MACHINING TECHNOLOGY
Machine Tools and Operations

MACHINING TECHNOLOGY
Machine Tools and Operations

Helmi A. Youssef

Hassan El-Hofy

CRC Press
Taylor & Francis Group
Boca Raton London New York

CRC Press is an imprint of the
Taylor & Francis Group, an **informa** business

FIRST INDIAN REPRINT, 2012

CRC Press
Taylor & Francis Group
6000 Broken Sound Parkway NW, Suite 300
Boca Raton, FL 33487-2742

© 2008 by Taylor & Francis Group, LLC
CRC Press is an imprint of Taylor & Francis Group, an Informa business

No claim to original U.S. Government works

Printed at Chennai Micro Print (P) Ltd, Chennai - 29.

International Standard Book Number-13: 978-1-4200-4339-6

This book contains information obtained from authentic and highly regarded sources Reasonable efforts have been made to publish reliable data and information, but the author and publisher cannot assume responsibility for the validity of all materials or the consequences of their use. The Authors and Publishers have attempted to trace the copyright holders of all material reproduced in this publication and apologize to copyright holders if permission to publish in this form has not been obtained. If any copyright material has not been acknowledged please write and let us know so we may rectify in any future reprint

Except as permitted under U.S. Copyright Law, no part of this book may be reprinted, reproduced, transmitted, or utilized in any form by any electronic, mechanical, or other means, now known or hereafter invented, including photocopying, microfilming, and recording, or in any information storage or retrieval system, without written permission from the publishers.

For permission to photocopy or use material electronically from this work, please access www.copyright.com (http://www.copyright.com/) or contact the Copyright Clearance Center, Inc. (CCC) 222 Rosewood Drive, Danvers, MA 01923, 978-750-8400. CCC is a not-for-profit organization that provides licenses and registration for a variety of users. For organizations that have been granted a photocopy license by the CCC, a separate system of payment has been arranged.

Trademark Notice: Product or corporate names may be trademarks or registered trademarks, and are used only for identification and explanation without intent to infringe.

Library of Congress Cataloging-in-Publication Data

Youssef, Helmi A.
   Machining technology : machine tools and operations / Helmi A. Youssef, Hassan El-Hofy.
      p. cm.
   Includes bibliographical references and index.
   ISBN 978-1-4200-4339-6 1. Machining. 2. Machine-tools. I. El-Hofy, Hassan.
II. Title.

TJ1185.Y68 2008
671.3′5--dc22
                                                    2008008302

Visit the Taylor & Francis Web site at
http://www.taylorandfrancis.com

and the CRC Press Web site at
http://www.crcpress.com

FOR SALE IN SOUTH ASIA ONLY

*To our grandsons and granddaughters,
Omar, Youssef, Nour, Anourine, Fayrouz, and Yousra*

Contents

Preface xix
Acknowledgments xxiii
Editors xxv
List of Symbols xxvii
List of Acronyms xxxiii

Chapter 1 Machining Technology 1

1.1 Introduction 1
1.2 History of Machine Tools 1
1.3 Basic Motions in Machine Tools 5
1.4 Aspects of Machining Technology 5
    1.4.1 Machine Tool 6
    1.4.2 Workpiece Material 9
    1.4.3 Machining Productivity 9
    1.4.4 Accuracy and Surface Integrity 10
    1.4.5 Product Design for Economical Machining 10
    1.4.6 Environmental Impacts of Machining 10
1.5 Review Questions 10
References 10

Chapter 2 Basic Elements and Mechanisms of Machine Tools 11

2.1 Introduction 11
2.2 Machine Tool Structures 13
    2.2.1 Light- and Heavy-weight Constructions 17
2.3 Machine Tool Guideways 18
    2.3.1 Sliding Friction Guideways 18
    2.3.2 Rolling Friction Guideways 21
    2.3.3 Externally Pressurized Guideways 22
2.4 Machine Tool Spindles 23
    2.4.1 Spindle Bearings 23
    2.4.2 Selection of Spindle-Bearing Fit 25
    2.4.3 Sliding Friction Spindle Bearing 27
2.5 Machine Tool Drives 28
    2.5.1 Stepped Speed Drives 28
        2.5.1.1 Belting 28
        2.5.1.2 Pick-Off Gears 30
        2.5.1.3 Gearboxes 30
        2.5.1.4 Stepping of Speeds According to Arithmetic Progression 31
        2.5.1.5 Stepping of Speeds According to Geometric Progression 32
        2.5.1.6 Kinetic Calculations of Speed Gearboxes 35
        2.5.1.7 Application of Pole-Changing Induction Motors 35
        2.5.1.8 Feed Gearboxes 37
        2.5.1.9 Preselection of Feeds and Speeds 39

	2.5.2	Stepless Speed Drives	40
		2.5.2.1 Mechanical Stepless Drives	40
		2.5.2.2 Electrical Stepless Speed Drive	42
		2.5.2.3 Hydraulic Stepless Speed Drive	43
2.6	Planetary Transmission		44
2.7	Machine Tool Motors		45
2.8	Reversing Mechanisms		45
2.9	Couplings and Brakes		46
2.10	Reciprocating Mechanisms		48
	2.10.1	Quick-Return Mechanism	48
	2.10.2	Whitworth Mechanism	50
	2.10.3	Hydraulic Reciprocating Mechanism	50
2.11	Material Selection and Heat Treatment of Machine Tool Components		51
	2.11.1	Cast Iron	51
	2.11.2	Steels	52
2.12	Testing of Machine Tools		53
2.13	Maintenance of Machine Tools		55
	2.13.1	Preventive Maintenance	56
	2.13.2	Corrective Maintenance	56
	2.13.3	Reconditioning	56
2.14	Review Questions		57
References			57

Chapter 3 General-Purpose Machine Tools 59

3.1	Introduction		59
3.2	Lathe Machines and Operations		59
	3.2.1	Turning Operations	59
	3.2.2	Metal Cutting Lathes	60
		3.2.2.1 Universal Engine Lathes	60
		3.2.2.2 Other Types of General-Purpose Metal Cutting Lathes	69
3.3	Drilling Machines and Operations		70
	3.3.1	Drilling and Drilling Allied Operations	70
		3.3.1.1 Drilling Operation	70
		3.3.1.2 Drilling Allied Operations	71
	3.3.2	General-Purpose Drilling Machines	74
		3.3.2.1 Bench-Type Sensitive Drill Presses	74
		3.3.2.2 Upright Drill Presses	74
		3.3.2.3 Radial Drilling Machines	76
		3.3.2.4 Multispindle Drilling Machines	76
		3.3.2.5 Horizontal Drilling Machines for Drilling Deep Holes	77
	3.3.3	Tool Holding Accessories of Drilling Machines	77
	3.3.4	Work-Holding Devices Used on Drilling Machines	79
3.4	Milling Machines and Operations		82
	3.4.1	Milling Operations	82
		3.4.1.1 Peripheral Milling	82
		3.4.1.2 Face Milling	84
	3.4.2	Milling Cutters	84

	3.4.3	General-Purpose Milling Machines	86
		3.4.3.1 Knee-Type Milling Machines	86
		3.4.3.2 Vertical Bed-Type Milling Machines	88
		3.4.3.3 Planer-Type Milling Machine	88
		3.4.3.4 Rotary-Table Milling Machines	89
	3.4.4	Holding Cutters and Workpieces on Milling Machines	90
		3.4.4.1 Cutter Mounting	90
		3.4.4.2 Workpiece Fixturing	91
	3.4.5	Dividing Heads	94
		3.4.5.1 Universal Dividing Heads	94
		3.4.5.2 Modes of Indexing	95
3.5	Shapers, Planers, and Slotters and Their Operations		99
	3.5.1	Shaping, Planing, and Slotting Processes	99
		3.5.1.1 Determination of v_{cm} in Accordance with the Machine Mechanism	101
	3.5.2	Shaper and Planer Tools	102
	3.5.3	Shapers, Planers, and Slotters	103
		3.5.3.1 Shapers	103
		3.5.3.2 Planers	105
		3.5.3.3 Slotters	107
3.6	Boring Machines and Operations		107
	3.6.1	Boring	107
	3.6.2	Boring Tools	108
		3.6.2.1 Types of Boring Tools	108
		3.6.2.2 Materials of Boring Tools	109
	3.6.3	Boring Machines	109
		3.6.3.1 General-Purpose Boring Machines	109
		3.6.3.2 Jig Boring Machines	110
3.7	Broaching Machines and Operations		111
	3.7.1	Broaching	111
		3.7.1.1 Advantages and Limitations of Broaching	112
	3.7.2	The Broach Tool	113
		3.7.2.1 Tool Geometry and Configuration	113
		3.7.2.2 Broach Material	116
		3.7.2.3 Broach Sharpening	116
	3.7.3	Broaching Machines	116
		3.7.3.1 Horizontal Broaching Machines	117
		3.7.3.2 Vertical Broaching Machines	118
		3.7.3.3 Continuous Horizontal Surface Broaching Machines	118
3.8	Grinding Machines and Operations		119
	3.8.1	Grinding Process	119
	3.8.2	Grinding Wheels	122
		3.8.2.1 Manufacturing Characteristics of Grinding Wheels	122
		3.8.2.2 Grinding Wheel Geometry	127
		3.8.2.3 Mounting and Balancing of Grinding Wheels and Safety Measures	127
		3.8.2.4 Turning and Dressing of Grinding Wheels	130
	3.8.3	Grinding Machines	131
		3.8.3.1 Surface Grinding Machines and Related Operations	132
		3.8.3.2 External Cylindrical Grinding Machines and Related Operations	133
		3.8.3.3 Internal Grinding Machines and Related Operations	136
		3.8.3.4 Centerless Grinding Machines and Related Operations	137

3.9 Microfinishing Machines and Operations 141
    3.9.1 Honing 141
        3.9.1.1 Process Capabilities 142
        3.9.1.2 Machining Parameters 144
        3.9.1.3 Honing Machines 145
    3.9.2 Superfinishing (Microhoning) 145
    3.9.3 Lapping 147
        3.9.3.1 Machining Parameters 147
        3.9.3.2 Lapping Machines 148
3.10 Review Questions 154
References 156

Chapter 4 Thread Cutting 157

4.1 Introduction 157
4.2 Thread Cutting 159
    4.2.1 Cutting Threads on the Lathe 160
    4.2.2 Thread Chasing 163
    4.2.3 Thread Tapping 164
    4.2.4 Die Threading 168
        4.2.4.1 Die Threading Machines 169
        4.2.4.2 Die Threading Performance 172
    4.2.5 Thread Milling 172
    4.2.6 Thread Broaching 175
4.3 Thread Grinding 175
    4.3.1 Center-Type Thread Grinding 175
    4.3.2 Centerless Thread Grinding 177
4.4 Review Questions 178
References 179

Chapter 5 Gear Cutting Machines and Operations 181

5.1 Introduction 181
5.2 Forming and Generating Methods in Gear Cutting 183
    5.2.1 Gear Cutting by Forming 184
        5.2.1.1 Gear Milling 184
        5.2.1.2 Gear Broaching 188
        5.2.1.3 Gear Forming by a Multiple-Tool Shaping Head 189
        5.2.1.4 Straight Bevel Gear Forming Methods 190
    5.2.2 Gear Cutting by Generation 190
        5.2.2.1 Gear Hobbing 190
        5.2.2.2 Gear Shaping with Pinion Cutter 198
        5.2.2.3 Gear Shaping with Rack Cutter 202
        5.2.2.4 Cutting Straight Bevel Gears by Generation 202
5.3 Selection of Gear Cutting Method 207
5.4 Gear Finishing Operations 207
    5.4.1 Finishing Gears Prior to Hardening 207
        5.4.1.1 Gear Shaving 207
        5.4.1.2 Gear Burnishing 211
    5.4.2 Finishing Gears After Hardening 212
        5.4.2.1 Gear Grinding 212
    5.4.3 Gear Lapping 214

5.5	Review Questions and Problems	215
References		215

Chapter 6 Turret and Capstan Lathes 217

6.1	Introduction	217
6.2	Difference Between Capstan and Turret Lathes	217
6.3	Selection and Application of Capstan and Turret Lathes	219
6.4	Principal Elements of Capstan and Turret Lathes	219
	6.4.1 Headstock and Spindle Assembly	220
	6.4.2 Carriage/Cross-Slide Unit	221
	6.4.3 Hexagonal Turret	221
	6.4.3.1 Manually Controlled Machines	222
	6.4.3.2 Automatically Controlled Headstock Turret Lathes	222
	6.4.4 Cross-Sliding Hexagonal Turret	223
6.5	Turret Tooling Setups	223
	6.5.1 Job Analysis	223
	6.5.2 Tooling Layout	226
6.6	Review Questions	232
References		232

Chapter 7 Automated Lathes 233

7.1	Introduction	233
7.2	Degree of Automation and Production Capacity	234
7.3	Classification of Automated Lathes	235
7.4	Semiautomatic Lathes	237
	7.4.1 Single-Spindle Semiautomatics	237
	7.4.2 Multispindle Semiautomatics	239
7.5	Fully Automatic Lathes	241
	7.5.1 Single-Spindle Automatic	241
	7.5.1.1 Turret Automatic Screw Machine	241
	7.5.1.2 Swiss-Type Automatic	252
	7.5.2 Horizontal Multispindle Bar and Chucking Automatics	256
	7.5.2.1 Special Features of Multispindle Automatics	256
	7.5.2.2 Characteristics of Parallel- and Progressive-Action Multispindle Automatic	258
	7.5.2.3 Operation Principles and Constructional Features of a Progressive Multispindle Automatic	260
7.6	Design and Layout of Cams for Fully Automatics	266
	7.6.1 Planning a Sequence of Operation and a Tooling Layout	267
	7.6.2 Cam Design	268
7.7	Review Questions and Problems	283
References		284

Chapter 8 Numerical Control and Computer Numerical Control Technology 285

8.1	Introduction	285
8.2	Coordinate System	290
	8.2.1 Machine Tool Axes for NC	290
	8.2.2 Quadrant Notation	292
	8.2.3 Point Location	292
	8.2.4 Zero Point Location	293

	8.2.5	Setup Point	293
	8.2.6	Absolute and Incremental Positioning	293
8.3	Machine Movements in Numerical Control Systems		294
8.4	Interpolation		296
8.5	Control of Numerical Control Machine Tools		297
8.6	Components of Numerical Control Machine Tools		299
8.7	Tooling for Numerical Control Machines		302
8.8	Numerical Control Machine Tools		305
8.9	Input Units		308
8.10	Forms of Numerical Control Instructions		310
8.11	Program Format		311
8.12	Feed and Spindle Speed Coding		312
	8.12.1	Feed Rate Coding	312
	8.12.2	Spindle Speed Coding	314
8.13	Features of Numerical Control Systems		314
8.14	Part Programming		316
8.15	Programming Machining Centers		320
	8.15.1	Planning the Program	320
	8.15.2	Canned Cycles	322
8.16	Programming Turning Centers		328
	8.16.1	Planning the Program	328
	8.16.2	Canned Turning Cycles	331
8.17	Computer-Assisted Part Programming		334
	8.17.1	Automatically Programmed Tools Language	334
	8.17.2	Programming Stages	337
8.18	CAD/CAM Approach to Part Programming		339
	8.18.1	Computer-Aided Design	339
	8.18.2	Computer-Aided Manufacturing	339
8.19	Review Questions		340
References			343

Chapter 9 Hexapods and Machining Technology 345

9.1	Introduction		345
9.2	Historical Background		345
9.3	Hexapod Mechanism and Design Features		348
	9.3.1	Hexapod Mechanism	348
	9.3.2	Design Features	349
		9.3.2.1 Hexapods of Telescopic Struts (Ingersoll System)	349
		9.3.2.2 Hexapods of Ball Screw Struts (Hexel and Geodetic System)	352
9.4	Hexapod Constructional Elements		354
	9.4.1	Strut Assembly	354
	9.4.2	Sphere Drive	354
	9.4.3	Bifurcated Balls	356
	9.4.4	Spindles	357
	9.4.5	Articulated Head	359
	9.4.6	Upper Platform	359
	9.4.7	Control System	361
9.5	Hexapod Characteristics		362
9.6	Manufacturing Applications		366

9.7	Review Questions	368
References		369

Chapter 10 Machine Tool Dynamometers 371

- 10.1 Introduction 371
- 10.2 Design Features of Dynamometers 371
  - 10.2.1 Rapier Parameters for Dynamometer Design 372
  - 10.2.2 Main Requirements of a Good Dynamometer 373
- 10.3 Dynamometers Based on Displacement Measurements 374
  - 10.3.1 Two-Channel Cantilever (Chisholm) Dynamometer 374
  - 10.3.2 Two-Channel-Slotted Cantilever Dynamometer 374
- 10.4 Dynamometers Based on Strain Measurement 375
  - 10.4.1 Strain Gauges and Wheatstone Bridges 375
  - 10.4.2 Cantilever Strain Gauge Dynamometers 377
  - 10.4.3 Octagonal Ring Dynamometers 378
    - 10.4.3.1 Strain Rings and Octagonal Ring Transducers 378
    - 10.4.3.2 Turning Dynamometer 382
    - 10.4.3.3 Surface Plunge-Cut Grinding Dynamometer 384
    - 10.4.3.4 Milling Dynamometers 384
- 10.5 Piezoelectric (Quartz) Dynamometers 384
  - 10.5.1 Principles and Features 384
  - 10.5.2 Typical Piezoelectric Dynamometers 386
- 10.6 Review Questions 389
- References 390

Chapter 11 Nontraditional Machine Tools and Operations 391

- 11.1 Introduction 391
- 11.2 Classification of Nontraditional Machining Processes 392
- 11.3 Jet Machines and Operations 392
  - 11.3.1 Abrasive Jet Machining 392
    - 11.3.1.1 Process Characteristics and Applications 392
    - 11.3.1.2 Work Station of Abrasive Jet Machining 395
    - 11.3.1.3 Process Capabilities 396
  - 11.3.2 Water Jet Machining (Hydrodynamic Machining) 397
    - 11.3.2.1 Process Characteristics and Applications 397
    - 11.3.2.2 Equipment of WJM 399
    - 11.3.2.3 Process Capabilities 401
  - 11.3.3 Abrasive Water Jet Machining 402
    - 11.3.3.1 Process Characteristics and Applications 402
    - 11.3.3.2 Abrasive Water Jet Machining Equipment 405
    - 11.3.3.3 Process Capabilities 409
- 11.4 Ultrasonic Machining Equipment and Operation 410
  - 11.4.1 Definitions, Characteristics, and Applications 410
  - 11.4.2 USM Equipment 413
    - 11.4.2.1 Oscillating System and Magnetostriction Effect 413
    - 11.4.2.2 Tool Feeding Mechanism 418
  - 11.4.3 Design of Acoustic Horns 419
    - 11.4.3.1 General Differential Equation 419
    - 11.4.3.2 Design of the Cylindrical Stepped Acoustic Horns ($A(x) = C$) 421
    - 11.4.3.3 Design of Exponential Acoustic Horns ($A(x) = A_0 e^{-2hx}$) 421

		11.4.4	Process Capabilities	430
			11.4.4.1 Stock Removal Rate	430
			11.4.4.2 Accuracy and Surface Quality	432
		11.4.5	Recent Developments	433

11.5 Chemical Machining 434
    11.5.1 Chemical Milling 435
    11.5.2 Photochemical Machining (Spray Etching) 441
11.6 Electrochemical Machines and Operations 445
    11.6.1 Process Characteristics and Applications 445
    11.6.2 Elements of Electrochemical Machining 447
        11.6.2.1 Tool 447
        11.6.2.2 Workpiece 449
        11.6.2.3 Electrolyte 449
    11.6.3 ECM Equipment 449
    11.6.4 Process Capabilities 451
11.7 Electrochemical Grinding Machines and Operations 453
11.8 Electrical Discharge Machines and Operations 454
    11.8.1 Process Characteristics and Applications 454
    11.8.2 ED Sinking Machine 458
    11.8.3 EDM-Spark Circuits (Power Supply Circuits) 460
        11.8.3.1 Resistance-Capacitance Circuit 460
        11.8.3.2 Transistorized Pulse Generator Circuits 462
    11.8.4 EDM-Tool Electrodes 463
    11.8.5 Process Capabilities 464
    11.8.6 Electrical Discharge Milling 465
    11.8.7 Electrodischarge Wire Cutting 468
11.9 Electron Beam Machining Equipment and Operations 470
    11.9.1 Process Characteristics and Applications 470
    11.9.2 Electron Beam Machining Equipment 471
    11.9.3 Process Capabilities 474
11.10 Laser Beam Machining Equipment and Operations 475
    11.10.1 Process Characteristics 475
    11.10.2 Types of Lasers 477
        11.10.2.1 Pyrolithic and Photolithic Lasers 477
        11.10.2.2 Industrial Lasers 477
        11.10.2.3 Laser Beam Machining Operations 478
    11.10.3 LBM Equipment 481
    11.10.4 Applications and Capabilities 483
11.11 Plasma Arc Cutting Systems and Operations 485
    11.11.1 Process Characteristics 485
    11.11.2 Plasma Arc Cutting Systems 486
    11.11.3 Applications and Capabilities of Plasma Arc Cutting 486
11.12 Review Questions 488
References 492

Chapter 12 Environment-Friendly Machine Tools and Operations 495

12.1 Introduction 495
12.2 Traditional Machining 498
    12.2.1 Cutting Fluids 501
        12.2.1.1 Classification of Cutting Fluids 501

| | | 12.2.1.2 Selection of Cutting Fluids | 502 |
| | | 12.2.1.3 Evaluation of Cutting Fluids | 502 |

- 12.2.2 Hazard Ranking of Cutting Fluids 503
- 12.2.3 Health Hazards of Cutting Fluids 504
- 12.2.4 Cryogenic Cooling 504
- 12.2.5 Ecological Machining 505

12.3 Nontraditional Machining Processes 510
- 12.3.1 Chemical Machining 510
- 12.3.2 Electrochemical Machining 512
- 12.3.3 Electrodischarge Machining 514
  - 12.3.3.1 Protective Measures 516
- 12.3.4 Laser Beam Machining 516
- 12.3.5 Ultrasonic Machining 519
  - 12.3.5.1 Electromagnetic Field 520
  - 12.3.5.2 Ultrasonic Waves 520
  - 12.3.5.3 Abrasives Slurry 520
  - 12.3.5.4 Contact Hazards 521
  - 12.3.5.5 Other Hazards 521
- 12.3.6 Abrasive Jet Machining 521

12.4 Review Questions 523

References 524

Chapter 13 Design for Machining 525

13.1 Introduction 525
- 13.1.1 General Design Rules 525

13.2 General Design Recommendations 526

13.3 Design for Machining by Cutting 528
- 13.3.1 Turning 528
  - 13.3.1.1 Economic Production Quantities 529
  - 13.3.1.2 Design Recommendations for Turning 530
  - 13.3.1.3 Dimensional Control 535
- 13.3.2 Drilling and Allied Operations 535
  - 13.3.2.1 Economic Production Quantities 536
  - 13.3.2.2 Design Recommendations for Drilling and Allied Operations 536
  - 13.3.2.3 Dimensional Control 539
- 13.3.3 Milling 539
  - 13.3.3.1 Design Recommendations 539
  - 13.3.3.2 Dimensional Factors and Tolerances 542
- 13.3.4 Shaping, Planing, and Slotting 542
  - 13.3.4.1 Design Recommendations 542
  - 13.3.4.2 Dimensional Control 543
- 13.3.5 Broaching 544
  - 13.3.5.1 Design Recommendations 544
  - 13.3.5.2 Dimensional Factors 549
  - 13.3.5.3 Recommended Tolerances 550
- 13.3.6 Thread Cutting 550
  - 13.3.6.1 Design Recommendations 550
  - 13.3.6.2 Dimensional Factors and Tolerances 551
- 13.3.7 Gear Cutting 552
  - 13.3.7.1 Design Recommendations 552
  - 13.3.7.2 Dimensional Factors 554

13.4	Design for Grinding	554
	13.4.1 Surface Grinding	554
	13.4.1.1 Design Recommendations	554
	13.4.1.2 Dimensional Control	556
	13.4.2 Cylindrical Grinding	556
	13.4.2.1 Design Recommendations	556
	13.4.2.2 Dimensional Factors	557
	13.4.3 Centerless Grinding	557
	13.4.3.1 Design Recommendations	558
	13.4.3.2 Dimensional Control	559
13.5	Design for Finishing Processes	559
	13.5.1 Honing	559
	13.5.2 Lapping	560
	13.5.3 Superfinishing	561
13.6	Design for Chemical and Electrochemical Machining	561
	13.6.1 Chemical Machining	561
	13.6.1.1 Design Recommendations	561
	13.6.1.2 Dimensional Factors and Tolerances	563
	13.6.2 Electrochemical Machining	563
	13.6.2.1 Design Recommendations	564
	13.6.2.2 Dimensional Factors	566
	13.6.3 Electrochemical Grinding	566
	13.6.3.1 Design Recommendations	566
	13.6.3.2 Dimensional Factors	567
13.7	Design for Thermal Machining	567
	13.7.1 Electrodischarge Machining	567
	13.7.1.1 Design Recommendations	567
	13.7.1.2 Dimensional Factors	568
	13.7.2 Electron Beam Machining	568
	13.7.3 Laser Beam Machining	569
13.8	Design for Ultrasonic Machining	570
13.9	Design for Abrasive Jet Machining	571
13.10	Review Questions	572
References		573

Chapter 14 Accuracy and Surface Integrity Realized by Machining Processes ... 575

14.1	Introduction	575
14.2	Surface Texture	575
14.3	Surface Quality and Functional Properties	577
14.4	Surface Integrity	579
14.5	Surface Effects by Traditional Machining	582
	14.5.1 Chip Removal Processes	582
	14.5.2 Grinding	583
14.6	Surface Effects by Nontraditional Machining	587
	14.6.1 Electrochemical and Chemical Machining	590
	14.6.2 Thermal Nontraditional Processes	591
	14.6.2.1 Electrodischarge Machining	591
	14.6.2.2 Laser Beam Machining	596
	14.6.2.3 Electron Beam Machining	597
	14.6.2.4 Plasma Beam Machining (PBM)	598

		14.6.2.5 Electroerosion Dissolution Machining	598

           14.6.2.5 Electroerosion Dissolution Machining..598
           14.6.2.6 Electrochemical Discharge Grinding...598
      14.6.3 Mechanical Nontraditional Processes 599
14.7 Reducing Distortion and Surface Effects in Machining..599
14.8 Review Questions..601
References...601

Chapter 15 Automated Manufacturing System...603

15.1 Introduction..603
15.2 Manufacturing Systems 605
15.3 Flexible Automation–Flexible Manufacturing Systems...609
      15.3.1 Elements of Flexible Manufacturing System 610
      15.3.2 Limitations of Flexible Manufacturing System...611
      15.3.3 Features and Characteristics...611
      15.3.4 New Developments in Flexible Manufacturing System Technology 611
15.4 Computer Integrated Manufacturing..612
      15.4.1 Computer-Aided Design 615
      15.4.2 Computer-Aided Process Planning...616
      15.4.3 Computer Aided Manufacturing 617
15.5 Lean Production–Just-in-Time Manufacturing Systems 617
      15.5.1 Steps for Implementing the IMPS Lean Production 618
      15.5.2 Just-in-Time and Just-in-Case Production 619
15.6 Adaptive Control..620
15.7 Smart Manufacturing and Artificial Intelligence 622
      15.7.1 Expert Systems 622
      15.7.2 Machine Vision...623
      15.7.3 Artificial Neural Networks 623
      15.7.4 Natural-Language Systems...624
      15.7.5 Fuzzy Logic (Fuzzy Models)...624
15.8 Factory of the Future 624
15.9 Concluding Remarks Related to Automated Manufacturing 625
15.10 Review Questions..625
References...626

Index..627

Preface

This book provides a comprehensive description of machining technologies related to metal shaping by material removal techniques, from the basic to the most advanced, in today's industrial applications. It is a fundamental textbook for undergraduate students enrolled in production, materials and manufacturing, industrial, and mechanical engineering programs. Students from other disciplines can also use this book while taking courses in the area of manufacturing and materials engineering. It should be also useful to graduates enrolled in high-level machining technology courses and professional engineers working in the field of manufacturing industry. The book covers the technologies, machine tools, and operations of several machining processes. The treatment of the different subjects has been developed from the basic principles of machining processes, machine tool elements, and control systems, and extends to ecological machining and the most recent machining technologies, including nontraditional methods and hexapod machine tools. Along with the fundamentals of the conventional and modern machine tools and processes, the book presents environmental-friendly machine tools and operations; design for machining, accuracy, and surface integrity realized by machining operations; machining data; and solved examples, problems, and review questions, which are very useful for undergraduate students and manufacturing engineers facing shop floor problems.

The book is written in 15 chapters, describing for the first time in one book the fundamentals, basic elements, and operations of general-purpose machine tools used for the production of cylindrical and flat surfaces by turning, drilling and reaming, shaping and planing, and milling processes. Special-purpose machines and operations used for thread cutting, gear cutting, and broaching processes are also dealt with. Semiautomatic, automatic, NC and CNC machine tools, operations, tooling, mechanisms, accessories, and work fixation are discussed. Abrasion and abrasive finishing machine tools and operations such as grinding, honing, superfinishing, and lapping are described. Modern machine tools and operations, dynamometers, and hexapod machine tools and processes are described. Design for accurate and economic machining, ecological machining, levels of accuracy, and surface finish attained by machining methods are also presented.

OUTLINE OF THE BOOK

In Chapter 1, the history and progress of machining, aspects of machining technology, and the basic motions of machine tools are introduced. Classification of machine tools and operations in addition to the basic motions of machining operations are also given.

Chapter 2 introduces the design considerations and requirements of machine tools, including basic elements such as beds, structures, frames, guideways, spindles and shafts, stepped and stepless drives, planetary transmission, machine tool motors, couplings, and brakes. Material selection and heat treatment of machine tool elements, and the testing and maintenance of machine tools are also discussed.

Chapter 3 covers general-purpose metal cutting machine tools including lathes, drilling, reaming, jig boring machines, milling machines, and the machine tools of a reciprocating nature such as shapers, planers, and slotters. Machine tool elements, mechanisms, tooling, accessories, and operations are also explained. Chapter 3 also presents abrasion machine tools, including grinding and surface finishing machines and processes.

Chapter 4 describes the different types and applications of commonly used screw threads. Thread machining by cutting and grinding methods are described, together with thread cutting machines and cutting tools.

In Chapter 5, common types of gears are listed and their applications described. Gear production by machining methods that include cutting, grinding, and lapping are described, together with their corresponding machine tools and operations.

Chapter 6 describes the capstan and turret lathes. Machine components, features, and applications are described. Tool layouts for bar-type capstan lathes and chucking-type turret lathes are described and solved examples are given.

Semiautomatic and automatic lathes are discussed in Chapter 7. Machine tool features, components, operation, tooling, and industrial applications are described. Solved examples for typical products that show process layout and cam design are given for turret-type and long-part automatics.

Chapter 8 presents computer numerical controlled machine tools, their merits, and their industrial applications. The basic features of such machines, tooling arrangements, and programming principles and examples are illustrated in case of machining and turning centers. An introduction to computer-assisted and CAD/CAM applications in part programming is also covered.

Hexapod mechanisms, design features, constructional elements, characteristics, control, and their applications in traditional and nontraditional machining, manufacturing, and robotics are covered in Chapter 9.

Chapter 10 describes the fundamentals, instrumentation, and operation of machine tool dynamometers used for cutting force measurements. Examples of turning, drilling, milling, and grinding dynamometers are explained.

Chapter 11 presents modern machine tools and operations for mechanical nontraditional machining processes, such as ultrasonic and jet machining. Chemical milling, electrochemical machining, and electrochemical grinding machine tools are also described, along with the machine tools for thermal processes such as electrodischarge, laser beam, electron beam, and plasma arc machining. Machine tools, basic elements, accessories, operations, removal rate, accuracy, and surface integrity are covered for each case.

Environment-friendly machine tools and operations are described in Chapter 12; these tend to detect the source of hazards and minimize their effect on the operator, machine tools, and environment.

An introduction to design recommendations for economic machining and sources of dimensional variations by traditional and nontraditional processes is covered in Chapter 13.

Dimensional accuracy and surface integrity by traditional and nontraditional machining processes are discussed in Chapter 14. Sources of surface alterations, their effects on the functional properties of machined parts, and recommendations for minimizing surface effects are also given.

Chapter 15 covers the fundamentals and applications of computer-integrated manufacturing, lean production, adaptive control, just-in-time manufacturing systems, smart manufacturing, artificial intelligence, and the factory of the future.

ADVANTAGES OF THE BOOK

This book provides several advantages to the reader since it:
1. Presents a wide spectrum of the machining technologies, machine tools, and operations used in manufacturing industries
2. Covers a wide range of abrasive machining and finishing technologies
3. Presents the nontraditional machine tools and processes
4. Provides coverage for CNC, hexapod technologies, and computer-aided manufacturing
5. Introduces the principles of ecological machining
6. Discusses the economics of design for machining, machining accuracy, and surface integrity aspects by the different machining techniques
7. Presents very useful technical data that help in solving and analysis of day-to-day shop floor problems
8. Presents solved examples, review questions, and problems related to the various machining topics

Preface

This book is intended to help the following readers:

1. Undergraduate students enrolled in mechanical, industrial, manufacturing, materials, and production engineering programs
2. Professional engineers
3. Industrial companies
4. Postgraduate students

WHY DID WE WRITE THE BOOK?

This book presents several years of the authors' experience in research and teaching of different machining technologies and related topics at many universities and institutions around the world. Although many aspects of the machining subject have been covered in detail through various books, the authors believe that this is the first attempt to cover such topics at this level in one book. The book follows the two books by Professor El-Hofy: *Advanced Machining Processes: Nontraditional and Hybrid Processes* that covered the principles of advanced machining process published by McGraw Hill (2005) and the book entitled *Fundamentals of Machining Processes: Conventional and Nonconventional Processes* by CRC Press (2007).

Helmi A. Youssef and Hassan El-Hofy
Alexandria, Egypt

Acknowledgments

Many individuals have contributed to the development of this book. It is a pleasure to express our deep gratitude to Professor Dr. Ing. A. Visser, Bremen University, Germany, for supplying valuable materials during the preparation of this book. The assistance of Dr. A. Khalil and I. Bayoumi of Alexandria University and I. El-Naggar of Lord Alexandria Razor Company for their valuable Auto-CAD drawings is highly appreciated.

Heartfelt thanks are due to our families for their great patience, support, encouragement, enthusiasm, and interest during the preparation of the manuscript.

We would like to acknowledge the dedication and continued help of the editorial and production staff of CRC Press for their efforts in ensuring that the book is accurate and as well-designed as possible.

We appreciate very much the permissions from all publishers to reproduce many illustrations from a number of authors as well as the courtesy of many industrial companies that provided photographs and drawings of their products to be included in this book. Their generous cooperation is a mark of sincere interest in enhancing the level of engineering education. The credits for all such great help are given in the captions under the corresponding illustrations.

Editors

Professor Helmi A. Youssef, born in August 1938 in Alexandria, Egypt, acquired his BSc degree with honors in production engineering from Alexandria University in 1960. He completed his scientific career in Carolo-Wilhelmina, TH Braunschweig, Germany during 1961–1967. In June 1964, he acquired his Dipl.-Ing. degree; then, in December 1967, he completed his Dr.-Ing. degree in the domain of nontraditional machining. In 1968, he returned to the Alexandria University Production Engineering Department as an assistant professor. In 1973, he was promoted to associate professor, and in 1978 to full professor. From 1995–1998, Professor Youssef was the chairman of the Production Engineering Department, Alexandria University. Since 1989, he has been a member of the scientific committee for promotion of professors in Egyptian universities.

Based on several research and educational laboratories that he has built, Professor Youssef founded his own scientific school in both traditional and nontraditional machining technologies. In the early 1970s, he established the first NTM-research laboratory in Alexandria University (and maybe in the whole region). Since that time, he has carried out intensive research in his fields of specialization, and supervised many PhD and MSc theses.

Between 1975 and 1995, Professor Youssef was a visiting professor in Arabic universities, such as El-Fateh University, Tripoli; the Technical University, Baghdad; King Saud University (KSU), Riyadh; and Beirut Arab University (BAU), Beirut. Besides his teaching activities in these universities, he established laboratories and supervised many MSc theses. Moreover, he was a visiting professor in different academic institutions in Egypt and abroad.

Professor Youssef has organized and participated in many international conferences. He has published numerous scientific papers in specialized journals. He authored many books in his fields of specialization. Currently, he is an emeritus professor at Alexandria University.

Professor Hassan El-Hofy was born in February 1953 in Egypt. He received a BSc honors degree in production engineering from Alexandria University, Egypt, in 1976, and then served as a teaching assistant in the same department and received an MSc degree in production engineering from Alexandria University in 1979. Professor El-Hofy has had a successful university career in education, training, and research. Following his MSc degree, he worked as an assistant lecturer until October 1980, when he left for Aberdeen University, Scotland, and began his PhD work with Professor J. McGeough in hybrid machining processes. He won the Overseas Research Student (ORS) award during pursuit of his doctoral degree, which he completed in 1985. He came back to Alexandria University and resumed his work as an assistant professor. In 1990, he was promoted an associate professor. He was on leave as a visiting professor for Al-Fateh University, Tripoli, between 1989 and 1994.

In July 1994, Professor El-Hofy returned to Alexandria University, and in November 1997 he was promoted to full professor. In September 2000, he was selected to work as a professor in the University of Qatar. He chaired the accreditation committee for mechanical engineering program toward ABET Substantial Equivalency Recognition that has been granted to the College of Engineering programs in 2005. Due to his role in that event, he received the Qatar University Award and a certificate of appreciation. Professor El-Hofy wrote his first book, entitled *Advanced Machining Processes: Nontraditional and Hybrid Processes*, which was published by McGraw-Hill in March 2005. His second book, *Fundamentals of Machining Processes: Conventional and Nonconventional Processes* was published in 2007 by CRC Press, Taylor & Francis. He has published more than 50 scientific and technical papers and supervised many graduate students in the area of machining by nontraditional methods. He is a consulting editor to many international journals and is a regular participant in international conferences.

Since August 2007, Professor El-Hofy has been the chairman of the Production Engineering Department of Alexandria University, College of Engineering, where he teaches advanced machining and related courses.

List of Symbols

Symbol	Definition	Unit
A	Included thread angle	Degree
$A(x)$	Area of acoustic horn at position x	mm^2
A_c	Uncut chip cross-sectional area	mm^2
a_c	Acme thread crest width	mm
A_0	Area of acoustic horn at position 0	mm^2
a_r	Acme thread root width	mm
B	Chip-tool contact length	mm
C	Acoustic speed in horn material	m/s
C	Capacitance	µF
$\acute{C}$	Modified acoustic speed in horn material	m/s
$c1$	Specific heat of workpiece material	N m/kg °C
C_d	Coefficient of thermal diffusivity	m^2/s
c_i	Constraints	
D	Diameter	mm
D	Diameter of grinding wheel	mm
$D(x)$	Diameter of acoustic horn at position x	mm
D_1	Depth removal per pass	mm
d_a	Addendum diameter of gear	mm
D_a	Burnishing gear addendum diameter	mm
d_c	Fixation hole diameter	mm
d_d	Dedendum diameter	mm
d_f	Electron beam focusing diameter	mm
d_g	Abrasive grain diameter	µm
D_0	Diameter of acoustic horn at position 0	mm
D_0	Depth of thread	mm
d_p	Pitch circle diameter	mm
d_R	Diameter of regulating wheel	mm
D_t	Depth of cut in the first pass	mm
d_u	Chemical undercut	
E	Young's modulus	MPa
E_d	Energy of individual discharge	J
EF	Etch factor	
EI	Flexural rigidity	N mm^2
e_m	Hydraulic motor eccentricity	mm
e_p	Hydraulic pump eccentricity	mm
F	Feed rate	mm/rev
F	Force	N
F_a	Axial force	N
f_b	Bending strength	N/mm^2
F_c	Main cutting force	N
f_e	Frequency of exciting vibration	s^{-1}
F_f	Feed force	N
f_n	Natural frequency	s^{-1}

Symbol	Definition	Unit
F_r	Radial force	N
F_R	Resultant cutting force	N
f_r	Frequency	s^{-1}
F_x	Horizontal (passive) force	N
F_y	Vertical force	N
F_z	Feed force in drilling	
G	Gravitational acceleration or deceleration	m/s^2
H	Ascent factor of exponential horn	m/s
h_a	Addendum	mm
h_d	Dedendum	mm
h_g	Frontal gap thickness in EDM	
h_o	Height of thread fundamental triangle	mm
h_t	Tooth height	mm
h_w	Working depth	mm
i	Ratio	—
I	Moment of inertia	mm^4
i_b	Electron beam current	A
i_c	Charging current	A
i_d	Discharging current	A
i_f	Transmission ratio of feed gear	—
I_p	Premagnetizing current	A
i_r	Transmission ratio of speed change gear	—
i_x	Transmission ratio of indexing gear	—
I_x	Depth of cut in the X-axis	mm
i_y	Transmission ratio of differential gear	—
J	Radius of hole to be milled minus cutter radius	mm
k	Static stiffness	N/mm^2
K	Spring constant	M/N m
K_1	Depth of peck	mm
k_g	Gauge factor	
k_r	Coefficient of magneto-mechanical coupling	
K_t	Thread height	mm
k_t	Thermal conductivity	N/s °C
K_Z	Depth of cut on the Z-axis	mm
L	Stroke length in shaper	mm
L	Length	mm
l_d	Cantilever length	mm
l_g	Position of displacement gauge	mm
l_o	Maximum stock available for sharpening	mm
l_r	Length of ring transducer	mm
L_t	Tool travel	mm
M	Mobility	—
m	Mass	kg
m_g	Module of gear	mm
m_n	Normal module of gear	mm
M_s	Moment at position s	N mm
M_x	Bending moment distribution due to horizontal force	N mm
M_y	Bending moment distribution due to vertical force	N mm
M_z	Drilling torque	N mm

List of Symbols

Symbol	Description	Unit
N	Rotational speed	rpm or stroke/min
N	Number of threads per inch	
n_{aux}	Auxiliary shaft rotational speed	rpm
n_{cam}	Camshaft rotational speed	rpm
n_e	Number of elements in the hexapod system	
n_m	Motor speed	rpm
n_{max}	Maximum rotational speed	rpm
n_{min}	Minimum rotational speed	rpm
n_r	Reverse spindle speed	rpm
n_s	Spindle speed	rpm
p	Pitch	mm
P_e	Power of electron beam	N m/s
P_h	Honing stone pressure	kgf/cm^2
Q	Cutting/return speed ratio in shaping	—
Q_s	Number of steps in peck drilling	
R	Radius	mm
R	Resistance	Ω
ΔR	Change in resistance	Ω
R_0	Initial level position	mm
R_1	Radius of ring transducer	mm
R_a	Average surface roughness	μm
R_d	Diameter range	—
r_d	Displacement ratio	—
R_g	Speed ratio	
R_m	Magnification factor	μm
R_n	Rotational speed range	—
RPT	Rise per tooth (superelevation)	mm
R_t, R_{max}	Peak-to-valley surface roughness	μm
R_v	Cutting speed range	—
S	Tooth thickness	mm
T	Depth of cut (time)	mm (s)
$T(x)$	Thickness function	m
T_1	Input torque	N mm
T_1	Plate thickness	mm
T_2	Output torque	N mm
t_a	Auxiliary (idle or nonproductive) time	min
t_c	Charging time	μs
T_{cyc}	Cycle time	min
t_d	Discharging time	μs
T_e	Chemical etch depth	mm
t_e	Etching time	min
t_f	Floor-to-floor time	min
t_{hel}	Lead of helical groove	mm
t_i	Pulse duration	°C
t_{ls}	Pitch of lead screw of the lathe	mm
t_m	Machining (production) time	min
t_{mh}	Machine handling time	min
t_o	Thickness	mm
t_r	Thickness of ring transducer	mm

Symbol	Definition	Unit
T_s	Total depth of material removed in one stroke in broaching	mm
t_s	Setup time	min
t_{th}	Pitch of thread to be cut on the lathe	mm
t_{wh}	Work handling time	min
U	Allowance for finishing in X-axis	mm
u	Feed in milling	mm/min
v	Cutting speed	m/min
v_A	Anodic dissolution rate	mm/min
V_b	Electron beam accelerating voltage	V
V_c	Cutting speed in shaper	m/min
V_c	Capacitor voltage	V
V_{cm}	Mean cutting speed of cutting stroke in shaper	m/min
v_f	Feed rate in ECM	mm/min
v_g	Peripheral speed of grinding wheel	m/s—m/min
V_l	Lower speed	m/min
V_{max}	Maximum cutting speed	m/min
V_{min}	Minimum cutting speed	m/min
V_o	Open circuit voltage	V
v_p	Peripheral speed of regulating wheel	m/s
V_r	Return speed in shaper	m/min
v_r	Reverse speed	m/min
v_{rc}	Reciprocation speed in honing	m/min
V_{rm}	Mean return speed of return stroke in shaper	m/min
v_{rt}	Surface rotation speed in honing	m/min
V_s	Breakdown voltage	V
v_t	Traverse speed	m/min
V_u	Economical speed	m/min
v_w	Peripheral speed of workpiece	m/s
W	Allowance for finishing in Z-axis	mm
W_{ave}	Average power	W
w_o	Width	mm
X	Axial position	m
x_n	Nodal point location	m
Y	Distance between shaper crank and link pivot/displacement	mm
Z	Number of speeds	—
Z	Number of teeth	—
Z'	Modified number of teeth	—
Z_1	End position of the groove/thread	mm
z_g	Number of speed steps	

Chapter	Symbol (Greek Letters)	Definition	Unit
2	α	Angle of cutting stroke of shaper	Degree
2	β	Angle of return (non cutting) stroke of shaper	Degree
2	δ	Deflection	mm
2	σ_e	Elastic limit	N/mm^2
2	δ_5	Elongation	mm
2	δ_n	Increase in speed	%

List of Symbols

2	ω_o	Natural frequency	Hz
2	φ_p	Progression ratio of pole change motor	
2	φ	Progression ratio	—
2	σ_u	Ultimate tensile strength	N/mm^2
3	α_h	Half cross-hatch angle (honing)	Degree
3	ω_h	Helix angle of spiral groove	Degree
3	α_1	Inclination angle of regulating wheel	Degree
3	χ	Setting (approach) angle	Degree
3	φ	Diameter notation	mm
4	α_t	Thread helix angle	Degree
4	φ_c	Threading tap chamfer angle	Degree
5	β_g	Helix angle (gear)	Degree
5	α_h	Helix angle of the hob	Degree
5	γ	Setting angle (hob)	Degree
9	μ	Coefficient of friction	
10	c_s	Elastic strain	
10	θ	Location angle	Degree
11	β_m	Abrasive/air weight mixing ratio	%
11	ξ	Oscillation amplitude	μm
11	ω	Angular speed	radian/s
11	ε	Chemical equivalent	
11	ε_{ms}	Coefficient of magnetostrictive elongation	
11	η	Current efficiency	%
11	ρ	Density of the magnetostriction material	kg/m^3
11	θ_m	Melting point of workpiece material	°C
11	σ	Stress	kg/mm^2
11	λ	Wavelength	μm

List of Acronyms

Abbreviation	Description
ac	Alternating current
AC	Adaptive control
ACC	Adaptive control with constraints
ACO	Adaptive control with optimization
AFM	Abrasive flow machining
AGMA	American Gear Manufacturing Association
AI	Artificial intelligence
AISI	American Iron and Steel Institute
AJECM	Abrasive jet electrochemical machining
AJM	Abrasive jet machining
AMZ	Altered material zone
ANN	Artificial neural network
ANSI	American National Standards Institute
APT	Automatically programmed tools
ASA	American Standards Association
ASCII	American Standard Code for Information Interchange
ASME	American Society of Mechanical Engineers
ASTM	American Society for Testing and Materials
ATM	Atmosphere
AWJ	Abrasive water jet
AWJD	Abrasive water jet deburring
AWJM	Abrasive water jet machining
BA	British association
BCD	Binary coded decimal
BHN	Brinell hardness number
BSW	British standard Whitworth
BUE	Built-up edge
CAD	Computer-aided design
CAI	Computer-aided inspection
CAM	Computer-aided manufacturing
CAPP	Computer-aided process planning
CBN	Cubic boron nitride
CCP	Conventional computer program
CCW	Counterclockwise
CFG	Creep feed grinding
CHM	Chemical machining
CH-milling	Chemical milling
CI	Cast iron
CIM	Computer-integrated manufacturing
CLDATA	Cutter location data
CNC	Computer numerical control
CPC	Computerized part changer
CRT	Cathode ray tube
CW	Continuous wave

CY	Cyaniding
DB	Database
DFM	Design for manufacturing
DIN	Deutsches Institut für Normung
DNC	Direct numerical control
DOF	Degrees of freedom
DOT	Department of Transportation
DXF	Drawing exchange file
EB	Electron Beam
EBM	Electron beam machining
ECA	Electrochemical abrasion
ECAM	Electrochemical arc machining
ECD	Electrochemical dissolution
ECDB	Electrochemical deburring
ECDG	Electrochemical discharge grinding
ECDM	Electrochemical discharge machining
ECG	Electrochemical grinding
ECH	Electrochemical honing
ECM	Electrochemical machining
ECS	Electrochemical sharpening
ECUSM	Electrochemical ultrasonic machining
EDG	Electrodischarge grinding
EDM	Electrodischarge machining
EDS	Electrodischarge sawing
EDT	Electrodischarge texturing
EDWC	Electrodischarge wire cutting
EEDM	Electroerosion dissolution machining
EF	Etch factor
EHS	Environmental health and safety
EIA	Electronics Industry Alliance
ELP	Electropolishing
EMF	Electromagnetic field
EMS	Environmental Management System
EOB	End of block
EP	Extreme pressure
EPA	Environmental Protection Agency
ES	Expert system
FEA	Finite element analysis
FFT	Floor-to-floor time
FL	Fuzzy logic
FMC	Flexible manufacturing cell
FMS	Flexible manufacturing system
FOF	Factory of the future
FRP	Fiber-reinforced plastics
GAC	Geometric adaptive control
GT	Group technology
GW	Grinding wheel
HAZ	Heat-affected zone
HB	Hardness Brinell
HF	High frequency
HMIS	Hazardous Material Identification System

List of Acronyms

HMP	Hybrid machining processes
HP	Hybrid process
HRC	Hardness Rockwell
HSS	High-speed steel
HT	High temperature
IBM	Ion beam macining
ICE	Internal combustion engine
IGA	Intergranular attack
IMPS	Integrated manufacturing production system
ipr	Inches per revolution
IR	Infrared
ISO	International Organization for Standardization
JIC	Just-in-case
JIT	Just-in-time
KB	Knowledge base
L and T	Laps and tears
Laser	Light amplification by stimulated emission of radiation
LAT	Laser-assisted turning
LBM	Laser beam machining
LBT	Laser beam torch
LECM	Laser-assisted electrochemical machining
LSG	Low-stress grinding
LVDT	Linear variable displacement transducer
MA	Mechanical abrasion
MCD	Machine control data
MCK	Microcracks
MCU	Machine control unit
MDI	Manual data input
MIT	Massachusetts Institute of Technology
MPE	Maximum permissible exposure
MQL	Minimum quantity lubrication
MRP	Material requirements planning
MRR	Material removal rate
MS	Manufacturing system
MSDS	Material safety data sheets
NASA	National Aeronautics and Space Administration
NC	Numerical control
Nd	Neodymium
Nd:YAG	Neodymium-doped yttrium aluminum garnet
NFPA	National Fire Protection Association
NHZ	Nominal hazard zone
NTD	Nozzle-tip distance
NTM	Nontraditional machining
OA	Overaging
OSHA	Occupational Safety and Health Administration
OTM	Overtempered martensite
PAC	Plasma arc cutting
PAH	Polycyclic aromatic hydrocarbons
PAM	Plasma arc machining
PBM	Plasma beam machining
PCB	Printed circuit board

PCD	Polycrystalline diamond
PCM	Photochemical machining
PD	Plastic deformation
PEO	Polyethylene oxide
PIV	Positive infinitely variable
PKM	Parallel kinematic mechanism
PKS	Parallel kinematic system
PLC	Programmable logic controller
PTP	Point-to-point
PVD	Physical vapor deposition
RC	Recast
RETAD	Rapid exchange of tooling and dies
RPT	Rise per tooth
RUM	Rotary ultrasonic machining
RW	Regulating wheel
SAE	Society of Automotive Engineers
SB	Sand blasting
SE	Selective etching
SI	Surface integrity
SM	Smart manufacturing
SMED	Single-minute exchange of die
SOD	Stand-off distance
SP	Special precision
SRR	Stock removal rate
TEM	Transverse excitation mode
TIR	Total indicator reading
UAW	United Auto Workers
UNC	Unified coarse
UNF	Unified fine
UP	Ultraprecision
US	Ultrasonic
USM	Ultrasonic machining
UTM	Untempered martensite
UV	Ultraviolet
VDU	Visual display unit
VESP	Vibratory-enhanced shear processing
VRR	Volumetric removal rate
WC	Tungsten carbide
WHO	World Health Organization
WIP	Work in progress
WJM	Water jet machining
WP	Workpiece
YAG	Yttrium aluminum garnet

# 1 Machining Technology

1.1 INTRODUCTION

Manufacturing is the industrial activity that changes the form of raw materials to create products. The derivation of the word *manufacture* reflects its original meaning: to make by hand. As the power of the hand tool is limited, manufacturing is done largely by machinery today. Manufacturing technology constitutes all methods used for shaping the raw metal materials into a final product. As shown in Figure 1.1, manufacturing technology includes plastic forming, casting, welding, and machining technologies. Methods of plastic forming are used extensively to force metal into the required shape. The processes are diverse in scale, varying from forging and rolling of ingots weighing several tons to drawing of wires less than 0.025 mm in diameter. Most large-scale deformation processes are performed at high temperatures so that a minimum of force is needed and the consequent recrystallization refines the metallic structure. Cold forming is used when smoother surface finish and high-dimensional accuracy are required. Metals are produced in the form of bars or plates. On the other hand, casting produces a large variety of components in a single operation by pouring liquid metals into molds and allowing them to solidify. Parts manufactured by plastic forming, casting, sintering, and molding are often finished by subsequent machining operations, as shown in Figure 1.2.

Machining is the removal of the unwanted material (machining allowance) from the workpiece (WP), so as to obtain a finished product of the desired size, shape, and surface quality. The practice of removal of machining allowance through cutting techniques was first adopted using simple handheld tools made from bone, stick, or stone, which were replaced by bronze or iron tools. Water, steam, and later electricity were used to drive such tools in power-driven metal cutting machines (machine tools). The development of new tool materials opened a new era for the machining industry in which machine tool development took place. Nontraditional machining techniques offered alternative methods for machining parts of complex shapes in hard, stronger, and tougher materials that are difficult to cut by traditional methods. Figure 1.3 shows the general classification of machining methods based on the material removal mechanism.

Compared to plastic forming technology, machining technology is usually adopted whenever part accuracy and surface quality are of prime importance. The technology of material removal in machining is carried out on machine tools that are responsible for generating motions required for producing a given part geometry. Machine tools form around 70% of operating production machines and are characterized by their high production accuracy compared with metal forming machine tools. Machining activities constitute approximately 20% of the manufacturing activities in the United States.

This book covers the different technologies used for material removal processes in which traditional and nontraditional machine tools and operations are employed. Machine tool elements, drives, and accessories are introduced for proper selection and understanding of their functional characteristics and technological requirements.

1.2 HISTORY OF MACHINE TOOLS

The development of metal cutting machines (once briefly called machine tools) started from the invention of the cylinder, which was changed to a roller guided by a journal bearing. The ancient Egyptians used these rollers for transporting the required stones from a quarry to the building site. The use of rollers initiated the introduction of the first wooden drilling machine, which dates back

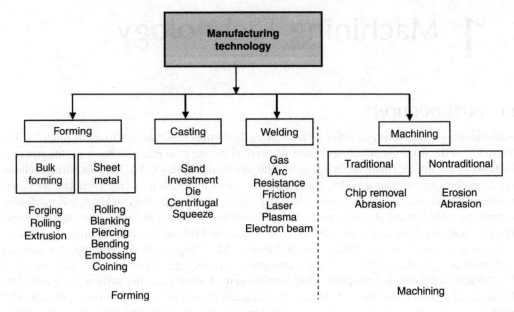

FIGURE 1.1 Classification of manufacturing processes.

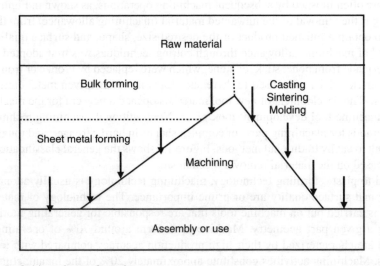

FIGURE 1.2 Definition of manufacturing.

to 4000 BC. In such a machine, a pointed flint stone tip acted as a tool. The first deep hole drilling machine was built by Leonardo da Vinci (1452–1519). In 1840, the first engine lathe was introduced. Maudslay (1771–1831) added the lead screw, back gears, and the tool post to the previous design. Later, slide ways for the tailstock and automatic tool feeding systems were incorporated. Planers and shapers have evolved and were modified by Sellers (1824–1905). Fitch designed the first turret lathe in 1845. That machine carried eight cutting tools on a horizontally mounted turret for producing screws. A completely automatic turret lathe was invented by Spencer in 1896. He was also credited with the development of the multispindle automatic lathe. In 1818, Whitney built the first milling machine; the cylindrical grinding machine was built for the first time by Brown and Sharpe in 1874. The first gear shaper was introduced by Fellows in 1896. In 1879, Pfauter invented the gear hobber, and the gear planers of Sunderland were developed in 1908. Figures 1.4 and 1.5 show the first wooden lathe and planer machine tools.

Machining Technology

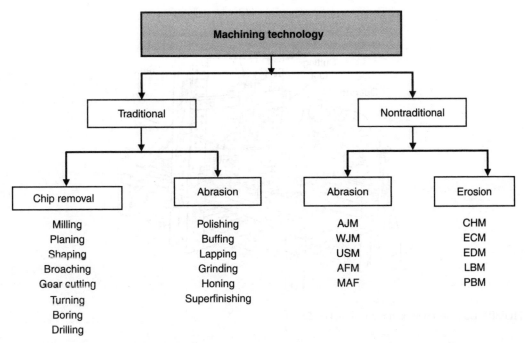

FIGURE 1.3 Classification of machining processes. AJM, abrasive jet machining; WJM, water jet machining; USM, ultrasonic machining; AFM, abrasive flow machining; MAF, magnetic abrasive finishing; CHM, chemical machining; ECM, electrochemical machining; EDM, electrodischarge machining; LBM, laser beam machining; PBM, plasma beam machining.

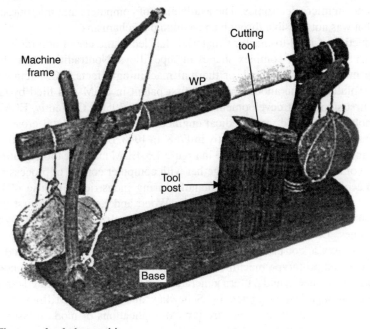

FIGURE 1.4 First wooden lathe machine.

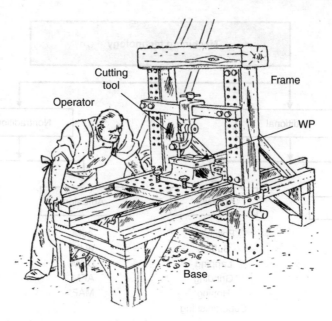

FIGURE 1.5 Wooden planer machine (1855).

Further developments for these conventional machines came via the introduction of copying techniques, cams, attachments, and automatic mechanisms that reduced manual labor and consequently raised product accuracy. Machine tool dynamometers are used with machine tools to measure, monitor, and control forces generated during machining operations. Such forces determine the method of holding the tool or WP and are closely related to product accuracy and surface integrity. In 1953, the introduction of numerical control (NC) technology opened doors to computer numerical control (CNC) and direct numerical control (DNC) machining centers that enhanced product accuracy and uniformity. Machine tools have undergone major technological changes through various developments in microelectronics. The availability of computers and microprocessors brought in flexibility that was not possible through conventional mechanisms.

The introduction of hard-to-machine materials has led to the use of nontraditional machining technology for production of complex shapes in superalloys. Nontraditional machining removes material using mechanical, chemical, or thermal machining effects. ECM removes material by electrolytic dissolution of the anodic WP. The first patent in ECM was filed by Gussef in 1929. However, the first significant development occurred in the 1950s. Currently, ECM machines are used in automobile, die, mold, and medical engineering industries. Metal erosion by spark discharges was first noted by Sir Joseph Priestly in 1768. In 1943, B. R. Lazerenko and N. I. Lazerenko introduced their first EDM machine, shown in Figure 1.6. EDM machine tools continued to develop through the use of novel power supplies together with computer control of process parameters that made EDM machines widespread in the manufacturing industries. The use of high-frequency sound waves in machining was noted in 1927 by Wood and Loomis. The first patent for USM appeared in 1945 by Balamuth. The benefits of USM were realized in the 1950s by the production of related machines. USM machines tackle a wide range of materials including glass, ceramic, and diamond. The earliest work on using electron beam machining (EBM) was attributed to Steigerwald, who designed the first prototype machine in 1947. Modern EBM machines are now available for drilling, perforation of sheets, and pattern generation associated with integrated circuit fabrication. Laser phenomenon was first predicted by Schawlaw and Townes. Drilling, cutting, scribing, and trimming of electronic components are typical applications of modern laser machine tools. The use of NC, CNC, computer-aided design or computer-aided manufacturing (CAD/CAM), and

Machining Technology

FIGURE 1.6 First industrial EDM machine in the world. Presentation of the Eleroda D1 at the EMO exhibition in Milan Italy, 1955. (Courtesy of Charmilles, 560 Bond St., Lincolnshire, IL.)

computer-integrated manufacturing (CIM) technologies provided robust solutions to many machining problems and made nontraditional machine tools widespread in industry. Table 1.1 summarizes the historical background of machine tools.

1.3 BASIC MOTIONS IN MACHINE TOOLS

In conventional machine tools, a large number of product features are generated or formed via the variety of motions given to the tool or the WP. The shape of the tool plays a considerable role in the final surface obtained. Basically, there are two types of motions in a machine tool. The primary motion, generally given to the tool or WP, constitutes the cutting speed. While the secondary motion feeds the tool relative to the WP. In some instances, combined primary motion is given either to the tool or to the WP. A classification of machine tool movements used for traditional machining is given in Table 1.2. Table 1.3 gives a classification for nontraditional machining technology. It may be concluded that movements of nontraditional machine tools are simple and mainly in the Z direction, while traditional machine tools have a minimum of two axes, that is, X and Y directions in addition to rotational movements.

1.4 ASPECTS OF MACHINING TECHNOLOGY

Machining technology covers a wide range of aspects that should be understood for proper understanding and selection of a given machining technology. Tooling, accessories, and the machine tool itself determine the nature of machining operation used for a particular material. As shown on the

TABLE 1.1
Developments of Machine Tools

1200–1299	Horizontal bench lathe appears, using foot treadle to rotate object
1770	Screw-cutting lathe invented: first to get satisfactory results (Ramsden, Britain)
1810	Lead screw adapted to lathe, leading to large-quantity machine-tool construction (Maudslay, Britain)
1817	Metal planing machine (Roberts, Britain)
1818	Milling machine invented (Whitney, United States)
1820–1849	Lathes, drilling, boring machines, and planers (most primary machine tools) refined
1830	Gear-cutting machine with involute cutters and geared indexing improved (Whitworth, Britain)
1830–1859	Milling machines, shapers, and grinding machines (United States)
1831	Surface-grinding machine patented (J. W. Stone, United States)
1834	Grinding machine developed: perhaps first (Wheaton, United States)
1836	Shaping machine invented; Whitworth soon added crank mechanism (Nasmyth, Britain)
1840 ca.	Vertical pillar drill with power drive and feed in use (originated in 1750)
1842	Gear-generating machine for cutting cycloidal teeth developed (Saxton, United States)
1850	Commercially successful universal milling machine designed (Robbins and Lawrence, Howe, and Windsor, United States)
1853	Surface grinder patented (Darling, United States)
1854 ca.	Commercial vertical turret lathe built for Robbins and Lawrence by Howe and Stone (Stone, Howe, Lawrence, United States)
1857	Whitney gauge lathe built (Whitney, United States)
1860–1869	First cylindrical grinder made; replaces single-point tool of engine lathe (United States)
1860–1879	Universal milling (1861–1865) and universal grinding machines (1876) produced (Brown and Sharpe, United States)
1873	Automatic screw machine invented (1893, produced finished screws from coiled wire—A2) (Spencer, United States)
1887	Spur-gear hobbing machine patented (Grant, United States)
1895	Multispindle automatic lathe introduced for small pieces (United States)
1896–1940	Heavy-duty precision, high production rate grinding machine introduced at Brown and Sharpe (Norton, United States)
1921	First industrial jig borer made for precision machining: based on 1912 single-point tool (Société Genevoise, Switzerland)
1943	Electrodischarge machining (spark erosion) developed for machine tool manufacturing
1944–1947	Centerless thread-grinding machine patented (Scrivener, Britain; United States)
1945	The USM was patented by Balamuth
1947	The first prototype of EBM was designed by Steigerwald
1950	Electrochemical machines introduced into industry
1952	Alfred Herbert Ltd.'s first NC machine tool operating
1958	Laser phenomenon first predicted by Schawlaw and Townes

Source: ASME International, 3 Park Ave., New York. With permission.

right-hand side of Figure 1.7, the main objective of the technology adopted is to utilize the selected machining resources to produce the component economically and at high rates of production. Parts should be machined at levels of accuracy, surface texture, and surface integrity that satisfy the product designer and avoid the need for postmachining treatment, which, in turn, maintains acceptable machining costs. The general aspects of machining technology include:

1.4.1 MACHINE TOOL

Each machine tool is capable of performing several machining operations to produce the part required at the specified accuracy and surface integrity. Machining is performed on a variety of

TABLE 1.2
Tool and WP Motions for Machine Tools Used for Traditional Machining

Machining Process	v (↻ →)		f (→)		Remarks
Chip removal					
Turning	WP	↻	Tool	→	WP stationary
Drilling	Tool	↻	Tool	→	
Milling	Tool	↻	WP	→	
Shaping	Tool	→	WP	⇢	Intermittent feed
Planing	WP	→	Tool	⇢	
Slotting	Tool	→	WP	⇢	
Broaching	Tool	•	WP	•	Feed motion is built in the tool
	WP	→	Tool	•	
Gear hobbing	Tool	↻	WP	↻	
			Tool	→	
Abrasion					
Surface grinding	Tool	↻	WP	→	
Cylindrical grinding	Tool	↻	WP	↻	
			Tool or WP	→	
Honing	Tool	↻ →		•	WP stationary
Superfinishing	WP	↻	Tool	→	

Note: ↻, Rotation; •, stationary; →, linear motion; ⇢, intermittent.

general-purpose machine tools that in turn perform many operations, including chip removal and abrasion techniques, by which cylindrical and flat surfaces are produced. Additionally, special-purpose machine tools are used to machine gears, threads, and other irregular shapes. Finishing technology for different geometries includes grinding, honing, lapping, and superfinishing techniques.

Figure 1.8 shows general-purpose machine tools used for traditional machining in chip removal and abrasion techniques. Typical examples of general-purpose machine tools include turning, drilling, shaping, milling, grinding, broaching, jig boring, and lapping machines intended for specific tasks. Gear cutting and thread cutting are examples of special-purpose machine tools. During the use of general- or special-purpose manual machine tools, product accuracy and productivity depend on the operator's participation during operation. Capstan and turret lathes are typical machines that somewhat reduce the operator's role during machining of bar-type or chucking-type WPs at higher rates and better accuracy. Semiautomatic machine tools perform automatically controlled movements, while the WP has to be hand loaded and unloaded. Fully automatic machine tools are those machines in which WP handling and cutting and other auxiliary activities are performed automatically. Semiautomatic and automatic machine tools are best suited for large production lots where the operator's interference is minimized or completely eliminated, and parts are machined more accurately and economically.

NC machine tools utilize a form of programmable automation by numbers, letters, and symbols using a control unit and tape reader, while CNC machine tools utilize a stored computer program to perform all the basic NC functions. NC and CNC have added many benefits to machining technology, since small and large numbers of parts can now be produced. Part geometry can be

TABLE 1.3
Tool and WP Motions for Nontraditional Machine Tools

	WP		Tool	
Machining Process	Stationary ●	Feed Movement ↓	Stationary ●	Remarks
Chemical (erosion)				In the slitting processes (plate cutting), a relative motion between tool and WP (traverse speed v_t) is imparted in horizontal directions (X,Y).
CHM	●		●	
ECM (sinking)	●	↓		
Thermal (erosion)				
EDM (sinking)	●	↓		
EBM (drilling)	●		●	
LBM (drilling)	●		●	
PBM (drilling)	●		●	
Mechanical (abrasion)				
USM	●		↓	
AJM	●		●	
WJM	●		●	
Abrasive water jet machining (AWJM)	●		●	

Note: ↻, Rotation; ●, stationary; →, linear motion; --→, intermittent.

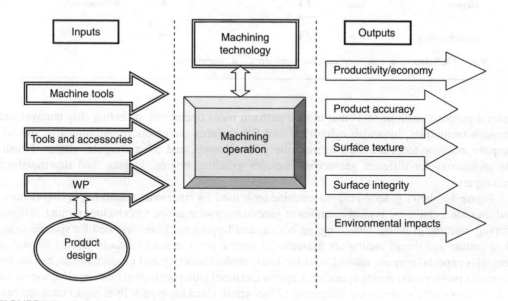

FIGURE 1.7 General aspects of machining technology.

changed through the flexible control of the part programs. The integration of CAD/CAM systems to machining technology has created new industrial areas in die, mold, aerospace, and automobile industries.

Hexapods have added a new area to the machining technology in which complicated parts can be machined using a single tool that is capable of reaching the WP from many sides. The hexapod has six degrees of movement and is very dexterous like a robot, but also offers the machine tool rigidity and accuracy generally beyond a robot's capability. The hexapods are used to help develop

Machining Technology

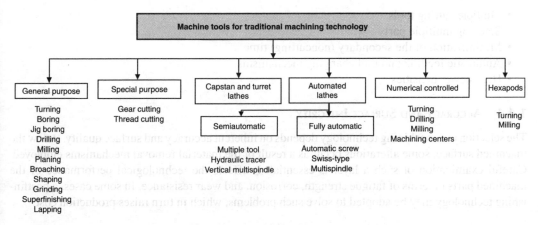

FIGURE 1.8 Classification of machine tools for traditional machining technology.

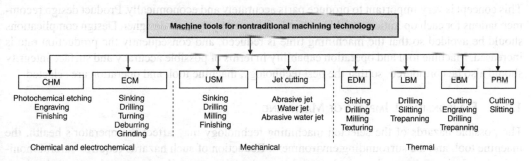

FIGURE 1.9 Classification of machine tools for nontraditional machining technology.

machining processes for WPs that need the dexterity offered by the hexapod design. For general-purpose machining, the hexapod is an ideal machine tool for mold-and-die machining applications. Its ability to keep a cutting tool normal to the surface being machined promotes use of larger radii ball nose end mills, which can cut more material with very small stepovers. In some applications, a flat nose end mill can be used very effectively for smooth surface finishes with little or no cusp. Nontraditional machining uses a wide range of machine tools such as ECM, USM, EDM, and LBM. Each machine tool is capable of performing a variety of operations, as shown in Figure 1.9. Nontraditional machining technology tackles materials that range from glass, ceramics, hard alloys, heat-resistant alloys, and other materials that are difficult to machine by traditional machining technology.

1.4.2 Workpiece Material

The WP material specified for the part influences the selection of the adopted machining method. Most materials can be machined by a range of processes, some by a very limited range. In any particular case, however, the choice of the material depends on the desired shape and size, the dimensional tolerances, the surface finish, and the required quantity. It must not depend only on technical suitability, but also on economy and environmental considerations.

1.4.3 Machining Productivity

The choice of any machining method should take into consideration a rate of production that is inversely proportional to machining time. Methods of raising productivity include the use of the following:

- High machining speeds
- High feed rates

- Multiple cutting tools
- Stacking multiple parts
- Minimization of the secondary (noncutting) time
- Automatic feeding and tool changing mechanisms
- High power densities

1.4.4 Accuracy and Surface Integrity

The selection of a machining technology depends on inherent accuracy and surface quality. Below the machined surface, some alterations occur as a result of the material removal mechanisms employed. Careful examination of such a layer is essential. It affects the technological performance of the machined parts in terms of fatigue strength, corrosion, and wear resistance. In some cases, a postfinishing technology may be adopted to solve such problems, which in turn raises production cost.

1.4.5 Product Design for Economical Machining

This concept is very important to produce parts accurately and economically. Product design recommendations for each operation should be strictly followed by the part designer. Design complications should be avoided so that the machining time is reduced, and consequently the production rate is increased. Machine tool and operation capability in terms of possible accuracy and surface integrity should also be considered, so that the best technology, machine tool, and operation are selected.

1.4.6 Environmental Impacts of Machining

The possible hazards of the selected machining technology may affect the operator's health, the machine tool, and the surrounding environment. Reduction of such hazards requires careful monitoring, analysis, understanding, and control toward environmentally clean machining technology. The hazards generated by the cutting fluids have led to the introduction of the minimum quantity lubrication (MQL), cryogenic machining, and dry machining techniques. Strict precautions are followed during laser beam machining (LBM) and abrasive jet machining (AJM), and these processes are covered in Chapter 11.

1.5 REVIEW QUESTIONS

1. Explain what is meant by *manufacturing*.
2. What are the different manufacturing methods used for metal shaping?
3. Explain the different mechanisms of material removal in machining technology.
4. List the main categories of machine tools used for traditional machining.
5. Classify the different nontraditional machine tools based on the material removal process.
6. Show basic motions of machine tools used for traditional and nontraditional processes.
7. Explain the different aspects of machining technology.
8. Explain what is meant by product design for economic machining.
9. Explain the importance of adopting an environmentally friendly machining technology.
10. What are the main objectives behind selecting a machining technology?

REFERENCES

ASME International, 3 Park Ave., New York.
Charmilles, 560 Bond St., Lincolnshire, IL.

2 Basic Elements and Mechanisms of Machine Tools

2.1 INTRODUCTION

Metal cutting machines (machine tools) are characterized by higher production accuracy compared with metal forming machines. They are used for the production of relatively smaller number of pieces; conversely, metal forming machines are economical for producing larger lots. Machine tools constitute about 70% of the total operating production machines in industry. The percentage of the different type of operating machine tools is shown in Table 2.1.

The successful design of machine tool requires the following fundamental knowledge:

1. Mechanics of the machining processes to evaluate the magnitude and direction and to control the cutting forces
2. The machinability of the different materials to be processed
3. The properties of the materials used to manufacture the different parts of the machine tool
4. The manufacturing techniques that are used to produce each machine tool part economically
5. The durability and capability of the different tool materials
6. The principles of engineering economy

The productivity of a machine tool is measured either by the number of parts produced in a unit of time, by the volumetric removal rate, or by the specific removal rate per unit of power consumed. Productivity levels can be enhanced using the following methods:

1. Increasing the machine speeds and feed rates
2. Increasing the machine tool available power
3. Using several tools or several WPs machined simultaneously
4. Increasing the traverse speed of the operative units during the nonmachining parts of the production time
5. Increasing the level of automation for the machine tool operative units and their switching elements
6. Adopting modern control techniques such as NC and CNC
7. Selecting the machining processes properly based on the machined part material, shape complexity, accuracy, and surface integrity
8. Introducing jigs and fixtures that locate and clamp the work parts in the minimum possible time

Machine tools are designed to achieve the maximum possible productivity and to maintain the prescribed accuracy and the degree of surface finish over their entire service life. To satisfy these requirements, each machine tool element must be separately designed to be as rigid as possible and

TABLE 2.1
Percentage of Different Types of Operating Machine Tools

Type of Machine Tool	Percentage
Lathes including automatics	34
Grinding	30
Milling	15
Drilling and boring	10
Planers and shapers	4
Others	7

then checked for resonance and strength. Furthermore, the machine tool, as whole, must have an adequate stability and should possess the following general requirements:

1. High static stiffness of the different machine tool elements such as structure, joints, and spindles
2. Avoidance of unacceptable natural frequencies that cause resonance of the machine tool
3. Acceptable level of vibration
4. Adequate damping capacity
5. High speeds and feeds
6. Low rates of wear in the sliding parts
7. Low thermal distortion of the different machine tool elements
8. Low design, development, maintenance, repair, and manufacturing cost

Machine tools are divided according to their specialization into the following categories:

- General-purpose (universal) machines, which are used to machine a wide range of products
- Special-purpose machines, which are used for machining articles similar in shape but different in size
- Limited-purpose machines, which perform a narrow range of operations on a wide variety of products

Machine tools are divided according to their level of accuracy into the following categories:

1. Normal-accuracy machine tools, which includes the majority of general-purpose machines
2. Higher accuracy machine tools, which are capable of producing finer tolerances and have more accurate assembly and adjustments
3. Machine tools of super-high accuracy, which are capable of producing very accurate parts

The main functions of a machine tool are holding the WPs to be machined, holding the tool, and achieving the required relative motion to generate the part geometry required.

Machine tools include the following elements:

1. A structure that is composed of bed, column, or frame
2. Slides and tool attachments
3. Spindles and spindle bearings
4. A drive system (power unit)
5. Work holding and tool holding elements

Basic Elements and Mechanisms of Machine Tools

6. Control systems
7. A transmission linkage

Stresses produced during machining, which tend to deform the machine tool or a WP, are usually caused by one of the following factors:

1. Static loads that include the weight of the machine and its various parts
2. Dynamic loads that are induced by the rotating or reciprocating parts
3. Cutting forces generated by the material removal process

Both the static and the dynamic loads affect the machining performance in the finishing stage, while the final degree of accuracy is also affected by the deflection caused by the cutting forces.

2.2 MACHINE TOOL STRUCTURES

The machine tool structure includes a body, which carries and accommodates all other machine parts. Figure 2.1 shows a typical machine tool bed of the lathe and a frame of the drilling machines. The main functions of the machine structure include the following:

1. Ability of the structure or the bed to resist distortion caused by static and dynamic loads
2. Stability and accuracy of the moving parts
3. Wear resistance of the guideway
4. Freedom from residual stresses
5. Damping of vibration

Machine tool structures are classified by layouts into open (C-frames) and closed frames. Open frames provide excellent accessibility to the tool and the WP. Typical examples of open frames are found in turning, drilling, milling, shaping, grinding, slotting, and boring machines (Figure 2.2). Closed frames find application in planers, jig boring, and double-spindle milling machines (Figure 2.3). A machine tool structure mounts and guides the tool and the WP and maintains their specified relative position during the machining process. Machine tool structures must therefore be designed to withstand and transmit, without deflection, the cutting forces and weights of the moving parts of the machine onto the foundation. For a multiunit structure, the unit must be designed to locate and guide each other in accordance with the required position between the tool and the WP.

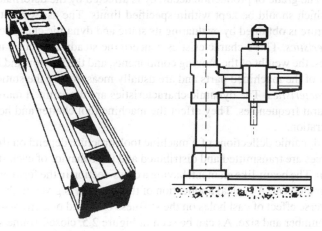

Lathe bed   Frame of radial drill

FIGURE 2.1 Typical bed of center lathe and frame of a drilling machine.

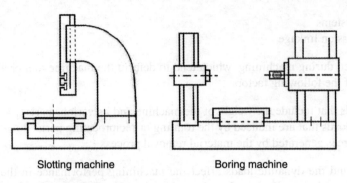

FIGURE 2.2 Examples of open frames (C-frames).

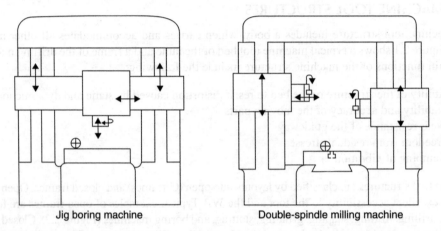

FIGURE 2.3 Examples of closed frames.

The configuration of machine tool structure is governed by the arrangement of the necessary cutting and feed movements and their stroke lengths, as well as the size and capacity of the machine. In this regard, chip disposal, transport, erection, and maintenance are also considered. The rate of material removal determines the power capacity of the machine tool and hence the magnitude of the cutting forces. The grade of production accuracy is affected by the deformation and deflections of the structure, which should be kept within specified limits. The assessment of the behavior of machine tool structure is obtained by evaluating its static and dynamic characteristics.

Static characteristics. These characteristics concern the steady deflection under steady operational cutting forces, the weight of the moving components, and the friction and inertia forces. They affect the accuracy of the machined parts and are usually measured by the static stiffness.

Dynamic characteristics. The dynamic characteristics are determined mainly by the dynamic deflection and natural frequencies. They affect the machine tool chatter and hence the stability of the machining operation.

The static and dynamic deflections of a machine tool structure depend on the manner by which the operational forces are transmitted and distributed and the behavior of each structural unit under operating condition. The beam-like element, having a cross-section in the form of a hollow rectangle, is the most superior element. A typical application of this concept is given in the lathe bed shown in Figure 2.4; the adverse effect of cast holes on the stiffness of closed box cross-section is minimized by reducing their number and size. As can be seen in Figure 2.5, closed-frame structures, although deformed under load, keep the alignment of their centerline axes unchanged. This, in turn, results in an axial (not lateral) displacement of the tool relative to the WP, which does not affect the accuracy

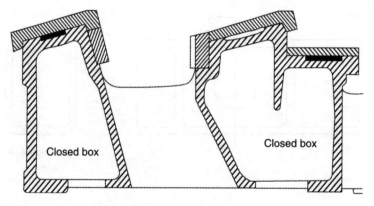

FIGURE 2.4 Hollow box sections of the lathe bed.

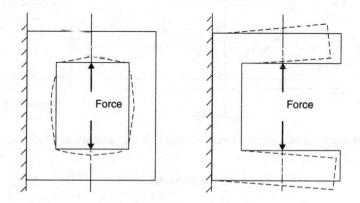

FIGURE 2.5 Deformation in open and closed frames.

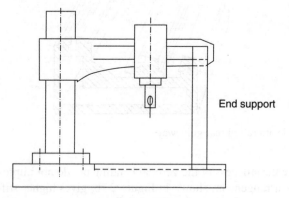

FIGURE 2.6 Radial drilling machine with end support.

of machined parts. Open frame can, therefore, be supplemented with a supporting element to close its frame during the machining operation, as shown in the radial drilling machine in Figure 2.6.

Machine tool stiffness and damping of its structure depend on the number and type of joint used to connect the different units of the structure. As a rule, the fewer the joints, the greater the stiffness of the structure and smaller its damping capability. The ribbing system is an effective method for increasing the stiffness of the machine tool structures. In this regard, simple vertical stiffeners, seen

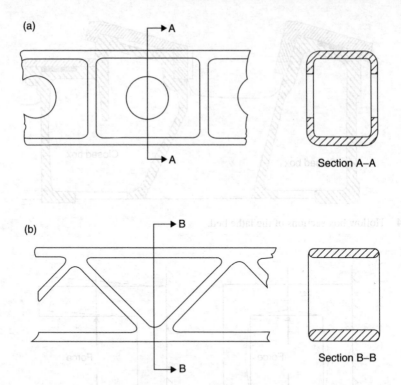

FIGURE 2.7 Arrangement of stiffeners in machine tool beds: (a) vertical and (b) diagonal stiffeners.

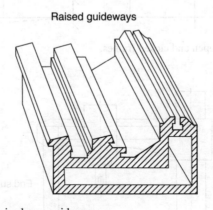

FIGURE 2.8 Lathe bed with raised rear guideways.

in Figure 2.7a, increase the stiffness of the vertical bending but do not improve horizontal bending. The diagonal stiffness arrangement, shown in Figure 2.7b, gives higher stiffness in both bending and torsion. In some cases, to eliminate the tilting movement that usually acts on the tailstock of the lathe machine, raised rear guideways are introduced, as shown in Figure 2.8. Machine tool frames can be produced as cast or welded construction. Welded structures ensure great saving of the material and the pattern costs. Figure 2.9 shows typical cast and fabricated machine tool structures. A cast iron (CI) structure ensures the following advantages:

- Better lubricating property (due to the presence of free graphite); most suitable for beds in which rubbing is the main criterion
- High compressive strength

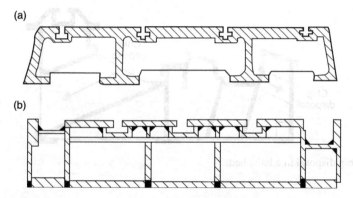

FIGURE 2.9 Cast and fabricated structures: (a) cast and (b) welded machine tool bases.

- Better damping capacity
- Easily cast and machined

2.2.1 Light- and Heavy-weight Constructions

Machine tool structures are classified according to their natural frequency as light- or heavy-weight construction. The natural frequency ω_0 of a machine tool can be described by

$$\omega_0 = \sqrt{\frac{k}{m}} \qquad (2.1)$$

where
k = structure static stiffness
m = mass

$$k = \frac{F}{\delta} \qquad (2.2)$$

where
F = force (N)
δ = deflection (mm)

To avoid resonance and thus reduce the dynamic deflection of the machine tool structure, ω_0 should be far below or far above the exciting frequencies, which is equal to a multiple of the rotational speed of the machine.

If the natural frequency of the machine structure is kept far below the speed working range of the machine tool then

$$\omega_0 < \text{exciting frequency}$$

or

$$\sqrt{\frac{k}{m}} < \text{exciting frequency}$$

This requirement is achieved by the increase of the mass m, which, in turn, leads to a heavyweight construction. On the other hand, lightweight constructions are made when

$$\omega_0 > \text{exciting frequency}$$

or

$$\sqrt{\frac{k}{m}} > \text{exciting frequency}$$

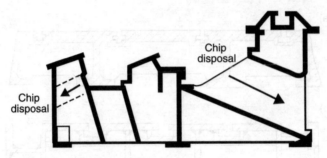

FIGURE 2.10 Chip disposal in a lathe bed.

Chip disposal, in the case of high-production machine tools, affects the construction of the machine tool frame as shown in Figure 2.10.

2.3 MACHINE TOOL GUIDEWAYS

Machining occurs as a result of a relative motion between the tool and the WP. Such a motion is a rotary, linear, or rectilinear one. Guideways are required to perform the necessary machine tool motion at a high level of accuracy under severe machining conditions. Generally guideways, therefore, control the movement of the different parts of the machine tool in all positions during machining and nonmachining times. Besides the accuracy requirements, ease of assembly, and economy in manufacturing guideways, the following features should be provided:

- Accessibility for effective lubrication
- Wear resistance, durability, and rigidity
- Possibility of wear compensation
- Restriction of motion to the required directions
- Proper contact all over the sliding area

Guideways are classified as sliding friction, rolling friction, and externally pressurized (Figure 2.11).

2.3.1 SLIDING FRICTION GUIDEWAYS

Sliding friction guideways consist of any one of or a combination of the flat, vee, dovetail, and cylindrical guideway elements. Flat circular guideways are used for guiding the rotating table of the vertical turning and boring machines. Figure 2.12 shows the different types of guideways that are normally used to guide sliding parts in the longitudinal directions. Holding strips may be provided to prevent the moving part from lifting or tilling by the operational forces. Scraping and the introduction of thin shims are used for readjustments that may be required to compensate wear of the sliding parts.

Vee-shaped guideways are either male or female type, which are self-adjusting under the weight of the guided parts. Practically, a vee guideway is usually combined with a flat one, as the case of the carriage guides of the center lathe to ensure proper contact all over the sliding surfaces. The combination of two vee guideways has an unfavorable effect on the machining accuracy and is limited to guideways of relatively small distance between the two vees. Circular vee guideways carry the operational loads and provide self location for the rotating table. Dovetail guideways, shown in Figure 2.12c, are used separately or in a combination of half dovetail and flat guideways. Cylindrical guides, shown in Figure 2.12d, are either male or female type that must be accurately manufactured. They require special devices to adjust their working clearances. The column of the drilling machine is a

Basic Elements and Mechanisms of Machine Tools

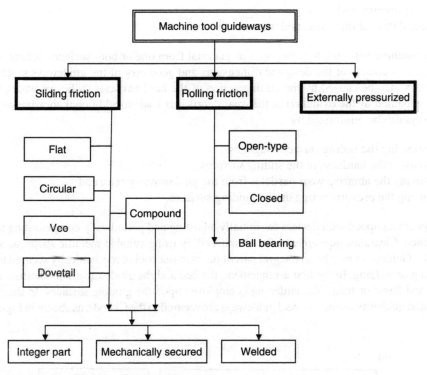

FIGURE 2.11 Classification of machine tool guideways.

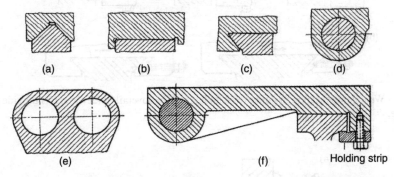

FIGURE 2.12 Types of guideways: (a) vee, (b) flat, (c) dovetail, (d) cylinder, (e) cylindrical–cylindrical, and (f) cylindrical–flat.

typical example of the male type, while the sleeve of the drilling machine spindle is a female type. The combinations of cylindrical guideways are shown in Figures 2.12e (cylindrical–cylindrical) and 2.12f (cylindrical–flat).

For the sliding surfaces, the bulk of the load is carried on the metal-to-metal contact. The load carried by the lubricating oil film is very small. The localized pressures cause elastic or plastic deformation to the supporting asperities of the surface, which in turn results in an instability of the sliding motion usually known as the *stick–slip effect*. This phenomenon can be reduced or eliminated by the use of proper lubricants or through the introduction of externally pressurized guideways. Friction condition and, consequently, the wear of the guideways are affected by

1. material properties of the fixed and moving element,
2. surface dimensions of the guideways,

3. acting pressure, and
4. accumulation of dirt, chip, and wear debris.

When the machine parts rub together, loss of material from one or both surfaces occurs, which in turn results in a change of the designed dimensions and geometry of the guideways system. Wear of guideways may be caused by the cutting action of the hard particles (adhesive wear), which is often accompanied by the oxidation of the wear debris that leads to additional abrasive wear. Wear of guideways can be minimized by

1. minimizing the sliding surface roughness,
2. increasing the hardness of the sliding surfaces,
3. removing the abrasive wear particles from the guideways system, and
4. reducing the pressure acting on the guiding surfaces.

Guideways are equipped with devices for initially adjusting and periodically compensating the working clearance. Clearance adjustment is accomplished by using suitable metallic strips, as shown in Figure 2.13. Guideways may be an integral part of the machine tool or mechanically secured to the bed by fastening or welding. In the first arrangement, the bed and the guideway are made from the same material, and flame or induction hardening is employed upon the guiding surfaces. In the mechanically secured guideways, separate steel guideways are secured to the CI beds, as shown in Figure 2.14a.

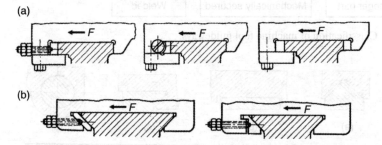

FIGURE 2.13 Wear compensation in guideways: (a) flat and (b) dovetail guideways. F is the side force acting on the carriage.

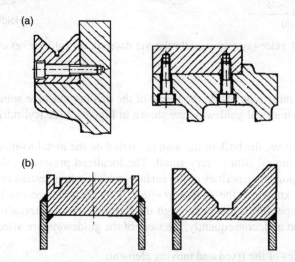

FIGURE 2.14 (a) Mechanically secured and (b) welded guideways.

Basic Elements and Mechanisms of Machine Tools

In plastic guideways, plates of phenolic resin bonded fiber are inserted into one of the sliding surfaces. These guideways reduce friction and stick–slip effect. They also reduce the danger of seizure when lubricant is inadequate and minimize vibrations. The design and arrangement of the guideways must prevent chip and dirt accumulation, which promotes the rate of wear. Methods of protecting guideways against foreign matter include

1. extending the length of the moving parts using cover plates that protect the guideways and
2. providing covering belts or a telescopic plate that surrounds the guideways and seals them from external materials.

2.3.2 Rolling Friction Guideways

In rolling friction guideways, rollers, needles, or balls are inserted between the moving parts to minimize the frictional resistance, which is kept constant irrespective of the traveling speed. Rolling friction guideways find wide applications in numerically controlled and medium-size machine tools in which the setting accuracy is decisive. Their expensive manufacturing, complicated construction, and the short life of the rolling elements create problems. Rolling friction guideways are either open or closed. The open type (Figure 2.15) is used when the load acts downward, which makes this type self-adjusting for wear in the guideways. In the closed type, wear compensation requires adjusting elements. For long strokes, recirculating rolling elements (as shown in Figure 2.16) or ball or roller bearing guideways (Figure 2.17) are used to shorten the length of the slider.

Circular rolling friction guideways find applications in high-speed vertical lathes. The size and the distribution of the load on the rolling elements and the deformation of the guideways are affected by:

1. magnitude, distribution, and type of loading,
2. stiffness of the rolling elements,
3. manufacturing errors of the rolling elements,
4. form error of the guideways,
5. magnitude of preloading, and
6. stiffness of the table, bed, fixture, and WP.

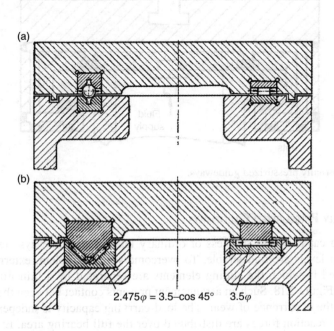

FIGURE 2.15 Open-type rolling friction guideways: (a) flat and (b) vee–flat guideways.

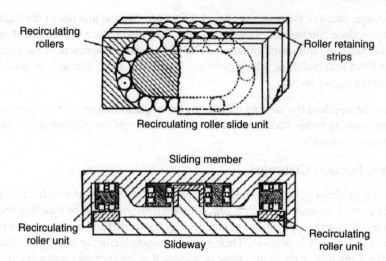

FIGURE 2.16 Recirculating rolling friction guideways.

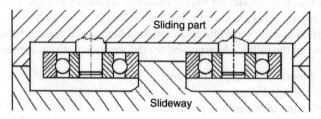

FIGURE 2.17 Ball bearing guideway.

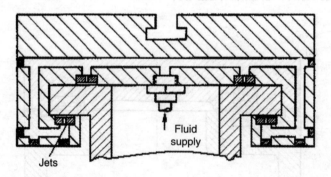

FIGURE 2.18 Externally pressurized guideways.

2.3.3 EXTERNALLY PRESSURIZED GUIDEWAYS

The load-carrying capacity and stiffness of ordinary lubricated guideways are excellent; however, their friction levels are undesirable. To overcome such a problem, externally pressurized guideways are used in which the sliding elements are separated by a thin film of pressurized fluid, as shown in Figure 2.18. Such an arrangement prevents contact between the sliding surfaces and hence avoids the occurrence of wear. The load-carrying capacity is independent of the sliding speed, and the reaction forces are distributed over the full bearing area. Externally pressurized guideways are ideal guideways in terms of stiffness, uniformity of travel, low friction, large damping, and better heat dissipation capacity. Generally, the service properties of machine tool guideways can be improved by

1. providing favorable frictional conditions, which can be achieved by using
   a. combined sliding and rolling guides,
   b. proper lubricants and materials for guideways, and
   c. hydrostatic ways with high-rigidity oil film and automatic control systems,
2. providing adequate protection of guideways,
3. using optimal cross-section of slideways, and
4. using optimal surface finishes.

2.4 MACHINE TOOL SPINDLES

Machine tool spindles are used to locate, hold, and drive the tool or the WP. These spindles possess a high degree of rigidity, rotational accuracy, and wear resistance. Spindles of the general-purpose machine tools are subjected to heavier loads compared with precision ones. In the former class of spindles, rigidity is the main requirement; in the second, the manufacturing accuracy is of the prime consideration. Spindles are normally made hollow and provided with an internal taper at the nose end to accommodate the center or the shank of the cutting tool (Figure 2.19). A thread can be added at the nose end to fix a chuck or a face plate. Medium-carbon steel containing 0.5% C is used for making spindles in which hardening is followed by tempering to produce a surface hardness of about 40 Rockwell (HRC). Low-carbon steel containing 0.2% C can also be carburized, quenched, and tempered to produce a surface hardness of 50–60 HRC. Spindles for high-precision machine tools are hardened by nitriding, which provides a sufficient hardness with the minimum possible deformation. Manganese steel is used for heavy-duty machine tool spindles.

2.4.1 SPINDLE BEARINGS

Machine tool spindles are supported inside housings by means of ball, roller, or antifriction bearings. Precision bearings are used for a precision machine tool. The geometrical accuracy and surface finish of the machined components depend on the quality of the spindle bearings. The considerable attention paid to the spindle design, selection and proper mounting of its bearings, and the construction of the housing of bearings makes the spindle system one of the most expensive parts of the machine tools. Drive shafts, which are subjected to bending and tensional stresses, are designed on the basis of strength while spindles are designed on the basis of stiffness. Generally, machine tool spindle bearings must provide the following requirements:

1. Minimum deflection under varying loads
2. Accurate running under loads of varying magnitudes and directions
3. Adjustability to obtain minimum axial and radial clearances
4. Simple and convenient assembly
5. Sufficiently long service

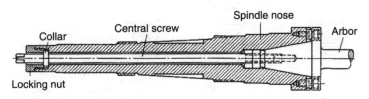

FIGURE 2.19 Typical milling machine spindle.

6. Maximum temperature variation throughout the speed ranges
7. Sufficient wear resistance

The forces acting on a machine tool spindle are the cutting force, which acts at the spindle nose, and the driving force, which acts in between the spindle bearings (Figure 2.20). The cutting force can be resolved into two components with respect to the spindle. The spindle bearings have to take radial and axial components of the cutting and driving forces. In this manner, when the machine tool spindle is mounted at two points, the bearing at one point takes the axial component besides the reaction of the radial component, while the other takes only the reaction of the radial component. The bearings that carry the axial component should prevent the axial movement of the spindle under the effect of the cutting and driving forces (fixed bearing). The other bearing (floating bearing) provides only a radial support and provides axial displacement due to differential thermal expansion of the spindle shaft and the housing.

The arrangement shown in Figure 2.21a is used in most high-speed machine tools because the free length of the spindle (from nose to the fixed bearing) is limited, which minimizes nose deflection. Additionally, the effect of differential thermal expansion of the spindle and spindle housing acts toward the floating (rear) end, which in turn reduces the axial displacements of the spindle nose. Figure 2.22 shows typical spindle bearing mounting arrangements. Figure 2.23 presents a

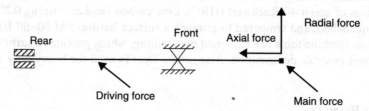

FIGURE 2.20 Forces acting on machine tool spindles.

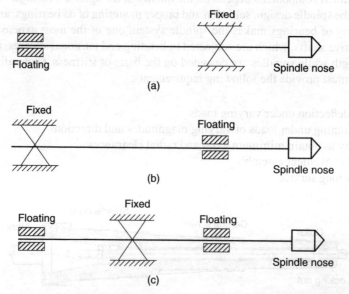

FIGURE 2.21 Fixed and floating bearing arrangements: (a) fixed front, (b) fixed rear, and (c) fixed middle.

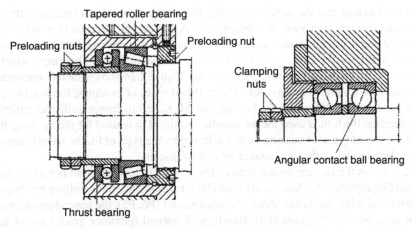

FIGURE 2.22 Typical spindle-bearing arrangements. (From Browne, J. W., *The Theory of Machine Tools*, *Book-1*, Cassell and Co. Ltd., London, 1965.)

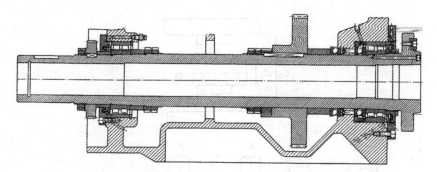

FIGURE 2.23 Typical machine tool spindle. (From Koenigsberger, F., *Berechnungen, Konstruktiosgrundlagen und Bauelemente spanender Werkzeugmaschinen*, Springer, Berlin, 1961. With permission.)

machine tool spindle with the fixed front bearing while the rear end axially slides at the outer race of the roller bearing. The various considerations in the selection of bearings are

1. direction of load relative to the bearing axial,
2. intensity of load,
3. speed of rotation,
4. thermal stability,
5. stiffness of the spindle shaft, and
6. class of accuracy of the machine.

Ball bearings sustain considerable loads; roller bearings are preferred for severe conditions and shock loads. Tapered roller bearings are suitable for high axial and radial forces (combined loads). To increase the accuracy of ball and roller bearings, these are fitted with very high interference fits, which eliminate the radial play between the bearing and the spindle. Angular contact ball bearings or roller bearings, installed in pairs, are preloaded by the adjustments made during their assembly.

2.4.2 Selection of Spindle-Bearing Fit

The high accuracy requirements of a machine tool have direct implications on the method of bearing mounting and the type of fit in the spindle assembly. To prevent creep, roll, or excessive interference fitting of bearing on either the spindle or the housing, it is important to select the correct fit between the bearing and the seats. A bearing fit (inner race on the spindle or outer race in the housing) is interference, transition, or clearance fit. A correct interference fit provides proper support around the whole circumference and, hence, provides a correct load distribution; moreover, the load-carrying ability of the bearing is fully utilized. In case of floating bearings, which are made free to move axially, interference fit is unacceptable. Figure 2.24 shows the recommended types of fit for machine tool spindle bearings. Apart from thrust types of bearings, the fixed bearing on the spindle has j or k types of fit, while the housing has M, K, or J to ensure sufficient stiffness. In case of floating bearing, the h fit is used for the spindle and the H fit is used for the housing. Because the stiffness of the thrust types of bearings is not affected by the type of fit, the spindle has a transition fit, while the housing has either a clearance or a transition fit.

Table 2.2 shows the recommended type of fits applied to machine tool bearings. According to International Organization for Standardization (ISO) recommendations, rolling bearings are manufactured in normal tolerance grade, close tolerance grades (P6, P5), the special precision (SP) grade, and the ultraprecision (UP) grade (P4). Bearings of normal tolerance grades are of general use, while SP and UP grades are used in spindles of high-precision machine tools. For comparison of the fits for machine tool spindles, see Table 2.3.

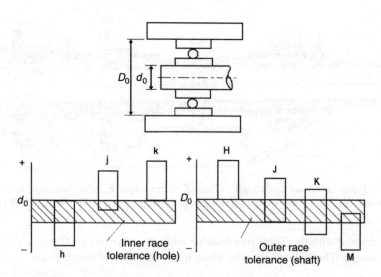

FIGURE 2.24 Recommended bearing fits.

TABLE 2.2
Recommended Fits for Machine Tool Spindle Bearings

	Spindle				Housing		
Bearing Type	P6	P5-SP	P4-UP	Working Conditions	P6	P5-SP	P4-UP
Deep groove ball bearing	j5	j4	j3	Point load	J5	J4	J5
Angular contact ball bearing				Rotating load	M6	M5	M4
Cylindrical roller bearing	k5	k4	—	Point load	K6	K5	K4
				Rotating load	M6	M5	M4
Tapered roller bearing	k5	k4	—	Loose adjustable	J6	J5	—
				Tight adjustable	K6	K5	—
				Rotating load	M6	M5	—
Angular contact ball thrust bearing	—	h5	h4		K6	K5	K4
Ball thrust bearing	h6	h5	—		H8	H8	—

Basic Elements and Mechanisms of Machine Tools

TABLE 2.3
Bearing Tolerances (Microns)

Tolerance Grade	Bore (80 mm)			Outer Diameter (80 mm)		
	Bore	Width	Radial Runout	Diameter	Width	Radial Runout
Normal	5–20	25	25	5–20	25	25
P4	0–8	4	5	0–8	5	6
SP	0–10	7	5	0–10	7	6
UP	0–8	3	3	0–8	3	3

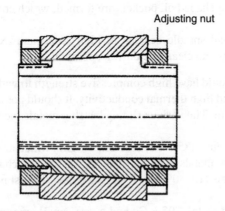

FIGURE 2.25 Sliding friction bearing.

2.4.3 SLIDING FRICTION SPINDLE BEARING

Rolling bearings are used at a speed and diameter range of $n \cdot d \leq 2 \times 10^5$ where n is the spindle rotational speed in revolutions per minute and d is the diameter of the spindle in millimeters. At higher running speeds, the bearing life is reduced due to the gyratory action, especially in bearings that take combined loads. At high spindle speeds, as in case of grinding, sliding friction (journal) bearings that have high damping capacity compared with rolling bearings are used. Their load-carrying capacity increases as the spindle speed increases due to the hydrodynamic action created within the bearing. For an optimum performance, the radial clearance between journal and bearing should be properly maintained, as it affects bearing friction, load-carrying capacity, and the efficiency of heat dissipation of the bearing. The main types of sliding friction bearings include the following:

1. Sliding bearing with radial play adjustment using segments that can be adjusted radically to control the clearance.
2. Bearing with axial play adjustment, in which a bush with a cylindrical bore and external taper has a slot along its length and is made to fit in a taper hole in the housing. When the bush slides axially, through two opposing nuts, on the two ends of the bush, radial play can be finely adjusted and controlled (Figure 2.25).
3. Mackensen bearing is used in highly accurate machine tool spindles, running at extremely high speeds, under limited applied load. As shown in Figure 2.26, an elastic bearing bush is supported at three points in the housing. This bush has nine equally spaced axial slots along its circumference. When the shaft is running, the bush deforms into a triangular

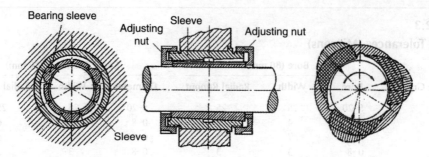

FIGURE 2.26 Mackensen bearing.

shape, and three wedge-shaped oil pockets are formed, which constitute the load-carrying parts of the bearing.
4. Hydrodynamic multipad spindle bearing of high radial and axial thrust capacity, high stiffness, and practically no clearance during operation.

Sliding bearing materials should have high compressive strength to withstand the bearing pressure, low coefficient of friction, and high thermal conductivity. It should possess high wear resistance and maintain a continuous oil film. The various sliding bearing metals include

1. copper base bearing metals (85% Cu, 10% Sn, 5% Zn), which are used for heavy loads,
2. tin base bearing (babbit) metals (85% Sn, 10% Sb, 5% Cu), which are used for higher loads,
3. lead base bearing metals (10–30% Pb, 10–15% Sb, and the rest is copper), which are used for light loads, and
4. cadmium base bearing metals (95% Cd and a very small amount of iridium) which have higher compressive strength and more favorable properties at higher temperatures.

2.5 MACHINE TOOL DRIVES

To obtain a machined part by a machine tool, coordinated motions must be imparted to its working members. These motions are either primary (cutting and feed) movements, which removes the chips from the WP or auxiliary motions that are required to prepare for machining and ensure the successive machining of several surfaces of one WP or a similar surface of different WPs. Principal motions may be either rotating or straight reciprocating. In some machine tools, this motion is a combination of rotating and reciprocating motions. Feed movement may be continuous (lathes, milling machine, drilling machine) or intermittent (shapers, planers). As shown in Figure 2.27, stepped motions are obtained using belting or gearing. Stepless speeds are achieved by mechanical, hydraulic, and electrical methods.

2.5.1 Stepped Speed Drives

2.5.1.1 Belting

The belting system, shown in Figure 2.28, is used to produce four running rotational speeds n_1, n_2, n_3, and n_4. It is cheap and absorbs vibrations. It has the limitation of the low-speed changing, slip, and the need for more space. Based on the driver speed n_1, the following speeds can be obtained in a decreasing order:

$$n_1 = n\frac{d_1}{d_5} \qquad (2.3)$$

$$n_2 = n\frac{d_2}{d_6} \qquad (2.4)$$

Basic Elements and Mechanisms of Machine Tools

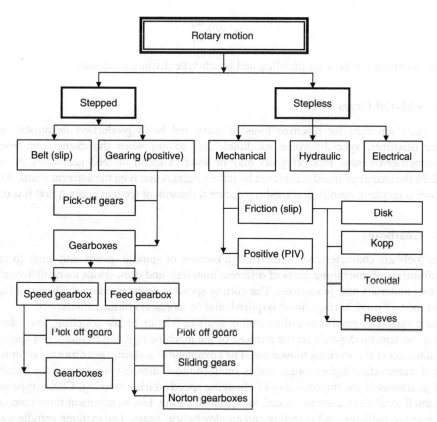

FIGURE 2.27 Classification of transmission of rotary motion.

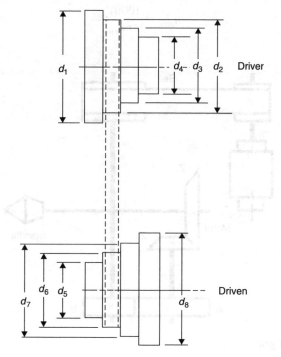

FIGURE 2.28 Belting transmission.

$$n_3 = n\frac{d_3}{d_7} \tag{2.5}$$

This type is commonly used for grinding and bench-type drilling machines.

2.5.1.2 Pick-Off Gears

Pick-off gears are used for machine tools of mass and batch production (automatic and semi-automatic machines, special-purpose machines, and so on) when the changeover from job to job is comparatively rare. Pick-off gears may be used in speed or feed gearboxes. As shown in Figure 2.29, the change of speed is achieved by setting gears A and B on the adjacent shafts. As the center distance is constant, correct gear meshing occurs if the sum of teeth of gears A and B is constant.

2.5.1.3 Gearboxes

Machine tools are characterized by their large number of spindle speeds and feeds to cope with the requirements of machining parts of different materials and dimensions using different types of cutting tool materials and geometries. The cutting speed is determined on the bases of the cutting ability of the tool used, surface finish required, and economical considerations.

A wide variety of gearboxes utilize sliding gears or friction or jaw coupling. The selection of a particular mechanism depends on the purpose of the machine tool, the frequency of speed change, and the duration of the working movement. The advantage of a sliding gear transmission is that it is capable of transmitting higher torque and is small in radial dimensions. Among the disadvantages of these gearboxes is the impossibility of changing speeds during running. Clutch-type gearboxes require small axial displacement needed for speed changing, less engagement force compared with sliding gear mechanisms, and therefore can employ helical gears. The extreme spindle speeds of a machine tool main gearbox n_{max} and n_{min} can be determined by

$$n_{max} = \frac{1000 V_{max}}{\pi d_{min}} \tag{2.6}$$

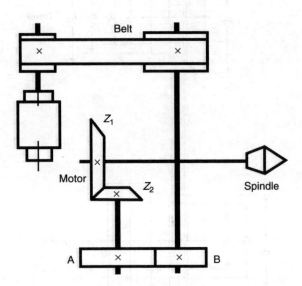

FIGURE 2.29 Pick-off gears.

Basic Elements and Mechanisms of Machine Tools

$$n_{min} = \frac{1000 V_{min}}{\pi d_{max}} \quad (2.7)$$

where

V_{max} = maximum cutting speed (m/min) used for machining the most soft and machinable material with a cutting tool of the best cutting property

V_{min} = minimum cutting speed (m/min) used for machining the hardest material using a cutting tool of the lowest cutting property or the necessary speed for thread cutting

d_{max}, d_{min} = maximum and minimum diameters (mm) of WP to be machined

The speed range R_n becomes

$$R_n = \frac{n_{max}}{n_{min}} = \frac{V_{max}}{V_{min}} \cdot \frac{d_{max}}{d_{min}} = R_v \cdot R_d \quad (2.8)$$

where

R_v = cutting speed range
R_d = diameter range

In case of machine tools having rectilinear main motion (planers and shapers), the speed range R_n is dependent only on R_v. For other machine tools, R_n is a function of R_v and R_d, large cutting speeds and diameter ranges are required. Generally, when selecting a machine tool, the speed range R_n is increased by 25% for future developments in the cutting tool materials. Table 2.4 shows the maximum speed ranges in modern machine tools.

2.5.1.4 Stepping of Speeds According to Arithmetic Progression

Let $n_1, n_2, \ldots, n_z$ be arranged according to arithmetic progression. Then

$$n_1 - n_2 = n_3 - n_2 = \text{constant} \quad (2.9)$$

The sawtooth diagram in such a case is shown in Figure 2.30. Accordingly, for an economical cutting speed v_0, the lowest speed v_1 is not constant; it decreases with increasing diameter. Therefore, the arithmetic progression does not permit economical machining at large diameter ranges.

TABLE 2.4
Speed Range for Different Machine Tools

Machine	Range
Numerically controlled lathes	250
Boring	100
Milling	50
Drilling	10
Surface grinding	4

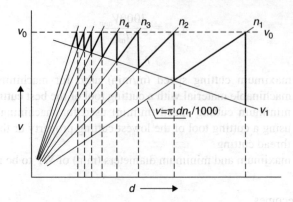

FIGURE 2.30 Speed stepping according to arithmetic progression.

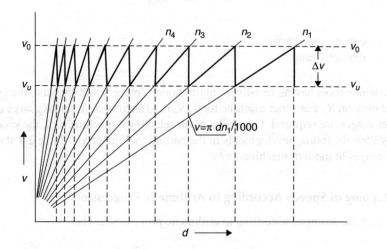

FIGURE 2.31 Speed stepping according to geometric progression.

The main disadvantage of such an arrangement is that the percentage drop from step to step δ_n decreases as the speed increases. Thus the speeds are not evenly distributed and more concentrated and closely stepped, in the small diameter range than in the large one. Stepping speeds according to arithmetic progression are used in Norton gearboxes or gearboxes with a sliding key when the number of shafts is only two.

2.5.1.5 Stepping of Speeds According to Geometric Progression

As shown in Figure 2.31, the percentage drop from one step to the other is constant, and the absolute loss of economically expedient cutting speed Δv is constant all over the whole diameter range. The relative loss of cutting speed $\Delta v_{max}/v_0$ is also constant. Geometric progression, therefore, allows machining to take place between limits v_0 and v_u independent of the WP diameter, where v_0 is the economical cutting speed and v_u is the allowable minimum cutting speed. Now suppose that $n_1, n_2, n_3, \ldots, n_z$ are the spindle speeds. According to the geometric progression,

$$\frac{n_2}{n_1} = \frac{n_3}{n_2} = \varphi \qquad (2.10)$$

TABLE 2.5
Standard Values of Progression Ratio φ According to ISO/R229 and Deutsches Institüt für Normung (DIN) 323

Basic and Derived Series	Standard Value	Accurate Value	Percentage Drop	Application
R20	$\sqrt[20]{10} = 1.12$	1.1221	10	Seldom used
R20/2	$(\sqrt[20]{10})^2 = 1.26$	1.258	20	Machines of large z
R20/3	$(\sqrt[20]{10})^3 = 1.4$	1.4125	30	Machines of large R_n
R20/4	$(\sqrt[20]{10})^4 = 1.6$	1.5849	40	and small z
R20/6	$(\sqrt[20]{10})^6 = 2.0$	1.9953	50	Drilling machines

Note: z, Number of speeds; R_n, speed range.

where φ is the progression ratio. The spindle speeds can be expressed in terms of the minimal speed n_1 and progression ratio φ.

n_1	n_2	n_3	n_4	n_z
n_1	$n_1\varphi$	$n_1\varphi^2$	$n_1\varphi^3$	$n_1\varphi^{z-1}$

Hence, the maximum spindle speed n_z is given by

$$n_z = n_1 \varphi^{z-1} \qquad (2.11)$$

where z is the number of spindle speeds, therefore,

$$\varphi = \sqrt[z-1]{\frac{n_z}{n_1}} = \sqrt[z-1]{R_n} = (R_n)^{1/(z-1)} \qquad (2.12)$$

from which

$$z = \frac{\log R_n}{\log \varphi} + 1 \qquad (2.13)$$

Progression ratios are standardized according to ISO standards in such a way as to allow standard speeds and feeds, including full load induction motor speeds of 2800, 1400, and 710 rpm to be used. Table 2.5 shows the standard values of φ according to ISO/R229. Similarly, machine tool speeds are standardized according to ISO/R229. Such speeds enable the direct drive of machine tool spindles using induction motors with changing poles. The full load speeds of induction motors are 236, 280, 322, 472, 200, 710, 920, 1400, and 2800 rpm. Tables 2.6 and 2.7 show the standard speeds and feeds according to ISO/R229.

ILLUSTRATIVE EXAMPLE

The following speeds form a geometric progression. Find the progression ratio and the percentage increase in the speed series.

Solution

n_1 (rpm)	n_2 (rpm)	n_3 (rpm)	n_4 (rpm)	n_5 (rpm)	n_6 (rpm)
14	18	22.4	28	35.2	45

TABLE 2.6
Standard Speeds According to ISO/R229 and DIN 804

Accurate Value (rpm)	Basic Series			Derived Series			Limiting Values	Considering 2% Mechanical Tolerance
	R20	R20/2	R20/3	R20/4 1400–800	R20/6 2800			
	$\varphi = 1.12$	$\varphi = 1.25$	$\varphi = 1.4$	$\varphi = 1.6$	$\varphi = 2.0$		-2%	$+2\%$
100	100						98	102
112.2	112	112	11.2		112	11.2	110	114
162.89	125			125			123	128
141.25	140	140		1400	140		1400	138	144
158.49	160		16				155	162
177.83	180	160		180		180	180	174	181
199.52	200			2000			193	204
223.87	224	224	22.4		224	22.4	219	228
251.19	250			250			246	256
281.84	280	280		2800	280		2800	276	287
316.23	315		31.5				310	323
854.81	355	355		355	355		355	348	368
398.11	400			4000			390	406
446.68	450	450	45		450	45	448	456
501.19	500			500			491	511
562.34	560	560		5600	560		5600	551	574
630.96	630		63				618	643
707.95	710	710		710		710	710	694	722
794.33	800			8000			778	810
891.25	900	900	90		900	90	873	909
1000	1000			1000			980	1020

or

$$\varphi = \frac{n_2}{n_1} = \frac{18}{14} = 1.25$$

$$\varphi = \sqrt[5]{\frac{45}{14}} = 1.25$$

The percentage increase in speed δ_n

$$\delta_n = \frac{n_2 - n_1}{n_1} = \frac{\varphi n_1 - n_1}{n_1} = (\varphi - 1) \times 100$$

hence, $\delta_n = (1.25 - 1) \times 100 = 25\%$.

ILLUSTRATIVE EXAMPLE

Given $n_1 = 2.8$ rpm, $n_z = 31.50$ rpm, and $\varphi = 1.41$, calculate the speed range R_n and the number of speeds z.

Solution

$$R_n = \frac{n_z}{n_1} = \frac{31.50}{2.8} = 11.2$$

since

$$\varphi = (R_n)^{1/(z-1)}$$

Basic Elements and Mechanisms of Machine Tools

TABLE 2.7
Standard Feeds According to ISO/R229 and DIN 803

		Nominal Values		
R20	R20/2	R20/3 ...1...	R20/4	R20/6 ...1...
$\varphi = 1.12$	$\varphi = 1.25$	$\varphi = 1.4$	$\varphi = 1.6$	$\varphi = 2.0$
1.00	1.0	1.0	1.0	1.0
1.12		11.2		
1.25	1.25	0.125		0.125
1.40		1.4		
1.60	1.6	16	1.6	16
1.80		0.18		
2.00	2.0	2.0		2.0
2.24		20		
2.50	2.5	0.25	2.5	0.25
2.80		2.8		
3.15	3.15	31.5		31.5
3.55		0.355		
4.00	4.0	4.0	4	4.0
4.50		45		
5.00	5.0	0.5		0.5
5.60		5.6		
6.30	6.3	63	6.3	63
7.10		0.71		
8.00	8.0	8.0		8.0
9.00		90		
10.00	10.0	1000	10	

or

$$z = \frac{\log R_n}{\log \varphi} + 1$$

hence,

$$z = \frac{\log 11.2}{\log 1.41} + 1 = 8$$

2.5.1.6 Kinetic Calculations of Speed Gearboxes

Consider the six-speed gearbox shown in Figure 2.32. There are two possibilities of the kinematic diagrams for this gearbox, $z = 6 = 3 \times 2$ or $z = 6 = 2 \times 3$. The structural diagram for the first arrangement is shown in Figure 2.33, and the speed chart for the structural diagram $z = 3 \times 2$ is shown in Figure 2.34. In the first group, the motor speed n_m is taken as 1000 rpm, thus allowing a speed reduction of 1:φ^4(1:2.5). In the second group, the ratio of speed R_g is $\varphi^3 = 2$, which is less than the permissible speed reduction in machine tools (1:4). Based on the transmission ratios shown in Figure 2.34, the number of gear teeth Z_1 through Z_{10} can be calculated.

2.5.1.7 Application of Pole-Changing Induction Motors

The use of the multispeed induction motors through pole changing in the machine tool simplifies the machine tool gearboxes. The possibility of changing speeds while the machine is running is an

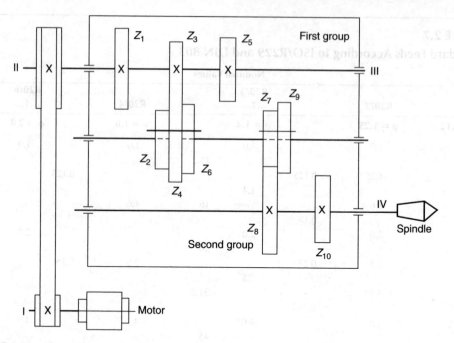

FIGURE 2.32 Six-speed gearbox.

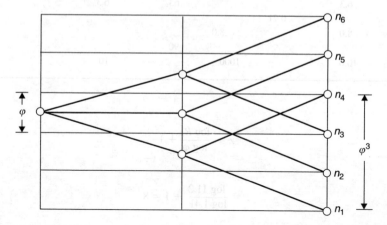

FIGURE 2.33 Six-speed gearbox structural diagram.

advantage of pole-changing motors. It reduces the auxiliary time and enables the automatic change of spindle speeds and feeds during operation in automatic machine tools. The pole-changing motor with its standard speeds replaces one of the transmission groups depending on its speed ratio φ_p, p the progression ratio of the gearbox to be constructed. For example, if a two-speed pole-changing motor of 1500 and 3000 rpm (full load speeds 1400 and 2800 rpm) is used, it can be used as the main group of a number of steps 2 and a progression ratio $\varphi_p = 2$. If the gearbox to be designed has a progression ratio $\varphi = 1.25$, then this motor is used as the first extension group of $\varphi_p = \varphi^3 = 2$. The number of speed steps of the main group Z_g following the electrical group is given by

$$z_g = \frac{\log \varphi_p}{\log \varphi} = 3 \tag{2.14}$$

Basic Elements and Mechanisms of Machine Tools

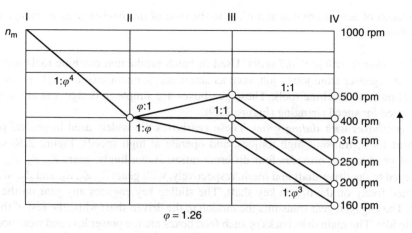

FIGURE 2.34 Speed chart for six-speed gearbox.

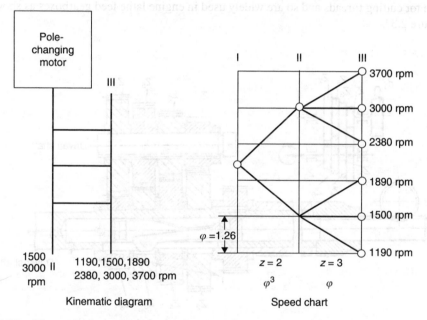

FIGURE 2.35 Kinematic diagram and speed chart for six-speed gearbox driven by pole-changing induction motor of two speeds.

Figure 2.35 shows the kinematic diagram and the speed chart for a six-speed gearbox driven by two-speed pole-changing motor. Accordingly, it is clear that the gearbox has been simplified by using the two-speed induction motor in which the number of shafts and gears has been reduced (2 shafts instead of 3 and 6 gears instead of 10).

2.5.1.8 Feed Gearboxes

Feed gearboxes are designed to provide the feed rates required for the machining operation. The values of feed rates are determined by the specified surface finish, tool life, and the rate of material removal.

The classification of feed gearboxes according to the type of mechanism used to change the rate of feed is as follows:

1. *Feed gearboxes with pick-off gears.* Used in batch-production machine tools with infrequent changeover from job to job, such as automatic, semiautomatic, single-purpose, and special-purpose machine tools. These gearboxes are simple in design and are similar to those used for speed changing (Figure 2.29).
2. *Feed gearboxes with sliding gears.* These gearboxes are widely used in general-purpose machine tools, transmit high torques, and operate at high speeds. Figure 2.36 shows a typical gearbox that provides four different ratios. Accordingly, gears Z_2, Z_4, Z_6, and Z_8 are keyed to the drive shaft and mesh, respectively, with gears Z_1, Z_3, Z_5, and Z_7, which are mounted freely on the driven key shaft. The sliding key engages any gear on the driven shaft. The engaged gear transmits the motion to the driven shaft while the rest of the gears remain idle. The main drawbacks of such feed boxes are the power loss and wear occurring due to the rotation of idle gears and insufficient rigidity of the sliding key shaft. Feed boxes with sliding gears are used in small- and medium-size drilling machines and turret lathes.
3. *Norton gearboxes.* These gearboxes provide an arithmetic series of feed steps that is suitable for cutting threads and so are widely used in engine lathe feed gearboxes as shown in Figure 2.37.

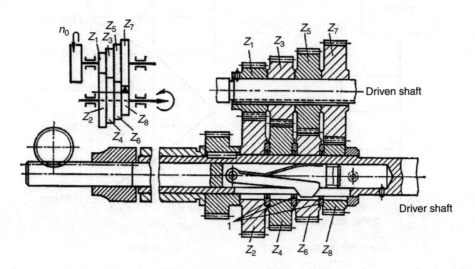

FIGURE 2.36 Feed gearbox with sliding gears.

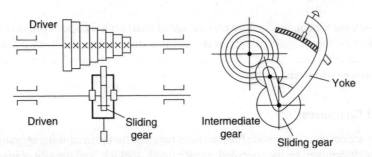

FIGURE 2.37 Norton gearbox.

2.5.1.9 Preselection of Feeds and Speeds

Preselection mechanisms in machine tools are used to select the speeds and feeds for the next machining operation during the machining time of the current operation. Once the current operation is finished, the selected speed and feed are automatically switched on with the press of a button. The main advantage of such a system in machine tools is to save the significant secondary time normally used for selecting the speeds and feeds at the end of each machining operation. Consequently, the total production time is reduced. The adoption of preselection mechanisms is justified whenever the speeds and feeds of the machine tool are frequently changed. Preselection has the following three steps:

1. Positioning the switching elements in the required position corresponding to the required (next) feeds and speeds without actuation, which is carried out during the cutting time of the current operation.
2. Switching on is achieved directly after the current machining operation is finished by bringing the corresponding coupling and shiftable gears in mesh.
3. Returning the switching elements to the original position automatically to be ready for the following preselection.

Preselection may be carried out mechanically, electrically, and hydraulically. Figure 2.38 shows an example of mechanical preselection for a nine-speed gearbox. The process is carried out by adjusting the preselection dial (a) to the required speed. Hence, the preselection drum (b) is rotated to the required position. Once the current machining operation is finished, the drums (b) and (c) are shifted axially against each other by pulling lever 1 in the switching-on position. The shifting forks (k_1) and (k_2) are moved using fingers (d) and (e) to the required position. Consequently, the blocks r_1 and r_2 are switched to mesh, giving the required speed.

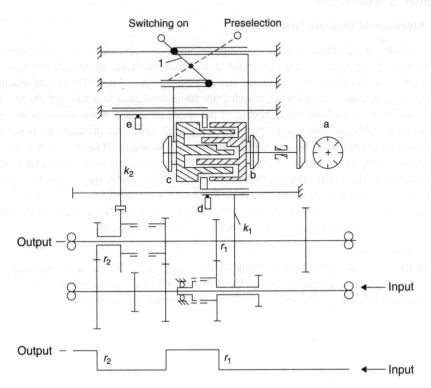

FIGURE 2.38 Preselection of spindle speeds. (From Youssef, H. et al., *Design and Construction of Machine Tool Elements*, Dar El-Maaref Publishing Co., Alexandria, Egypt, 1976. With permission.)

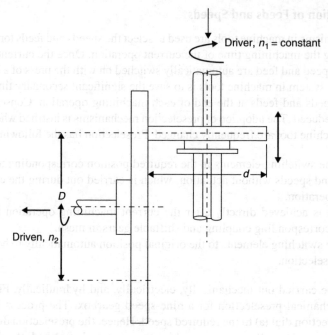

FIGURE 2.39 Disk-type friction stepless drive.

2.5.2 Stepless Speed Drives

2.5.2.1 Mechanical Stepless Drives

Infinitely variable speed (stepless) drives provide output speeds, forming infinitely variable ratios to the input ones. Such units are used for main as well as feed drives to provide the most suitable speed or feed for each job, thereby reducing the machining time. They also enable machining to be achieved at a constant cutting speed, which leads to an increased tool life and ensures uniform surface finish. The easy and smooth changing of the speed or feed, without stopping the machine, results in an appreciable reduction in the production time that raises the productivity of the machine tool. Stepless speed drives may be mechanical, hydraulic, or electric. The selection of the suitable drive depends on the purpose of the machine tool, power requirements, speed range ratio, mechanical characteristics of the machining operation, and cost of the variable speed unit. In most stepless drives, the torque transmission is not positive. Their operation involves friction and slip losses. However, they are more compact, less expensive, and quieter in operation than the stepped speed control elements. Mechanical stepless drives include the following types.

Friction Stepless Drive
Figure 2.39 shows the disk-type friction stepless mechanism. Accordingly, the drive shaft rotates at a constant speed n_1 as well as the friction roller of diameter d. The output speed of the driven shaft rotates at a variable speed n_2 that depends on the instantaneous diameter D.
Because

$$n_1 d = n_2 D \tag{2.15}$$

hence

$$n_2 = n_1 \frac{d}{D} \tag{2.16}$$

The diameter ratio d/D can be varied in infinitely small steps by the axial displacement of the friction roller. If the friction force between the friction roller and the disk is F,

$$F = \frac{\text{input torque }(T_1)}{\text{input radius }(d/2)} = \frac{\text{output torque }(T_2)}{\text{output radius }(D/2)} \quad (2.17)$$

If the power, contact pressure, transmission force, and efficiency are constant, the output torque T_2 is inversely proportional to the speed of the output shaft n_2.

$$T_2 \alpha\, T_1 \frac{n_1}{n_2} \quad (2.18)$$

Due to the small contact area, a certain amount of slip occurs, which makes this arrangement suitable for transmitting small torques and is limited to reduction ratios not more than 1:4.

Kopp Variator

In the Kopp variator, shown in Figure 2.40, the drive balls (4) mounted on inclinable axes (3) run in contact with identical, effective radii $r_1 = r_2$, and drive cones (1 and 2) are fixed on coaxial input and output shafts. When the axes of the drive balls (3) are parallel to the drive shaft axes, the input and output speeds are the same. When they are tilted, r_1 and r_2 change, which leads to the increase or decrease of the speed. Using Kopp mechanism, a speed range of 9:1, efficiency of higher than 80% and 0.25–12 hp capacity are obtainable.

Toroidal and Reeves Mechanisms

Figure 2.41 shows the principle of toroidal stepless speed transmission. Figure 2.42 shows the Reeves variable speed transmission, which consists of a pair of pulleys connected by a V-shaped belt; each pulley is made up of two conical disks. These disks slide equally and simultaneously along the shaft and rotate with it. To adjust the diameter of the pulley, the two disks on the shaft are made to approach each other so that the diameter is increased or decreased. The ratio of the driving diameter to the driven one can be easily changed and, therefore, any desired speed can be obtained without stopping the machine. Drives of this type are available with up to 8:1 speed range and 10 hp capacity.

Positive Infinitely Variable Drive

Figure 2.43 shows a positive torque transmission arrangement that consists of two chain wheels, each of which consists of a pair of cones that are movable along the shafts in the axial direction.

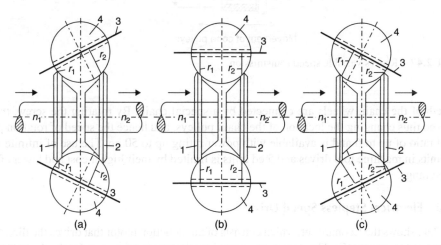

FIGURE 2.40 Kopp stepless speed mechanism: (a) $n_2 < n_1$, (b) $n_2 = n_1$, and (c) $n_2 > n_1$.

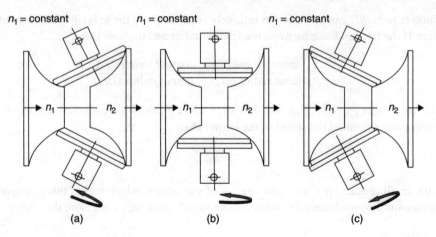

FIGURE 2.41 Toroidal stepless speed transmission: (a) $n_2 < n_1$, (b) $n_2 = n_1$, and (c) $n_2 > n_1$.

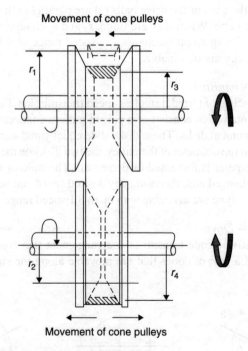

FIGURE 2.42 Reeves variable speed transmission.

The teeth of the chain wheels are connected by a special chain. By rotating the screw, the levers get moved thus changing the location of the chain pulleys, and hence the speed of rotation provides a speed ratio of up to 6 and is available with power rating up to 50 hp. The use of infinite variable speed units in machine tool drives and feed units is limited by their higher cost and lower efficiency or speed range.

2.5.2.2 Electrical Stepless Speed Drive

Figure 2.44 shows the Leonard set, which consists of an induction motor that drives the direct current generator and an exciter (E). The dc generator provides the armature current for the dc motor, and the

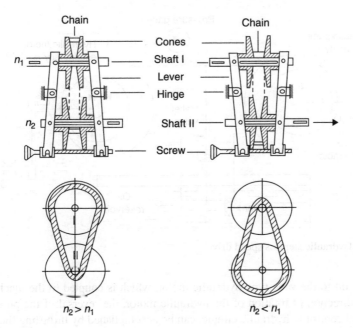

FIGURE 2.43 Positive infinitely variable drive.

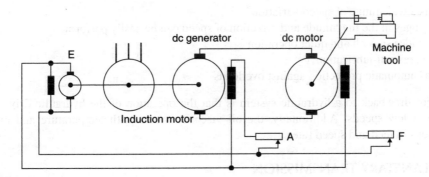

FIGURE 2.44 Leonard set (electrical stepless speed drive).

exciter provides the field current; both are necessary for the dc motors that drive the machine tool. The speed control of the dc motor takes place by adjusting both the armature and the field voltages by means of the variable resistances A and F, respectively. By varying the resistance A, the terminal voltage of the dc generator and hence the rotor voltage of the dc motor can be adjusted between zero and a maximum value. The Leonard set has a limited efficiency: it is large, expensive, and noisy. Nowadays, dc motors and thyrestors that permit direct supply to the dc motors from an alternating current (ac) mains are available and, therefore, the Leonard set can be completely eliminated. Thyrestor feed drives can be regulated such that the system offers infinitely variable speed control.

2.5.2.3 Hydraulic Stepless Speed Drive

The speeds of machine tools can be hydraulically regulated by controlling the oil discharge circulated in a hydraulic system consisting of a pump and hydraulic motor, both of the vane type, as shown in Figure 2.45. This is achieved by changing either the eccentricity of the pump e_p or the eccentricity of the hydraulic motor e_m or both. The vane pump running approximately at a constant speed delivers

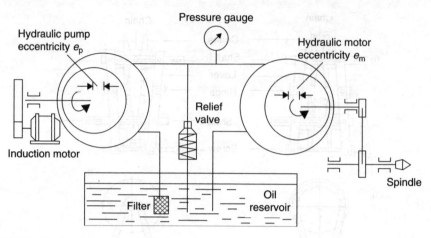

FIGURE 2.45 Hydraulic stepless speed drive.

the pressurized oil to the vane type hydraulic motor, which is coupled to the machine tool spindle. To change the direction of rotation of the hydraulic motor, the reversal of the pump eccentricity is preferred. Speed control in hydraulic circuits can be accomplished by throttling the quantity of fluid flowing into or out of the hydrocylinders or hydromotor. The advantages of the hydraulic systems are as follows:

1. Has a wide range of speed variation
2. Changes in the magnitude and direction of speed can be easily performed
3. Provides smooth and quiet operation
4. Ensures self-lubrication
5. Has automatic protection against overloads

The major drawback to a hydraulic system is that the operation of the hydraulic drive becomes unstable at low speeds. Additionally, the oil viscosity varies with temperature and may cause fluctuations in feed and speed rates.

2.6 PLANETARY TRANSMISSION

Figure 2.46 shows a planetary transmission with bevel gears that is widely used in machine tools. Accordingly, any two members may be the driving members, while the third one is the driven member. The differential contains central gears Z_1 and Z_4, and satellites Z_2 and Z_3 (an additional wheel) rotated by worm gear 2. The differential can operate as follows (Chernov, 1984):

1. Z_4 is a driving member, the carrier is a driven member, and worm gear 2 is stationary.
2. The carrier is a driving member, gear Z_4 is a driven member, and worm gear 2 is stationary.
3. Gear wheel Z_1 is a driving member (rotated by worm gear 2), gear wheel Z_4 is a driven member, and the carrier is fixed.
4. The carrier is a driving member, so is gear Z_1, and gear wheel Z_4 is a driven member.
5. Gear wheels Z_1 and Z_4 are driving members and the carrier is a driven member.

The principal relationship between axes speed is described by Willis formula, with $Z_2 = Z_3$ and $Z_1 = Z_4$, as follows:

$$i = \frac{n_4 - n_0}{n_1 - n_0} = \frac{Z_2 Z_1}{Z_4 Z_3} = -1$$

Basic Elements and Mechanisms of Machine Tools

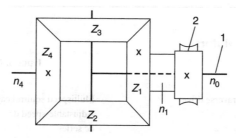

FIGURE 2.46 Planetary transmission.

where

i = conversion ratio
n_0 = speed of carrier rotation
n_1, n_2 = rotational speeds of Z_1 and Z_4, respectively.

The minus sign in the previous equation indicates that gear wheels Z_1 and Z_4 rotate in opposite direction when the carrier is stationary. Willis also suggested the following relations:

$$n_0 = \frac{n_4}{2}, \quad \text{i.e., } i = \frac{1}{2} \tag{2.19}$$

$$n_4 = 2n_0, \quad \text{i.e., } i = 2 \tag{2.20}$$

$$n_4 = n_1, \quad \text{i.e., } i = 1 \tag{2.21}$$

$$n_0 = \frac{n_1}{2} \pm \frac{n_4}{2} \tag{2.22}$$

The plus sign in equation (2.22) indicates opposite rotational directions, and the minus sign indicates the same direction of the differential driving members.

2.7 MACHINE TOOL MOTORS

Most of machine tool drives operate on standard three-phase 50 Hz, 400/440 V ac supply. The selection of motors for machine tools depends on the following:

1. Motor power
2. The power supply used (ac/dc)
3. Electrical characteristics of the motor
4. Mechanical features that include mounting, transmission of drive, noise level, and the type of cooling
5. Overload capacity

Squirrel-cage induction motors are the most popular due to their simplicity, robustness, availability with a wide range of operating characteristics, and low cost. Alternating current (ac) motors can provide infinitely variable speed over a wide range; however, their cost is high. Direct current (dc) shunt motors with field and armature control are commonly used for the main drives. For traverse drives, dc series or compound wound motors are preferred. Table 2.8 shows the different machine tool motors recommended for machine tools (Nagpal, 1996).

2.8 REVERSING MECHANISMS

Movements of machine tool elements can be reversed by mechanical, electrical, and hydraulic devices. Among these are the mechanisms with spur gears and bevel gears. Figure 2.47 shows the reversing

TABLE 2.8
Machine Tool Motors

Machine Tool	Types of Motor
Lathe	
Main drive and traverse drive	Multispeed squirrel cage
	Adjustable-speed dc
Traverse drive	dc series
	High-slip squirrel cage
Shapers and slotters	Constant-speed squirrel cage
Planers	Multispeed squirrel cage
	dc adjustable voltage
Drilling machines	Constant-speed squirrel cage
	dc shunt motor
Milling machines	Squirrel cage
	dc shunt motor
Power saws	Constant-speed squirrel cage
Grinding machines	
Wheel	Constant-speed squirrel cage
	Adjustable-speed dc
Traverse	Constant-speed squirrel cage

Source: Nagpal, G.R., in *Machine Tool Engineering*, Khanna Publishers, Delhi, India, 1999.

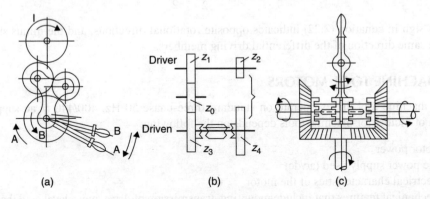

FIGURE 2.47 Reversing mechanisms: (a) tumbler yoke gear, (b) spur gear with clutch, and (c) bevel gear with clutch.

mechanisms with sliding spur gears (a) and those with fixed gears and clutches (b). Figure 2.47 also shows the reversing mechanism with bevel gears and a double-claw clutch (c). Hydraulic reversal of motion is effected by redirection of the oil delivered to an operative cylinder using a directional control valve, and electrical reversal is achieved by changing the direction of the drive motor rotation.

2.9 COUPLINGS AND BRAKES

Shaft couplings are used to fasten together the ends of two coaxial shafts. Permanent couplings cannot be disengaged while clutches engage and disengage shafts in operation. Safety clutches avoid the breakdown of the engaging mechanisms due to sharp increase in load, while overrunning clutches

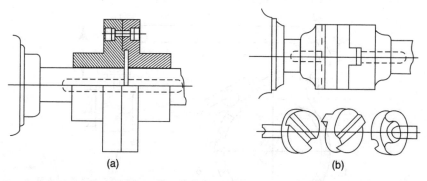

FIGURE 2.48 (a) Flanged coupling and (b) Oldham coupling.

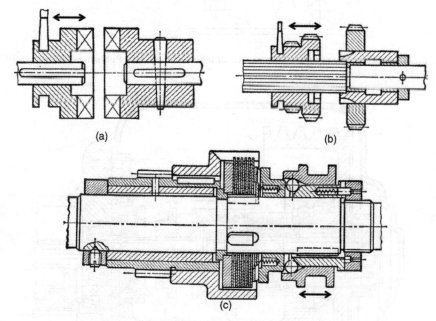

FIGURE 2.49 (a) Claw clutch, (b) toothed clutch, and (c) friction clutch. (From Chernov, N., *Machine Tools*, Mir Publishers, Moscow, 1975. With permission.)

transmit the motion in only one direction. Figure 2.48 shows permanent couplings. Figure 2.49 shows a typical claw clutch (a) and a toothed clutch (b). These two clutches cannot be engaged when the difference between the speeds of shafts is high. However, a friction clutch (c) can be engaged regardless of the speeds of its two members. Additionally, they can slip in case of overloading. Other types of clutches include friction multidisk, contactless magnetic, or hydraulic clutch (Chernov, 1984).

Brakes are used in machine tools to quickly slow or completely stop their moving parts. This step can be performed using mechanical, electrical, or hydraulic (or a combination of these) devices. Figure 2.50 shows the shoe brake in which shoes (1 and 6) are connected by a rod (3), whose length is controlled by a nut (2) that controls the clearance between the shoes and the pulley (7). Braking is achieved by pressing the shoe against the pulley by an arm (4) driven by the brake actuator (5). Band brakes operate frequently by electromagnetic or solenoid actuators.

In a multiple-disk friction brake, shown in Figure 2.51, when the shaft sleeve (3) is moved to the left, it engages with its lever (2), which, in turn, compresses the clutch disks, thereby engaging the clutch. For braking, the sliding sleeve (3) is moved to the right, disengaging the clutch (1) and engaging the friction brake (4).

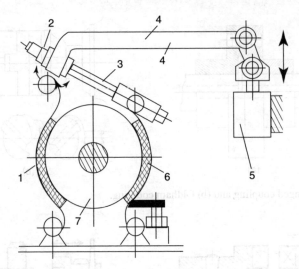

FIGURE 2.50 Shoe brake. (From Chernov, N., *Machine Tools*, Mir Publishers, Moscow, 1975. With permission.)

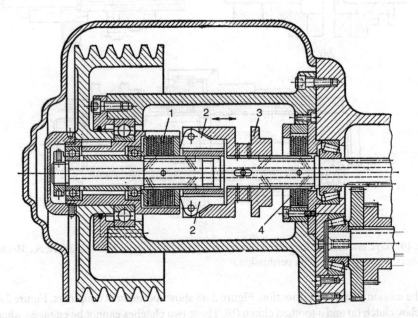

FIGURE 2.51 Friction brake. (From Chernov, N., *Machine Tools*, Mir Publishers, Moscow, 1975. With permission.)

2.10 RECIPROCATING MECHANISMS

2.10.1 QUICK-RETURN MECHANISM

Ruled flat surfaces are machined on the shaping or planing machines by the combined reciprocating motion and the side feed of the tool and WP. Figure 2.52 shows the quick-return mechanism of the shaper machine. Accordingly, the length of the stroke is controlled by the radial position of the crank pin and sliders A and B. The time taken for the crank pin to move through the angle corresponding to the cutting stroke α is less than that of the noncutting stroke β (the usual ratio is 2:1). Velocity curves for the cutting and reverse strokes are shown in Figure 2.52. The maximum speed occurs when the link is vertical.

Basic Elements and Mechanisms of Machine Tools

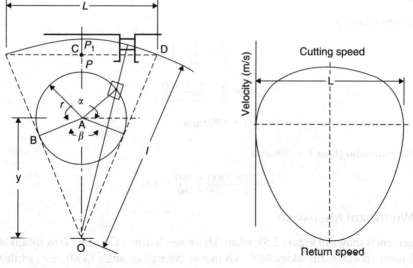

FIGURE 2.52 The quick-return mechanism.

The speed of the link at point P for a given stroke length L will be that at the corresponding crank radius r, hence, the cutting speed v_c at point P_1 is

$$v_c = 2\pi r n \frac{l}{y+r} \text{ m/min} \tag{2.23}$$

where

- n = number of strokes per minute
- l = length of crank arm (constant)

Similarly, the maximum reverse speed v_r is given by the following equation:

$$v_r = 2\pi r n \frac{l}{y-r} \text{ m/min} \tag{2.24}$$

In terms of the stroke length for maximum radius using similar triangles OBA and OCD

$$\frac{OD}{OA} = \frac{DC}{AB}$$

$$\frac{l}{y} = \frac{L}{2r} \tag{2.25}$$

hence

$$v_c = \pi n \left[\frac{lL}{l + L/2} \right] \tag{2.26}$$

and

$$v_r = \pi n \left[\frac{lL}{l - L/2} \right] \tag{2.27}$$

therefore, the speed ratio, Q

$$Q = \frac{V_r}{V_c} = \frac{2l + L}{2l - L} \tag{2.28}$$

EXAMPLE

In the slotted arm quick-return mechanism of the shaping machine, the maximum quick-return ratio is 3/2 and the stroke length is 400 mm. Calculate the length of the slotted arm. Calculate the maximum quick-return ratio if the stroke length is 180 mm.

Solution

The quick-return ratio Q

$$Q = \frac{V_r}{V_c} = \frac{2l + L}{2l - L}$$

$$Q = \frac{3}{2} = \frac{2l + 400}{2l - 400}$$

$$l = 1200 \text{ mm}$$

The quick-return ratio Q, for $L = 180$ mm

$$Q = \frac{2 \times 1200 + 180}{2 \times 1200 - 180} = 1.11$$

2.10.2 Whitworth Mechanism

This arrangement is shown in Figure 2.53; when AB rotates, it drives CE about D by means of the slider F so that G moves horizontally along MN. AB moves through an angle $(360° - \alpha)$ while CE moves through 180°, which is less than $360° - \alpha$. Also, the crank moves through α while CE moves through 180°, which is greater than α. Hence, with a uniformly rotating crank, the link moves through one-half of its revolution more quickly than the other. The angle α is used for the return stroke. Hence

$$\frac{\text{Time for cutting stroke}}{\text{Time for return stroke}} = \frac{360 - \alpha}{\alpha} \qquad (2.29)$$

The stroke can be changed by altering the radius DE, with the angle α being unchanged. Provided that the fixed center D lies on the line of movement of G, the ratio of the cutting speed to the return speed lies between 1:2 and 1:2.5.

2.10.3 Hydraulic Reciprocating Mechanism

As shown in Figure 2.54, the electrically driven pump supplies the fluid under pressure to the operating cylinder through the solenoid operated valve. The piston is connected to the machine table. At the end of the forward stroke, the direction control valve reverses the direction of the flow through limit switches set at the stroke limits and the table moves backward.

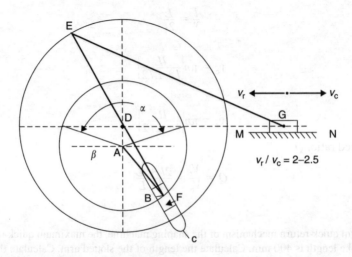

FIGURE 2.53 Whitworth quick-return mechanism.

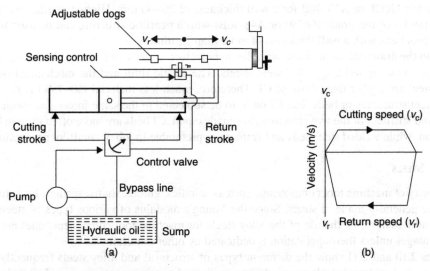

FIGURE 2.54 Reciprocating mechanism (a) and velocity diagram (b) of hydraulic shaper.

TABLE 2.9
Grades of Gray CI According to DIN 1691, American Iron and Steel Institute (AISI), Society of Automotive Engineers/American Society for Testing and Materials (SAE/ASTM)

DIN 1691	AISI, SAE/ASTM	C (%)	Brinell Hardness Number (BHN) (kg/mm²)	Applications	Approximate Composition (%)
GG 12	A48-20B	3.5	160	No acceptance test for parts of no special requirements	C = 3.2–3.6, Si = 1.7–3, Mn = 0.5, P = 0.5, S = 0.12
GG 14	A48-26B	3.4	180		
GG 18	A48-30B	3.3	200		
GG 22	A48-30B	3.3	210	Machine parts and frames to withstand high stresses	
GG 26	A48-40B	3.2	230		
GG 30	A48-50B	2.8	240	Machine parts and frames of special quality	C = 2.8–3.0, Si = 1.5–1.7, Mn = 0.8–1.8, P = 0.3, S = 0.12

2.11 MATERIAL SELECTION AND HEAT TREATMENT OF MACHINE TOOL COMPONENTS

The operating characteristics of a machine tool component depend on the proper choice of the material of each component. The most extensively used materials in machine tool components include CI and steels.

2.11.1 Cast Iron

In the majority of cases, machine tool beds and frames are made of gray CI (see Table 2.9) because of its good damping characteristics. If the guideways are cast as an integral part of the bed, frame, column, and so on, the high wear resistance grade CI (GG22 or A48-30B) with pearlitic matrix is recommended for medium-size machine tool beds and frames for a wall thickness of 10–30 mm and the grade GG26 or A48-40B for a wall thickness of 20–60 mm. High-strength, wear-resistant special gray CI of the grade (GG30 or A48-50B) with a pearlitic structure can be used for heavy machine tool beds with a wall thickness of more than 20 mm.

Due to the drawbacks associated with the manufacture of beds and frames by casting, beds and frames are made by welding rolled steel sheets. The elastic limit and the mechanical properties of such steel are higher than those of CI. Therefore, much less material (50–75%) is required for welded steel structures or beds than CI ones, to be subjected to the same forces and torques, if the rigidity and stiffness of the two structures are made equal. CI beds are more often used in large-lot production, while welded steel beds and frames are preferable in job or small-lot production.

2.11.2 Steels

The majority of machine tool components, such as spindles, guides, shafts, springs, keys, forks, and levers, are generally made of steels. Since the Young's modulus of various types of steels cannot vary by more than ±3%, the use of the alloy steels for machine tool components does not provide any advantages unless their application is dedicated by other requirements.

Tables 2.10 and 2.11 show the different types of structural and alloy steels frequently used in machine tools. Structural steels are used when no special requirements are needed. Case hardening steels of carbon content <0.25%, phosphorous (P) or sulfur (S) should not exceed 0.40% are used

TABLE 2.10
Structural Steel According to DIN 17100 and AISI, SAE/ASTM

DIN 17100	AISI, SAE/ASTM	C (%)	Mechanical Properties			Hardening Temperature (°C)	Properties	Applications
			σ_u (kg/mm²)	σ_e (kg/mm²)	δ_5 (%)			
St 34	—	0.17	34–42	18	30	920	Case hardenable and weldable	Case hardened parts
St 37	—	0.20	37–45	—	25	920	Low grade, low weldability T* or M*	General machine constructions
St 42	—	0.25	42–50	23	25	880–900	Case hardenable, hard core, machinable, not weldable	Machine elements and shafts withstanding variable loads
St 50	A570Cr50	0.35	50–60	27	22	820–850	Not case hardenable, not weldable, may be hardened, machinable	Machine elements and shafts withstanding heavy loads, not hardened gears
St 52	—	0.17	52–64	35	22	920	High strength, weldable	Welded steel construction in bridges and automotives
St 60	—	0.45	60–70	30	17	800–820	Can be hardened and toughened	Same applications like St 50 but for higher loads, keys, gears, worms
St 70	—	0.60	70–85	35	12	780–800	Can be hardened and toughened	For parts in which wear resistance is recommended

Note: T, Thomas; M, Martin.

TABLE 2.11
Case Hardened Steels According to DIN 17210 and AISI, SAE/ASTM

DIN 17210	Quenching	AISI, SAE/ASTM	Composition (%)				Mechanical Properties			Applications
			C	Mn	Cr	Ni	σ_u (kg/mm)	σ_e (kg/mm^2)	δ_5 (%)	
C 10	Water	1010	0.06–0.12	0.25–0.5	—	—	50	29	—	Typewriter parts
C 15		1015	0.12–0.18	0.25–0.5	—	—	55	35	—	Levers, bolts, sleeves
CK 10*		1010	0.06–0.12	0.25–0.5	—	—	50	30	20	Levers, bolts, pins of
CK 15*		1015	0.12–0.18	0.25–0.5	—	—	55–60	35	15	good surface finish
15Cr3		—	0.12–0.18	0.4–0.6	0.5–0.8	—	70–90	49	12	Spindles, cam shafts, piston pins, bolts, measuring tools
16MnCr3	Oil	5115	0.14–0.19	1–1.3	0.8–1.1	—	85–110	60	20–10	Pinions, automotive shafts, machine shafts
15CrNi6		—	0.12–0.17	0.4–0.6	1.4–1.7	1.4–1.7	95–120	70–90	15–6	Highly stressed small gears
20MnCr5		5120	0.17–0.22	1.1–1.4	1.0–1.3	—	110–145	75	12–7	Medium-size gears, automotive shafts, machine shafts
18CrNi8		—	0.15–0.22	0.4–0.6	1.8–2.1	1.8–2.1	120–145	90–110	14–7	Highly stressed gears, shafts, spindles, differential gears
41Cr4	Cy	5140	0.38–0.40	0.5–0.8	0.9–1.2	—	160–190	130–140	12–7	Cyanided gears

Note: CK 10* and CK 15* are carbon steels of quality better than C10 and C15 due to smaller contents of S and P; Cy, cyaniding.

when the surface hardness of the component should be very high while the core remains tough. Typical applications of case-hardening steels are in gears, shafts, and spindles. Tempered steels, shown in Table 2.12, contain higher carbon content than case-hardened steels. They are used when high strength and toughness are required. Nonalloy tempered steels are used for machine components that are not heavily loaded. For components that are heavily loaded, such as gears, spindles, and shafts, the alloy type is recommended.

Nitriding steels (see Table 2.13) contain aluminum as the main alloying element. After nitriding, the components possess an extraordinary surface hardness and therefore are used for machine parts subjected to wear such as spindles, guideways, and gears. The main advantage of the nitriding steel is minimum distortion after nitriding.

2.12 TESTING OF MACHINE TOOLS

After manufacture or repair of any machine tool, a machine tool test (usually called an *acceptance test*) should be performed according to the approved general specification. Such tests are essential because the accuracy and the surface quality of parts produced depend on the performance of the machine tool used. Testing machine tools has the following general advantages:

1. Determines the precision class and the accuracy capabilities of the machine tool
2. Prepares plans for preventive maintenance
3. Determines the actual condition and hence the expected life of the machine tool

Machine tool tests are classified into two categories: geometrical alignment tests and performance tests.

TABLE 2.12
Tempered Steels According to DIN 17100, AISI, SAE/ASTM

DIN 17100	AISI, SAE/ASTM	Composition (%)						Mechanical Properties			
		C	Si	Mn	Cr	Mo	Others	BHN	σ_u (kg/mm²)	σ_e (kg/mm²)	δ_5 (%)
C22	1020	0.18–0.25	0.15–0.36	0.3–0.6	—	—	—	155	50–60	30	22
C35	1035	0.32–0.40	0.15–0.36	0.4–0.7	—	—	—	172	60–72	37	18
C45	1045	0.42–0.50	0.15–0.36	0.5–0.8	—	—	—	206	65–80	40	16
C60	1060	0.57–0.65	0.15–0.36	0.5–0.8	—	—	—	243	75–90	40	14
CK22	1020–1023	0.18–0.25	0.15–0.36	0.3–0.6	—	—	—	155	50–60	30	22
CK35	1035	0.32–0.40	0.15–0.36	0.4–0.7	—	—	—	172	60–72	37	18
CK45	1045	0.42–0.50	0.15–0.36	0.5–0.8	—	—	—	206	65–80	49	16
CK60	1055	0.57–0.65	0.15–0.36	0.5–0.8	—	—	—	243	75–90	40	14
40Mn4	1039	0.36–0.44	0.25–0.50	0.8–1.1	—	—	—	217	80–95	55	14
30Mn5	—	0.27–0.34	0.15–0.35	1.2–1.5	—	—	—	217	88–95	55	14
37MnSi5	1330	0.38–0.41	1.1–1.4	1.1–1.4	—	—	—	217	90–105	56	12
42MnV7	—	0.38–0.45	0.15–0.35	1.6–1.9	—	—	0.07–0.12 V	217	100–120	80	11
34Cr4	—	0.30–0.37	0.15–0.55	0.5–0.8	0.9–1.2	—	—	217	90–105	65	12
41Cr4, 42Cr4	5140	0.38–0.44	0.15–0.55	0.5–0.8	0.9–1.2	—	—	217	90–105	65	12
25CrMo4	4130	0.22–0.29	0.15–0.55	0.5–0.8	0.9–1.2	0.15–0.25	—	217	80–95	55	14
34CrMo4	4135–4137	0.30–0.37	0.15–0.55	0.5–0.8	0.5–0.15		—	217	90–105	65	12
42CrMo4	4140–4142	0.38–0.45	0.15–0.55	0.5–0.8	0.9–1.2		—	217	100–120	80	11
50CrMo4	4150	0.46–0.54	0.15–0.55	0.5–0.8	0.9–1.2		—	235	110–130	90	10
30CrMoV9	—	0.26–0.34	0.15–0.55	0.4–0.7	2.3–2.7		0.1–0.2 V	248	125–145	105	9
36CrNiMo4	9840	0.32–0.40	0.15–0.55	0.5–0.8	0.9–1.2		0.9–1.2 Ni	217	100–120	80	11
34CrNiMo6	4340	0.30–0.38	0.15–0.55	0.4–0.7	1.4–1.7		1.4–1.7 Ni	235	110–130	90	10
30CrNiMo8	—	0.26–0.34	0.15–0.55	0.3–0.6	1.8–2.1		1.8–2.1 Ni	248	125–145	105	9
27NiCrV4	—	0.24–0.30	0.15–0.55	1.0–1.3	0.6–0.9		0.07–0.12 V	217	80–95	55	14
36Cr6	—	0.32–0.40	0.15–0.55	0.3–0.6	1.4–1.7		—	217	100–105	65	12
42CrV6	—	0.38–0.46	0.15–0.55	0.5–0.8	1.4–1.7		0.07–0.12 V	217	100–120	80	11
50CrV4	6150	0.47–0.56	0.15–0.55	0.8–1.1	0.9–1.12		0.07–0.12 V	235	110–130	90	10

TABLE 2.13
Nitriding Steels

Not Specified in DIN	AISI, SAE/ASTM	Composition (%)					Mechanical Properties			Applications
		C	Cr	Al	Mn	Others	σ_u (kg/mm²)	σ_e (kg/mm²)	δ_5 (%)	
27CrAl6	—	0.27	1.5	1.1	0.6	—	85–80	45	16	Valve stems
34CrAl6	A355Cl.D	0.34	1.5	1.1	0.6	—	80–100	60	12	Shafts, measuring instruments
32AlCrMo4	—	0.32	1.1	1.1	0.6	0.2 Mo	80–95	60	12	Steam machinery shafts
32AlNi7	—	0.33	0.7	1.7	0.5	1.0 Ni	88–100	60	14	Piston rods, shafts
31CrMoV9	—	0.31	2.3	—	0.6	0.15Mo/0.1Ni	90–115	75	12	Cam- and crankshaft
30CrAlNi7	—	0.30	0.3	0.9	0.5	0.5 Ni	65–80	45	14	Spindles and shafts

Geometrical tests cover the manufactured accuracy of machine tools. These tests are carried out to determine the various relationships between the various machine tool elements when idle and unloaded (static test). They include checking parallelism of the spindle and a lathe bed, squareness of the table movement to the milling machine spindle, straightness of guideways, and so on. Static tests are inadequate to judge the machine tool performance, because they do not reveal the machine tool rigidity or the accuracy of machining. The normal procedure for acceptance tests is made through the following steps:

1. Checking the principal horizontal and vertical planes and axes using a spirit level
2. Checking the guiding and bearing surfaces for parallelism, flatness, and straightness, using dial gauge, test mandrel, straight edge, and squares
3. Checking the various movements in different directions using dial gauges, mandrels, straight edges, and squares
4. Testing the spindle concentricity, axial slip, and accuracy of axis
5. Conducting working tests to check whether the accuracy of machined parts are within the specified limits
6. Preparing acceptance charts for the machine tool that specify the type of test and the range of allowable limits of deformation, deflection, error in squareness, flatness eccentricity, parallelism, and amplitude of vibrations

In contrast, dynamic tests are used to check the working accuracy of machine tools through the following steps:

1. Performing an idle run test and operation check mechanisms
2. Checking for geometrical accuracy and surface roughness of the machined parts
3. Performing rigidity and vibration tests

Standards for testing machine tools are covered by Schlesinger (1961).

2.13 MAINTENANCE OF MACHINE TOOLS

Machine tools cannot produce accurate parts throughout their working life if there is excessive wear in their moving parts. Machine tool maintenance delays the possible deterioration in machine tools and avoids the machine stoppage time that leads to lower productivity and higher production cost. Maintenance is classified under the following schemes.

2.13.1 Preventive Maintenance

Preventative maintenance is mainly carried out to reduce wear and prevent disruption of the production program. Lubrication of all the moving parts that are subjected to sliding or rolling friction is essential. A regular planned preventive maintenance consists of minor and medium repairs as well as major overhaul. The features of a well-conceived preventive maintenance scheme include

1. adequate records covering the volume of work,
2. inspection frequency schedule,
3. identification of all items to be included in the maintenance program, and
4. well-qualified personnel.

Preventive maintenance of machine tools ensures reliability, safety, and the availability of the right machine at the right time. Figure 2.55 shows preventive maintenance of a machine tool.

2.13.2 Corrective Maintenance

When a machine tool is in use, it should be regularly checked to determine whether wear has reached the level when corrective maintenance should be carried out to avoid machine tool failure. A record of all previous repairs shows those elements of the machine tool that need frequent inspection. Additionally, such records are used for decisions regarding the need for machine tool reconditioning and replacement.

2.13.3 Reconditioning

The need for machine tool recondition is determined by the frequency of the corrective maintenance repairs. Every machine tool component has a certain life span beyond which it becomes unserviceable despite the best preventive maintenance. A major overhaul or reconditioning is required.

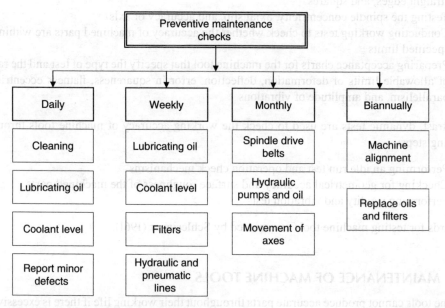

FIGURE 2.55 Preventive maintenance scheme.

Inspection reports of the machine indicate the components to be replaced, labor time, and the cost estimate. As a general rule, it is undesirable to recondition the machine if the cost exceeds 50% of buying new equipment.

2.14 REVIEW QUESTIONS

1. State the main requirements of a machine tool.
2. Give examples for open and closed machine tool structures.
3. Explain why closed box elements are best suited for machine tool structures.
4. Sketch the different types of ribbing systems used in machine tool frames.
5. Explain what is meant by light- and heavyweight construction in machine tools.
6. Sketch the different types of machine tool guideways.
7. Show how wear is compensated for in machine tool guideways.
8. Differentiate between cast and welded structures.
9. Distinguish among the kinematic, structural, and speed diagrams of gearboxes.
10. Show an example of externally pressurized and rolling friction guideways.
11. Show the different schemes of spindle mounting in machine tools.
12. What are the main applications of pick-off gears, feed gearboxes with a sliding gear, and Norton gearboxes?
13. Compare between toroidal and disk-type stepless speed mechanisms.
14. Give examples for speed-reversing mechanisms in machine tools.
15. Derive the relationship between the cutting and the reverse speeds of the quick-return mechanism used in the mechanical shaper.
16. State the main objectives behind machine tool testing.
17. Compare between corrective and preventive maintenance of machine tools.

REFERENCES

Browne, J. W. (1965) *The Theory of Machine Tools*, Book-1, Cassell and Co. Ltd., London.
Chernov, N. (1975) *Machine Tools*, Mir Publishers, Moscow.
DIN 1691—Grades of gray cast iron.
DIN 17100—Tempered and structural steels.
DIN 17210—Case hardened steels.
DIN 323—Standard values of progression ratio.
DIN 803—Standard feeds.
DIN 804—Standard speeds.
ISO/R229—Standard feeds and speeds.
ISO/R229—Standard values of progression ratio.
Koenigsberger, F. (1961) *Berechnungen, Konstruktiosgrundlagen und Bauelemente spanender Werkzugmaschinen*, Springer, Berlin.
Nagpal, G. R. (1996) *Machine Tool Engineering*, Khanna Publishers, Delhi, India.
Schlesinger, G. (1961) *Testing Machine Tools*, The Machine Publishing Company, London.
Youssef, H., Ragab, H., and Issa, S. (1976) *Design and Construction of Machine Tool Elements*, Dar Al-Maaref Publishing Company, Alexandria.

3 General-Purpose Machine Tools

3.1 INTRODUCTION

Machine tools are factory equipment used for producing machines, instruments, tools, and all kinds of spare parts. Therefore, the size of a country's stock of machine tools, and their technical quality and condition, characterize its industrial and technical potential fairly well. Metal cutting machine tools are mainly grouped into the following categories:

- General-purpose machine tools. These are multipurpose machines used for a wide range of work.
- Special-purpose machine tools. These are machines used for making one type of part of a special configuration, such as screw thread and gear cutting machines.
- Capstan, turret, and automated lathes.
- Numerical and computer numerical controlled machine tools.

In this chapter, the general-purpose machine tools are characterized and dealt with in brief. This group of machine tools comprises lathes, drilling machines, milling machines, shapers, planers, slotters, boring machines, jig boring machines, broaching machines, and microfinishing machines.

3.2 LATHE MACHINES AND OPERATIONS

Lathes are generally considered to be the oldest machine tools still used in industry. About one-third of the machine tools operating in engineering plants are lathe machines. Lathes are employed for turning external cylindrical, tapered, and contour surfaces; boring cylindrical and tapered holes, machining face surfaces, cutting external and internal threads, knurling, centering, drilling, counterboring, countersinking, spot facing and reaming of holes, cutting off, and other operations. Lathes are used in both job and mass production.

3.2.1 Turning Operations

In operations performed on lathes (turning operations), the primary cutting motion v (rotary) is imparted to the WP, and the feed motion f (in most cases straight along the axis of the WP) is imparted to a single-point tool. The tool feed rate f is usually very much smaller than the surface speed v of the WP. Figure 3.1 visualizes the basic machining parameters in turning that include:

1. Cutting speed v

$$v = \frac{\pi D n}{1000} \text{ m/min} \tag{3.1}$$

where
D = initial diameter of the WP (mm)
n = rotational speed of the WP (rpm)

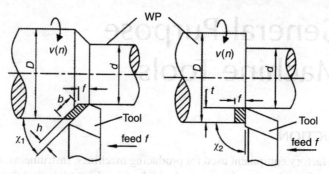

FIGURE 3.1 Basic machining parameters in turning.

2. Rotational speed n

$$n = \frac{1000v}{\pi D} \text{ rpm} \quad (3.2)$$

3. Feed rate f, which is the movement of the tool cutting edge in millimeters per revolution of the WP (mm/rev).
4. Depth of cut t, which is measured in a direction perpendicular to the WP axis, for one turning pass.

$$t = \frac{D - d}{2} \text{ mm} \quad (3.3)$$

where d is the diameter of the machined surface.

5. Undeformed chip cross-section area A_c

$$A_c = f \cdot t = h \cdot b \text{ mm}^2 \quad (3.4)$$

where

h = chip thickness in millimeters ($h = f \sin \chi$ mm)
b = contact length in millimeters
χ = cutting edge angle (setting angle)

Different types of turning operations using different tools together with cutting motions v, f are illustrated in Table 3.1.

3.2.2 Metal Cutting Lathes

Every engine lathe provides a means for traversing the cutting tool along the axis of revolution of the WP and at right angles to it. Beyond this similarity, the lathe may embody other characteristics common to several classifications according to fields of application that ranges from manual to full automatic machining. Metal cutting lathes may differ in size and construction. Among these are the general-purpose machines that include universal engine lathes, plain turning lathes, facing lathes, and vertical turning and boring mills.

3.2.2.1 Universal Engine Lathes

Universal engine lathes are widely employed in job and lot production, as well as for repair work. Parts of very versatile forms may be machined by this lathe. Its size varies from small bench

General-Purpose Machine Tools

TABLE 3.1
Lathe Operations and Relevant Tools

Lathe Operation and Relevant Tool	Sketch and Directions of Cutting Movements
1. Cylindrical turning with a straight-shank turning tool 2. Taper turning with a straight-shank turning tool	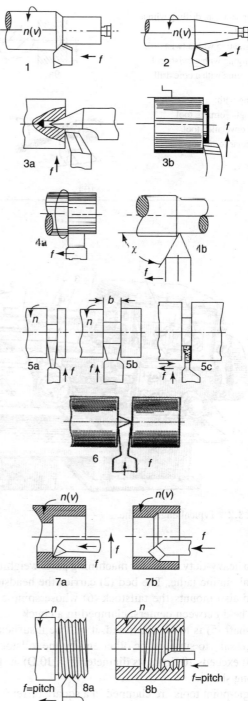

3. Facing of a WP with:
    a. Facing tool while the WP is clamped by a half center
    b. Facing tool while the WP is mounted in a chuck	

4. Finish turning with:
    a. Broad-nose finishing tool
    b. Straight finishing tool with a nose radius	

5. Necking or recessing with:
    a. Recessing tool
    b. Wide recessing tool
    c. Wide recessing using narrow recessing tool	

6. Parting off with parting-off tool	

7. Boring of cylindrical hole with:
    a. Bent rough-boring tool
    b. Bent finish-boring tool	

8. Threading with:
    a. External threading tool
    b. Internal threading tool	

(Continued)

TABLE 3.1 Continued
Lathe Operations and Relevant Tools

Lathe Operation and Relevant Tool	Sketch and Directions of Cutting Movements

9. Drilling and core drilling with a twist drill:
  a. Originating with a twist drill
  b. Enlarging with a twist drill
  c. Enlarging with a core drill
	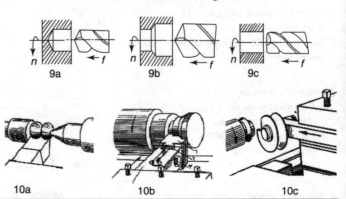
	9a 9b 9c

10. Forming with:
  a. Straight forming tool
  b. Flat dovetailed tool
  c. Circular form tool
	10a 10b 10c

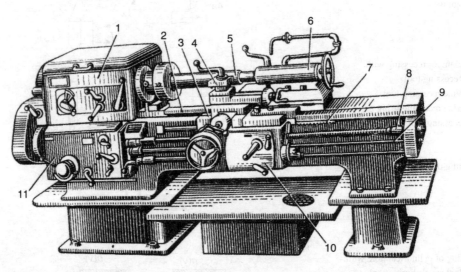

FIGURE 3.2 Typical engine lathe.

lathes to heavy-duty lathes for machining parts weighing many tons. Figure 3.2 illustrates a typical universal engine lathe. The bed (2) carries the headstock (1), which contains the speed gearbox. The bed also mounts the tailstock (6) whose spindle usually carries the dead center. The work may be held between centers, clamped in a chuck, or held in a fixture mounted on a faceplate. If a long shaft (5) is to be machined, it will be insufficient to clamp one end in a chuck; therefore, it is necessary to support the other end by the tailstock center. In many cases when the length of the shaft exceeds 10 times its diameter ($\ell > 10\,D$), a steady rest or follower rest is used to support these long shafts.

Single-point tools are clamped in a square turret (4) mounted on the carriage (3). Tools such as drills, core drills, and reamers are inserted in the tailstock spindle after removing the center. The carriage (3), to which the apron (10) is secured, may traverse along the guideways either manually or powered. The cross slide can also be either manually or power traversed in the cross direction.

General-Purpose Machine Tools

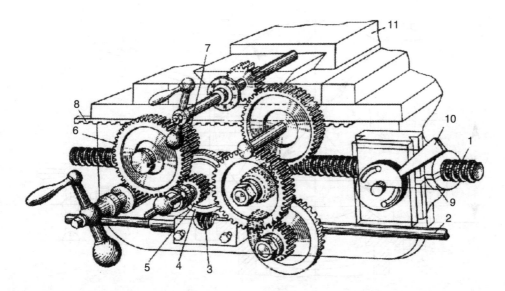

FIGURE 3.3 Lathe apron mechanism.

Surfaces of revolution are turned by longitudinal traverse of the carriage. The cross slide feeds the tool in the cross direction to perform facing, recessing, forming, and knurling operations. Power traverse of the carriage or cross slide is obtained through the feed mechanism. Rotation is transmitted from the spindle through change gears and the quick change feed gearbox (11) to either the lead screw (8) or feed rod (9). From either of these, motion is transmitted to the carriage. Powered motion of the lead screw is used only for cutting threads using a threading tool. In all other cases, the carriage is traversed by hand or powered from the feed rod. Carriage feed is obtained by a pinion and rack (7) fastened to the bed. The pinion may be actuated manually or powered from the feed rod. The cross slide is powered by the feed rod through a gearing system in the apron (10). Figure 3.3 shows an isometric view of the apron mechanism. During thread cutting, the half nuts (9) are closed by the lever (10) over the lead screw (1).

Specifications of an Engine Lathe

Figure 3.4 shows the main dimensions that indicate the capacity of an engine lathe. These are:

- Maximum diameter D of work accommodated over the bed (swing over bed). According to most of national standards, D varies from 100 to 6300 mm, arranged in geometric progression $\varphi = 1.26$.
- Maximum diameter D_1 of work accommodated over the carriage.
- Distance between centers, which determines the maximum work length. It is measured with the tailstock shifted to its extreme right-hand position without overhanging.
- Maximum bore diameter of spindle, which determines the bar capacity (maximum bar stock).

In addition to these dimensions, other important specifications are:

- Number of spindle speeds and speed range
- Number of feeds and feed range
- Motor power and speed
- Overall dimensions and net weight

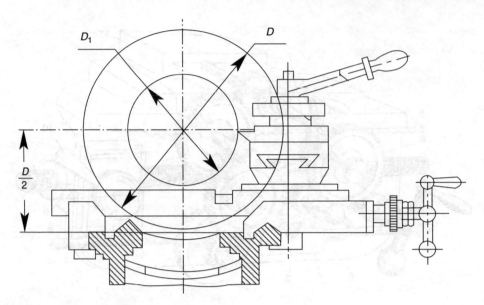

FIGURE 3.4 Main dimensions of an engine lathe.

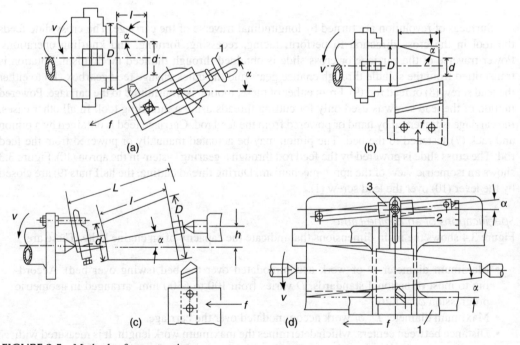

FIGURE 3.5 Methods of taper turning.

Setting Up the Engine Lathe for Taper Turning
Tapered surfaces are turned by employing one of the following methods (Figure 3.5):

a. *By swiveling the compound rest to the required angle α.* Before performing the operation, the compound rest is to be clamped in this position. The tool is fed manually by rotating handle (1). This method is used for turning short internal and external tapers with large taper angles, while the work is commonly held in a chuck and a straight turning tool is used (Figure 3.5a).

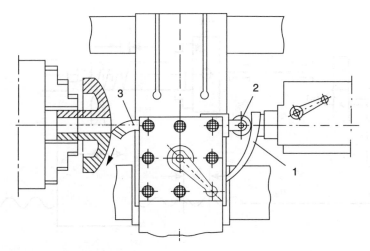

FIGURE 3.6 Turning of a spherical surface.

b. *By using a straight-edge broad-nose tool.* The tool of width that exceeds the taper being turned is cross-fed. The work is held in a chuck or clamped on a faceplate (Figure 3.5b).
c. *By setting over the tailstock.* The angle of taper α should not exceed 8°. Since the turned surface is parallel to the spindle axis, the powered feed of the carriage can be used while the work is to be mounted between centers as shown in Figure 3.5c. Before turning cylindrical surfaces, it is a good practice to check whether the tailstock is not previously set over for taper turning; otherwise, tapered surfaces are produced.
d. *By using a taper-turning attachment.* This is best suited for long tapered work. The cross slide (1) is disengaged from the cross feed screw and is linked through the tie (2) to the slide (3) (Figure 3.5d).

Setting Up the Engine Lathe for Turning Contoured Surfaces with a Tracer Device
Longitudinal contoured surfaces are produced using a tracer device similar to the taper-turning attachment, except that the template of the required profile is substituted by the guide bar. Disadvantages of such mechanical duplicating are the difficulties in making a template sufficiently accurate and strong enough to withstand the cutting force and the rapid wear of such templates. A mechanical tracer for turning spherical surfaces, shown in Figure 3.6, operates by similar principles. Accordingly, the template (1) is clamped in the tailstock spindle and a roller (2) is clamped in the square turret opposite the tool (3) and in contact with the template. If the cross feed is transmitted to the cross slide, the profile of the template will be produced on the WP. When much contour turning work is to be done with longitudinal feeds, a hydraulic tracer slide is often installed on engine lathe where the stylus sliding on the template does not carry the cutting force.

Setting Up the Engine Lathe for Cutting Screw Threads
In some cases when the machine has not a quick-change gearbox, or when the thread pitch to be cut is nonstandard, change gears must be used and setup on the quadrant as shown in Figure 3.7. Because one revolution of the spindle provides the pitch t_{th} of the screw thread to be produced, the kinematic linkage is given by the following equation:

$$t_{th} = t_{ls} \cdot i_{cg} \tag{3.5}$$

or

$$i_{cg} = \frac{t_{th}}{t_{ls}} = \frac{a}{b} \times \frac{c}{d} \tag{3.6}$$

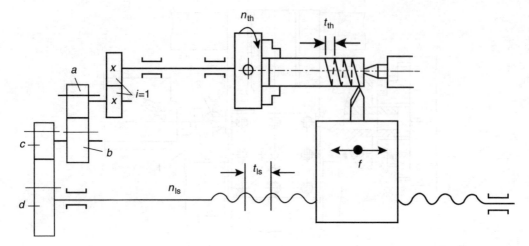

FIGURE 3.7 Setting up the engine lathe for thread cutting.

where

t_{ls} = pitch of the lead screw of the lathe
i_{cg} = gearing ratio of the quadrant
a, b, c, d = number of teeth of change gears

Holding the Work on Engine Lathe
WP fixation on an engine lathe depends mainly upon the geometrical features of the WP and the precision required. The WP can be held between centers, on a mandrel, in a chuck, or on a faceplate:

1. *Holding the WP between centers.* A dog plate (1) and a lathe dog (2) are used (Figure 3.8a). It is an accurate method for clamping a long WP. The tailstock center may be a dead center (Figure 3.8b), or a live center (Figure 3.8c), when the work is rotating at high speed. In such a case, rests are used to support long WPs to prevent their deflection under the action of the cutting forces. The steady rest (Figure 3.9a) is mounted on the guideways of the bed while the follower rest (Figure 3.9b) is mounted on the saddle of the carriage.
2. *Clamping hollow WPs on mandrels.* Mandrels are used to hold WPs with previously machined holes. The WP to be machined (2) is tightly fitted on a conical mandrel, tapered at 0.001, and provided with center holes to be clamped between centers using a dog plate and a lathe dog (Figure 3.10a). The expanding mandrel (Figure 3.10b) consists of a conical rod (1), a split sleeve (2), and nuts (3 and 4). The work is held by expansion of a sleeve (2), as the latter is displaced along the conical rod (1) by nut (3). Nut (4) removes the work from the mandrel. There is a flat (5) on the left of the conical rod used for the setscrew of the driving lathe dog.
3. *Clamping the WP in a chuck.* The most commonly employed method of holding short work is to clamp it in a chuck (Figure 3.11a). If the work length is considerably large relative to its diameter, supporting the free end with the tailstock dead or live center (Figure 3.11b) is also used. Chucks may be universal (self-centering) of three jaws, which are expanded and drawn simultaneously (Figure 3.11c); or they may be independent of four jaws (Figure 3.11d). The three-jaw chucks are used to clamp circular and hexagonal rods, whereas the independent four-jaw chucks are especially useful in clamping

General-Purpose Machine Tools

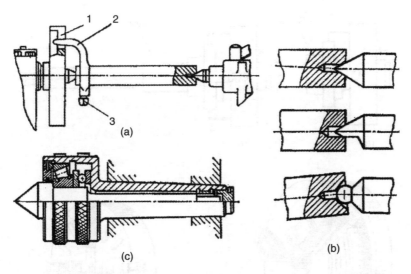

FIGURE 3.8 Holding the work between centers: (a) Dog plate, (b) dead centers, (c) live centers.

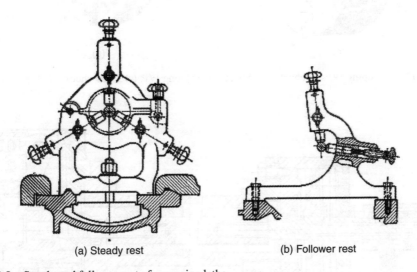

FIGURE 3.9 Steady and follower rest of an engine lathe.

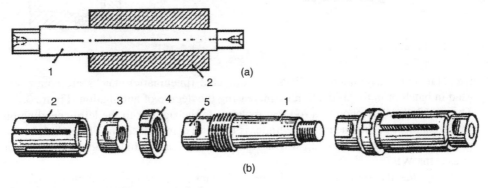

FIGURE 3.10 Mounting WPs on a mandrel.

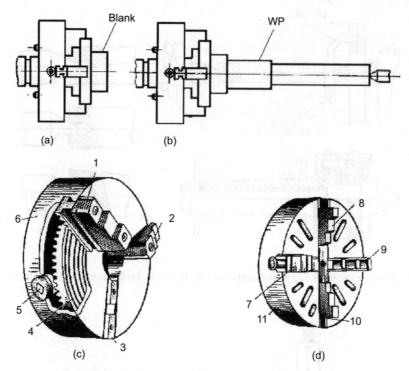

FIGURE 3.11 Clamping WPs in chucks: (a) Short WP, (b) long WP, (c) 3-jaw chuck, (d) 4-jaw chuck.

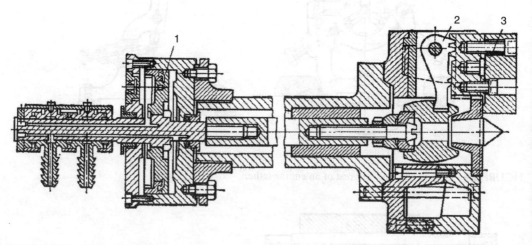

FIGURE 3.12 Pneumatic chuck.

irregular and nonsymmetrical WPs. Air-operated (pneumatic) chucks are commonly used in batch or mass production by increasing the degree of automation (Figure 3.12). The piston (1) is attached to a rod that moves it to the right or to the left depending on which chamber of the pneumatic cylinder is fed with compressed air. The end of the rod is connected to three levers (2), which expand jaws (3) in a radial direction to clamp or release the WP.

4. *Clamping the WP on a faceplate.* Large WPs cannot be clamped in a chuck and are, therefore, mounted either directly on a faceplate (Figure 3.13a), or mounted on a plate

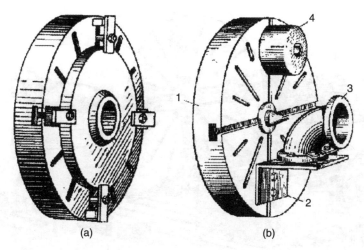

FIGURE 3.13 Mounting WPs on faceplates.

fixture (2) that is attached to faceplate (1) (Figure 3.13b). The work (3) and angle plate (2) must be counterbalanced by using the counterweight (4) mounted at the opposite position on the faceplate. The plate fixture has been proved to be highly efficient in machining asymmetrical work of complex and irregular shape.

3.2.2.2 Other Types of General-Purpose Metal Cutting Lathes

These include plain turning lathes, facing lathes, and vertical turning and boring mills. Facing lathes, vertical turning and boring mills, and heavy-duty plain turning lathes are generally used for heavy work. They are characterized by low speeds, large feeds, and high cutting torques.

1. *Plain turning lathes.* Plain turning lathes differ from engine lathes in that they do not have a lead screw. They perform all types of lathe work except threading and chasing. The absence of the lead screw substantially simplifies the kinematic features and the construction of the feed gear trains. Their dimensional data are similar to those of engine lathes. Plain turning lathes are available in three different size ranges: small, medium, and heavy duty. Heavy-duty plain turning lathes have several common carriages that are powered either from a common feed rod, linked kinematically to the lathe spindle, or powered from a variable speed dc motor mounted on each carriage. The tailstock traverses along the guideway by a separate drive.
2. *Facing lathes.* These are used to machine work of large diameter and short length in single-piece production and for repair jobs. These machines are generally used for turning external, internal, and taper surfaces, facing, boring, and so on. Facing lathes have relatively small length and large diameter of faceplates (up to 4 m). Sometimes, they are equipped with a tailstock. Its construction differs, to some extent, from the center lathe. It consists of the base plate (1), headstock (4) with faceplate (5), bed (2), carriage (3), and tailstock (6) (Figure 3.14). The work is clamped on the faceplate using jaws, or clamps, and T-slot bolts. It may be additionally supported by the tailstock center. The feed gear train is powered from a separate motor to provide the longitudinal and transverse feeds. Facing lathes have been almost superseded by vertical turning and boring mills; however, because of their simple construction and low cost, they are still employed.

FIGURE 3.14 Facing lathe.

3. *Vertical turning and boring mills*. These machines are employed in machining heavy pieces of large diameters and relatively small lengths. They are used for turning and boring of cylindrical and tapered surfaces, facing, drilling, countersinking, counterboring, and reaming. In vertical turning and boring mills, the heavy work can be mounted on rotating tables more conveniently and safely as compared to facing lathes. The horizontal surface of the worktable excludes completely the overhanging load on the spindle of the facing lathes. This facilitates the application of high-velocity machining and, at the same time, enables high accuracy to be attained. These small machines are called vertical turret lathes. As their name implies, they are equipped with turret heads, which increase their productivity.

3.3 DRILLING MACHINES AND OPERATIONS

3.3.1 Drilling and Drilling Allied Operations

3.3.1.1 Drilling Operation

Drilling is a process used extensively by which through or blind holes are originated or enlarged in a WP. This process involves feeding a rotating cutting tool (drill) along its axis of rotation into a stationary WP (Figure 3.15). The axial feed rate f is usually very small when compared to the peripheral speed v. Drilling is considered a roughing operation and, therefore, the accuracy and surface finish in drilling are generally not of much concern. If high accuracy and good finish are required, drilling must be followed by some other operation such as reaming, boring, or grinding.

The most commonly employed drilling tool is the twist drill, which is available in diameters ranging from 0.25 to 80 mm. A standard twist drill (Figure 3.16) is characterized by a geometry in which the normal rake and the velocity of the cutting edge are a function of their distance from the center of the drill. Referring to the terminology of twist drill shown in Figure 3.17, the helix angle of the twist drill is the equivalent of the rake angle of other cutting tools. The standard helix is 30°, which, together with a point angle of 118°, is suitable for drilling steel and CI (Figure 3.17a). Drills with a helix angle of 20°, known as slow-helix drills, are available with a point of 118° for cutting brass and bronze (Figure 3.17b), and with a point of 90° for cutting plastics. Quick helix drills, with

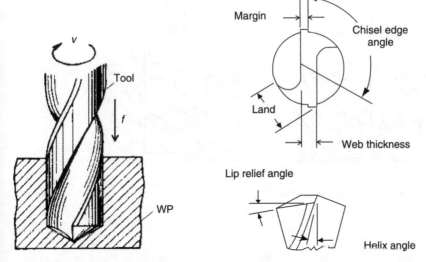

FIGURE 3.15 Drilling operation.

FIGURE 3.16 Terminology of a standard point twist drill.

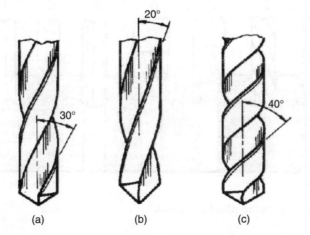

FIGURE 3.17 Helix drills of different helix angles: (a) Standard, (b) slow, (c) quick.

a helix angle of 40° and a point of 100°, are suitable for drilling softer materials such as aluminum alloys and copper (Figure 3.17c). Figure 3.18 visualizes the basic machining parameters in drilling and enlarging holes.

3.3.1.2 Drilling Allied Operations

Drilling allied or alternative operations such as core drilling, center drilling, counterboring, countersinking, spot facing, reaming, tapping, and other operations can also be performed on drilling machines as shown in Figure 3.19. Accordingly, the main and feed motion are the same as in drilling; that is, the drill rotates while it is fed into the stationary WP. In these processes, the tool shape and geometry depend upon the machining process to be performed.

The same operations can be accomplished in some other machine by holding the tool stationary and rotating the work. The most general example is performing these processes on a center lathe, in

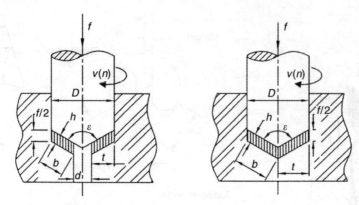

FIGURE 3.18 Basic machining parameters in drilling.

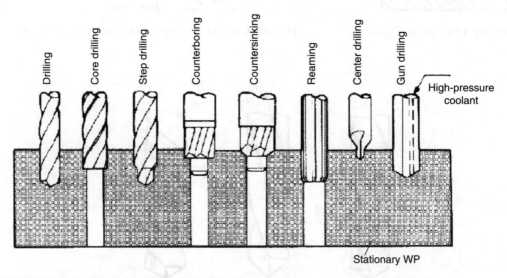

FIGURE 3.19 Drilling and drilling allied operations.

which the tool (drill, counterbore, reamer, tap, and so on) is held in the tailstock and the work is held and rotated by a chuck (Figure 3.20). The most important drilling allied processes are as follows:

1. Core drilling, which is performed for the purpose of enlarging holes, as shown in Figure 3.19. Higher dimensional and form accuracy and improved surface quality can be obtained by this operation. It is usually an intermediate operation between drilling and reaming. Similar allowances should be considered for both reaming and core drilling. Core drills are of three- or four-flutes; they have no web or chisel edge and consequently provide better guidance into the hole than ordinary twist drills, which produces better and accurate performance (third and fourth grade of accuracy). Core drilling is a more productive operation than drilling, since at the same cutting speeds, the feeds used may be two to three times larger. It is recommended to enlarge holes with core drills wherever possible, instead of drilling with a larger drill. This process is much more efficient than boring large diameter holes with a single drill.

General-Purpose Machine Tools

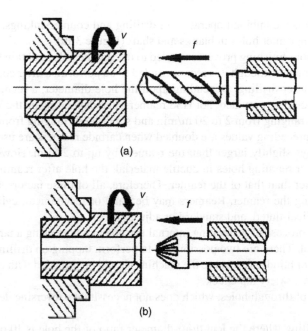

FIGURE 3.20 Drilling and drilling allied operations as performed on an engine lathe: (a) Drilling and (b) Countersinking.

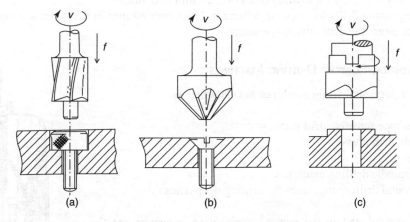

FIGURE 3.21 Counterboring, countersinking, and spot facing operations.

2. Counterboring, countersinking, and spot facing, which are performed with various types of tools. Counterboring and countersinking (Figure 3.19) are used for machining cylindrical and tapered recesses in previously drilled holes. Such recesses are used for embedding the heads of screws and bolts, when these heads must not extend over the surface (Figures 3.21a and 3.21b). Spot facing is the process of finishing the faces of bosses for washers, thrust rings, nuts, and other pieces (Figure 3.21c). Spot facing tools cut only to a very limited depth. The tools used in these processes are made of HSS and have a guide or pilot, which is usually interchangeable. For these processes, cutting speeds and feeds are similar to those of core drilling.

3. Center drilling is a combined operation of drilling and countersinkings. Center drills are used for making center holes in blanks and shaft (Figure 3.19).
4. Reaming is a hole-finishing process intended to true up the hole to obtain high dimensional and form accuracy. Although it is recommended to be performed after core drilling, it may be performed after drilling. Depending upon the hole diameter, a reaming allowance of 40–400 μm should be provided. For HSS reamers, and depending on the WP material, low cutting speeds ranging from 2 to 20 m/min and small feeds ranging from 0.1 to 1 mm/rev are used. The preceding values are doubled when carbide reamers are used. The produced holes are always slightly larger than the reamers by up to 20 μm. However, when using worn reamers or reaming holes in ductile material, the hole after reaming may have to a smaller diameter than that of the reamer. Therefore, all of these factors should be considered in selecting the reamer. Reamers may be hand or mechanical, cylindrical or taper, straight- or helical-fluted, and standard or adjustable.
5. Tapping is the process of generating internal threads in a hole using a tap that is basically a threading tool. There are two possibilities to perform tapping on drilling machines:
   - The tapping of blind holes where the machine should be provided with a reversing device together with a safety tap chuck.
   - The tapping of through holes, which does not necessitate a reversing device and a safety tap chuck.
6. Deep-hole drilling where the length-to-diameter ratio of the hole is 10 or more, the work is rotated by a chuck and supported by a steady rest, while the drill is fed axially. The following special types of drills are used:
   - Gun drills for drilling holes up to 25 mm in diameter.
   - Half-round drills for drilling holes over 25 mm in diameter.
   - Trepanning drills for annular drilling of holes over 80 mm in diameter, leaving a core that enters the drill during operation.

3.3.2 General-Purpose Drilling Machines

The general-purpose drilling machines are classified as

- Bench-type sensitive drill presses
- Upright drill presses
- Radial drills
- Multispindle drilling machines
- Horizontal drilling machines for drilling deep holes

The most widely used in the general engineering industries are the upright drill presses and radial drills.

3.3.2.1 Bench-Type Sensitive Drill Presses

These drill presses are used for machining small diameter holes of 0.25–12 mm diameter. Manual feeding characterizes this machine and that is why they are called "sensitive." High speeds are typical for bench-type sensitive drill presses.

3.3.2.2 Upright Drill Presses

These machines are used for machining holes up to 50 mm in diameter in relatively small-size work. Figure 3.22 shows a typical drilling

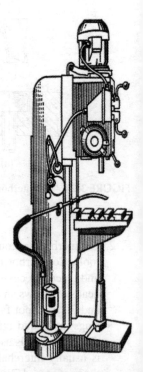

FIGURE 3.22 Typical upright drill press.

General-Purpose Machine Tools

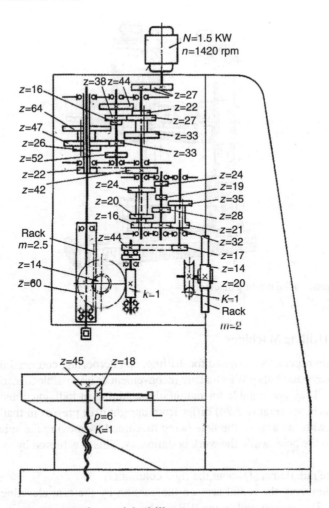

FIGURE 3.23 Kinematic diagram of an upright drill press.

machine. It has a wide range of spindle speeds and feeds. Therefore, they are employed not only for drilling from solid material, but also for core drilling, reaming, and tapping operations. Figure 3.23 illustrates the gearing diagram of the machine.

Cutting movements. As shown in the gearing diagram (Figure 3.23), the kinematic chain equations for the maximum spindle speed and feed are given by

$$n_{max} = 1420 \cdot \frac{27}{27} \cdot \frac{33}{33} \cdot \frac{52}{26} = 2840 \text{ rpm} \tag{3.7}$$

and

$$f_{max} = 1 \cdot \frac{22}{42} \cdot \frac{24}{24} \cdot \frac{32}{21} \cdot \frac{17}{44} \cdot \frac{1}{60} \times \pi \times 2.5 \times 14 = 0.56 \text{ mm/rev} \tag{3.8}$$

Auxiliary movements. The drill head, housing the speed and feed gearboxes, moves along the machine column through the gear train: worm gearing 1/20-rack and pinion ($z = 14$, $m = 2$). The machine table can be moved vertically by hand through bevels 18/45 and an elevating screw driven by means of a handle (Figure 3.23).

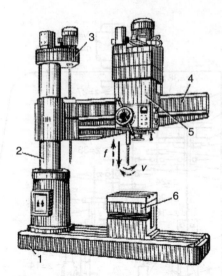

FIGURE 3.24 Typical radial drilling machine.

3.3.2.3 Radial Drilling Machines

These machines are especially designed for drilling, counterboring, countersinking, reaming, and tapping holes in heavy and bulky WPs that are inconvenient or impossible to machine on the upright drilling machines. They are suitable for multitool machining in individual and batch production. Radial drilling machines (Figure 3.24) differ from upright drill presses in that the spindle axis is made to coincide with the axis of the hole being machined by moving the spindle in a system of polar coordinate to the hole, while the work is stationary. This is achieved by

1. Swinging the radial arm (4) about the rigid column (2)
2. Raising or lowering the radial arm on the column by the arm-elevating and -clamping mechanism (3) to accommodate the WP height
3. Moving the spindle head (5) along the guideways of the radial arm (4)

Accordingly, the tool is located at any required point on the stationary WP, which is set either on detachable table (6) or directly on base (1). After the maneuvering tasks performed by the radial arm and spindle head, they are held in position using power-operated clamping devices. The spindle head gearing diagram of the radial drilling machine is very similar to that of the upright drill press.

3.3.2.4 Multispindle Drilling Machines

These are mainly used in lot production for machining WPs requiring simultaneous drilling, reaming, and tapping of a large number of holes in different planes of the WP. A single spindle drilling machine is not economical for such purposes, as not only a considerably large number of machines and operators are required but also the machining cycle is longer. There are three types of multiple-spindle drilling machines:

a. *Gang multispindle drilling machines.* The spindles (2–6) are arranged in a row, and each spindle is driven by its own motor. The gang machine is in fact several upright drilling machines having a common base and single worktable (Figure 3.25). They are used for consecutive machining of different holes in one WP, or for the machining of a single hole with different cutting tools.

General-Purpose Machine Tools

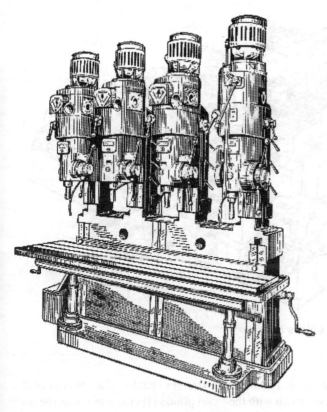

FIGURE 3.25 Gang, multiple-spindle drilling machine.

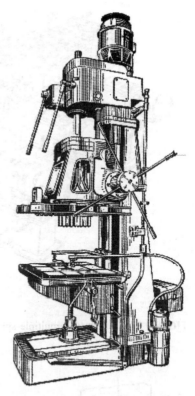

FIGURE 3.26 Multiple-spindle drilling machine.

b. *Adjustable-center multispindle vertical drilling machines.* These differ from gang-type machines in that they have a common drive for all working spindles. The spindles are adjusted in the spindle head for drilling holes of varying diameters at random locations on the WP surface (Figure 3.26).

c. *Unit-type multispindle drilling machines.* These are widely used in mass production. They are, as a rule, chiefly built of standard units. Such machines are designed for machining a definite component held in a jig and are frequently built into an automatic transfer machine (Figure 3.27).

3.3.2.5 Horizontal Drilling Machines for Drilling Deep Holes

Such machines are usually equipped with powerful pumps, which deliver cutting fluid under high pressure, either through the hollow drilling tool or through the clearance between the drill stem and the machined hole. The cutting fluid washes out the chips produced by drilling.

In deep hole drilling, the work is rotated by chuck and supported by a steady rest while the drill is fed axially. This process reduces the amount by which the drill departs from the drilled hole-center. Deep-hole drilling machines (also called drill lathes) are intended for drilling hole having a length-to-diameter ratio of 10 or more.

3.3.3 Tool Holding Accessories of Drilling Machines

Twist drills are either of a straight shank (for small sizes) or of a tapered shank (for medium to large sizes). A self-centering, three-jaw drill chuck (Figure 3.28) is used to hold small drilling

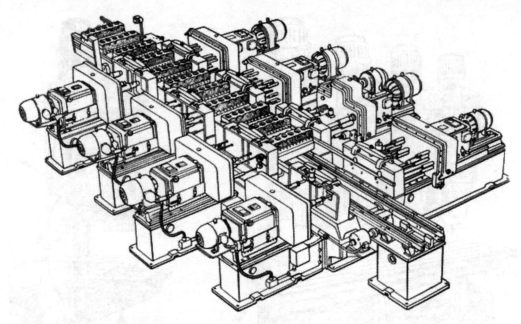

FIGURE 3.27 Unit-type multiple-spindle drilling machine.

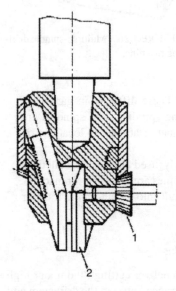

FIGURE 3.28 Three-jaw drilling chuck.

tools (up to 15 mm) with straight shanks. The rotation of the chuck wrench with the bevel pinion (1) closes or opens the jaws (2). The chuck itself is fitted with a Morse-taper shank, which fits into the spindle socket. Tapered sleeves (Figure 3.29a) are used for holding tools with taper shanks in the spindle socket (Figure 3.29b). The size of a Morse-taper shank is identified from smallest to largest by the numbers 1–6 and depends on the drill diameter (Table 3.2).

The included angle of the Morse taper is in the range of 3°. If the two mating tapered surfaces are clean and in good condition, such a small taper is sufficient to provide a frictional drive between the two surfaces. At the end of the taper shank of the tool or the taper shank of the sleeve, two flats are machined, leaving a tang. The purpose of the tang is to remove the tool from the spindle socket by a drift, as shown in Figure 3.29c. When the cutting tool has a Morse taper smaller than that of the spindle socket, the difference is made up by using one or two tapered sleeves (Figure 3.29d).

If a single hole is to be machined consecutively by several tools in a single operation, quick-change chucks are used for reducing the handling times in operating drilling machines. They enable tools to be changed rapidly without stopping the machine.

A quick-change chuck for tapered-shank tools is shown in Figure 3.30. The body (1) has a tapered shank inserted in the machine spindle. A sliding collar (2) may be raised (for releasing) or lowered (for chucking). Interchangeable tapered sleeves (3) into which various tools have been secured are inserted in the chuck. When the collar (2) is lowered, it forces the balls (4) into recess b, and the torque is transmitted. The sleeve is rapidly released by raising the collar upward.

General-Purpose Machine Tools

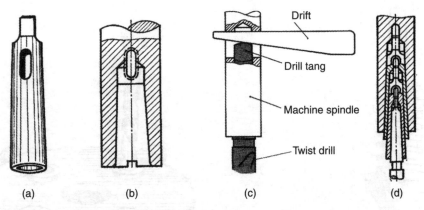

FIGURE 3.29 Holding drills in spindle socket or sleeves and drifting out from a socket or sleeve. (a) Tapered sleeve, (b) spindle socket, (c) drifting out, and (d) holding by different sleeves.

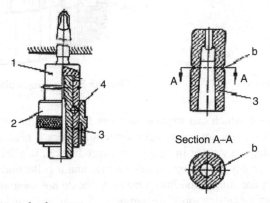

FIGURE 3.30 Quick-change chuck.

TABLE 3.2
Morse Taper Sizes

Morse Taper Number	1	2	3	4	5	6
Drill diameter (mm)	up to 14	14.25–23	23.25–31.75	32–50.5	51–76	77–100

A safety-tap chuck (Figure 3.31) is used in tapping blind holes on machines having a reversing device. It is difficult to time the reversal at the proper moment; if a safety chuck is not used, consequently the tap may run up against the bottom of the hole and break. The safety-tap chuck is secured in the machine spindle by the taper shank of the central shaft (4). Clutch member (2) is keyed on shaft (4) and the second clutch member (3) is mounted freely on shaft 4. Both members are held in engagement by the action of the spring (1). The compression of the spring is adjusted by a nut (6). Rotation is transmitted to the sleeve (5) through clutch member (3). When the actual torque exceeds the preset value, clutch member (2) begins to slip, the tap stops rotating and the spindle is then reversed.

3.3.4 Work-Holding Devices Used on Drilling Machines

The type of work-holding device used depends upon the shape and size of the WP, the required accuracy, and the production rate. It should be stressed that the work being drilled should never be held by hand. High torque is transmitted by a revolving drill, especially when the drill is breaking

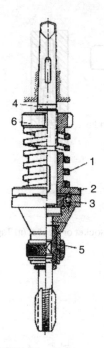
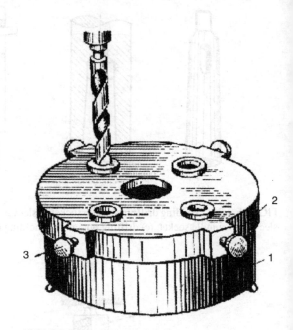

FIGURE 3.31 Safety tap chuck. **FIGURE 3.32** Simple plate jig.

through the bottom surface, which can wrench the work from the hand. The resulting injuries can vary from a small hand cut to the loss of a finger. Generally, work is held on a drilling machine by clamping to the worktable, in a vise, or in case of mass production, in a drilling jig.

Standard equipment in any workshop includes a vise and a collection of clamps, studs, bolts, nuts, and packing, which are simple and inexpensive. Vises do not accurately locate the work and provide no means for holding cutting tools in alignment. A small WP can be held in a vise, whereas larger work and sheet metal are best clamped on to the worktable surface that is provided with standard tee slots for clamping purposes.

Drilling jigs are special devices designed to hold a particular WP and guide the cutting tool. Jigs enable work to be done without previously laying out the WP. Drilling using jigs is, therefore, accurate and quicker than standard methods. However, larger quantities of WPs must be required to justify the additional cost of the equipment.

Jigs are provided with jig bushings to ensure that the hole is machined in the correct location. Jig bushings are classified as press-fit bushings for jigs used in small-lot production for machining holes using a single tool. Slip renewable bushings are used for mass production. Bushings are made of hardened steels to ensure the required hardness to resist the wear. The drilling jigs are generally produced on jig boring machines.

According to Figure 3.32, the plate jig (2) is mounted on the surface of the WP (1), where the holes are to be drilled. The WP is clamped under the plate jig with screws (3).

Figure 3.33 shows the jig used for drilling three holes in thin gauge components. The press-fit drill bushings are pressed into a separate top plate that is doweled and screwed to the jig body and the base plate. A post jig used to make eight holes (up and down) in the flanges of cylindrical component is shown in Figure 3.34. Accordingly, clamping is achieved by the finger nut. The previously drilled holes are located by the spring-loaded location pin in the jig base to enable the holes to be drilled in line. Figure 3.35 illustrates a jig design that enables a hole to be drilled at an angle to the component centerline. A special drill bushing is used to take the drill as close to the component as possible. Figure 3.36 shows an inverted post jig with four legs, and Figure 3.37 presents an indexing jig used for drilling six equally spaced holes around the periphery of the component.

General-Purpose Machine Tools

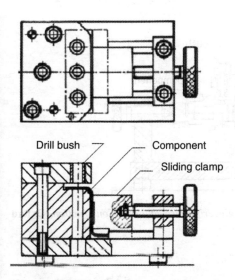

FIGURE 3.33 A thin plate drilling jig. (From Mott, L. C., *Engineering Drawing and Construction*, Oxford University Press, Oxford, 1976. With permission.)

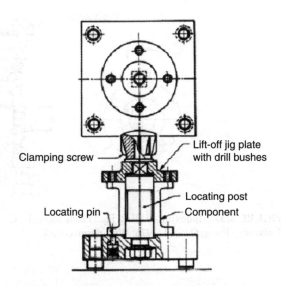

FIGURE 3.34 A post jig to drill holes into flanged, cylindrical WP. (From Mott, L. C., *Engineering Drawing and Construction*, Oxford University Press, Oxford, 1976. With permission.)

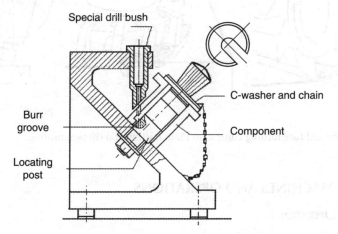

FIGURE 3.35 Angle drilling jig. (From Mott, L. C., *Engineering Drawing and Construction*, Oxford University Press, Oxford, 1976. With permission.)

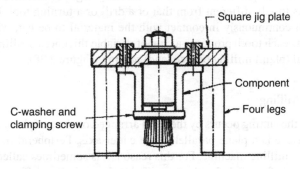

FIGURE 3.36 Inverted post jig. (From Mott, L. C., *Engineering Drawing and Construction*, Oxford University Press, Oxford, 1976. With permission.)

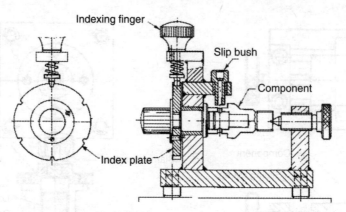

FIGURE 3.37 Indexing drilling jig. (From Mott, L. C., *Engineering Drawing and Construction*, Oxford University Press, Oxford, 1976. With permission.)

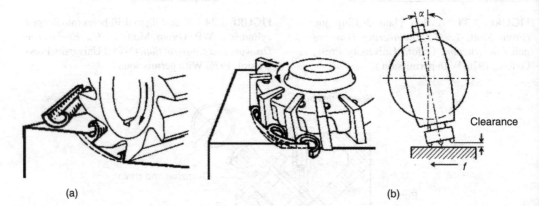

FIGURE 3.38 Plain and face milling cutters: (a) Plain milling and (b) face milling.

3.4 MILLING MACHINES AND OPERATIONS

3.4.1 MILLING OPERATIONS

Milling is the removal of metal by feeding the work past a rotating multitoothed cutter. In this operation the material removal rate (MRR) is enhanced as the cutter rotates at a high cutting speed. The surface quality is also improved due to the multicutting edges of the milling cutter. The action of the milling cutter is totally different from that of a drill or a turning tool. In turning and drilling, the tools are kept continuously in contact with the material to be cut, whereas milling is an intermittent process, as each tooth produces a chip of variable thickness. Milling operations may be classified as peripheral (plain) milling or face (end) milling (Figure 3.38).

3.4.1.1 Peripheral Milling

In peripheral milling, the cutting occurs by the teeth arranged on the periphery of the milling cutter, and the generated surface is a plane parallel to the cutter axis. Peripheral milling is usually performed on a horizontal milling machine. For this reason, it is sometimes called horizontal milling. The appearance of the surface and also the type of chip formation are affected by the direction of

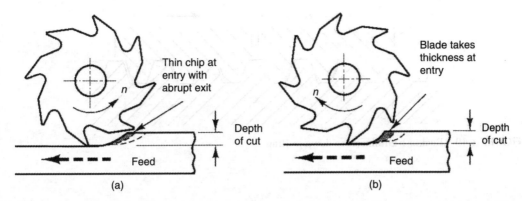

FIGURE 3.39 Up-milling and down-milling: (a) Up-milling (conventional cut) and (b) down-milling (climb cut).

cutter rotation with respect to the movement of the WP. In this regard, two types of peripheral milling are differentiable, namely, up-milling and down-milling.

Up-Milling (Conventional Milling)
Up-milling is accomplished by rotating the cutter against the direction of the feed of the WP (Figure 3.39a). The tooth picks up from the material gradually; that is, the chip starts with no thickness and increases in size as the teeth progress through the cut. This means that the cycle of operation to remove the chip is first a sliding action at the beginning and then a crushing action takes place, which is followed by the actual cutting action. In some metals, up-milling leads to strain hardening of the machined surface, and also to chattering and excessive teeth blunting.

Advantages of up-milling include the following:

- It does not require a backlash eliminator.
- It is safer in operation (the cutter does not climb on the work).
- Loads on teeth are acting gradually.
- Built-up edge (BUE) fragments are absent from the machined surface.
- The milling cutter is not affected by the sandy or scaly surfaces of the work.

Down-Milling (Climb Milling)
Down-milling is accomplished by rotating the cutter in the direction of the work feed, as shown in Figure 3.39b. In climb milling, as implied by the name, the milling cutter attempts to climb the WP. Chips are cut to maximum thickness at initial engagement of cutter teeth with the work, and decrease to zero at the end of its engagement.

The cutting forces in down milling are directed downward. Down-milling should not be attempted if machines do not have enough rigidity and are not provided with backlash eliminators (Figure 3.40). Under such circumstances, the cutter climbs up on the WP and the arbor and spindle may be damaged.

Advantages of down-milling include the following:

- Fixtures are simpler and less costly, as cutting forces are acting downward.
- Flat WPs or plates that cannot be firmly held can be machined by down-milling.
- Cutter with higher rake angles can be used, which decreases the power requirements.
- Tool blunting is less likely.
- Down-milling is characterized by fewer tendencies of chattering and vibration, which leads to improved surface finish.

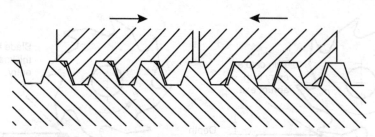

FIGURE 3.40 Backlash eliminator in down-milling.

3.4.1.2 Face Milling

In face milling, the generated surface is at a right angle to the cutter axis. When using cutters of large diameters, it is a good practice to tilt the spindle head slightly at an angle of 1–3° to provide some clearance, which leads to an improved surface finish and eliminate tool blunting (Figure 3.38b). Face milling is usually performed on vertical milling machines; for this reason, the process is called vertical milling, which is more productive than plain milling.

3.4.2 Milling Cutters

The milling cutters are selected for each specified machining duty. The milling cutter may be provided with a hole to be mounted on the arbor of the horizontal milling machines, or provided with a straight or tapered shank for mounting on the vertical or horizontal milling machine. Figure 3.41 visualizes commonly used milling cutters during their operation. These include the following:

1. Plain milling cutters are either straight or helical ones. Helical milling cutters are preferred for large cutting widths to provide smooth cutting and improved surface quality (Figure 3.41a). Plain milling cutters are mainly used on horizontal milling machines.
2. Face milling cutters are used for the production of horizontal (Figure 3.41b), vertical (Figure 3.41c), or inclined (Figure 3.41d) flat surfaces. They are used on vertical milling machines, planer type milling machines, and vertical milling machines with the spindle swiveled to the required angle α, respectively.
3. Side milling cutters are clamped on the arbor of the horizontal milling machine and are used for machining of the vertical surface of a shoulder (Figure 3.41e) or cutting a keyway (Figure 3.41f).
4. Interlocking (staggered) side mills (Figure 3.41g) mounted on the arbor of the horizontal milling machines are intended to cut wide keyways and cavities.
5. Slitting saws (Figure 3.41h) are used on horizontal milling machines.
6. Angle milling cutters, used on horizontal milling machines, for the production of longitudinal grooves (Figure 3.41i) or for edge chamfering.
7. End mills are tools of a shank type, which can be mounted on vertical milling machines (or directly in the spindle nose of horizontal milling machines). End mills may be employed in machining keyways (Figure 3.41j) or vertical surfaces (Figure 3.41k).
8. Key-cutters are also of the shank type that can be used on vertical milling machines. They may be used for single-pass milling or multipass milling operations (Figures 3.41l and 3.41m).
9. Form-milling cutters are mounted on horizontal milling machines. Form cutters may be either concave as shown in Figure 3.41n or convex as in Figure 3.41o.

General-Purpose Machine Tools

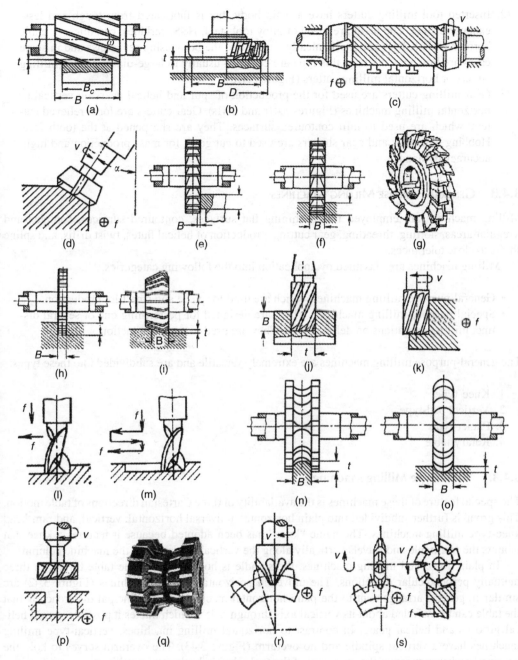

FIGURE 3.41 Different types of milling cutters during operation.

10. T-slot cutters are used for milling T-slots and are available in different sizes. The T-slot is machined on a vertical milling machine in two steps:
    - Slotting with end mill (Figure 3.41j)
    - Cutting with T-slot cutter (Figure 3.41p)
11. Compound milling cutters are mainly used to produce compound surfaces. These cutters realize high productivity and accuracy (Figure 3.41q).

12. Inserted tool milling cutters have a main body that is fabricated from tough and less-expensive steel. The teeth are made of alloy tool steel, HSS, carbides, ceramics, or cubic boron nitride (CBN) and mechanically attached to the body using set screws and in some cases are brazed. Cutters of this type are confined usually to large-diameter face milling cutters or horizontal milling cutters (Figure 3.41q).
13. Gear milling cutters are used for the production of spur and helical gears on vertical or horizontal milling machines (Figures 3.41r and 3.41s). Gear cutters are form-relieved cutters, which are used to mill contoured surfaces. They are sharpened at the tooth face. Hobbing machines and gear shapers are used to cut gears for mass production and high-accuracy demands.

3.4.3 General-Purpose Milling Machines

Milling machines are employed for machining flat surfaces, contoured surfaces, complex and irregular areas, slotting, threading, gear cutting, production of helical flutes, twist drills, and spline shafts to close tolerances.

Milling machines are classified by application into the following categories:

- General-purpose milling machines, which are used for piece and small-lot production.
- Special-purpose milling machines, which are designed for performing one or several distinct milling operations on definite WPs. They are used in mass production.

The general-purpose milling machines are extremely versatile and are subdivided into these types:

1. Knee-type
2. Vertical bed-type
3. Planer-type
4. Rotary-table

3.4.3.1 Knee-Type Milling Machines

The special feature of these machines is the availability of three Cartesian directions of table motion. This group is further subdivided into plain horizontal, universal horizontal, vertical, and ram-head knee-type milling machines. The name "knee" has been adopted because it features a knee that mounts the worktable and travels vertically along the vertical guideway of the machine column.

In plain horizontal milling machines, the spindle is horizontal and the table travels in three mutually perpendicular directions. The universal horizontal milling machines (Figure 3.42) are similar in general arrangement to the plain horizontal machines. The principal difference is that the table can be swiveled about its vertical axis through ±45°, which makes it possible to mill helical grooves and helical gears. In contrast to horizontal milling machines, vertical-type milling machines have a vertical spindle and no overarm (Figure 3.43). The overarm serves to hold the bearing bracket supporting the outer end of the tool arbor in horizontal machines.

The ram-head milling machines (Figure 3.44) differ from the universal type in that they have an additional spindle that can be swiveled about both the vertical and horizontal axes. In ram-head milling machines, the spindle can be set at any angle in relation to the WP being machined. In modern machines, a separate drive for the principal movement (cutter), feed movement (WP), rapid traversal of the worktable in all directions, and a single lever control for changing speeds and feeds are provided. Units and components of milling machines are widely unified. Horizontal knee-type milling machine specifications are as follows:

- Dimensions of table working surface
- Maximum table travel in the three Cartesian directions

General-Purpose Machine Tools

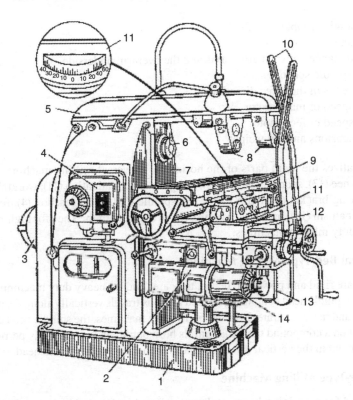

FIGURE 3.42 Universal horizontal-spindle milling machine.

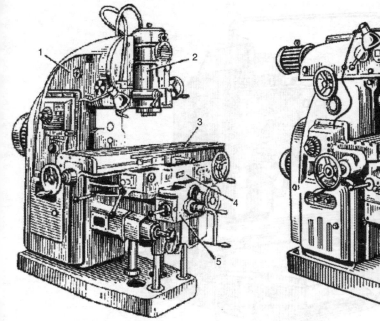

FIGURE 3.43 Vertical milling machine.

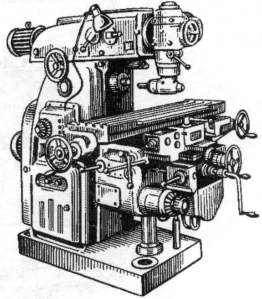

FIGURE 3.44 Ram-head milling machine.

- Maximum angle of table swivel
- Arbor diameter
- Maximum distance between arbor axis and the overarm underside
- Number of spindle speeds
- Number of feeds in the three directions
- Power and speed of main motor
- Power and speed of feed motor
- Overall dimensions and net weight

Figure 3.42 visualizes the main parts of the horizontal universal milling machine. These are Base (1), column (7), knee (13), saddle (12), table swivel plate with graduation (11), worktable (9), overarm (5), holding bearing bracket (8), main motor (3), spindle (6), speed gearbox (4), feed gearbox (2), feed control mechanism (14), braces (10) to link the overarm with the knee for high-rigidity requirements in heavy-duty milling machines.

3.4.3.2 Vertical Bed-Type Milling Machines

These machines are rigid and powerful; hence, they are used for heavy duty machining of large WPs (Figure 3.45). The spindle head containing a speed gearbox travels vertically along the guideways of the machine column and has a separate drive motor. In some machines, the spindle head can be swiveled. The work is fixed on a compound table that travels horizontally in two mutually perpendicular directions. The adjustment in the vertical direction is accomplished by the spindle head.

3.4.3.3 Planer-Type Milling Machine

They are intended for machining horizontal, vertical, and inclined planes as well as form surfaces by means of face, plain, and form milling cutters. These machines are of single or double housing, with one or several spindles; each has a separate drive.

Figure 3.46 shows a single-housing machine with two spindle heads traveling vertically and horizontally.

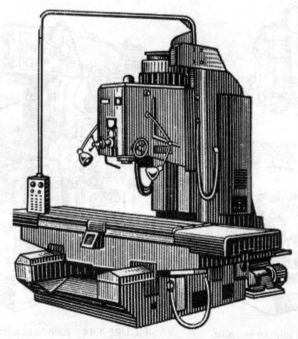

FIGURE 3.45 Vertical-bed general-purpose milling machine.

General-Purpose Machine Tools

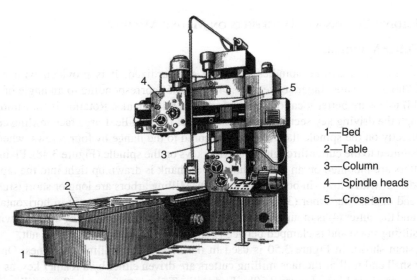

1—Bed
2—Table
3—Column
4—Spindle heads
5—Cross-arm

FIGURE 3.46 Planer-type general-purpose milling machine.

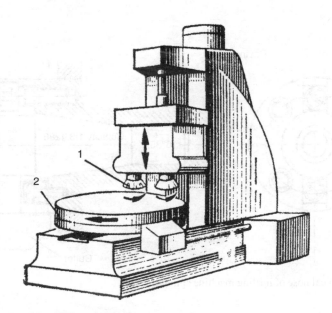

FIGURE 3.47 Rotary-table milling machine.

3.4.3.4 Rotary-Table Milling Machines

These are also called continuous milling machines, as the WPs are set up without stopping the operation. Rotary-table machines are highly productive; consequently, they are frequently used for both batch and mass production. The WPs being machined are clamped in fixtures installed on the rotating table (2) (Figure 3.47). The machines may be equipped with one or two spindle heads (1).

When several surfaces are to be machined, the WPs are indexed in the fixtures after each complete revolution of the table. The machining cycle provides as many table revolutions as the number of surfaces to be machined.

3.4.4 Holding Cutters and Workpieces on Milling Machines

3.4.4.1 Cutter Mounting

The nose of milling machine spindles has been standardized. It is provided with a locating flange ϕ H7/h6 and a steep taper socket of 7:24 (1:3.4286) corresponding to an angle of 16° 35.6′ (Figure 3.48) to ensure better location of arbor and end mill shanks. Rotation is transmitted to the cutter through the driving key secured to the end face of the spindle. Large face milling cutters are mounted directly on the spindle flange and are secured to the flange by four screws, whereas rotation is transmitted to the cutter through the driving keys on the spindle (Figure 3.48). Plain and side milling cutters are mounted on an arbor whose taper shank is drawn up tight into the taper socket of the spindle (2) with a draw-in bolt 1 (Figure 3.49). Milling arbors are long or short (stub arbors). The outer end of the long arbor (3) is supported by an overarm support (5) in horizontal milling machines, and the cutter (4) is mounted at the required position on the arbor by a key (or without key in case of slitting saws) and is clamped between collars or spacers (6) with a large nut.

The system shown in Figure 3.50 is used in the duplex bed milling machines. On the stub arbors, the shell end mill or the face milling cutters are driven either by a feather key, as shown in Figure 3.50a, or an end key (Figure 3.50b). End mills, T-slot cutters, and other milling cutters of tapered shanks are secured with a draw-in bolt directly in the taper socket of the spindle by means of adaptors (Figure 3.51a). Straight shank cutters are held in chucks (Figure 3.51b).

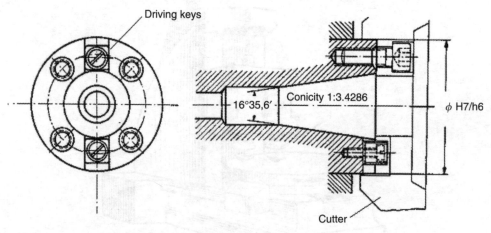

FIGURE 3.48 Typical nose of milling machine spindle.

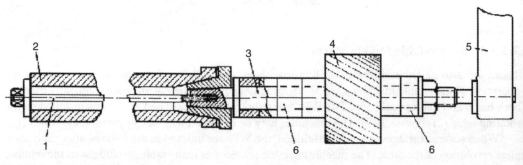

FIGURE 3.49 Milling machine arbor.

General-Purpose Machine Tools

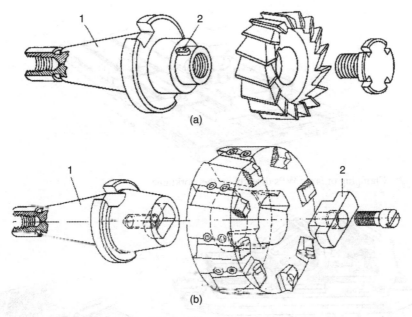

FIGURE 3.50 Mounting of end mills and face milling cutters on duplex-bed milling machine.

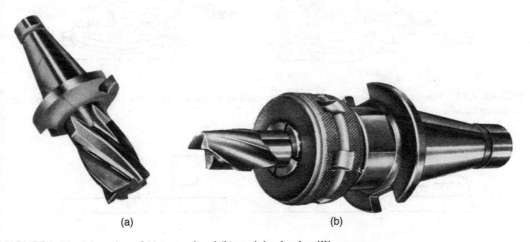

FIGURE 3.51 Mounting of (a) tapered and (b) straight-shank milling cutters.

3.4.4.2 Workpiece Fixturing

Large WPs and blanks that are too large for a vise are clamped directly on the worktable using standard fastening elements such as strap clamps, support blocks, and T-bolts (Figure 3.52). Small WPs and blanks are clamped most frequently in general-purpose plain, swivel, or universal milling vises fastened to the worktable (Figure 3.53). Shaped jaws are sometimes used instead of the flat type to clamp parts of irregular shapes. For more accurate and productive work, expensive milling fixtures are frequently used.

Figure 3.54 shows a simple milling fixture for a bearing bracket. A full-form and a flatted locator, firmly fitted into the base plate, are used to locate the WP from two previously machined holes. The clamping is effected by two solid clamps. To achieve correct alignment and, hence, increased accuracy, a tool-setting block is used to locate the cutter with respect to the WP.

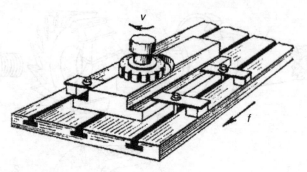

FIGURE 3.52 Clamping of large WPs directly on the worktable.

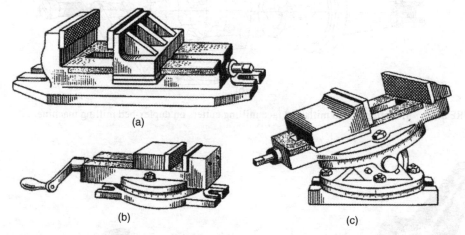

FIGURE 3.53 Vises for clamping of small WPs on milling machines: (a) plain vise, (b) swivel vise, and (c) universal vise.

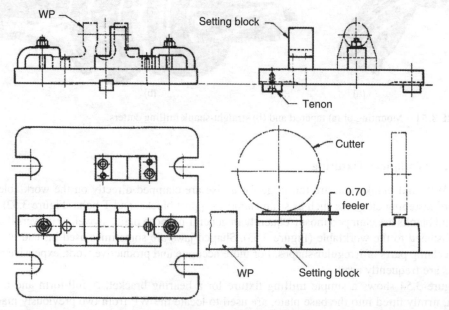

FIGURE 3.54 Simple milling fixture for a bearing bracket. (From Mott, L. C., *Engineering Drawing and Construction*, Oxford University Press, Oxford, 1976. With permission.)

General-Purpose Machine Tools

Figure 3.54 illustrates how the height of the cutter is setup using setting blocks and 0.7 mm feeler. The main body of the fixture is frequently made of CI because of its ability to absorb vibrations initiated by the milling operation. However, welded and other steel constructions are also used for various specialized purposes.

Figure 3.55 illustrates a vise used as a fixture for milling six cylindrical WPs in one clamp. The setting block is designed for a feeler gage of 0.025 mm, the thickness of which should be stamped on the setting block in some suitably prominent position. In this type of fixture, it is essential that when the components are unloaded, all the swarf must be removed; otherwise, the component subsequently loaded into the fixture will not seat correctly.

Figure 3.56 shows a WP and fixture of more specialized nature designed by the U.S. Naval Gun Factory. Two rectangular components are to be milled together. They are located and clamped

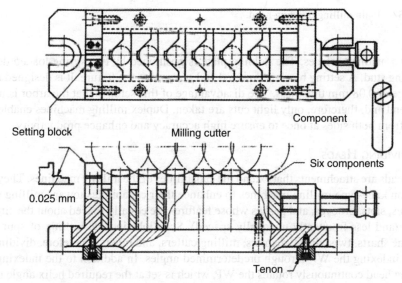

FIGURE 3.55 Special fixture for milling six cylindrical WPs. (From Mott, L. C., *Engineering Drawing and Construction*, Oxford University Press, Oxford, 1976. With permission.)

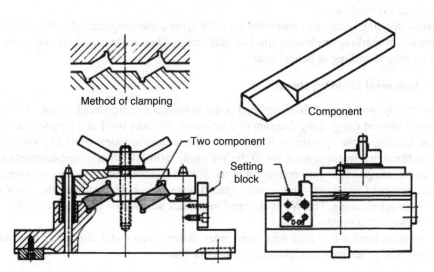

FIGURE 3.56 A special milling fixture for mounting two rectangular components. (From Mott, L. C., *Engineering Drawing and Construction*, Oxford University Press, Oxford, 1976. With permission.)

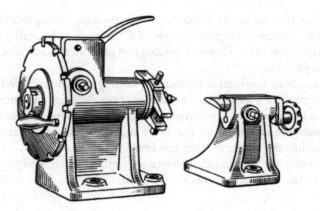

FIGURE 3.57 Plain milling dividing head.

between two mating surfaces. The holding plate is positioned by two spring-loaded dowels and a central fixing stud. A setting block is doweled and screwed to the fixture. It is designed for use with a feeler gage of 0.08 mm thickness. The disadvantage of this setup is that the arbor is unsupported at its free end and, therefore, only light cuts are taken. Duplex milling machines enable WPs to be machined from both sides at once to ensure high accuracy and enhance productivity.

3.4.5 Dividing Heads

Dividing heads are attachments that extend the capabilities of the milling machines. They are mainly employed on knee-type milling machines to enhance their capabilities toward milling straight and helical flutes, slots, grooves, and gashes whose features are equally spaced about the circumference of a blank (and less frequently unequally spaced). Such jobs include milling of spur and helical gears, spline shafts, twist drills, reamers, milling cutters, and others. Therefore, dividing heads are capable of indexing the WP through predetermined angles. In addition to the indexing operation, the dividing head continuously rotates the WP, which is set at the required helix angle during milling of helical slots and helical gears. There are several versions of dividing heads:

- Plain dividing heads (Figure 3.57) are mainly used for indexing milling fixtures.
- Universal dividing heads.
- Optical dividing heads are commonly used for precise indexing, and also for checking the accuracy of marking graduation lines on dial scales. Their main drawback is that they cannot be used in milling of helical gears.

3.4.5.1 Universal Dividing Heads

The most widely used type of dividing head is the universal dividing head. Figure 3.58 illustrates an isometric view of the gearing diagram of a universal dividing head in a simple indexing mode. Periodical turning of the spindle (3) is achieved by rotating the index crank (2), which transmits the motion through a worm gearing 6/4 to the WP (gear ratio 1:40; that is, one complete revolution of the crank corresponds to 1/40 revolution of the WP). The index plate (1), having several concentric circular rows of accurately and equally spaced holes, serves for indexing the index crank (2) through the required angle. The WP is clamped in a chuck screwed on the spindle (3). It can also be clamped between two centers.

The dividing head is provided with three index plates (Brown and Sharpe) or two index plates (Parkinson). The plates have the following number of holes:

Brown and Sharpe
  Plate 1: 15, 16, 17, 18, 19, and 20

General-Purpose Machine Tools

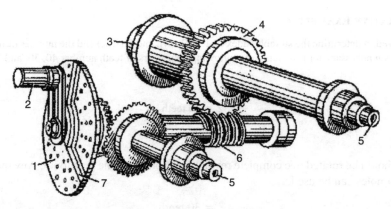

FIGURE 3.58 An isomeric gearing diagram of a universal dividing head.

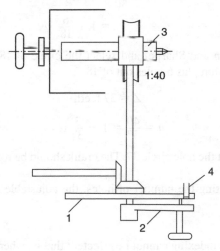

FIGURE 3.59 Simple indexing.

Plate 2: 21, 23, 27, 29, 31, and 33
Plate 3: 35, 37, 39, 41, 43, 47, and 49
Parkinson
Plate 1: 24, 25, 28, 30, 34, 37, 38, 39, 41, 42, and 43
Plate 2: 46, 47, 49, 51, 53, 54, 57, 58, 59, 62, and 66

3.4.5.2 Modes of Indexing

The universal dividing head can be set up for simple or differential indexing, or for milling helical slots.

1. Simple Indexing

The index plate (1) is fixed in position by a lock pin (4) to be motionless (Figure 3.59). The work spindle (3) is rotated through the required angle by rotating the index crank (2). For determining the number of index crank revolutions n to give the number of divisions Z on the job periphery (assuming a worm/worm gear ratio of 1:40), the kinematic balance equation is given by:

$$n = \frac{40}{Z} \tag{3.9}$$

ILLUSTRATIVE EXAMPLE 1

It is required to determine the suitable index plates (Brown and Sharpe) and the number of index crank revolutions n necessary for producing the following spur gears of teeth number 40, 30, and 37 teeth.

Solution

$$Z = 40 \text{ teeth}$$

$$n = \frac{40}{40} = 1 \text{ rev}$$

The crank should be rotated one complete revolution to produce one gear tooth. Any index plate and any circle of holes can be used.

$$Z = 30 \text{ teeth}$$

$$n = \frac{40}{30} = 1\tfrac{1}{3} \text{ rev}$$

$$= 1 + \frac{6}{18}$$

Then choose plate 1 (Brown and Sharpe) and select the circle of 18 holes. The crank should be rotated one complete revolution plus 6 holes out of 18.

$$Z = 37 \text{ teeth}$$

$$n = \frac{40}{37} = 1 + \frac{3}{37} \text{ rev}$$

Choose the plate 3 and select the hole circle 37. The crank should be rotated one complete revolution plus 3 holes out of 37.

To avoid errors in counting the number of holes, the adjustable selector (Figure 3.60) on the index plate should be used.

2. Differential Indexing

It is employed where simple indexing cannot be effected; that is, where an index plate with number of holes required for simple indexing is not available.

In differential indexing, a plunge (5) is inserted in the bore of the work spindle (Figure 3.61) while the index plate is unlocked. The spindle drives the plate through change and bevel gears while the crank through the worm is driving the spindle.

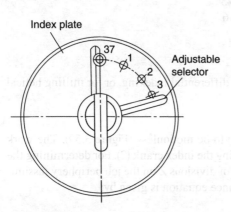

FIGURE 3.60 Counting with adjustable sector.

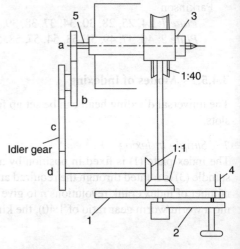

FIGURE 3.61 Differential indexing.

General-Purpose Machine Tools

Hence, the required turn of the work spindle is obtained as sum of two turns (Figure 3.61):

- A turn of the index crank (2) relative to the index plate (1)
- A turn of the index plate itself, which is driven from the work spindle through change gears $(a/b) \times (c/d)$ to provide the correction

Depending on the setup, the index plate rotates either in the same direction with the index crank or in the opposite direction. An idler gear should be used if the crank and plate move in opposite directions to each other (Figure 3.61).

To perform a differential indexing, the following steps are to be considered:

- The number of revolutions of index crank is set up in the same manner as in simple indexing, but not for the required number of divisions Z. Another number Z' nearest to Z makes it possible for simple indexing to be carried out.
- The error of such setup Z' is compensated for by means of a respective setting up of the differential change gears a, b, c, and d (Figure 3.61). The change gears supplied to match the three plate system (Brown and Sharpe) are 24(2), 28, 32, 40, 44, 48, 56, 64, 72, 86, and 100 teeth.
- The number of teeth of the change gears a, b, c, and d are determined from the corresponding kinematic balance equation:

$$\frac{40}{Z'} + \frac{1}{Z}\frac{a \cdot c}{b \cdot d} = \frac{40}{Z} \qquad (3.10)$$

from which

$$\frac{a \cdot c}{b \cdot d} = \frac{40}{Z'}(Z' - Z) \qquad (3.11)$$

It is more convenient to assume that $Z' > Z$ to avoid the use of an idler gear. If $Z' < Z$, then an idler gear must be used (Figure 3.61).

ILLUSTRATIVE EXAMPLE 2

Select the differential change gears and the index plate (Brown and Sharpe), and determine the number of revolutions of the index crank for cutting a spur of $Z = 227$ teeth.

Solution

Assume $Z' = 220$, $Z' < Z$, therefore, idler is required:

$$n = \frac{40}{Z'} = \frac{40}{220} = \frac{2}{11} = \frac{6}{33}$$

$$\frac{a}{b} \times \frac{c}{d} = \frac{40}{Z'}(Z' - Z)$$

$$= \frac{2}{11}(220 - 227) = -\frac{2 \times 7}{11}$$

$$= -\frac{8}{4} \times \frac{7}{11} = -\frac{64}{32} \cdot \frac{28}{44}$$

$a = 64$, $b = 32$, $c = 28$, and $d = 44$ teeth with an idler gear.

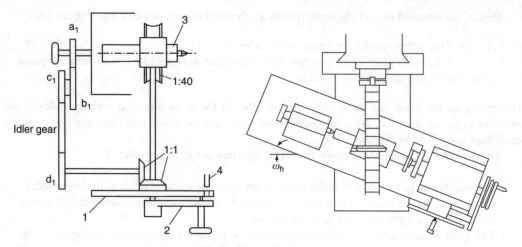

FIGURE 3.62 Setting the dividing head for milling helical grooves.

3. *Setting the Dividing Head for Milling Helical Grooves*

In milling helical grooves and helical gears, the helical movement is imparted to the WP through a reciprocating movement along its axis and rotation of the WP about the same axis. The WP receives reciprocation together with the worktable, and rotation from the worktable lead screw through a set of change gears. The table is set to the spindle axis at an angle ω_h equal to the helix angle of the groove being cut. The table is swiveled clockwise for left-hand grooves and counterclockwise for right-hand grooves (Figure 3.62).

$$\tan \omega_h = \frac{\pi D}{t_{hel}} \quad (3.12)$$

where
t_{hel} = lead of helical groove (mm)
D = diameter of the WP (mm)

The kinematic balance is based on the fact that for every revolution of the blank, it travels axially a distance equal to the lead of the helical groove to be milled. This balance is obtained by setting up the gear train that links the lead screw to the work spindle; therefore,

$$t_{hel} = t_{ls} \times \frac{d_1}{c_1} \cdot \frac{b_1}{a_1} \times 1 \times 40 \quad (3.13)$$

from which

$$\frac{a_1}{b_1} \cdot \frac{c_1}{d_1} = 40 \frac{t_{ls}}{t_{hel}} \quad (3.14)$$

where
t_{ls} = lead of worktable lead screw (mm)
$(a_1/b_1) \cdot (c_1/d_1)$ = change gears (Figure 3.62)

ILLUSTRATIVE EXAMPLE 3

It is required to mill six right-hand helical flutes with a lead of 600 mm; the blank diameter is 90 mm. If the pitch of the table lead screw is 7.5 mm, give complete information about the setup.

Solution

Indexing:

$$n = \frac{40}{6} = 6\frac{2}{3} = 6\frac{12}{18} \text{ crank revolution}$$

Choose plate 1 (Brown and Sharpe) and select the hole circle 18. The crank should be rotated six complete revolutions plus 12 holes out of 18.

Helix:

$$\tan \omega_h = \frac{\pi D}{t_{hel}} = \frac{\pi \times 90}{600} = 0.471$$

then

$$\omega_h = 25.23°$$

The milling table should be set counterclockwise at an angle of 25.23°.

Change gears:

$$\frac{a_1}{b_1} \cdot \frac{c_1}{d_1} = 40 \times \frac{t_{ls}}{t_{hel}}$$

$$= 40 \times \frac{7.5}{600} = \frac{24}{56} \cdot \frac{28}{24}$$

Then, $a_1 = 24$, $b_1 = 56$, $c_1 = 28$, and $d_1 = 24$ teeth.

3.5 SHAPERS, PLANERS, AND SLOTTERS AND THEIR OPERATIONS

3.5.1 SHAPING, PLANING, AND SLOTTING PROCESSES

These processes are used for machining horizontal, vertical, and inclined flat and contoured surfaces, slots, grooves, and other recesses by means of special single-point tools. The difference between these three processes is that in planing, the work is reciprocated and the tool is fed across the work, while in shaping and slotting, the tool is reciprocating and the work is fed across the cutting tool. Moreover, the tool travel is horizontal in shaping and planing and vertical in case of slotting (Figure 3.63).

The essence of these processes is the same as of turning, where metals are removed by single-point tools similar in shape to lathe tools. A similarity also exists in chip formation. However, these operations differ from turning in that the cutting action is intermittent, and chips are removed only during the forward movement of the tool or the work. Moreover, the conditions under which shaping, planing, and slotting tools are less favorable than in turning, even though the tools have the opportunity to cool during the return stroke, when no cutting takes place. That is because these tools operate under severe impact conditions. For these conditions, the related machine and tools are designed to be more rigid and strongly dimensioned, and the cutting speed in most cases does not exceed 60 m/min. Consequently, tools used in these processes should not be shock-sensitive, such as ceramics and CBN. It is sufficient to use low-cost and easily sharpened tools such as HSS and carbides.

The limited cutting speed and the time lost during the reverse stroke are the main reasons behind the low productivity of shaping, planing, and slotting compared to turning. However, in planing, not only the productivity but also the accuracy are enhanced due to the possibility of using multiple tooling in one setting. Figure 3.64 illustrates the kinematics and machining parameters in shaping, planing, and slotting.

The basic machining parameters are the average speed during the cutting stroke v_{cm}, the feed f, the depth of cut t, and the uncut cross-section area A_c. The feed is the intermittent relative movement of the tool (in planing) or the WP (in shaping and slotting), in a direction perpendicular to the

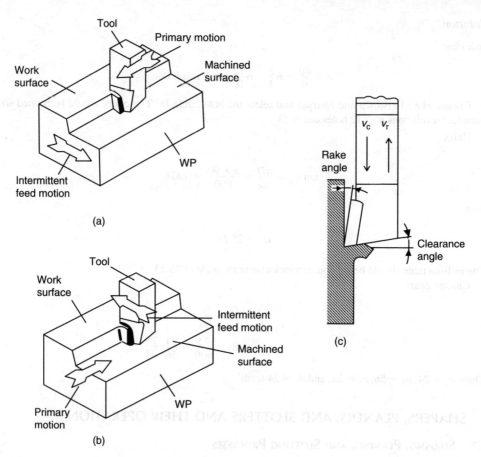

FIGURE 3.63 (a) Shaping, (b) planing, and (c) slotting operations.

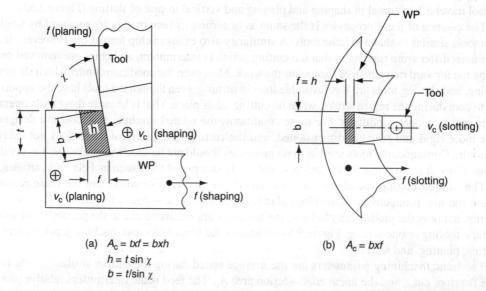

(a) $A_c = t \times f = b \times h$
$h = t \sin \chi$
$b = t/\sin \chi$

(b) $A_c = b \times f$

FIGURE 3.64 Kinematics and machining parameters in (a) shaping and planning and (b) slotting.

General-Purpose Machine Tools

cutting motion and expressed in millimeters per stroke. The feed movement is always actuated at the end of the return stroke when the tool is not engaged with the work.

The depth of cut is the layer removed from the WP in millimeters in a single pass and is measured perpendicular to the machined surface. The uncut chip cross-section in square millimeters is given by the following equation for shaping and planing:

$$A_c = b \cdot h = t \cdot f \, \text{mm}^2 \tag{3.15}$$

where
b = chip contact length (mm)
    = $t/\sin x$
h = chip thickness (mm)
    = $f \sin x$
x = setting angle (frequently $x = 75°$)

and the following equation for slotting

$$A_c = b \cdot f \, \text{mm}^2 \tag{3.16}$$

where b is the slot width (mm).

v_{cm} in meters per minute can be calculated depending on the type of machine mechanism. It should not exceed the permissible cutting speed which depends upon:

- Machining conditions (depth of cut, feed, tool geometry, and related conditions)
- Tool material used
- Properties of WP material

3.5.1.1 Determination of v_{cm} in Accordance with the Machine Mechanism

1. *Machines Equipped by the Quick Return Motion (QRM) Mechanisms ($v_{cm} < v_{rm}$)*

$$v_{cm} = \frac{nL(1+Q)}{1000} \, \text{m/min} \tag{3.17}$$

where
L = selected stroke length (mm)
n = selected number of strokes per minute
$Q = v_{cm}/v_{rm} < 1$ (v_{cm} and v_{rm} are the mean cutting and the mean reverse speeds, respectively)

Equation 3.17 is applicable for:

- Hydraulic shapers, planers, and slotters, where v_c and v_r are constant (Figure 3.65a)
- Shapers and slotters (Figure 3.65b) of lever arm mechanism
- Planers of rack and pinion mechanism (Figure 3.65c)

2. *Machines of Crank Mechanism ($v_{cm} = v_{rm}$)*
These machines are applicable for small size slotters (Figure 3.65d).

$$v_{cm} = v_{rm} \frac{2nL}{1000} \, \text{m/min} \tag{3.18}$$

where
L = selected stroke length (mm)
n = selected number of strokes per minute

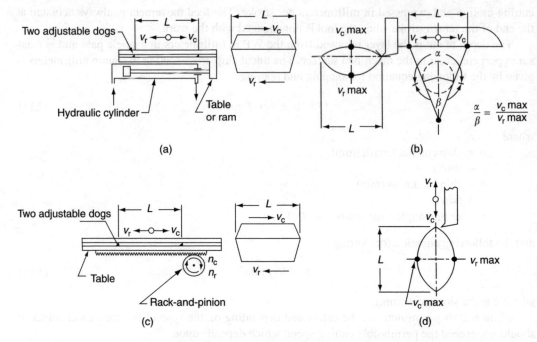

FIGURE 3.65 Shaper, planer, and slotter mechanisms. (a) Hydraulic shapers, planers, and slotters ($v_c < v_r$), (b) shapers and slotters of lever arm mechanism ($v_c < v_r$), (c) planers of rack and pinion mechanism ($v_c < v_r$), (d) small slotters of simple crank mechanism ($v_c < v_r$).

Hydraulic shapers, planers, and slotters are becoming increasingly popular for the following characteristics:

- Greater flexibility of speed (infinite variable)
- Smoother in operation
- Ability to slip in case of overload, thus eliminating tool and machine damage
- Possibility of changing speeds and feeds during operation
- Providing a constant speed all over the stroke

3.5.2 Shaper and Planer Tools

Shaper and planer tools are strongly dimensioned single-point tools designed to withstand the operating impact loads. Figure 3.66 shows typical tools that are used for different machining purposes. These include the following:

  a. Straight-shank roughing tool
  b. Bent-type roughing tool
  c. Side-cutting tool
  d. Finishing tool
  e. Broad-nose finishing tool
  f. Slotting tool
  g. Tee-slot tool
  h. Gooseneck tool

Figure 3.67 compares straight and gooseneck tools. The gooseneck tools are used to reduce digging in; scoring the WP and thus better surface quality is thereby achieved. The tendency to gouge will

General-Purpose Machine Tools

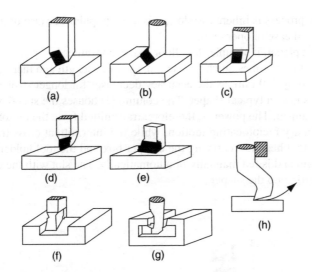

FIGURE 3.66 Shaper and planer tools.

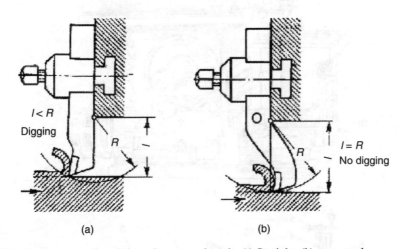

FIGURE 3.67 Performance of straight and gooseneck tools. (a) Straight, (b) gooseneck.

be lessened if the tool nose is leveled up with the base of the tool shank. For eliminating chatter, and accordingly achieving an acceptable surface quality, the tool overhang should be kept as small as possible. Shaper and planer tools have rake angles of 5–10° for HSS tools and between 0 and –15° for carbide tools. The cutting edge inclination angle is normally 10°, while a nose radius of 1–2 mm is used in case of roughing tools.

3.5.3 Shapers, Planers, and Slotters

3.5.3.1 Shapers

A shaper machine is commonly used in single-piece and small-lot production as well as in repair shops and tool rooms. Owing to its limited stroke length, it is conveniently adapted to small jobs and best suited for surfaces comprising straight-line elements and contoured surfaces when the shaper is equipped with a tracing attachment. It is also applicable for cutting keyways and splines on shafts.

Although the shaping process is inherently slow, it is quite popular because of its short setup time, inexpensive tooling, and ease of operation.

In comparison to a planer, it occupies less floor space, consumes less power, costs less, is easier to operate, and is about three times quicker in action, as stroke length and inertia forces are less. Its stroke length is limited by 750 mm, as the accuracy decreases for longer strokes due to ram overhanging. Figure 3.68 shows a typical shaper. The column (1) houses the speed gearbox, the crank, and slotted arm mechanism. The power is, therefore, transmitted from the motor (2) to the ram (3). Ram travel is the primary reciprocating motion, while the intermittent cross travel of the table is the feed motion. The tool head (5), carrying the clapper box and the tool holder (6), is mounted at the front end of the ram and is fed manually or automatically. The slot with the clamp (7) serves to position the ram in setting up the shaper.

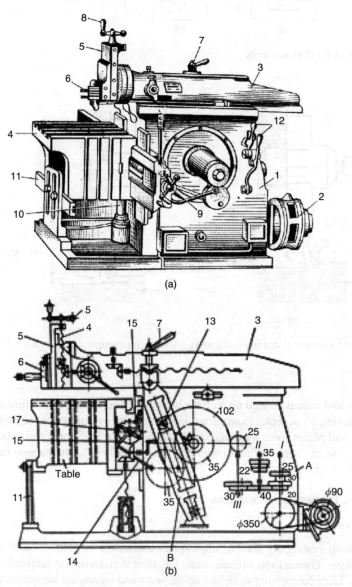

FIGURE 3.68 Typical mechanical shaper: (a) General view and (b) gearing diagram.

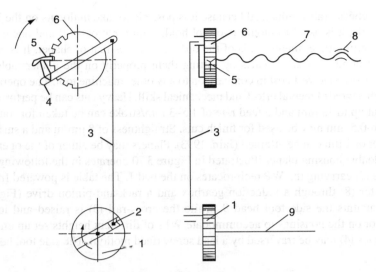

FIGURE 3.69 Table feed mechanism of a mechanical shaper.

The tool head has a tool slide and feed screw rotated by a ball crank handle (8) for raising and lowering the tool to adjust the depth of the cut. A swivel motion of the tool head enables it to take angular cuts to machine inclined surfaces. The WP is clamped either directly on the table or is held in a machine vise. By means of a ratchet and pawl mechanism (9) driven from the crank and slotted arm mechanism, the table is fed cross-wise in a horizontal plane. The table is raised or lowered by the elevating screw (10). A support bracket (11) is provided to clamp the table rigidly during operation. The number of ram strokes per minute is set by shifting levers 12.

Figure 3.68b shows a simplified gearing diagram of a mechanical shaper. Rotation of the main motor is transmitted to the six-speed gearbox (A). The pinion, $Z = 25$, drives the bull gear, $Z = 102$. The rotation movement of the bull gear is converted to reciprocating motion of the slotted arm (B) linked to the ram (3). The stroke length of the ram can be varied by adjusting the radius of the crankpin on the bull gear. This adjustment alters the speed of the ram (see Equation 3.17). Figure 3.69 visualizes the table feed mechanism of the mechanical shaper. Rotation of the crank gear (1), mounted on the driving shaft (9) (driven by the gear, $Z = 35$ [Figure 3.68b]), rocks the pawl carrier (4) and pawl (5) of the ratchet and pawl arrangement through the connection (3). In its forward stroke, the pawl engages a tooth of the ratchet wheel (6), which is fastened to the table lead screw (7). This causes the ratchet wheel and lead screw to rotate a fraction of a revolution. On the return stroke, the pawl slips over the ratchet teeth. Accordingly, the table is fed upon rotation of the lead screw. Radial adjustment of the cluster (2) on the crank gear (1) varies the amount by which the ratchet wheel is moved at each stroke and, consequently, changes the table displacement per stroke (feed). Maximum feed is obtained when the cluster is adjusted to its maximum radius. The table feed direction can be reversed by revering the ratchet (5).

3.5.3.2 Planers

Planers are intended for machining large-size WPs because of their capacity for long table travel (1–15 m) and robust construction. They are used to machine plane surfaces that may be horizontal, vertical, or at an angle. Angular surfaces are often easier to machine on planers. Some of the work formerly done on planers is done now on planer-type milling machines using large face milling cutters. However, it is found that milling cutters tend to be glazed and the machined component is work-hardened and hence becomes difficult to be hand-scraped. Therefore, plane surfaces that required hand-scraping are preferably machined on planers. Both the productivity and accuracy of planers are considerably enhanced because it is possible to take multicuts on the WP in a single stroke. Generally, it is usual to mount two tool holders on the cross-rail and one each side of the column. The setting time, therefore, is of the order 5–6 times that of shaper. It is also possible to machine large number of small parts by setting them properly on the planer table. The planers produce large work at lowest cost in comparison to any other machine tool. The operator of a planer requires a high degree of mental effort and mechanical skill. Heavy cuts can be performed on planers. A depth of cut up to 18 mm and a feed rate of 1.5–3 mm/stroke can be taken for roughing, while a depth of 0.25–0.5 mm may be used for finish cuts. Straightness of 8 μm/m and a surface roughness R_a of the order of 1 μm can be attained (Jain, 1993). Planers may be either of the open-side or housing type. A double housing planer illustrated in Figure 3.70 operates in the following manner.

The table (2) carrying the WP reciprocates on the bed 1. The table is powered from a variable-speed dc motor (8) through a reduction gearbox and a rack-and-pinion drive (Figure 3.71). The housing (6) mounts the side tool head 9, while the cross-rail (3) is raised and lowered from a separate motor on the housings to accommodate WPs of different heights set up on the table. The upper tool heads (4) may be traversed by a lead screw (feed motion). The side tool head is traversed

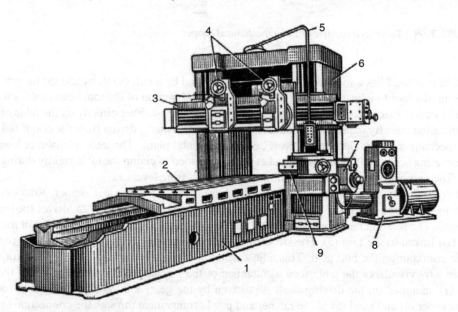

FIGURE 3.70 Double housing planer.

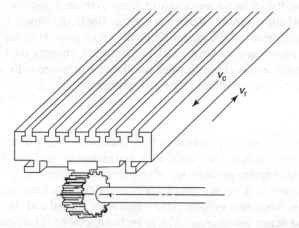

FIGURE 3.71 Rack-and-pinion mechanism of a planer.

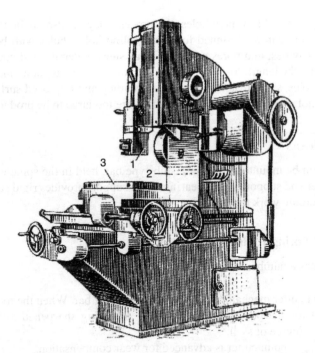

FIGURE 3.72 Typical slotter.

vertically (feed motion) by the feed gearbox (7) to machine vertical surfaces. All tool heads operate independently of each other. The control panel and the suspended cable (5) are shown in Figure 3.70. The tool heads (4) may be swiveled to machine an inclined surface. Like all reciprocating machine tools, planers are equipped with a clapper box to raise the tools on the return stroke. As the tool and holder are quite heavy, air cylinders are employed to lift the tool from the WP on the return stroke.

3.5.3.3 Slotters

Slotters are commonly used for internal machining of blind holes, or vertical machining of complicated shapes that are difficult to machine on horizontal shapers. They are useful for machining keyways, and cutting of internal and external teeth on large gears.

As illustrated in Figure 3.72, the job is generally supported on a round table (3) that has a rotary feed in addition to the usual table movement in cross-directions. The ram (1) travels vertically along the ways of the column (2). The ram stroke of a slotter ranges from 300 to 1800 mm. The slotters are generally very robust machines and there is a possibility of tilting the ram up to $\pm 15°$ from vertical to permit machining of dies with relief.

The rams are either crank-driven or hydraulically driven. Ram speeds are usually from 2 to 40 m/min. Longitudinal and transverse feeds range from 0.05 to 2.5 mm/stroke. Cutting action takes place on downward stroke.

3.6 BORING MACHINES AND OPERATIONS

3.6.1 Boring

Boring is the machining process in which internal diameters are generated in true relation to the centerline of the spindle by means of single-point tools. It is the most commonly used process for enlarging and finishing holes or other circular contours. Although most boring operations are performed on simple straight-through holes, the process may be also applied to a variety of other configurations. Tooling can be designed for boring blind holes, holes with bottle configurations, circular-contoured cavities, and bores with numerous steps undercuts and counterbores. The process is not limited by the length-to-diameter ratio of holes. Boring is sometimes used after drilling to provide drilled holes with greater dimensional accuracy and improved surface finish. It is used for finishing large holes in castings and forgings that are too large to be produced by drilling.

3.6.2 Boring Tools

The boring tools can be mounted in either a stub-type bar, held in the spindle, or in a long boring bar that has its outer end supported in a bearing. Such support provides rigid support for the boring bar and permits accurate work to be done.

3.6.2.1 Types of Boring Tools

Figure 3.73 illustrates a number of typical boring tools:

a. A single-point cutter mechanically secured to a boring bar. When the tool becomes worn, it is removed for sharpening and reset again. Resetting sharpened tools is tedious and requires a fair degree of skill.
b. Adjustable single-point cutter is advanced for wear compensation.
c. Boring tools are clamped in a universal boring head that is attached to the end of the boring bar. The head is designed to accommodate a variety of tool configurations.
d. A fixed cutter, held by a stub boring bar, is simple and widely used.
e. A blade-type boring tool, where the cutter is inserted through the body, thus providing two cutting edges that enable a substantially higher increase of feed rate than that is possible when only one cutting edge is used. The main advantage of this tool is that it equalizes the

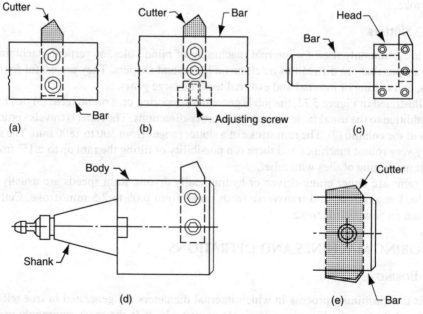

FIGURE 3.73 Typical boring tools.

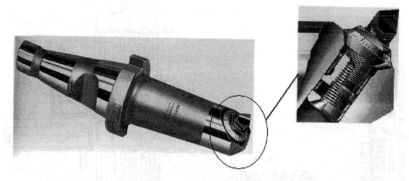

FIGURE 3.74 Adjustable tools in stub. (From DeVlieg Machine Co., Michigan, USA.)

force imposed on the bar during operation. Therefore, it is possible to maintain closer tolerances with bars having maximum unsupported length than when using a boring tool that has only one cutting edge. The main disadvantage is that the blade cannot be adjusted to compensate for wear, and therefore must be removed for sharpening and then reset. The boring bar illustrated in Figure 3.74 is the same as d in Figure 3.73, but the cutter is adjustable for wear compensation.

3.6.2.2 Materials of Boring Tools

1. For low cutting speeds, HSS is more suitable than carbide.
2. Carbides are used almost exclusively for precision boring when the maximum rigidity is maintained in the setup.
3. Ceramics are increasingly applied for precision boring operation at high cutting speeds.

Ceramic inserts are characterized by reduced tool wear. Ceramics have the ability to bore hard materials (steel of 60–65 HRC), thus eliminating the need of subsequent grinding. Ceramics are not recommended for boring, interrupted cuts and refractory metals. Also, they are not recommended for aluminum alloys, because they develop BUE.

3.6.3 Boring Machines

Boring is performed on almost every type of machine that has facilities for rotating a spindle or a WP. Most boring operations are done in conjunction with turning and cutting on NC and CNC machines, and so on discussed in other chapters of this book. However, in this section, the general-purpose horizontal boring machines and the jig boring machines are discussed.

3.6.3.1 General-Purpose Boring Machines

In these machines, the WP remains stationary; the tool rotates and may simultaneously perform a feed motion. The boring machine is designed to machine relatively large, irregular, and bulky WPs that cannot be easily rotated. Among the operations performed on this machine are boring, facing, drilling, counterboring, counterfacing, external and internal thread cutting, and milling. A horizontal boring machine is especially suitable for work where several parallel bores with accurate center distances are to be produced. Because of its flexibility, this machine is especially suited to work in which other machining operations are performed in conjunction with boring.

A typical general-purpose boring machine is shown in Figure 3.75. The cutting tool is mounted either in the spindle (13) or on the facing slide (8). The rotation of the spindle and faceplate (7) is the

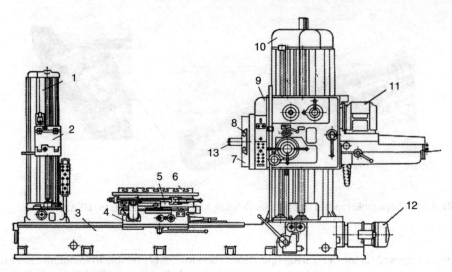

FIGURE 3.75 Typical general-purpose boring machine.

principal movement that is effected by the main motor (11) through the speed gearbox housed in the headstock (9). The spindle can also be fed axially so that drilling and boring can be done over a considerable distance without moving the work. The WP is installed either directly on the table (6) or in a fixture. The table is moved longitudinally or transversally on the cross-slide (5). The table and cross-slide are located on a saddle (4), which moves longitudinally on the bed (3).

The headstock (9) moves vertically along the column (10) simultaneously with the spindle rest (2), which is moving vertically along the end support column (1). The spindle travels axially when boring or cutting internal thread, and so on. The facing slide is moved radially on the faceplate to perform facing operations. The rotational speed of the faceplate is much less than that of the spindle. The table feed and its rapid reverse are powered by the motor (12). In some setups, the work is fed toward the tool, while in other cases, the tool is fed toward the work.

3.6.3.2 Jig Boring Machines

Jig borers are extra-precise vertical boring machines intended for precise boring, centering, drilling, reaming, counterboring, facing, spot facing, and so on in addition to lay out work. They are mainly designed for use in tool making, jigs and fixtures, and machining of other precisions parts. No jigs whatsoever are required in these machines. A jig boring machine contains similar features of a vertical milling machine, except that the spindle and its bearings are constructed with very high precision, and the worktable permits extra-precise movement and control.

As illustrated schematically in Figure 3.76, the jig borer is generally built lower to the floor and is of much more rigid and accurate construction than any other machine tool. The table and saddle ensure the longitudinal and cross movements, X and Y. The machine has a massive column that supports and accurately guides the spindle housing in the vertical direction, thus achieving the third position adjustment, Z.

The jig boring machines are rigid enough to perform heavy cuts and sensitive enough for précising. They are equipped with special devices ensuring accurate positioning of the machine operative units including a precision lead screw-and-nut and are supplemented by vernier dials and precision scales in combination with optical read-out devices, inductive transducers, and also optical and electrical measuring devices. To prevent the influence of ambient temperature changes on the machining accuracy, jig borers are installed in special environmental enclosures with temperature

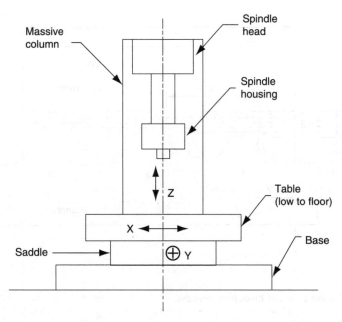

FIGURE 3.76 Schematic of a jig-boring machine.

maintained at a level of 20°C. Currently, jig boring machines are often replaced by CNC machining centers to do similar work.

3.7 BROACHING MACHINES AND OPERATIONS

3.7.1 Broaching

Broaching is a cutting process using a multitoothed tool (broach) having successive cutting edges, each protruding to a greater distance than the proceeding one in the direction perpendicular to the broach length. In contrast to all other cutting processes, there is no feeding of the broach or the WP. The feed is built into the broach itself through the consecutive protruding of its teeth. Therefore, no complex motion of the tool relative to the WP is required, where the tool is moved past the WP with a rectilinear motion v_c (Figure 3.77). Equally effective results are obtained if the tool is stationary and the work is moved. The total depth of the material removed in one stroke T is the sum of rises of teeth of the broach. T may be as deep as 6 mm broached in one stroke. If more depth is to be broached, two broaches may be used to perform the task. Broaching is generally used to machine through holes of any cross-sectional shape, straight and helical slots, external surfaces of various shapes, and external and internal toothed gears (Figure 3.78). To permit the broaching of spiral grooves and gun-barrel rifling, a rotational movement should be added to the broach. Broaching usually produces better accuracy and finish than drilling, boring, or reaming operations. A tolerance grade of IT6 and a surface roughness R_a of about 0.2 μm can be easily achieved by broaching.

Broaching dates back to the early 1850s, when it was originally developed for cutting keyways in pulleys and gears. However, its obvious advantages quickly led to its development for mass-production machining of various surfaces and shapes to tight tolerance. Today, almost every conceivable form and material can be broached.

Broaches must be designed individually for a particular job. They are very expensive to manufacture ($15,000–$30,000 per tool). It follows that the broaching can only be justified when a very

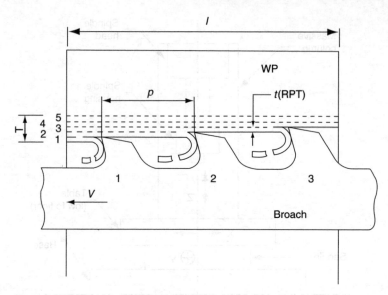

FIGURE 3.77 Cutting action of broaching process.

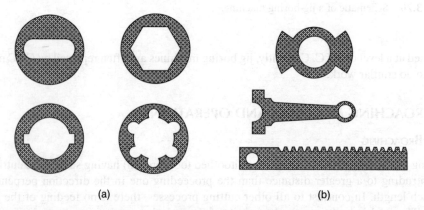

FIGURE 3.78 Typical parts produced by internal and external broaching: (a) internal broaching and (b) external broaching. (From Kalpakjian, S. and Schmid, S. R., *Manufacturing Processes for Engineering Materials*, Prentice-Hall, New York, 2003.)

large batch size (100,000–200,000) is to be machined. However, sometimes the WP is designed so that a nonexpensive standard broach can be used.

3.7.1.1 Advantages and Limitations of Broaching

1. *Advantages*
   - Broaching is a process in which both roughing and finishing operations are completed in one pass, giving a high rate of production.
   - It is a fast process; it takes only seconds to accomplish a task that would require minutes with any other method. Rapid loading and unloading of fixtures keeps the total production time to the minimum.
   - Automation is easily arranged.
   - Internal and external surfaces can be machined within close tolerance that is normally required for interchangeable mass production.

- As all the performance is built into the tooling, little skill is needed to operate a broaching machine.
- Broaches have an exceptionally long life (10,000–20,000 parts per each sharpening), as each tooth passes over the work only once per pass.

2. Limitations
   - Broaches are costly to make and sharpen. Hence, broaching is adopted only in cases of mass production.
   - Standard broaches are available; however, most broaches are expensive, as they are made especially to perform only one job.
   - Special precautions may be necessary when broaching cast and forged parts to control the variations in stock. Operations for removing excess stock may be necessary, which add to the overall cost of manufacturing.
   - Surfaces to be machined must be parallel to the axis of the broach.
   - Broaching is impractical in the following cases:
     a. A surface that has obstructions across the path of the broach travel.
     b. Blind holes and pockets.
     c. Fragile WPs, because they cannot withstand broaching forces without distortion or breakage.

3.7.2 THE BROACH TOOL

3.7.2.1 Tool Geometry and Configuration

Figure 3.79 illustrates the broach tooth terminology. Each individual tooth has the basic wedge form.

- Depending on the material being cut, the rake (hook) angle ranges from 0° to 20°. The small clearance (back-off) angle is usually 3°–4° for roughing teeth and 1°–2° for finishing teeth.
- The rise per tooth (RPT) (superelevation) is the difference in height of two consecutive teeth (Figure 3.77). It is selected depending upon the material to be machined and the type (form) of the broach (Table 3.3).
- The pitch is the distance between two consecutive teeth of a broach. It depends upon the following factors:
  - Length of cut l
  - Material of WP and its mechanical properties
  - RPT (superelevation)

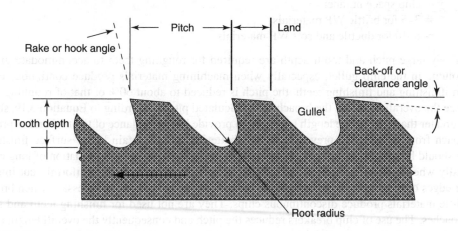

FIGURE 3.79 Broach tooth terminology.

TABLE 3.3
RPT of Broaches

Type of Broach	RPT (μm)			
	Steel	CI	Aluminum	Bronze–brass
Round	15–30	30–100	20–50	50–120
Spline	25–100	40–100	20–100	50–120
Square and hexagon	15–80	80–150	20–100	50–200
Keyway	50–200	60–200	50–80	80–200

Source: Arshinov, V. and Alekseev, G., *Metal Cutting Theory and Cutting Tool Design*, Mir Publishers, Moscow, 1970. With permission.

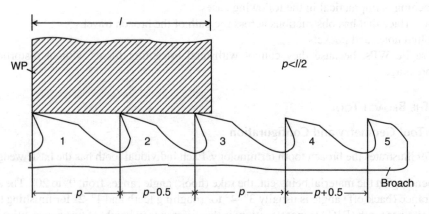

FIGURE 3.80 Nonuniform pitching to prevent chattering, and engagement of more than two teeth to ensure guidance.

The pitch P can be expressed empirically by

$$p = 3\sqrt{RPT \cdot l \cdot X} \qquad (3.19)$$

where
X = chip space number
    = 3–5 for brittle WP materials
    = 6–10 for ductile and soft WP materials

A relatively large pitch and tooth depth are required for roughing teeth to accommodate greater chip volume in the chip gullet, especially when machining materials produce continuous chips. For semifinishing and finishing teeth, the pitch is reduced to about 60% of that of roughing teeth to reduce the overall length of the broach. The calculated pitch, according to Equation 3.19, should not be greater than $l/2$ (l is the length to be cut) to provide better guidance of the tool and to prevent the broach from drifting. To prevent possible chattering and to obtain better surface finish, the pitch p should be made nonuniform as shown in Figure 3.80. To avoid the formation of long chips, especially when broaching profiles and circular shapes, chip breakers are uniformly cut into the cutting edges of the broach in a staggered manner. Chip breakers are not necessary when broaching brittle materials produce discontinuous chips. They are not used for finishing teeth and small size broaches. The use of chip breakers reduces the pitch and consequently the overall length of the broach. As a result, the productivity is enhanced and the tool cost may be reduced.

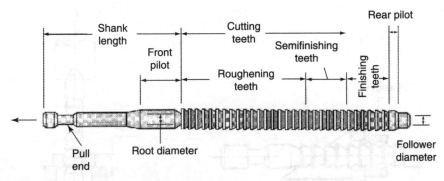

FIGURE 3.81 Solid-pull broach configuration.

FIGURE 3.82 Shell broach. (From Degarmo, E. P. et al., *Materials and Processes in Manufacturing*, 8th Edition, Prentice-Hall, New York, 1997.)

Broach Configuration

Figure 3.81 illustrates the terminology of a pull-type internal broach for enlarging circular holes. The cutting teeth on the broach have three regions: roughing, semifinishing, and finishing teeth. On some round broaches, burnishing teeth are provided for finishing or sizing. These teeth have no cutting edge, but are rounded. Their diameters are oversized by 25–30 μm larger than the finished hole. Irregular shapes are produced by starting from circular broaching in the WP originally provided with drilled, bored, cored, or reamed hole.

The pull end provides a means of quickly attaching the broach to the pulling mechanism. The front pilot aligns the broach in the hole before it begins to cut and the rear pilot keeps the tool square with the finished hole as it leaves the WP. It also prevents sagging of the broach. The follower end is ground to fit in the machine follower rest.

Internal broaches are also made of shells mounted on an arbor (Figure 3.82). Shell broaches are superior to solid broaches in that worn or broken shells can be replaced without discarding the entire broach. Shell construction, however, is initially more expensive than a solid broach of comparable size. The disadvantage of shell broaches is that some accuracy and concentricity are sacrificed.

Regarding the application of broaching force, two types of broaching are distinguished (Figure 3.83):

1. Pull-broaching, as the name implies, involves the broach being pulled through the hole (Figure 3.83a). In this case, the main cutting force is applied to the front of the broach, subjecting the body to tension. Most internal broaching is done with pull-broaches. Because there is no problem of buckling, pull-broaches can be longer than push-types for the same broaching depth. Pull-broaches can be made to long lengths, but cost usually limits the length to approximately 2 m. Broaches longer than 2 m are shell broaches, because the cost is less for replacing damaged or worn sections than for replacing the entire broach.
2. Push-broaching applies the main cutting force to the rear of the broach, thus subjecting the body to compression (Figure 3.83b). A push-broach should be shorter than a pull-broach and its length does not usually exceed 15 times its diameter to avoid buckling

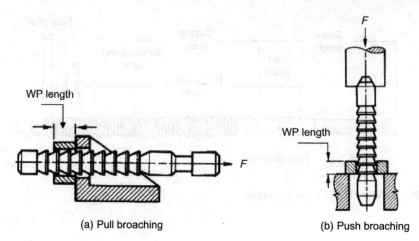

FIGURE 3.83 Pull- and push-broaching.

3.7.2.2 Broach Material

The low cutting speeds used in most broaching operations (2–12 m/min) do not lend themselves to the advantage of carbide tooling. Accordingly, most broaches are made of alloy steel and HSS of high grades (Cr-V-grade), which have less distortion during heat treatment. This is an important factor in the manufacture of long broaches. Titanium-coated HSS broaches are becoming more common due to their prolonged tool life.

Recently, carbide-tipped K-type (cobalt group) tools are employed to machine CI, thus allowing higher cutting speeds, increased durability, and improved surface finish. However, carbide-tipped broaches are seldom used for machining steels and forged parts, as the cutting edges tend to chip in the first stroke due to lack of rigidity of work fixture/tool combination.

3.7.2.3 Broach Sharpening

Broach sharpening is essential, as dull tools require more force, leading to less accuracy and broach damage. Dull internal broaches have the tendency to drift during cutting.

The clearance angle of the sizing teeth of a broach is made as small as possible (1°–2°) to minimize the loss of size when it is sharpened. Also, the finishing or sizing teeth are commonly provided with a land of a small width of 50–200 μm to limit the size loss due to sharpening. Most of broaches are sharpened by grinding the hook faces of the broach. The lands must not be reground because this would change the size of the broach (Figure 3.84). After sharpening, the tooth characteristics such as rake angle, clearance angles, tooth depth, root radius, RPT, and pitch should not be altered.

3.7.3 Broaching Machines

In comparison with other types of machine tools, broaching machines are notable for their simple construction and operation. This is due to the fact that the shape of the surface produced in broaching depends upon the shape and arrangement of the cutting edges on the broach. The only cutting motion of the broaching machine is the straight line motion of the ram. Broaching machines have no feed mechanisms, as the feed is provided by a gradual increase in the height of the broach teeth.

Hydraulic drives, developed in the early 1920s, offered pronounced advantages over the various early mechanical driving methods. Most broaching machines existing today are of hydraulic drive, accordingly characterized by smooth running and safe operation.

The choice between vertical and horizontal machines is determined primarily by the length of stroke required and the available floor space. Vertical machines seldom have strokes greater than

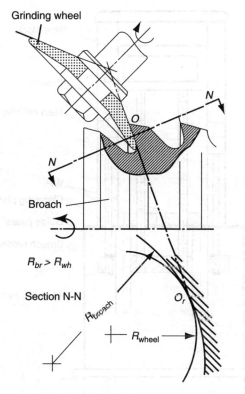

FIGURE 3.84 Sharpening of tool face of a round broach.

1.5 m because of ceiling limitation. Horizontal machines can have almost any stroke length; however, they require greater floor space.

The main specifications of a broaching machine are as follows:

- Maximum pulling or pushing force (capacity) (ton)
- Maximum stroke length (m)
- Broaching speed (m/min)
- Overall dimensions and total weight

3.7.3.1 Horizontal Broaching Machines

Currently, horizontal machines are finding increasing favor among users because of their long strokes and the limitation that ceiling height places on vertical machines. About 47% of all broaching machines are horizontal units (ASM International, 1989). Horizontal internal broaching machines are used mainly for some types of work such as automotive engine blocks. The pulling capacity ranges from 2.5 to 75 tons, strokes up to 3 m, and cutting speeds limited to less than 12 m/min. Broaching that requires rotation of the broach, as in rifling and spiral splines, is usually done on horizontal internal broaching machines. Horizontal machines are seldom used for broaching small holes.

Horizontal surface broaching machines may be hydraulically or electromechanically driven. In these machines, the broach is always supported in guides. The surface hydraulic broaching machines are built with capacities up to 40 tons, strokes up to 4.5 m, and cutting speeds up to 30 m/min. These machines are basically used in the automotive industry to broach a great variety of CI parts for nearly 30 years.

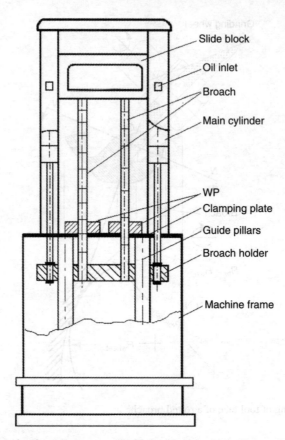

FIGURE 3.85 A schematic of a pull-down vertical broaching machine. (Adapted from ASM International, *Machining*, Vol. 16, *Metals Handbook*, ASM International, Materials Park, OH, 1989.)

On the other hand, the electromechanically driven horizontal surface broaching machines are available with higher capacities, stroke lengths, and cutting speeds (up to 100 tons, 9 m, and 30 m/min, respectively). Carbide-tipped broaches are used to machine CI blocks of internal combustion engines (ICEs).

3.7.3.2 Vertical Broaching Machines

These machines are almost all hydraulically driven. They are used in every major area of metal working. Depending on their mode of operation, they may be pull-up, pull-down, or push-down units. Figure 3.85 schematically illustrates a pull-down vertical broaching machine in which the work is placed on the worktable. These machines are capable of machining internal shapes to close tolerances by means of special locating fixtures. They are available with pulling capacities from 2 to 50 tons, strokes from 0.4 to 2.3 m, and cutting speeds up to 24 m/min. When cutting strokes exceed existing factory ceiling clearances, expensive pits must be dug for the machine so that the operator can work at the factory floor level.

3.7.3.3 Continuous Horizontal Surface Broaching Machines

In this type of machines, the broaches are usually stationary and mounted in a tunnel on the top of the machine, while the work is pulled past the cutters by means of a conveyor (Figure 3.86). Fixtures are usually attached to the conveyor chain, so that the WPs can be provided automatically by the loading chute at one end of the bed and removed at the other end. The key to the productivity of

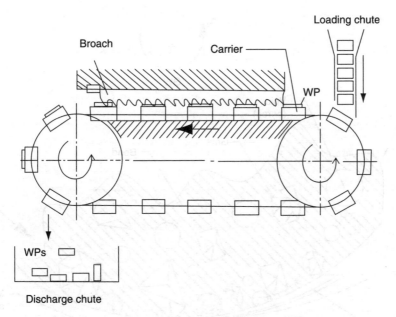

FIGURE 3.86 Continuous horizontal surface broaching machine. (Adapted from ASM International, *Machining*, Vol. 16, *Metals Handbook*, ASM International, Materials Park, OH, 1989.)

this type of machines is the elimination of the return stroke by mounting the WPs on continuous chain. In the rotary continuous horizontal broaching machines, the broaches are also stationary, while the work is passed beneath or between them. The work is held in fixtures on a rotary table. These machines are also used in mass production, as there is no loss of time due to the noncutting reciprocating strokes.

3.8 GRINDING MACHINES AND OPERATIONS

3.8.1 Grinding Process

Grinding is a metal removal process that employs an abrasive GW whose cutting elements are grains of abrasive materials of high hardness and high refractoriness. Grinding is generally among the final operations performed on manufactured products. It is not necessarily confined to small-scale material removal; it is also used for large-scale material removal operations and specifically compete economically in this domain with some machining processes such as milling and turning. The development of abrasive materials and better fundamental understanding of the abrasive machining have contributed in placing grinding among the most important basic machining processes.

Because the abrasives employed are very hard, abrasive machining is used for:

- Finishing hard materials and hardened steels
- Shaping hard nonmetallic materials such as carbides, ceramics, and glass
- Cutting-off hardened shafts, masonry, granite, and concrete
- Removing weld beads
- Cleaning surfaces

The sharp-edged and hard grains are held together by bonding material. Projecting grains (Figure 3.87) abrade layers of metal from the work in the form of very minute chips as the wheel rotes at high speeds of up to 60 m/s. Owing to the small cross-sectional area of the chip and the

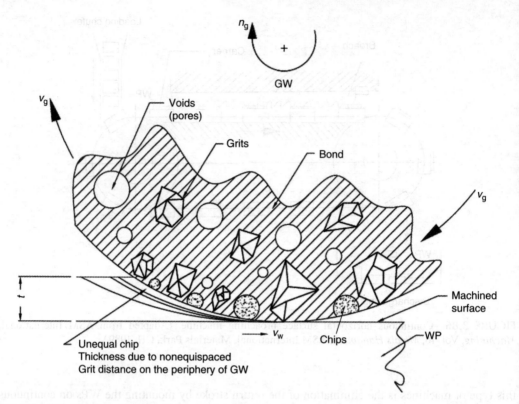

FIGURE 3.87 Cutting principles and main variables of a surface grinding process.

high cutting speed, grinding is characterized by high accuracy and good surface finish. Consequently, it is usually employed as a finishing operation. However, it is also used in snagging.

The chip formation in grinding is similar to milling. In spite of the small size of the layer being cut in grinding, the chip has the same comma form similar to that obtained by milling. However, in grinding, not all the grains participate equally in the metal removal as in milling.

Along with the general features of other typical methods of machining, the grinding process has certain specific features of its own, such as the following:

- In contrast to the teeth of a milling cutter, individual grains of a GW have an irregular and nondefinite geometry. They are randomly spaced along the periphery of the GW.
- The radial positions of the grains (protruding) on the wheel periphery vary, which make the grains cut layers of material in the form of chips of different volumes (Figure 3.87).
- The grains of the GW are characterized by high negative rake angles of −40° to −80°, consequently, the shear angles are very small (Figure 3.88).
- Owing to the minute chip thickness and the highly abrasive negative rakes of the grinding operation, the specific cutting energy in grinding is considerably larger than that of operations using tools of definite geometry. Grinding is thus not only time-consuming but also power-consuming and is hence a costly operation.
- The GW has a self-sharpening characteristic. As the grains wear during grinding, they either fracture or are torn off the wheel bond, exposing new sharp grains to the work.
- The cutting speeds of GWs are very high, typically 30 m/s, which together with the minute chip removal of the grains provide high dimensional and form accuracy along with high surface quality.

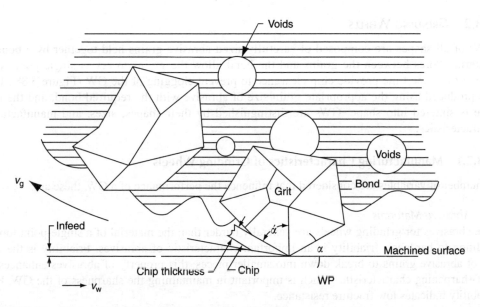

FIGURE 3.88 Schematic to illustrate the constituents of a GW.

These features make the grinding process more complicated than the other kinds of machining processes and offer considerable difficulties in both theoretical and experimental investigations. However, grinding possesses certain advantages over other metal cutting methods:

- It cuts hardened steels easily. Parts requiring hard surfaces are first machined to shape in annealed condition, with only a small amount left as the grinding allowance, considering the tendency of material to warp during hardening operation.
- Very accurate dimensions and smoother surfaces can be achieved in a very short time.
- Very little pressure is required, thus permitting very light work to be ground that would otherwise tend to spring away from the tool. This permits the use of magnetic chucks for holding the work in many grinding operations.

Machining variables of a surface grinding process. Figure 3.87 illustrates the main variables of a surface grinding process. The main rotary motion is performed by the GW (v_g), whereas the feed motion is performed by the WP (v_w). The depth of cut t (feed/stroke) is fed by the GW perpendicular to the machined surface.

The cutting speed v_g is given by

$$v_g = \frac{\pi D n_g}{1000} \text{ m/min} \qquad (3.20)$$

where
n_g = rotational speed of the GW (rpm)
D = outside diameter of the GW (mm)

The feed motion of the WP, v_w is considerably smaller than the main cutting speed of the GW, v_g. Typical values of the ratio v_w/v_g ranges from 1/20 (for rough grinding) to 1/120 (for finishing). The depth of cut t (feed/stroke) ranges from 10 μm (for finish cuts) to 100 μm (for roughing).

3.8.2 Grinding Wheels

GWs of all shapes are composed of carefully sized abrasive grains held together by a bonding material. Pores between the grains and the bond allow the grains to act as single-point tools, and at the same time provide chip clearance to prevent clogging of the GW (Figure 3.88). GWs are produced using the appropriate grain size of abrasive with the required bond, and the mixture is sintered into shape. GWs are distinguished by their shapes, sizes, and manufacturing characteristics.

3.8.2.1 Manufacturing Characteristics of Grinding Wheels

A number of variables are considered that influence the performance of a GW, these are:

1. Abrasive Materials

The abrasives for grinding wheels are generally harder than the material of a single-point tool. In addition to hardness, friability is an important characteristic of abrasives. Friability is the ability of abrasive grains to break down into smaller pieces; this property of abrasives enhances the self-sharpening characteristic, which is important in maintaining the sharpness of the GW. High friability indicates low fracture resistance.

Aluminum oxide (Al_2O_3) has lower friability than silicon carbide (SiC); thus, it has less tendency to fragment and self-sharpen. The shape and size of grain also affect its friability. Small grains of negative rakes are less friable than plate-like grains.

The four types of abrasive materials used in manufacturing of GWs are produced synthetically. They are classified into

a. *Conventional Abrasives* as follows:

- Aluminum oxide, Al_2O_3 (corundum), which has high hardness (Knoop number = 2100) and toughness and is mainly used for grinding metals and alloys of high tensile strength such as steel, malleable iron, and soft bronze.
- Silicon carbide, SiC (carborundum), which is harder than Al_2O_3 (Knoop number = 2500). It is more friable (more brittle) and is mainly used for grinding materials that have low strength like CI, aluminum, cemented carbides, and so on. Silicon carbides are available in black (95% SiC) or green (98% SiC). Carbide-tipped tools which are sharpened by SiC dull more rapidly than Al_2O_3 when grinding steels.

b. *Super Abrasives*:

- CBN, which has been manufactured by the General Electric Company since 1970 under the trade name of Borazon. Its properties are similar to those of diamond. CBN is very hard (Knoop number = 4500). It is used for manufacturing wheels intended for grinding extra-hard materials at high speeds. CBN is 10–20 times more expensive than Al_2O_3.
- Diamond, which is the hardest of all materials (Knoop number = 7500). It has been synthetically produced since 1955. Synthetic diamonds are friable. Diamond has a very high chemical resistance as well as a low coefficient of thermal expansion. Diamond abrasive wheels are extensively used for sharpening carbide and ceramic cutting tools. Diamonds are used for truing and dressing other types of abrasive wheels. Diamonds are best suited for nonferrous metal and are not recommended for machining steels.

Table 3.4 shows the characteristics of abrasive materials used in GW manufacturing and their applications.

TABLE 3.4
Characteristics of Abrasive Materials and Their Applications

Abrasive	Knoop Number	Uses
Al_2O_3	2100–3000	Safer and tougher than SiC, used for steels and high-strength materials
SiC	2500–3000	Nonferrous, nonmetallic materials, CI, carbides, hard metals, and good finish
CBN	4000–5000	Hard and tough tool steel, stainless steel, aerospace alloys, hard coating
Diamond	7000–8000	Nonferrous metals, sharpening carbide, and WC tools

Source: Raw, P. N., *Metal Cutting and Machine Tools*, Tata McGraw-Hill, New Delhi, 2000.

2. *Abrasive Grain Size*

The size of an abrasive grain is identified by the grit number, which is a function of sieve size. The smaller the sieve size, the larger the grit number. The sieve sizes (mesh number) of abrasives are grouped into four categories:

Coarse	10, 12, 14, 16, 20, and 24
Medium	30, 36, 46, 56, and 60
Fine	70, 80, 90, 100, 120, 150, and 180
Very fine	220, 240, 280, 320, 400, 500, and 600

The choice of the grain size is determined by the nature of the grinding operation, the material to be ground, and the relative importance of the stock removed rate to the finish required. Coarse and medium sizes are normally used for roughing and semifinishing operations. Fine and very fine grains are used for finishing operations and also used for making form GWs.

3. *Wheel Grade*

The wheel grade designates the force holding the grains. It is a measure of the strength of the bond. The wheel grade depends upon the type and amount of the bond, the structure of the wheel, and the amount of abrasive grains. Because strength and hardness are directly related, the grade is also referred to as the hardness of the bonded abrasive.

The grade is designated by letters, as follows:

Very soft	A, B, C, D, E, F, and G
Soft	H, I, J, and K
Medium	L, M, N, and O
Hard	P, Q, R, and S
Very hard	T, U, V, W, X, Y, and Z

Soft grades are generally used for machining hard materials and vice versa. When grinding hard materials, the grit is likely to become dull quickly, thus increasing the grinding force, and tends to knock off the dull grains easily. In contrast, when a hard grade is used to machine soft material, the grits are retained for a longer period of time, which prolongs the grinding wheel's life. Table 3.5 illustrates the recommended wheel grades for different materials and operations. During grinding, and depending on the machining variables, the wheel behaves as if it is harder or softer than its nominal or selected grade, as illustrated in Table 3.6.

TABLE 3.5
Grinding Wheel Hardnesses for Different Materials and Operations

	Wheel Hardness			
WP Material	Cylindrical Grinding	Surface Grinding	Internal Grinding	Deburring
Steel up to 80 kg/mm²	L, M, N	K, L	K, L	
Steel up to 140 kg/mm²	K	K, J	J	
Steel more than 140 kg/mm²	J	L, J	I	
Light alloys	J	I, K	I	O, P, Q, R
CI	K	I	J	
Bronze, brass, and copper	L, M	J, K	J	

Source: Raw, P. N., *Metal Cutting and Machine Tools*, Tata McGraw-Hill, New Delhi, 2000.

TABLE 3.6
Effect of Machining Variables on the Wheel Grade

Variable	Wheel Grade Appears
Increasing work speed (v_w)	Soft
Increasing wheel speed (v_g)	Harder
Increasing work diameter (d)	Harder
Increasing wheel diameter (D)	Harder

4. Wheel Structure

The structure of a GW is a measure of its porosity. Some porosity (Figure 3.87) is essential to provide clearances for the grinding chips; otherwise, they would interfere with the grinding process. The wheel loses its cutting ability due to loading by chips.

Wheels of open or porous structure are used for high metal removal rates that produce rough surfaces, whereas those of dense or compact structure are used for precision grinding at low MRRs. Wheel structure is designated by numbers from 1 (for extra-dense) to 15 (for extra-compact).

5. Wheel Bond

The wheel bond holds the grains together in the wheel with just the right strength that permits each grain on the cutting face to perform its work effectively. As the grains become dull, they may be either broken, forming new cutting edges, or torn out, leaving the bond. Thus, the bond acts like a tool post that supports the abrasive grains. When the amount of bond is increased, the size of the posts connecting each grain is increased. The seven standard GW bonds are:

Vitrified bond (V). This bond is of refractory clay, which vitrifies or fuses into glass. About 70% of the GWs are made of vitrified bond, as its strength and porosity yield high stock removal rates. Moreover, vitrified bonds are not affected by water, oils, or acids. However, they are brittle and sensitive to impact, but they can withstand velocities up to 2000 m/min.

Resinoid bond (B). This bond is stronger and more elastic than a vitrified bond. However, it is not resistant to heat and chemicals. Because the bond is an organic compound, wheels with resinoid bonds are also called "organic wheels." Resinoid bonded wheels can be used for rough grinding, parting off, and high- speed grinding at 3500 m/min.

Silicate bond (S). This is a soda silicate bond ($NaSiO_3$) that releases abrasive grains more rapidly than a vitrified bond. It is used to a limited extent where the heat generated in grinding must be kept to a minimum, as in a very large GW bond for tool sharpening.

Rubber bond (R). This is the most flexible bond, as the principal constituent is natural or synthetic rubber. It is not so porous and is widely used in thin cut-off large wheels, portable snagging wheels, and centerless regulating wheels (RWs).

Shellac bond (E). This bond is frequently used for strong, thin wheels having some elasticity. They tend to produce a high polish and thus have been used in grinding such parts as camshafts and mill rolls. Thin cut-off wheels may be shellac bonded.

Oxychloride bond (O). This magnesium oxychloride bond is used to a limited extent in certain wheels and segments used on disc grinders.

Metallic bond (M). These are made of Cu- or Al-alloys. Metallic bonds are used in diamond and CBN wheels, especially for electrochemical (EC) grinding applications. The depth of abrasive layer can be up to 6 mm.

6. *Grinding Wheel Marking*

A standard marking system has been adopted by the American National Standards Institute (ANSI). It is implemented by all GW manufacturers today. This system involves the use of numbers and letters, in the sequence indicated in Figure 3.89.

Abrasive type–grain size–grade–structure–bond.

The wheel selected in Figure 3.89 is, therefore, designated as:

51 (optional) A 36 L 5 V 23 (optional)

Moreover, the maximum allowable peripheral speed should be printed on the GW. Because GWs are brittle and operated at high speeds, precautions must be carefully followed in their handling,

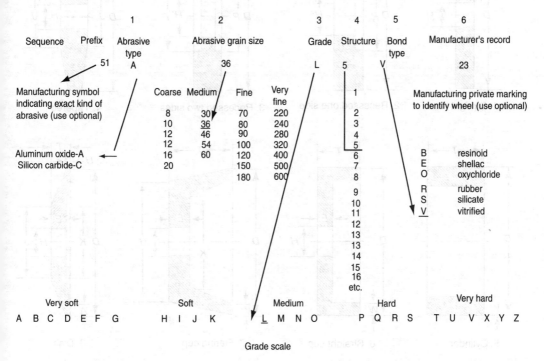

FIGURE 3.89 GW marking according to ANSI.

storage, and use. Failure to follow warnings and instructions printed on individual wheel labels may result in serious injury or even death. In general, the following guidelines are considered when selecting a GW marking:

- Choose Al_2O_3 for steels and SiC for CI, carbides, and nonferrous metals.
- Choose a hard grade for soft materials and a soft grade for hard materials.
- Choose a large grit size for soft ductile materials and a small grit for hard brittle materials.
- Choose a small grit for a good surface finish and a large grit for a maximum metal removal rate.
- Choose an open structure for rough cutting and a compact one for finishing.
- Choose a resinoid, rubber, or shellac bond for a good surface finish, and a vitrified bond for maximum removal rate.
- Do not choose vitrified bonded wheels for cutting speeds more than 32 m/s.
- Choose softer grades for surface and internal cylindrical grinding and harder grades for external cylindrical grinding.
- Choose harder grades on nonrigid grinding machines.
- Choose softer grades and friable abrasives for heat-sensitive materials.

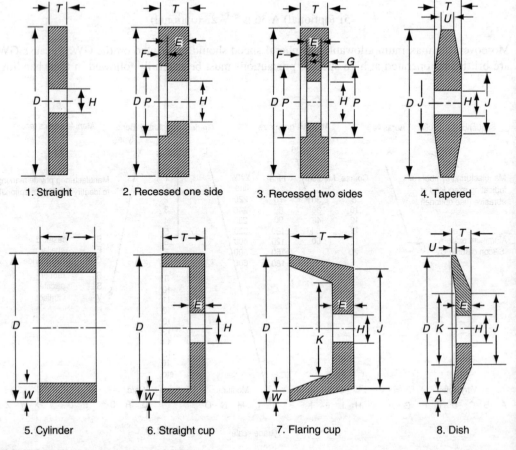

FIGURE 3.90 Standard shapes of GWs.

General-Purpose Machine Tools

3.8.2.2 Grinding Wheel Geometry

GW shapes must permit proper contact between the wheel and the surfaces to be ground. Figure 3.90 illustrates eight standard shapes of GWs, whose applications are as follows:

Shapes 1, 3 and 5 are intended for grinding external or internal cylindrical surfaces and for plain surface grinding.
Shape 2 is intended for grinding with the periphery or the side of the wheel.
Shape 4 is of a safely tapered shape to withstand breakage during snagging.
Shape 6 is a straight cup intended for surface grinding.
Shape 7 is a flaring cup intended for tool sharpening.
Shape 8 is a dish type intended for sharpening cutting tools and saws.

Each has a specific grinding surface. Grinding on other faces is improper and unsafe. Figure 3.91 shows a variety of standard face contours for the straight GWs.

3.8.2.3 Mounting and Balancing of Grinding Wheels and Safety Measures

A. *Mounting*: Proper and reliable clamping of the GW on its spindle is a prime requisite, both for operator safety and to ensure high accuracy and surface finish. Figure 3.92 shows

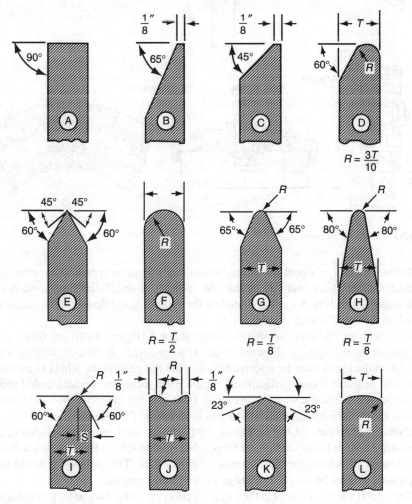

FIGURE 3.91 Standard face contours of a straight-shape GWs.

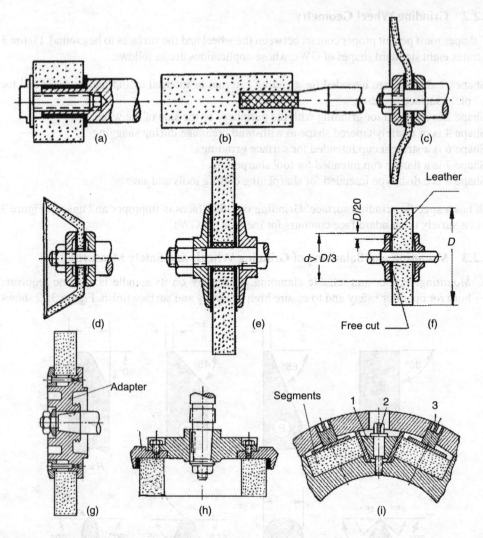

FIGURE 3.92 Methods of mounting GWs.

different methods of wheel mounting, which depends upon type and construction of the grinder and the shape and size of the GW. Wheels of small diameter, used in chucking type internal grinding, are either seated on the spindle nose (Figure 3.92a), or cemented or glued on the spindle stem (Figure 3.92b).

GWs with small bores (all shapes except shape 5 [Figure 3.90]) are directly clamped by flanges on the spindle (Figures 3.92c through 3.92e). Rubber or leather washers of 0.5–3 mm thickness must be inserted between the flanges and the wheel to assure that the clamping pressure is evenly disturbed. Figure 3.92f shows the recommended proportions of flanges relative to the wheel diameter D. GWs of large mounting holes are mounted on an adaptor (Figure 3.92g), which in turn is mounted on the spindle.

Cylindrical wheels (shape 5 [Figure 3.90]) are secured on a special chuck, either by cementing with bakelite varnish, or by pouring molten sulfur, Babbitt, or led into the gap between the wheel and the chuck flange (Figure 3.92h). The surfaces of wheel and chuck being jointed must be rough, cleaned of all dirt, and degreased.

Segmental wheels are held in their chucks either by cementing or by mechanical clamping using tapered keys (1) and screws (2 and 3, Figure 3.92i).

FIGURE 3.93 Revolving-disk wheel balancing stand.

B. *Balancing*: Because of the high rotational speeds involved, GWs must never be used unless they are in good balance. A slight imbalance produces vibrations that cause waviness errors and harm the machine parts. This may cause wheel breakage, leading to serious damage and injury. Static unbalance of a GW is necessary due to the lack of coincidence between its center of gravity and its axis of rotation. Lack of balance is measured at the manufacturing plant in special balancing machines and is eliminated. The user balances GWs either on a balancing stand or directly in the grinder. In the first case, and before mounting the wheel on the spindle, each wheel with its sleeve should be balanced on an arbor that is placed on the straight edges or revolving disks for a balancing stand (Figure 3.93). The wheel is balanced by shifting three balance weights (1) in an annular groove of the wheel sleeve (or mounting flange). The wheel is rotated until it no longer stops its rotation at a specific position. Certain grinders are equipped with a mechanism for balancing the wheel during operation without stopping the wheel spindle rotation.

C. *Safety measures*: Any unsafe practice in grinding can be hazardous for operation and deserves careful attention. Various important aspects in this respect are:
- Mounting of GWs. The wheel should be correctly mounted and enclosed by a guard. Wheel bore should not fit tightly on the sleeve.
- Wheel speed. The printed speed on the GW should not be exceeded.
- Wheel inspection. Before mounting the wheel, it should be checked for damage, cracks, and other defects. A ringing test should be performed. It is good enough for vitrified bonded wheels.
- Wheel storage. When not used, the wheels should be stored in a dry room and placed on their edges in racks.
- Wheel guards should always be used during grinding.
- Dust collection and health hazard precautions. When grinding dry, provisions for extracting grinding dust should be made. Operators should wear safety devices to protect themselves from abrasives and dust.
- Adequate power is necessary; otherwise the wheels slow down and develop flat spots, making the wheel to run out-of-balance.
- Wet grinding. The wheel should not be partly immersed, as this would seriously throw the wheel out of balance.

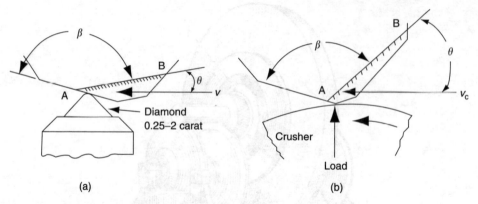

FIGURE 3.94 (a) Diamond truing and (b) crush dressing of GWs.

3.8.2.4 Turning and Dressing of Grinding Wheels

In the grinding process, the sharp grains of the GW become rounded and hence lose their cutting ability. This condition is termed GW-glazing. Along with grain wear (glazing), another factor that reduces the cutting ability is the loading of voids between the grains with the chips and waste of the grinding process, resulting in a condition known as wheel loading. Loading especially occurs when grinding ductile and soft materials.

A worn and loaded wheel ceases to cut. Its cutting ability can be restored by dressing or truing. Dressing is a sharpening operation, which removes the worn and dull grits and embedded swarf to improve the cutting action. Truing is an allied operation with the same tools done to restore the correct geometrical shape of the wheel that has been lost due to nonuniform wear. Truing makes the face of the wheel concentric and its sides plane and parallel, or forms the wheel true for grinding special contours. It also restores the cutting ability of a worn wheel as in dressing. Dressing a wheel does not necessarily make it true; however, the distinction between truing and dressing is a difficult one. There is some difference between diamond truing and crush dressing. The abrasive grit, being crystalline, tends to fracture along the most highly stressed crystallographic plane. Diamond truing tends to chip the grits along planes that make a small angle with respect to the direction of motion of the grit. Crush dressing may cause shear fractures along planes that make a large angle with respect to the direction of motion of the grit (Figure 3.94). As shown in Figure 3.94b, the crushed grit is likely to have more favorable cutting angles than diamond-trued grit (Figure 3.94a). A crush-dressed GW will have free-cutting properties but will not produce a finish on the work equal to that of a wheel that is diamond-trued.

In diamond truing and dressing, a single diamond (0.25–2 carat) is held in a steel holder. The GW is rotated at a normal speed and a small depth typically of 25 μm is given while moving the diamond across the face of the wheel in an automatic feed. The diamond tool is pointed in the same direction during wheel rotation to prevent gouging the wheel face. It is placed at the height of the wheel axis or 1–2 mm below it. Figure 3.95 shows the setting angle in two planes for truing and dressing operations. For best results in diamond truing and dressing, the maximum rate of traverse should be 0.05–0.4 m/min, the infeed 5–30 μm/pass, with 2 or 3 roughing and 1 or 2 finishing passes. The lower the rates of longitudinal traverse and infeed, the smoother the active surface of the wheel will be.

Wheel truing and dressing that do not require diamond make use of:

- Solid cemented carbide rollers
- Rollers of cemented carbide grains in a brass matrix

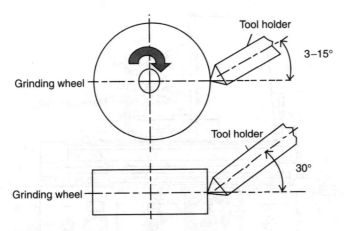

FIGURE 3.95 Diamond truing and dressing of GWs.

- Steel rollers and star-type dressers
- Abrasive wheels of black SiC with a vitrified bond of diameter 60–150 mm and width of 20–32 mm

Wheel truing and dressing without diamond is less efficient and does not require the expensive diamond tool. Of all the dressing tools that do not require diamond, the abrasive wheel dressers are the most widely employed. They have a grain size three to five steps coarser and five or six grades harder than the wheel that is to be dressed or trued. Three to five passes are made in dressing or truing; the traverse feed is 0.5–0.9 m/min and the infeed is 10–30 μm/pass. The last (finishing) passes are made without infeed and at reduced traverse feed (0.4–0.5 m/min). Ample coolant is applied in all dressing and truing methods that do not use diamond.

Diamond grinding wheels are not conventionally dressed; they are trued only when their shape is no longer sufficiently accurate. Metal-bonded diamond wheels are trued with a green SiC dressing stick having a vitrified bond, a grain size of 16 or 12, and of hard grades. Resinoid-bonded diamond wheels are trued with pumice. Truing is done at the working speed of the wheel with a coolant being applied. However, in more recent developments, metal-bonded diamond GWs can be dressed nonconventionally by electrodischarge (ED) and electrochemical (EC) techniques, which erode away a very small layer or a metal bond, thus exposing new diamond cutting edges.

3.8.3 Grinding Machines

A distinguishing feature of a grinding machine is the rotating GW. Grinding machines handle WPs that have been previously machined on other machine tools, which leaves a small grinding allowance. Such an allowance depends upon the required accuracy, size of work, and the preceding machining operations to which it has been subjected. Grinding machines are available for various WP geometries and sizes. Modern machines are computer-controlled, with features such as automatic WP loading, and unloading, clamping, cycling, gauging, and dressing, thereby reducing labor cost and producing parts accurately and repetitively. According to the shape of the ground surface, general-purpose grinding machines are classified into surface external cylindrical, internal cylindrical, and centerless grinding machines. In addition to the general-purpose grinding machines, there are other important types of single-purpose grinding machines, such as thread-gear, spline, contour, milling rolls, and tool grinders. The general-purpose machines and their related operations are discussed briefly in the following sections.

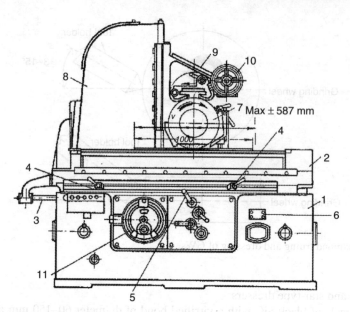

FIGURE 3.96 Typical surface grinder.

3.8.3.1 Surface Grinding Machines and Related Operations

These machines are used to finish flat surfaces. The most widely used types are:

1. *Horizontal-spindle reciprocating-table grinders.* Figure 3.96 illustrates a typical horizontal-spindle reciprocating-table grinder, on which a straight-shaped wheel (7) is commonly used. The bed (1) contains the drive mechanisms and the main table hydraulic cylinder. The table (2), actuated by the piston rod (3) of the hydraulic cylinder, reciprocates along ways on the bed to provide the longitudinal feed of the WP. T-slots are provided in the table surface for clamping WPs directly onto the table or for clamping grinding fixtures or a magnetic chuck. Nonmagnetic materials are held by a vise or special fixtures. The table stroke is set up by adjustable dogs (4). By means of a lever (5), the dogs reverse the table travel at the ends of the stroke. Push-button controls (6) start and stop the machine. A column (8) secured to the bed guides the vertical slide (9), which can be raised or lowered with the GW manually by the hand wheel (11). The vertical slide has horizontal ways to guide the wheel horizontally crosswise for traverse grinding. This slide is actuated by hand using a wheel (10) or by a hydraulic drive housed in the slide. The GW rotates at a constant speed; it is powered by a special built-in motor. Operations that can be performed on horizontal-spindle reciprocating-table grinders are:
    a. Traverse grinding, in which the table reciprocates longitudinally (v_w), and periodically fed laterally after each stroke at a rate f_2 that is less than the GW width. The wheel is fed down to provide the infeed f_1 after the entire surface has been ground (Figure 3.97a).
    b. Plunge grinding, in which the wheel is fed perpendicular to the work surface at a rate f_1, while the work reciprocates, as in grinding a groove (Figure 3.97b).
    c. Creep feed grinding (CFG), which is used for large-scale metal removal operations. The work is fed very slowly past the wheel and the tool depth ($d = 1-6$ mm) is accomplished in a single path. The wheels are mostly of a softer grade with a capability for continuously dressing using a diamond roll to improve the surface finish (Figure 3.97c). The machines used for CFG commonly have special features such as high power of up to 225 kW, high stiffness, high damping factor, variable and well-controlled spindle

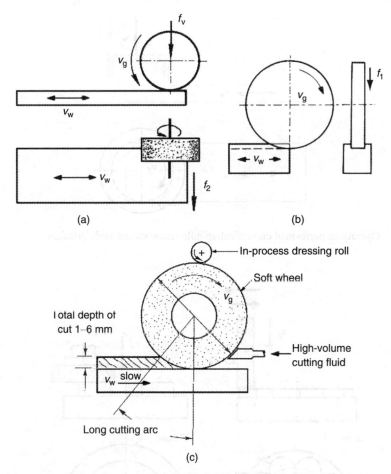

FIGURE 3.97 Operations performed on horizontal-spindle reciprocating table grinders. (a) Transverse grinding, (b) plunge grinding, (c) CFG.

and worktable speeds, and ample grinding fluids. Although a single pass generally is sufficient, a second pass may be necessary to improve surface finish.

2. *Vertical-spindle reciprocating-table grinders.* In these machines a cup, ring, or segmented wheel grinds the work over its full width using the end face of the wheel in one or several strokes of the table. The tool is fed down periodically at the infeed rate f (Figure 3.98).
3. *Horizontal-spindle rotary-table grinders.* The reciprocating cross-feed motion f_1 is transmitted in these machines to either the GW or the table unit, the feed f_2 is actuated per table revolution (Figure 3.99a). The worktable rotates at a speed v_w.
4. *Vertical-spindle rotary-table grinders.* These machines are similar to the previous type, except that the spindle is vertical. The configuration of these machines allows a number of pieces to be ground in one setup (Figure 3.99b).

3.8.3.2 External Cylindrical Grinding Machines and Related Operations

This type of machine is mainly used for grinding external cylindrical surfaces, which may be parallel or tapered, or filets, grooves, shoulders, or other formed surfaces of revolution. Typical applications include crankshaft bearings, spindles, shafts, pins, and rolls for rolling mills. The rotating cylindrical WP reciprocates laterally along its axis. However, in machines used for long shafts,

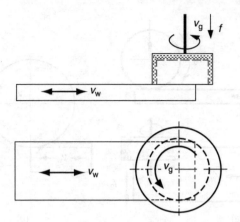

FIGURE 3.98 Operations performed on vertical-spindle reciprocating table grinders.

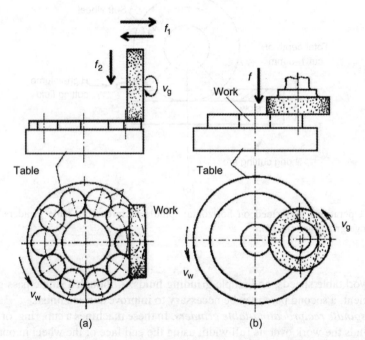

FIGURE 3.99 Operations performed on rotary-table grinders. (a) Horizontal spindle, (b) vertical spindle.

the GW reciprocates. The latter design configuration is called a roll grinder and is capable of grinding heavy rolls as large as 1.8 m in diameter.

Center-type cylindrical grinders are subdivided into:

1. Universal-type grinders make it possible to swivel the GW by swiveling the headstock. This enables steep tapers to be ground. Owing to their versatility, universal cylindrical grinders are best suited for tool room applications.
2. Plain-type grinders, in which the worktable can be swiveled through an angle of only ±6°. This type is basically designed for heavy repetitive single work. It is not very versatile, and is used for grinding tapers with small included angles.

General-Purpose Machine Tools

Similar to surface grinders, described before, the table assembly of cylindrical grinders, is reciprocated using a hydraulic drive. The table speed, therefore, is infinitely varied, and the stroke is controlled by means of adjustable trip dogs.

Infeed is provided by the movement of the wheel head crosswise to the table axis. Most grinders have automatic infeed with retraction when the desired size has been reached. Such machines are also equipped with an automatic diamond wheel truing device that dresses the wheel and resets the measuring element before grinding is started on each piece. Similar to engine lathes, cylindrical grinders are identified by the maximum diameter and length of the WP that can be ground.

These machines are generally equipped with computer control, simultaneously reducing labor cost and producing parts accurately and repetitively. Computer-controlled grinders are capable of grinding noncylindrical parts and cams. Moreover, in these machines the WP spindle speed could be synchronized such that the distance between the WP and wheel axes is varied continuously to produce accurate longitudinal profiles.

Two methods of cylindrical grinding are illustrated in Figure 3.100:

1. Traverse cylindrical grinding, in which the wheel has two movements: rotation about its axis and infeed into the work to remove the grinding allowance (usually an intermittent crosswise motion at ends of traverse stroke). The work rotates about its axis and also traverses longitudinally past the wheel so as to extend the grinding action over the full length of the work (Figure 3.100a). The longitudinal traverse should be about ¼ – ½ of the wheel width per revolution of the work. For a fine surface finish, it should be held to the smaller value of this range. The depth of cut (infeed) varies according to the finish required. It ranges from 50 to 100 μm for rough cut and 6–12 μm for finish cut. The grinding allowance ranges from 125 to 250 μm for short parts, and from 400 to 800 μm for long parts subjected to hardening treatment. About 70% of the grinding allowance is allocated for roughing and 30% for finish grinding.
2. Plunge-cut cylindrical grinding, in which there is no traversal motion of either the wheel or the work. The GW extends over the entire length of the surface being ground on the work ($B > l$), which rotates about its axis. The wheel rotates and, at the same time, is continuously fed into the work at a rate of 2.5–20 μm per revolution of the work (Figure 3.100b). This method is used in form grinding of relatively short work at high output. In any plunge-cut operation (cylindrical or surface), the wheel is fed normal to the work surface (infeed). The feed f, which is the depth of the layer of material removed during one work revolution or stroke in case of surface grinding or stroke, will initially be less than the nominal feed setting on the machine. The difference results from the machine-tool elements, the GW, and WP. It occurs due to the forces generated during the grinding operation. Thus, on

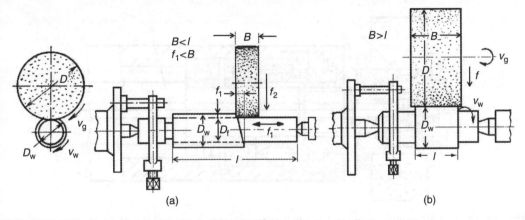

FIGURE 3.100 External cylindrical grinding operations.

completion of the estimated number of work revolutions or strokes required, some additional work material has to be removed. The removal of this material, called sparking-out, is achieved by continuing the grinding operation with no further application of feed until metal removal becomes insignificant (no further sparks appear). Therefore, when calculating grinding time, an additional time t_s should be considered for sparking out.

3.8.3.3 Internal Grinding Machines and Related Operations

In internal grinding, a small GW is used to grind the inside diameter of a WP, such as bushings, bearing races, and heavy housings. It is usually of the traverse type; however, the plunge-cut technique may also be used. Two difficulties in internal grinding are encountered:

1. The GW and consequently the machine spindle should be small to suit the small internal holes. The reduced rigidity of small spindles makes it impossible to take heavy cuts. Furthermore, the rotational speed of the small GW must be very high (up to 150,000 rpm) to operate at the recommended cutting speeds. Therefore, high-speed drives for the GW with special spindle mounting are required.
2. In internal grinding, conventional methods of coolant supply are not efficient. A method of internal coolant delivery is illustrated in Figure 3.101. The coolant is pumped to the GW through the axial hole A in the wheel spindle and the radial holes B and C are drilled in the spindle nose and the sleeve on which the wheel is mounted. Owing to the action of the centrifugal force, the coolant passes through the pores of the wheel to its periphery. The coolant is applied intensively in the grinding zone, where it also washes the waste products of grinding out of the wheel. This method raises the output by 10–20%, improves the surface finish by one class, avoids burning, and reduces the GW wear.

There are two types of internal cylindrical grinding machines:

a. *Chucking-type machine.* Used in grinding comparatively small WPs. In addition to the primary cutting motion of the GW v_g, the following feed motions are encountered (Figure 3.102a):
   - Work feed v_w due to the work rotation
   - Traverse feed f_1 as a reciprocating motion of the work or GW
   - Infeed f_2 as a periodic crosswise motion of the GW
b. *Planetary-type machine.* Designed to grind holes in large irregular parts that are difficult to mount and rotate (Figure 3.102b). In this case, the work is stationary while the wheel rotates not only around its own axis v_g but also around the axis (v_w) of the hole being

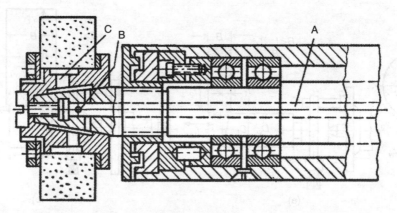

FIGURE 3.101 GW cooling for internal grinding operations.

General-Purpose Machine Tools

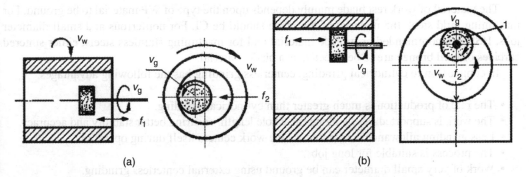

FIGURE 3.102 Internal cylindrical grinding operations: (a) chucking and (b) planetary.

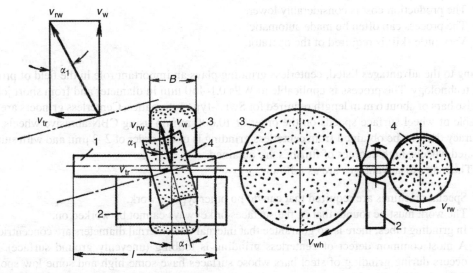

FIGURE 3.103 Centerless grinding operations.

ground. In addition to these two motions, traders feed f_1 and infeed f_2, as in the chucking-type, are affected.

3.8.3.4 Centerless Grinding Machines and Related Operations

As the name implies, the work is not supported between centers but is held against the face of GW, a supporting rest, and regulating wheel RW. Therefore, centerless grinding does not require center holes, a driver, and other fixtures for holding the WP. During cutting, the WP (1) is supported on the work rest blade (2) by the action of the GW (3). The RW (4) of infinite variable speed holds the WP against the horizontal force controlling its size and imparting the necessary rotational and longitudinal feeds of the WP (Figure 3.103).

The wheels rotate clockwise, and the work driven by the RW, having approximately the same peripheral speed of typically 20–30 m/min, rotates counterclockwise. To increase friction between the work and the RW, the latter has a fine grain size of mesh number 100–180 and is rubber bonded and of a sufficiently hard grade (R or S). Resinoid-bonded RWs are also employed. In comparison to the RWs, the GWs run at a much higher speed (2000 m/min) and accomplish the cutting action. To ensure that a true cylindrical surface is ground on the work, it is set above the centers of the RWs and GWs by 0.15–0.25 of the work diameter, but not over 10 mm, to avoid chattering.

The material of work rest blade mainly depends upon the type of WP material to be ground. For machining mild steel, the material of the blade should be CI. For nonferrous and small diameter jobs, HSS is recommended as a blade material and for machining stainless steels, either sintered carbides or hard bronze are to be selected as a blade material.

In comparison to cylindrical grinding, centerless grinding has the following advantages:

- The rate of production is much greater than cylindrical grinding.
- The work is supported rigidly along the whole length, ensuring better stability and accuracy.
- Less grinding allowance is required, as the work centers itself during operation.
- The process is suitable for long jobs.
- Work of very small diameter can be ground using external centerless grinding.
- Because centering is unnecessary, no time is lost in job setting and the cost to provide centers is eliminated.
- The machine is easy and economical to maintain.
- The production cost is considerably lower.
- The process can often be made automatic.
- Very little skill is required of the operator.

Owing to the advantages listed, centerless grinding plays an important role in the field of production technology. The process is applicable to WPs 0.1–150 mm in diameter and from short jobs to precise bars of about 6 m in length required for Swiss-type automatics. Centerless grinders are now capable of wheel surface speeds on the order of 10,000 m/min using CBN abrasive wheels. The accuracy that can be obtained from centerless grinding is of the order of 2–3 μm, and with suitable selected wheels, high degrees of finishes are obtained.

The major disadvantages are as follows:

- Special machines are required that can do no other type of work.
- The work must be round; that is, flat surfaces or keyways cannot be worked on.
- In grinding tubes, there is no guarantee that internal and external diameters are concentric.
- A most common defect of centerless grinding is lobbing (unevenly ground surface). It occurs during grinding of steel bars whose surfaces have some high and some low spots due to hot or cold rolling.

Centerless grinding may be external or internal.

1. External Centerless Grinding

Basically, three methods of centerless grinding are commonly used in practice.

a. Through-Feed Centerless Grinding

In this method, the axial traverse motion is imparted to the work by the RW because the latter is inclined at a small angle α_1 with respect to the axis of the GW (Figure 3.104a) or because the work-rest blade is inclined to an angle α_1 (Figure 3.104b).

The peripheral velocity v_p of the RW is resolved into peripheral speed of the work v_w and the rate of the work traverse v_{tr} (mm/min) (Figure 3.103), which can be calculated by the equation

$$v_{tr} = \pi d_R n_R \sin \alpha_1 \tag{3.21}$$

where

$\alpha_1 = 0.5°–1.5°$ for finish grinding
$ = 1.5°–6°$ for rough grinding
$d_R =$ diameter of the RW
$n_R =$ rotational speed of the RW

General-Purpose Machine Tools

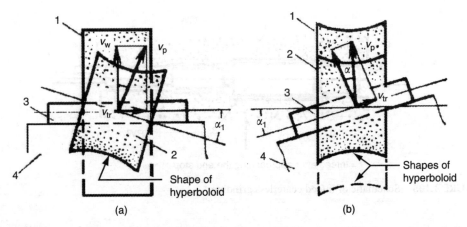

FIGURE 3.104 Ensuring linear contact between WP and wheels in through-feed centerless grinding by providing wheels of hyperboloid of revolution profiles: (a) RW inclination and (b) work rest inclination.

Equation 3.21 does not consider the effect of slip. The contact between the WP and the GWs and RWs must be linear in through-feed grinding. For this reason, the face of the RW (in case of RW inclination), or the faces of both RWs and GWs (in case of work-rest inclination) are trued by diamonds to the shape of hyperboloid of revolution (Figure 3.104).

Advantages of through-feed grinding are as follows:

- The method becomes automatic by employing magazine feeds for bars and hoppers for small jobs.
- Long bars can be ground easily without any deflection being produced.

Disadvantages of through-feed grinding are as follows:

- This method is used for straight cylindrical parts; if there is a head on the WP or it is tapered, then the process cannot be employed.
- Form grinding cannot be produced by this method.

b. Infeed (Plunge-Cut) Centerless Grinding

This method of grinding is used when the WP is of headed, stepped, or tapered form (Figure 3.105). In this case, there is no axial movement of the WP, which has a rotating movement only. The WP (3) is supported on work rest blade (4). After approaching the end stop (5), the cross feed is actuated by a method similar to plunge-cut grinding in which the grinding or RW is fed in a direction square to the WP axis by a precise feed movement.

In some cases, the infeed is ensured by the use of an RW 2 of special shape. Its periphery consists of sectors I, II, and III (Figure 3.106). Sectors I and III are circular arcs of different radii. Sector II is an Archimedean spiral, which enables infeed to be actuated without movement of the wheel heads. The whole grinding cycle takes place during one revolution of the RW. The WP (3) is automatically loaded between the wheels (1 and 2) from the top and the axis of the RW (or the work rest blade, 4) is inclined slightly at an angle $\alpha_1 = \frac{1}{2}°$ to provide for a fixed axial position of the WP by holding it against the locating stop (not shown in Figure 3.106). Infeed is done at various rates (sector II). At the beginning of the process, a large part of the allowance is removed with a high rate of infeed and then this rate is reduced. At the end of operation, the WP is ground for several revolutions without infeed for sparking out (sector III). The RW has a longitudinal slot (A) into which the finished WP drops after rolling off the work rest blade and is removed outside the grinding zone.

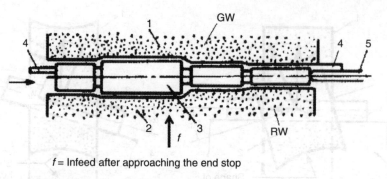

f = Infeed after approaching the end stop

FIGURE 3.105 Schematic of infeed centerless grinding.

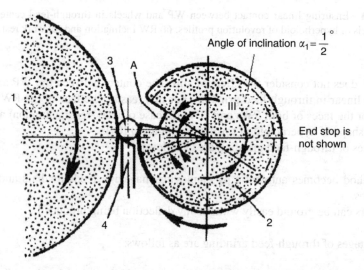

FIGURE 3.106 Automatic-infeed centerless grinding provided by a profiled RW with a loading recess.

c. *End-Feed Centerless Grinding*

This method is essentially an intermediate method between through-feed and infeed centerless grinding. It is employed for headed components that are too long to be ground by the infeed method. It is used when the length of WPs is greater than the width of the GW and not allowed to pass between the wheels for through-feed (Figure 3.107).

The work (3) is fed as in case of the infeed method, and after a certain portion of length has been ground, the axial movement takes place until the whole of length has been ground after approaching the end stop (5). In this method, the angle of inclination of the RW is typically 2–3°, which is larger than that used in the infeed method.

2. *Internal Centerless Grinding*

This process was recently developed for grinding internal surfaces of short or long tube work. The arrangement is schematically illustrated in Figure 3.108. The WP is supported by two steel rollers (1 and 2) and an RW (5). Roller 1 is a supporting roller and roller 2 is a pressure roller. The GW (4) and the WP (3) rotate in the same direction, while the RW (5) rotates in the opposite direction. The GW is generally smaller than the RW. The process may work either on the on-center principle (Figure 3.108a), or on the above-center principle (Figure 3.108b). The on-center method is used for thin-walled components; however, it tends to duplicate the errors of the outside diameter and those

General-Purpose Machine Tools

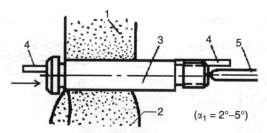

FIGURE 3.107 Schematic of end-feed centerless grinding.

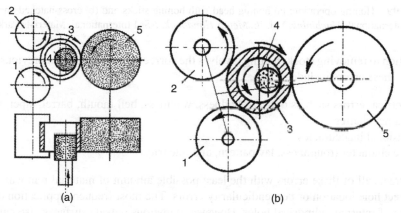

FIGURE 3.108 Internal centerless grinding: (a) on-center principle (used for thin-walled components) and (b) above-center principle.

of roundness and waviness, which can to some extent be corrected by the above-center arrangement. In internal centerless grinding, as the roundness of internal surface depends to a great extent upon the external surface, the latter must be ground first.

3.9 MICROFINISHING MACHINES AND OPERATIONS

These are operations by which a product receives the final machining stage that applies for the service for which it is intended. These remove a very small amount of metal, and hence the surface finish obtained is specified in the ranges of microfinishes. These operations include honing, microhoning (superfinishing), lapping, polishing, and buffing. The first three operations are discussed briefly in the following sections.

3.9.1 Honing

Honing is a controlled, low-speed sizing and surface-finishing process in which stock is abraded by the shearing action of a bonded abrasive honing stick.

In honing, simultaneous rotating and reciprocating action of the stick (Figure 3.109a) results in a characteristic cross-hatch lay pattern (Figure 3.109b). For some applications, such as cylinder bores, angles between cross-hatched lines are important and may be specified within a few degrees. Because honing is a low-speed operation, metal is removed without the increased temperature that accompanies grinding and thus any surface damage caused by heat (heat-affected zone [HAZ]) is avoided.

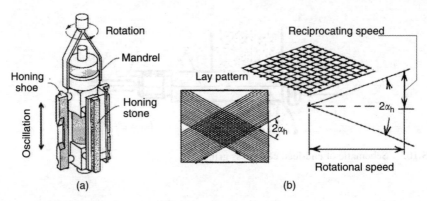

FIGURE 3.109 Honing operation: (a) honing head with honing sticks and (b) cross-hatched angle. (Adapted from ASM International, *Machining*, Vol. 16, *Metals Handbook*, ASM International, Materials Park, OH, 1989.)

In addition to removing stock, honing involves the correction of errors from previous machining operations. These errors include

- Geometrical errors such as out-of-roundness, waviness, bell mouth, barrel, taper, rainbow, and reamer chatter
- Dimensional inaccuracies
- Surface character (roughness, lay pattern, and integrity)

Honing corrects all of these errors with the least possible amount of material removal; however, it cannot correct hole location or perpendicularity errors. The most frequent application of honing is the finishing of internal cylindrical holes. However, numerous outside surfaces also can be honed. Gear teeth, valve components, and races for antifriction bearings are typical applications of external honing. The hone is allowed to float by means of two universal joints so that it follows the axis of the hole (Figure 3.110). Owing to the fact that the tool floats, the honing sticks are able to exert an equal pressure on all sides of the bore regardless of the machine vibrations, and therefore, round and straight bores are produced.

As the tool reciprocates through the bore, the pressure and the resulting penetration of grit is greatest at high spots and consequently the waviness crests are abraded, making the bore straight and round. After leveling high spots, each section of the bore receives equal abrading action. The hole axis is usually in the vertical position to eliminate gravity effects on the honing process; however, for long parts the axis may be horizontal.

Advantages of Honing
- It is characterized by rapid and economical stock removal with a minimum of heat and distortion.
- It generates round and straight holes by correcting form errors caused by previous operations.
- It achieves high surface quality and accuracy.

3.9.1.1 Process Capabilities

1. Materials
Although CI and steel are the most commonly honed materials, the process can also be used for finishing materials ranging from softer metals like Al- and Cu-alloys to extremely hard materials like case-nitrided steels or sintered carbides. The process can also be used for finishing ceramics and plastics.

2. Bore Size and Shape
Bores as small as 1.6 mm in diameter can be honed. The maximum bore diameter is governed by the machine power and its ability to accommodate the WP. Machines powered by motors of up to

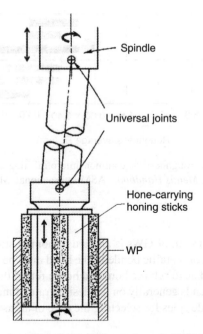

FIGURE 3.110 Floating hone using two universal joints to permit the bore and the tool to align.

37 kW are available that can hone bores up to about 1200 mm in diameter. Honing bores up to 760 mm in diameter is a common practice (ASM International, 1989). Although most internal honing is done on simple, straight-through holes, blind holes with a slight taper can also be honed. It is not feasible to hone the sides of a blind hole flush with the bottom. Bores having keyways can be honed and so can male or female splines (ASM International, 1989).

3. *Stock Removal*

In honing, a general rule is to remove twice as much stock as the existing error in the WP. For example, if a cylinder is 50 μm out-of-round or tapered, a removal of 100 μm will be required for complete cleanup.

The work in preceding operations is usually planned so that the amount of stock removed in honing is minimized. On the other hand, stock removal of up to 6.4 mm may be practical for rough honing in some applications. For instance, as much as 2.5 mm is honed from the inside diameter of hydraulic cylinders, because stock removal through honing is more practical and economical than attaining close preliminary dimensions by grinding or boring. Another example occurs in finishing bores of long tubes, where even larger amounts as much as 6 mm may be removed by honing, because it is the only practical method. Such tubes are finished by honing immediately after drawing. Honing is performed at a rate of 32 cm^3/min from soft steel tubes; for tubes steel-hardened to 60 HRC, the rate is reduced to 16 cm^3/min (ASM International, 1989).

Rough honing is employed before finish honing when large amounts of stock are to be removed and specific finishes are required. Sticks containing abrasives of 80 grit or even coarser are used for rough honing to maximize the removal rate. Finish honing is accomplished by abrasives of 180–320 grit or finer.

4. *Dimensional accuracy and surface finish*

Internal honing to tolerances of 2.5–25 μm is common. Surface roughness R_a of 0.25–0.38 μm can be easily obtained by rough honing and roughness of less than 0.05 μm can be achieved and reproduced in finish honing. Figure 3.111 compares typical ranges of surface roughness obtained by honing to other common microfinishing processes.

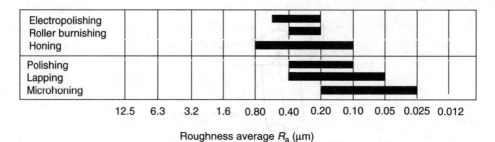

FIGURE 3.111 Average surface roughness of common microfinishing operations. (Adapted from ASM International, *Machining*, Vol. 16, *Metals Handbook*, ASM International, Materials Park, OH, 1989.)

5. *Honing Sticks*

The same ANSI-designation system of GWs is applied to honing sticks. Honing sticks commonly used may be vitrified, resinoid, or metallic bonded. The bond must be strong enough to hold the grit; however, it must not be so hard as to rub the bore and hence retard the cutting action.

The grit size selection depends generally on the desired rate of material removal and the degree of surface finish required. Guide rules for selecting the type of abrasive materials are as follows:

- Al_2O_3 is widely used for steels.
- SiC is generally used for CI and nonferrous materials.
- CBN is used for all steels (soft and hard), Ni- and Co-base super alloys, stainless steels, Br-Cu-alloy, and Zr.
- Diamonds are used for chromium plating, carbides, ceramics, glass, CI, brass, bronze, and surfaces nitrided to depths greater than 30 μm.

3.9.1.2 Machining Parameters

Parameters affecting the perforance of honing process are:

1. *Rotation speed*. The choice of the optimum surface speeds is influenced by:
    - *Material being honed*—higher speed can be used for metals that shear easily.
    - *Material hardness*—harder material requires lower speed.
    - *Surface roughness*—rougher surfaces that mechanically dress the abrasive stick permit higher speed.
    - *Number and width of sticks in the hone*—speed should be decreased as the area of abrasive per unit area to hone increases.
    - *Finish requirement*—higher speed usually results in finer surface finish.

    Depending on the material to be honed, the rotational surface speed typically varies from 15 to 90 m/min. Experience with a particular application may indicate advantages for higher or lower speeds. Rotation speeds as high as 183 m/min have been used successfully. However, a reduction of surface rotation speed can reduce the number of rejects (ASM International, 1989). Excessive speeds contribute to decreased dimensional accuracy, overheating of the WP, and glazing of the abrasive stick. Overheating causes breakdown of honing fluid and distortion of the WP.

2. *Reciprocation speed*. Reciprocation speed commonly ranges from 1.5 to 30 m/min for a variety of metals and alloys.

3. *Control of Cross-Hatch Angle*. The cross-hatch angle $2\alpha_h$ (Figure 3.109b) obtained on a honed surface is given by

$$\tan \alpha_h = \frac{v_{rc}}{v_{rt}} \qquad (3.22)$$

General-Purpose Machine Tools

where

v_{rc} = reciprocation speed (m/min)
v_{rt} = surface rotation speed (m/min)
α_h = half cross-hatch angle

When the rotation and reciprocation speeds are equal, the cross-hatch angle is 90°.

For some applications (engine cylinder bores), the cross-hatch angle is an important feature that should be noted in specifications. The cross-hatch scratch pattern left on the wall of cylindrical surfaces tends to retain lubricating fluids and thus reduce the wear in mating components. In the majority of applications, although an angle of 30° is commonly recommended, any angle within the range 20–45° is usually suitable.

4. *Honing pressure*. It is selected depending on hardness and toughness of the material, characteristics of honed surface (plain or interrupted by keyways), type of stick, and so on. Insufficient pressure results in a subnormal rate of metal removal and rough surface finish. Excessive removal rate and rough finish can cause an increased stick cost as well as decreased productivity due to time loss of frequent tool exchange.

5. *Honing fluids*. Lubrication is more critical in honing than in most other material removal operations. Honing fluids are necessary to act as lubricants, coolants, and to remove swarf. No single honing fluid possesses all requirements needed for honing process. Therefore, mixtures of two or more liquids are commonly used.

Water-based solutions are superior as coolants, but they are poor lubricants, have insufficient viscosity to prevent chatter, and cause rust. Because of this, water-base solutions are seldom used as honing fluids.

Mineral seal oil is effective and widely used for honing. It has a higher viscosity and flash point than kerosene. It is less likely to cause skin irritation. Mineral oils used for other machining operations have also proved satisfactory when one part oil is diluted with four parts kerosene.

3.9.1.3 Honing Machines

For the production of few parts, honing may be performed on drill presses or engine lathes on which arrangements can be made for simultaneous rotating and reciprocating motions. The stroking can be done manually or powered depending on the equipment capabilities. On the other hand, the production honing is done with machines built for the purpose. These vertical machines are available in a wide range of sizes and designs. Some horizontal machines operate by manual stroking. In power stroking, the WP is usually held stationary in a rigid fixture, while the hone is rotated and hydraulically powered for stroking, which is considered beneficial for heavier WPs.

3.9.2 SUPERFINISHING (MICROHONING)

Superfinishing (microhoning) is an abrading process that is used for external surface refining or cylindrical, flat, and spherical-shaped parts. It is not a dimension-changing process, but is mainly used for producing finished surfaces of superfine quality. Only a slight amount of stock is removed (2–30 µm), which represents the surface roughness (Figure 3.112). The process of honing involves two main motions, whereas superfinishing requires three or more motions. As a result of these motions, the abrasive path is random and never repeats itself.

The primary distinction between honing and superfinishing is that in honing, the tool rotates, while in superfinishing, the WP always rotates. The operating principle of the superfinishing process is illustrated in Figure 3.113. The bonded abrasive stone, whose operating face complies with the form of the WP surface, is subjected to very light pressure. A short, HF stroke, super-imposed on a reciprocating traverse, is used for superfinishing of long lengths.

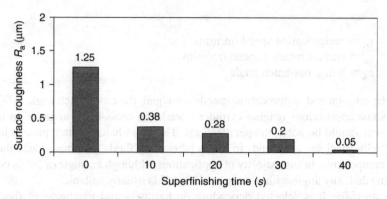

FIGURE 3.112 Gradual improving a rough surface by superfinishing. (Adapted from ASM International, *Machining*, Vol. 16, *Metals Handbook*, ASM International, Materials Park, OH, 1989.)

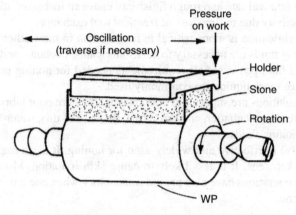

FIGURE 3.113 Principles of superfinishing process.

Machining Parameters

The following parameters affect the superfinishing process considerably

1. Abrasive stones
   Two types are mainly used: Al_2O_3 for carbon and alloy steels and SiC (more friable) for very soft and tough steels as well as for CI and most nonferrous metals.
   Grit size. The grit size is selected from a wide range (60–1000) to suit the machining situation, which varies from rough superfinishing to fine or extra-fine finishing.
   Grade. This varies from J (soft) used for extremely hard alloys to P (very hard) used for extremely soft materials, CI, and nonferrous metals.
   Width. This ranges from 60% to 80% of the part diameter, but not more than 25 mm.
   Number of stones. For parts over 150 mm diameter, several stones are arranged.
   Stone length. This length is somewhat less than the work length, but not more than three times the width of the stone. For superfinishing of longer work, an additional traverse movement is needed.
2. Work speed
   Roughing. 12–15 m/min.
   Finishing. 30–60 m/min (for very fine finish, higher speeds of 120 m/min may be applied).
   At lower work speeds, the superfinishing process generally develops a distinguished cross-hatched pattern, which may be desirable in many applications despite its low surface reflectivity. At higher speeds, this pattern disappears and a brighter surface is developed.

3. Stroke length and speed of stone reciprocation
   The fast reciprocation of stones in a short stroke is a main characteristic that sets superfinishing apart from honing. Some machines employ a single stroke length of 4.76 mm, while others provide a variable stroke length over a range of about 2–5 mm. The actual linear speed of oscillation is a function of the stroke length (amplitude) and the rate of reciprocation (frequency). Typical extreme reciprocation speeds are 3–20 m/min.
4. Stone pressure
   For normal work, $p_h = 1.5$–3.0 kg/cm^2
   For roughing, $p_h = 3.0$–6.0 kg/cm^2
   For extra fine finishing, $p_h = 1.0$–2.0 kg/cm^2

3.9.3 Lapping

The usual definition of lapping is the random rubbing of WP against a CI lapping plate (lap) using loose abrasives carried in an appropriate vehicle (oil) to improve fit and finish. It is a low-speed, low-pressure abrading process. In general, the surface quality that can be obtained by lapping is not easily or economically obtained by other processes. Moreover, the life of the moving parts that are subjected to wear can be increased by eliminating hills and valleys that create a maximum percentage at bearing area.

Lapping is a final machining operation that realizes the following major objectives:

- Extreme dimensional accuracies
- Mirror-like surface quality
- Correction of minor shape imperfections
- Close fit between mating surfaces

It does not require holding devices and consequently no WP distortion occurs. Also less heat is generated than in most of other finishing operations. Therefore, metallurgical changes are totally avoided. The temperature increase of the surface is only 1–2°C over ambient.

3.9.3.1 Machining Parameters

The following parameters have an effect on the lapping process.

1. *Abrasives type*:
   - Diamond is used for lapping tungsten carbide (WC) and precious stones.
   - B$_4$C is used for lapping dies and gauges. It is more expensive than SiC and Al$_2$O$_3$ (10–25 times).
   - SiC is intended for rapid stock removal. It is mainly used for lapping hardened steels, CI, and nonferrous metals.
   - Al$_2$O$_3$ is intended for improved finish. It is used for lapping soft steels.
2. *Grit size and abrasive grading*. Grit size (mesh number) generally ranges from 50 to 3800; however, more frequently, grit size from 100 to 1000 is used depending on the degree of surface finish required. Soft materials require finer grains to obtain a good finish. Commercially available abrasives of certain grit size may contain finer or coarser grit than the specified size. Abrasives increase in cost as their grading becomes closer. The use of a low-cost, loosely graded commercial abrasive is not recommended for reasons of economy.
3. *Vehicle*. This prevents scoring of the lapped surfaces and varies from clean water to heavy grease. It is selected to suit the work, method, and type of surface finish required. For machine lapping, an oil-base type is recommended; however, a commercial mixture of kerosene and machine oil can be used. Grease-based vehicles are recommended for lapping soft metals.

4. *Speed*. Speeds of 1.5–4 m/s are commonly used for machine lapping.
5. *Pressure*. A pressure of 0.1–0.2 kg/cm^2 is used for soft materials, while 0.7 kg/cm^2 is recommended for lapping hard materials. If the preceding values are exceeded, rapid breakdown and scoring of the WP results.

3.9.3.2 Lapping Machines

Lapping machines usually fall into one of the two categories: individual-piece and equalising lapping machines.

3.9.3.2.1 Individual-Piece Lapping

It is the most effective lapping method for hard metals and other hard materials. It is used to produce optically flat surfaces and accurate surface plates. When both sides of a flat WP are lapped simultaneously, extreme parallelism is achieved.

Individual-piece lapping is performed using a lap that is softer than the WP so that the abrasives get embedded in the lap. The lap is usually made of close-grain soft CI that is free from porosity and defects. When CI is not suitable as a lap material, steel, brass, copper, or aluminum may be used. Wood is sometimes used for certain applications.

The vast majority of individual piece lapping installations are of the following categories:

- Specialized single- or double-plate machines, such as ball or pin laps
- Single-sided flat or double-sided planetary laps
- A cup-lapping machine for lapping spherical surfaces

1. *Double-plate lapping machines for cylindrical WPs*

Figure 3.114 visualizes a typical vertical lapping machine used for lapping cylindrical surfaces in production quantities. The laps are two opposing CI circular plates that are held on vertical spindles of the machine. The WPs are retained between laps in a slotted-holder plate and rotate and slide in and out to break the pattern of motion by moving over the inside and outside edges of the laps that prevent grooving. The lower lap is usually rotated and drives the WPs. The upper one is held stationary but is free to float so that it can adjust to the variations in WP size. The lower lap regulates the speed of rotation. To avoid damage of the surface being lapped, the holder plate or carrier is made of soft material (copper, laminate fabricate base, and so on).

An alternative design of this machine is illustrated in Figure 3.115. Accordingly, the retainer is arranged eccentrically between the two laps and has a separate drive. In this design, both the upper and lower laps are rotating.

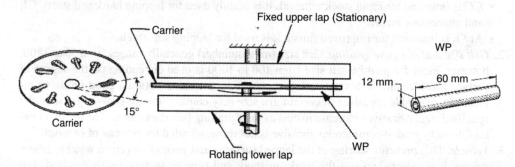

FIGURE 3.114 Typical vertical lapping machines for cylindrical surfaces. (Adapted from ASM International, *Machining*, Vol. 16, *Metals Handbook*, ASM International, Materials Park, OH, 1989.)

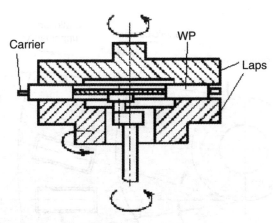

FIGURE 3.115 Two-plate lapping machine with two rotating laps and eccentrically rotating plate holder.

The abrasive with vehicle is provided to the laps before starting. Oil or kerosene is then added during the cycle to prevent drying of the vehicle, which could result in surface scratching. The best lapping practice is to load as many parts as possible to reduce the pressure applied on each part and slowdown the operation, which provides easier control on tolerances.

Achieved accuracy and surface finish. Fine surface finishes of 0.025 μm R_a and metal removal of 2.5–10 μm are feasible when CI laps are used. A diametral tolerance as low as ±0.5 μm, out-of-roundness of 0.13 μm, and taper less than 0.25 μm have been achieved. Such accuracies depend greatly on the accuracies achieved in prior machining operations.

Applications. Machine lapping between plates, as described earlier, is an economical and productive (100 parts/h) method of lapping cylindrical surfaces. The machine can be used for lapping parts such as plug gages, piston pins, hypodermic plungers, ceramic pins, small valve pistons, cylindrical valves, small engine pistons, roller and needle bearings, diesel injector valves, plungers, and miscellaneous cylindrical pins. Either hard or soft materials can be lapped, provided that they are rigid enough to accept pressure of laps. Because the hardness slows the operation, soft materials lap more rapidly than hard ones. Additionally, hard materials provide easier control of tolerances.

Limitations. A part with diameter greater than its length is difficult or impossible to machine lap between plates. Parts with shoulders require special fixtures. Parts with keyways, flats, or interrupted surface are difficult to lap because the variations in lapping pressure that occur are likely to fall out of round. If the relief extends over the entire length of the part, this method of lapping cannot be used at all.

Thin-wall tubing can be lapped, provided that the deflection due to lapping pressure is insignificant. Parts that are hollow on one end and solid on the other present problems in obtaining roundness and straightness. Plugging the hollow end of the part will sometimes solve the problem.

The outside edges of the laps lap at a faster rate than the inside edges. Therefore, it is expected that the cylindrical WP will become tapered. One method of overcoming this problem consists of using short cycles, while the WPs are reversed in their slots. In addition, they are mixed between slots. Taper can also be minimized by positioning the workholder so that parts in slots are at 15° angle to a radius, as illustrated in Figure 3.116.

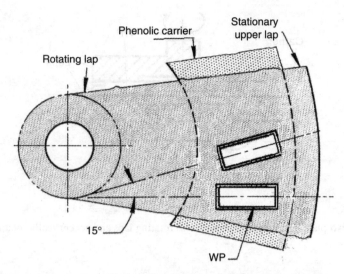

FIGURE 3.116 Lapping setup that minimizes taper for production quantity of cylindrical parts. (Adapted from ASM International, *Machining*, Vol. 16, *Metals Handbook*, ASM International, Materials Park, OH, 1989.)

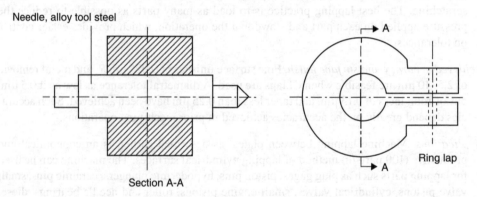

FIGURE 3.117 Lapping of valve needle using a ring lap. (Adapted from ASM International, *Machining*, Vol. 16, *Metals Handbook*, ASM International, Materials Park, OH, 1989.)

EXAMPLE

The valve needle (high-alloy tool steel of 60–65 HRC) shown in Figure 3.117 is to be lapped to achieve the accuracy requirements where $R_a = 0.05$ μm, tolerance = ±0.13 μm, out-of-roundness = 0.13 μm, and a taper = 0.25 μm. Discuss the possible alternatives to achieve the preceding requirement.

Solution

There are two alternatives for lapping:

1. For small quantities, a ring lap of CI is used (Figure 3.117). Each needle is chucked by its stem and rotated in a lathe at 650 rpm. The CI lap is stroked back and forth over the needle until grinding marks are vanished. The needle is coated with lapping compound (CrO mixed with spindle oil).
2. For lot and mass production, the part is finished on a two-plate lapping machine (Figure 3.115). Before being machine-lapped, parts are carefully ground for roundness and classified into groups according to their diametral variations of ±2.5 μm, ±5 μm, and so on. A laminated phenolic workholder is designed to be eccentric to the laps to provide an oscillating motion (Figure 3.115). The short cycles are stopped to measure the parts with an electro limit gauge. If the desired size has not yet been attained, more lapping compound is added and lapping proceeds until the required finish is achieved.

2. Lapping of Flat Surfaces

Flat surfaces can be lapped by either manual or mechanical methods.

a. Manual Lapping

Manual lapping is used only for limited quantities, or when special requirements must be met. Hand rubbing of a flat WP on a plate lap charged with an abrasive compound is the simplest method of flat lapping. The lap, usually made of iron, has regularly spaced grooves of about 1.6 mm depth to retain the lapping medium. The WP is rubbed on the lap in a figure eight or a similar pattern that covers almost the entire lap surface. The lap remains flat for a considerable amount of work. This method of lapping is time-consuming and tedious, and requires a high degree of labor skill.

Another somewhat faster method makes use of a vertical drilling press where the lap is fixed on the machine table and the work is held by the spindle. The WP rotates against the lap, while light pressure is applied by hand. However, this method violates one of the basic rules of lapping, namely, the random and the nonrepeated paths between the lap and the work.

b. Mechanical Lapping

Mechanical lapping is performed by flat lapping machines. The two general types are single- and dual-face lapping machines. However, dual-face lapping machines are preferred due to their enhanced accuracy.

Most of dual-face lapping machines are of the planetary type, with the workholder (carriers) nested between a center drive and a ring drive. These drives can be either gear- or pin-type configurations that must have positive engagement (Figures 3.118 and 3.119). The WP is propelled by the carrier in a serpentine path between lap plates on which abrasives have been charged or continuously fed in the form of slurry.

In the planetary fixed-plate machine (Figure 3.118), the bottom lap is fixed and the top lap is restrained from rotating. It is allowed to float to bear on the largest pieces and laps all the pieces to the same size. The part is dragged between the plates by the carrier and all the power is directed to the flat, thin carrier plates, exerting high forces on their thin teeth that may cause edge chipping on fragile parts.

FIGURE 3.118 Planetary fixed-plate double-face lapping machine for flat surfaces. (From Hoffman Co., Carlisle, PA, USA.)

Figure 3.119 illustrates another dual-face lapping machine, having two-bonded abrasive laps (400-grit SiC) that are rotated in opposite directions at 88 rpm. The head is air-actuated to provide the lapping pressure to the top lap. The WP carrier is eccentrically mounted over the bottom lap and rotates at 7.5 rpm. The viscous cutting oil is fed to the laps during operation. The laps are dressed two or three times during an eight-hour shift.

Figure 3.120 illustrates some typical shapes that can be machined on flat lapping machines. Symmetrical components (a) and (b) do not require workholders. Asymmetrical components (c) and (d) require workholders. Parts similar to (e) require holders to keep them from tipping.

Tolerance, roughness, flatness, and parallelism. Achieved tolerance of parts having parallel shapes can be ±2.5 μm (for small parts) to ±25 μm (for large parts). It is difficult to maintain accuracy for parts of uneven configuration. Such parts may require fixtures that determine the level of accuracy attainable. The flatness may attain a value of 0.3 μm and the achievable surface roughness R_a is 0.05 μm.

Flat parallel surfaces can be lapped on either double-lap machines, which lap both sides of the WP in a single operation, or the single-lap machines, which require two operations. In the

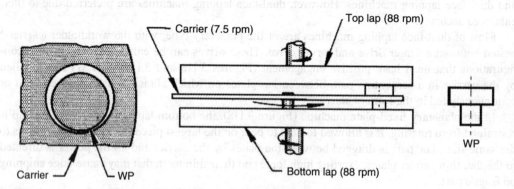

FIGURE 3.119 Dual-face lapping machine using two bonded abrasive laps. (Adapted from ASM International, *Machining*, Vol. 16, *Metals Handbook*, ASM International, Materials Park, OH, 1989.)

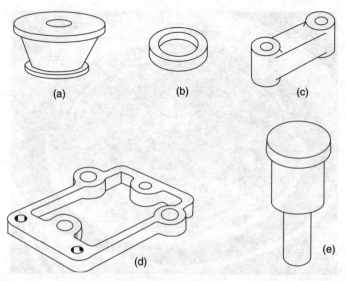

FIGURE 3.120 Typical shapes lapped on flat lapping machine. (Adapted from ASM International, *Machining*, Vol. 16, *Metals Handbook*, ASM International, Materials Park, OH, 1989.)

latter case, extraordinary attention is required to such details as cleanliness and lap flatness. Flatness of laps must be kept within the required flatness tolerance of the WP. In case of lot production, a parallelism of 0.2 μm/mm dictates the use of a dual machine; however, it is possible to produce parts with opposing faces parallel within 0.02 μm/mm. Allowance for stock removal in this operation should be 1.5–2 times the amount of the part out-of-parallelism plus the amount of the variation in part size (ASM International, 1989).

3. *Lapping Machines for Spherical Surfaces*

These are classified into two classes: single- and multiple-pieces lapping machines. Single-piece machines have the following two configurations:

a. *A single-spindle machine with a vertical spindle that rotates the lap.* Ferrous WPs are held stationary by a magnetic chuck; those of nonferrous materials are clamped in a fixture. A crank is held by the chuck of a lathe, is provided by a ball-end crankpin that fits in a drilled hole in the back of the lap (Figure 3.121a), rotates over the spherical surface of the WP. The WP is in line with the spindle of the lathe. The lap should be heavy enough to provide the required lapping pressure.
b. *A two-spindle machine.* One spindle holds and rotates the WP, while the other holds the lap in a floating position and oscillates it through an angle large enough to lap the required area of the surface (Figure 3.121b).

3.9.3.2.2 Equalizing Lapping

In this process, two WP surfaces are separated by a layer of abrasives mixed with a vehicle and rubbed against each other. Each piece drives the abrasive, so that the particles act on the opposing surface. Irregularities that prevent the two surfaces from fitting together precisely are thus lapped, and the surfaces are mated (equalized).

In many cases, a part is first lapped by individual-piece lapping and then mated with another part by equalizing lapping. Equalizing lapping enables mating parts such as cylinder heads and blocks of ICEs to be liquid- or gas-tight without the need for gaskets. It also eliminates the need for piston rings when fitting plungers to cylinders. Another common application is the equalizing lapping of tapered valve components (Figure 3.122).

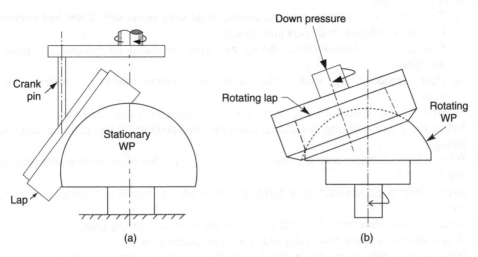

FIGURE 3.121 Lapping of spherical surfaces: (a) single-spindle machine and (b) two-spindle machine. (Adapted from ASM International, *Machining*, Vol. 16, *Metals Handbook*, ASM International, Materials Park, OH, 1989.)

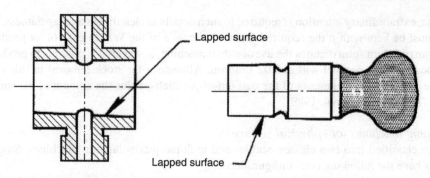

FIGURE 3.122 Tapered valve component finished by equalizing lapping. (Adapted from ASM International, *Machining*, Vol. 16, *Metals Handbook*, ASM International, Materials Park, OH, 1989.)

3.10 REVIEW QUESTIONS

1. The produced accuracy of planers are more superior than shapers. Explain.
2. What is the maximum table travel of a large planer? What is the maximum ram travel in a shaper?
3. What are the main features that limit the stroke of a shaper?
4. How does form turning differ from ordinary turning?
5. Why is the table spindle hollow?
6. What will happen to a WP held between centers if the centers are not exactly in line?
7. How does a steady rest differ from a follower rest?
8. Why should the projected length of a lathe tool be minimized?
9. List the methods of taper turning on a lathe.
10. Why it is desirable to use a heavy depth of cut and light feeds rather than the opposite?
11. Why is the cutting speed in planing, shaping, and slotting limited by 50–60 m/min?
12. On what diameter is the rpm of the work based for a facing cut, assuming given work and tool material?
13. Why are vertical boring machines better-suited than a facing lathe for machining large WPs?
14. Mark true or false.
    a. To enhance productivity, efficient cutting tools such as carbide, CBN, and ceramic tools are employed on shapers and planers.
    b. Surface to be hand-scraped should be better produced on planer-type milling machines.
    c. Planers produce large work at the lowest cost in comparison to any other machine tools.
    d. Broaching machines are simpler in basic design than other machine tools.
15. Define these terms: boring, broaching, counterboring, countersinking, reaming, and spot facing.
16. When a large-diameter hole is to be drilled, why is a smaller-diameter hole often drilled first?
17. Explain why a gooseneck tool is highly recommended to use as a shaping or planning tool.
18. What is unique about broaching compared to other basic machining operations?
19. Why broaching is more practically suited for mass production?
20. What is the main point to be considered in pitching the broach teeth that reduce chattering?

General-Purpose Machine Tools

21. Why are broaching speeds usually relatively low compared to that of other machining operations?
22. What are the advantages of a shell-type broach?
23. Can continuous broaching machines be used for broaching holes? Explain why or why not.
24. State some ways that improve the efficiency of a planer. Do any of these apply to the shaper?
25. How does the process of shaping differ from planing?
26. How does a gang drilling machine differ from a multiple-spindle drilling machine?
27. It is required to drill eight equally spaced holes φ 10 mm in a bolt-hole circle of 160 mm. The holes must be ±1° from each other around the bolt-hale circle.
    - Calculate the tolerance between hole centers.
    - Do you think a typical multiple-spindle drill set up could be used, or would a drilling jig be better in this situation?
28. What might happen when holding work by hand during drilling?
29. In a turning operation, a cutting speed of 55 m/min has been selected. At what rpm should a 15 mm diameter bar be rotated?
30. Describe the relative characteristics of climb milling and up-milling, mentioning the advantages of each.
31. Which type of milling (up or down) do you think uses less power under the same cutting conditions?
32. Why does the use of climb milling make it easier to design a milling fixture than up-milling?
33. What is the advantages of a helical-tooth cutter over a straight-tooth cutter for slab milling?
34. Explain the steps required to produce a T-slot by milling.
35. Why would a plain horizontal-knee milling machine be unsuitable for milling helical flutes?
36. What is the basic principle of a universal dividing head?
37. The input end of a universal dividing head can be connected to the lead screw of the milling machine table—for what purpose?
38. What is the purpose of indexing plates on a universal dividing head?
39. Explain how a standard universal dividing head having a hole circle 21, 24, 27, 30, and 32 would be operated to cut 18 gear teeth.
40. Why is friability an important grit property in abrasive machining?
41. What are the commonly used materials for binding GWs?
42. What are the differences between dressing and truing?
43. How is a WP controlled in centerless grinding?
44. Why should a grinding fluid be used in very copious quantities when performing wet grinding?
45. What is CFG?
46. How does CFG differ from conventional plunge surface grinding?
47. What are the common causes of grinding accidents?
48. What other machine tool does a cylindrical grinding resemble?
49. A set of granite or wooden stairs shows wear on the treads in the central region where people step when they climb or descend the stairs. The higher the stair step, the less wear on the tread. Explain.
    - Why do the stairs wear?
    - Why the lower stairs are more worn than the upper ones?
50. Explain the major differences between the specific energies involved in grinding and in machining.

51. Explain why the same GW might act soft or hard.
52. A soft-grade GW is generally recommended for hard materials. Explain.
53. Explain why CFG has become an important process.
54. The RW of a surface grinder is rotating at a surface speed of 20 m/min. It is inclined at an angle of 5% with respect to the GW axis for roughing. What is the feed rate of the WP past the GW? If the inclination angle is reduced to 2° for finishing, what would be the feed rate in this case?

REFERENCES

Arshinov, V. and Alekseev, G. (1970) *Metal Cutting Theory and Cutting Tool Design*, Mir Publishers, Moscow.
ASM International (1989) *Machining*, Vol. 16, *Metals Handbook*, ASM International, Materials Park, OH.
Degarmo, E. P., Black, J. T., and Kohser, R. A. (1997) *Materials and Processes in Manufacturing*, 8th Edition, Prentice-Hall, New York.
DeVlieg Machine Co.
Hoffman Co., Carlisle, PA.
Jain, R. K. (1993) *Production Technology*, 13th Edition, Khanna Publishes, Delhi, India.
Kalpakjian, S. and Schmid, S. R. (2003) *Manufacturing Processes for Engineering Materials*, 4th Edition, Prentice-Hall, New York.
Mott, L. C. (1976) *Engineering Drawing and Construction*, Oxford University Press, Oxford.
Raw, P. N. (2000) *Metal Cutting and Machine Tools*, Tata McGraw-Hill, New Delhi.

4 Thread Cutting

4.1 INTRODUCTION

Production of screw threads is of prime importance to engineers because nearly every piece of equipment has some form of screw thread. Machine parts are held together, adjusted, or moved by screw threads of many sizes and kinds. Screw threads are commonly used as fasteners, to transmit power or motion, and for adjustment. Different thread forms (V, square, acme) and thread series (coarse, fine, and so on) are available. The screw threads used in manufacturing should conform to an established standard to be interchangeable and replaceable. The following terms (Figure 4.1) are used to describe the geometry of a screw thread:

Major diameter. The outside diameter and the largest diameter of the thread.
Minor diameter. The inside diameter of the screw and the base of the thread. It is also called the *root diameter.*
Pitch. The distance from one point on a thread to the same point on the adjacent thread. The reciprocal of the pitch is the number of threads per inch (tpi).
Pitch diameter. The diameter of an imaginary cylinder whose surface would pass through the threads at such points as to make the width of the thread cut equal to the width of the spaces cut by the imaginary cylinder.
Crest. The top surface joining the flanks of the threads.
Root. The bottom surface joining the flanks of the thread.
Flank. The slanted surface joining the crest and the root; the surface that is in contact with the nut.
Lead. The amount by which the nut advances along the screw with one turn if the screw is held stationary. For single-pitch threads, the lead is equal to the pitch; for double-pitch threads the lead is double the pitch; and for triple-pitch threads, the lead is three times the pitch.

The standard and most widely used threads are as follows (Figure 4.2):

1. *The ISO metric thread.* This thread (Figure 4.2a) is based on the recommendation of the ISO technical committee and was published in British Standard Specification No. 3643 in 1963. It was intended that this thread become a British standard. As a part of the International System of Units (SI), pitches are in millimeters, and the system allows for coarse and fine pitches. The thread form is identical with the unified thread.
2. *The unified thread.* This thread standard was published in 1949 as a result of a conference held in Ottawa between the United States, Canada, and Britain in 1945. There are two subtypes of this thread: the unified coarse (UNC) and the unified fine (UNF). The pitches of this thread are in inches (Figure 4.2a).
3. *Whitworth thread.* This thread form was proposed by Sir J. Whitworth in 1840s. It has been used as the British standard Whitworth (BSW) ever since (Figure 4.2b).
4. *British association (BA) thread.* This thread has been used for screws of diameters less than 1/4 in. and for electrical fitting and accessories (Figure 4.2c).
5. *Square thread.* As shown in Figure 4.2d, this thread is used for power transmission. It is the most difficult to cut and is not compatible for using split nuts.

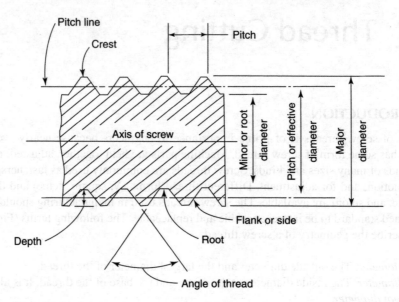

FIGURE 4.1 Thread terms.

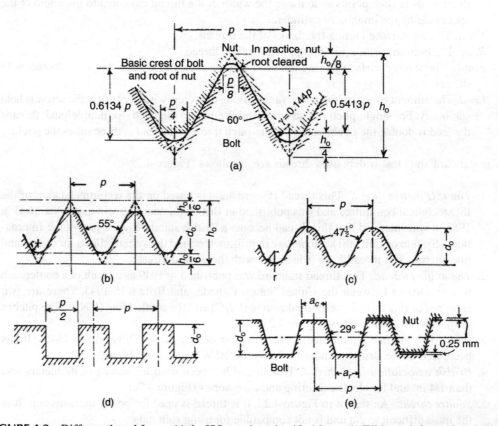

FIGURE 4.2 Different thread forms: (a) the ISO metric and unified form, (b) Whitworth form ($d_o = 0.64\,p$, $r = 0.137\,p$), (c) British association thread ($d_o = 0.6\,p$, $r = 2\,p/11$), (d) square thread ($d_o = 0.5\,p$), and (e) acme thread ($d_o = 0.5\,p + 0.25$ mm, $c_1 = 0.371\,p$, $a_r = a_c - 0.13$).

Thread Cutting

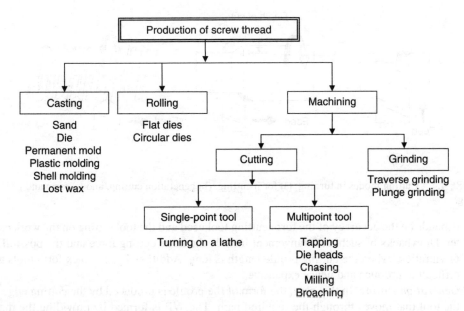

FIGURE 4.3 Methods of screw thread production.

6. *Acme thread.* This thread (Figure 4.2e) is often employed instead of the square thread because it is easier to cut; also it is easier to engage a split nut with an acme thread than with the square one. It has 29° thread angle.
7. *Trapezoidal metric thread.* Similar to the acme thread, except that it has a 30° thread angle. It may replace the acme thread for lead screws (Figure 4.2e).

Acme, square, and trapezoidal threads are used for power transmissions such as screw jacks, lead screws of lathes, vices, presses, and so on. Acme threads are cheaper to manufacture than square ones, but are less efficient than square threads. Acme threads are sometimes used with a split nut to facilitate the engagement and disengagement of the nut and to transmit power in any direction, while trapezoidal threads are used to transmit the power in one direction.

Threads can be produced in a number of ways. The manner of producing them depends on many factors such as the cost and use of the threaded WP, the equipment available, the number of parts to be made, the location of the threaded portion, the smoothness and accuracy desired, and the material to be used. Methods of thread manufacturing are shown in Figure 4.3. This chapter deals with thread machining methods; threads produced by rolling and casting methods are beyond the scope of this book.

4.2 THREAD CUTTING

During machining by cutting, the tool is penetrated into the WP by a depth of cut. Cutting tools have a definite number of cutting edges of known geometry. Moreover, the machining allowance is removed in the form of visible chips. The shape of the WP produced depends upon the tool and the relative motions of the WP. Three main arrangements that occur during machining by cutting are as follows:

1. *Form cutting.* The shape of the WP is obtained when the cutting tool completes the final contour of the WP. The WP profile is formed through the main WP rotary motion in addition to the tool feed at a specified depth (Figure 4.4a). For automatic machine tools, a circular form tool would be used, which has much greater work life, provides more regrinds, and is often easier to manufacture. The quality of the machined surface profile

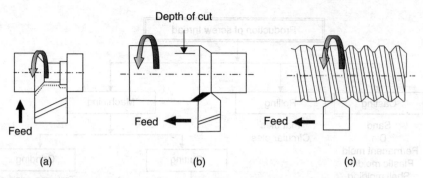

FIGURE 4.4 Cutting modes in turning: (a) form cutting, (b) generation cutting, and (c) form and generation cutting.

depends on the accuracy of the form cutting tool used and the tool setting on the work center. Drawbacks of such an arrangement include the large cutting force and the possibility of vibrations when the cutting profile length is long. Additionally, complex form tools are difficult to produce and hence expensive.

2. *Generation cutting.* In this case, the form of the profile is produced by the cutting edge of the tool that moves through the required path. The WP is formed by providing the main motion to the WP and moving the tool point in a feed motion. In the turning operation, shown in Figure 4.4b, the WP rotates around its axis while the tool is set at a feed rate to generate the required profile much longer profiles can be generated using this method than using the form cutting method. The work finish is better and it is easier to generate internal profiles than to form them.
3. *Form and generation cutting.* During thread cutting, the tool, having the thread form (form cutting), is allowed to feed (generation cutting) axially at the appropriate rate while the WP rotates around its axis (main motion) (Figure 4.4c).

Screw threads can be produced by a variety of cutting tools and processes. The simplest of these is the use of a single-point threading tool in an engine lathe, semiautomatics, and automatics. This method is widely used in piece and small lot production and for cutting coarse-pitch threads. Threads can be produced by hand and in a machine by means of taps and threading dies, which cut internal and external threads, respectively. Solid dies have a low production capacity and, therefore, self-opening dies are currently used. Die heads with radial chasers, circular chasers, and tangential chasers are available. Upon the completion of the thread, the chasers of the self-opening die head automatically withdraw from the work, making reversal of the machine to screw off the die head unnecessary.

Threads can also be produced by milling. Trapezoidal and acme threads with coarse pitch are milled with a disk-type milling cutter and short threads can be produced with a multithread milling cutter. The axis of this type of cutter is set parallel to the work axis and its length must be slightly greater than that of the threaded portion of the WP. In thread milling machines, employing this type of cutters, the whole length of the thread is milled in a slightly over one revolution of the WP.

4.2.1 Cutting Threads on the Lathe

For cutting an accurate screw thread on the lathe machine, it is necessary to control the relation between the feed movement of the cutting tool and the turns of the WP. This is done by means of the lead screw, which is driven by a train of gears from the spindle (Figure 4.5). Modern lathes are fitted with change gears boxes, by means of which any thread pitches can be cut without working out and setting up the change gears. However, there are some machines on which change gears must be fitted for screw cutting.

Thread Cutting

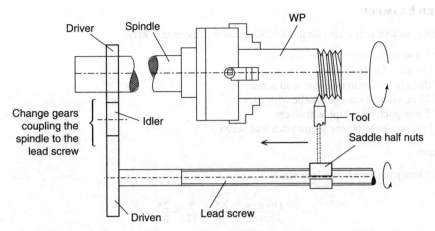

FIGURE 4.5 Diagrammatic representation of screw cutting on a lathe.

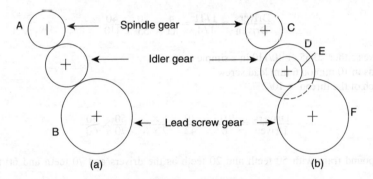

FIGURE 4.6 Gear trains for thread cutting on a lathe: (a) simple gearing and (b) compound gearing.

When cutting a screw thread, the tool is moved along the bed and is driven by a nut engaging with the lead screw. The lead screw is driven by a train of gears from the machine spindle. The gear train may be arranged in one of the following ways:

A. *Simple gear train*. In such a gear train, shown in Figure 4.6a, the following ratio holds:

$$\frac{\text{Turns of lead screw}}{\text{Turns of spindle}} = \frac{\text{Teeth on driver (A)}}{\text{Teeth on driven (B)}}$$

The intermediate gear has no effect on the ratio. It simply acts as a connection that makes the lead screw rotate in the same direction of the machine spindle.

B. *Compound gear train*. In this case, as shown in Figure 4.6b, the gear ratio becomes

$$\frac{\text{Turns of lead screw}}{\text{Turns of spindle}} = \frac{\text{Teeth on C}}{\text{Teeth on D}} \times \frac{\text{Teeth on E}}{\text{Teeth on F}} = \frac{\text{Teeth on drivers}}{\text{Teeth on driven}}$$

Gears supplied with lathes, generally, range from 20 to 120 teeth in steps of 5 teeth with two 40s or two 60s. The lead screw on lathes is always single-threaded and of a pitch varying from 5 to 10 mm depending on the size of the machine. For English lathes, the most common screw threads have 2, 4, or 6 tpi.

SOLVED EXAMPLE

Calculate suitable gear trains for the following cases (Chapman, 1981):

a. 2.5 mm pitch on a 6 mm lead screw
b. 11 tpi on a 4 tpi lead screw
c. 7 threads in 10 mm on 6 mm lead screw
d. 7/22 in. pitch, 3 start on a lathe with 2 tpi
e. 2.5 mm pitch on a 4 tpi lead screw
f. 12 tpi on a lathe having 6 mm pitch lead screw

Solution

a. 2.5 mm pitch on a 6 mm lead screw

$$\frac{\text{Drivers}}{\text{Driven}} = \frac{2.5}{6} = \frac{5}{12} = \frac{25}{60}$$

25 teeth driving 60 teeth in a simple train.

b. 11 tpi on a 4 tpi lead screw

$$\frac{\text{Drivers}}{\text{Driven}} = \frac{1/11}{1/4} = \frac{4}{11} = \frac{20}{55} = \frac{40}{110}$$

This gives either 20/55 or 40/110 in a simple train.

c. 7 threads in 10 mm on 6 mm lead screw
The pitch of the thread = 10/7

$$\frac{\text{Drivers}}{\text{Driven}} = \frac{10/7}{6} = \frac{10}{42} = \frac{5 \times 2}{7 \times 6} = \frac{50}{70} \times \frac{20}{60}$$

A compound train with 50 teeth and 20 teeth as the drivers and 70 teeth and 60 teeth as the driven.

d. 7/22 in. pitch, 3 start on a lathe with 2 tpi
Lead of the thread = 3 × 7/22 = 21/22 in.
Pitch of lead screw = 1/2 in.

$$\frac{\text{Drivers}}{\text{Driven}} = \frac{21/22}{1/2} = \frac{42}{22} = \frac{21}{11} = \frac{3 \times 7}{2 \times 5\frac{1}{2}} = \frac{30}{20} \times \frac{70}{55}$$

A compound train with 30 teeth and 70 teeth as the drivers and 20 teeth and 55 teeth as the driven.

e. 2.5 mm pitch on a 4 tpi lead screw
For cutting metric threads on English lathes:

$$\frac{\text{Drivers}}{\text{Driven}} = \frac{1}{25.4} = \frac{10}{254} = \frac{5}{127}$$

Cutting p mm pitch would require a ratio p as large as $5p/127$, and if the lead screw had N_t threads per inch instead of 1 thread per inch, it would need to turn faster still in the ratio N_t:1. That is (Chapman, 1981):

$$\frac{\text{Drivers}}{\text{Driven}} = \frac{5\,pN}{127}$$

where
  N = number of threads per inch of the lead screw
  p = required pitch in mm

Thread Cutting

hence,

$$\frac{\text{Drivers}}{\text{Driven}} = \frac{5 \times 4 \times 2\frac{1}{2}}{127} = \frac{50}{127}$$

50 teeth driving 127 teeth in a simple train.

f. 12 tpi on a lathe having 6 mm pitch lead screw
Dealing with English threads (in.) on lathes with metric leadscrew (mm)

$$\frac{\text{Drivers}}{\text{Driven}} = \frac{127}{5\,pN}$$

$$\frac{\text{Drivers}}{\text{Driven}} = \frac{127}{5 \times 6 \times 12} = \frac{127}{6 \times 60} = \frac{20}{60} \times \frac{127}{120}$$

A compound train with 20 teeth and 127 teeth as the drivers and 60 teeth and 120 teeth as the driven.

4.2.2 Thread Chasing

Thread chasing is the process of cutting a thread on a lathe with a chasing tool that comprises several single-point tools banked together in a single tool called a *chaser*. Thread chasers are shown in Figure 4.7. Chasing is used for the production of threads that are too large in diameter for a die head. It can be used for internal threads greater than 25 mm in diameter. Figures 4.7a and 4.7b show a tangential-type chaser for cutting external threads and Figure 4.7c shows a circular chaser for cutting internal threads. During internal and external chasing (Figure 4.8), the chaser moves from the headstock. The chaser is moved radially into the WP for each cut by means of the cross slide screw. Thread chasing reduces the threading time by 50% compared to single-point threading. However, thread chasing is a relatively slow method of cutting a thread, as a small depth of cut is used per pass. Depending on the size of the thread, 20–50 passes may be required to complete a thread. Multiple threads, square threads, threads on tapers, threads on diameters not practical to thread with a die, threads that are not standard or those that are so seldom cut that buying a tap or die would be impracticable, or threads with a quick lead are all cut by chasing.

Chasing lends itself better to nonferrous materials rather than ferrous ones. Multistart threads can be chased without any indexing of WP. Taper threads can be generated by chasing, if the chasing attachment is used in conjunction with taper attachment. For HSS cutters, a cutting speed of the order of 40 m/min and upward should be used. Feed varies from 5 to 7.5 cm/min for coarse threads in tough materials to 20–25 cm/min under more favorable conditions.

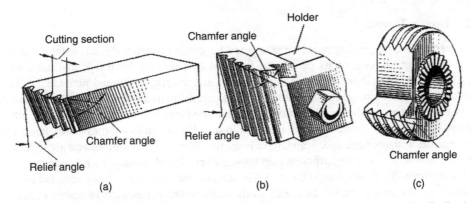

FIGURE 4.7 Thread chasers: (a) flat (shank type), (b) block, and (c) circular. (From Rodin, P., *Design and Production of Metal Cutting Tools*, Mir Publishers, Moscow, 1968. With permission.)

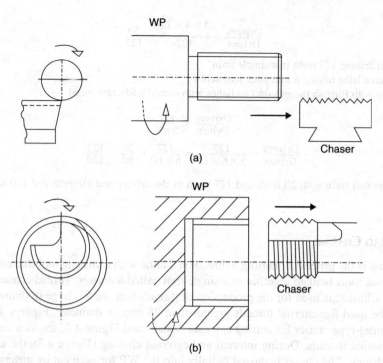

FIGURE 4.8 Thread chasing methods: (a) right-hand external and (b) right-hand internal.

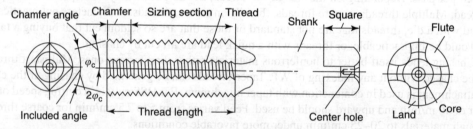

FIGURE 4.9 Tap nomenclature. (From Rodin, P., *Design and Production of Metal Cutting Tools*, Mir Publishers, Moscow, 1968. With permission.)

4.2.3 Thread Tapping

Thread tapping is a machining process that is used for cutting internal threads using a tap having threads of the desired form on its periphery (Figure 4.9). There are hand taps and machine taps, straight shank and bent shank taps, regular pipe taps and interrupted thread pipe taps, solid taps, and collapsible taps. A tap has cutting teeth and flutes parallel to its axis that act as channels to carry away the chips formed by the cutting action. Hand taps are furnished in three sets—taper, plug, and bottoming (Figure 4.10). These three are identical in size, length, and vital measurements, differing only in chamfer at the bottom end. Standard taps are furnished with four flutes and are used for iron and steel. These do not provide sufficient chip room for certain soft metals, such as copper, in which case two- or three-fluted taps should be used. The tap cuts threads through its combined rotary and axial motions. The cost of taping increases as the work material hardness becomes greater. Fine threads of 360 tpi in 0.33 mm diameter holes and coarse threads as 3 tpi in 619 mm diameter pipe fitting are possible (*Metals Handbook*, 1989).

Thread Cutting

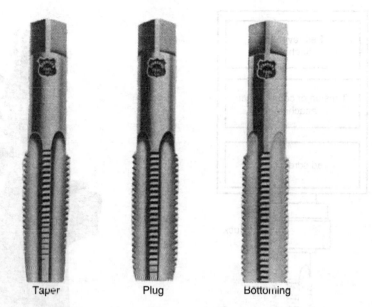

FIGURE 4.10 Straight flute hand taps. (Standard Tool Co., Athol, MA.)

Tapping machines are basically drill presses equipped with lead screws, tap holders, and reversing mechanisms. Lead screws convert the rotary motion into a linear one so that the axial motion of the tap into the hole to be threaded conforms with the pitch of the thread. Lead screw control is often used with larger tap sizes to ensure high-quality threads. However, such an arrangement has the following two major disadvantages:

- The need to return to the starting point to begin each cycle and to stop the rotation between cycles
- Changing the taps for different thread sizes requires time-consuming changes in the feed-controlling members

Tension or compression tapping spindles and attachments provide axial float and compensate for any differences between machine feed and correct tap feed. This provides the possibility to tap different thread pitches at the same time with a single machine feed rate. Self-reversing tapping attachments eliminate the need for reversing motors for tap retraction. Nonreversing tapping attachments are generally used with machines equipped with reversing motors. Figure 4.11 shows the components of a tapping attachment. Tapping machines include the following:

1. *Drill presses.* Simple to set up, easy to operate, and can be provided with lead-control devices that regulate the tap feed rates. When a solid tap is used, the drill press must be supplied with a self-reversing tapping attachment or a reversing motor having a tension compression tap holder. With a collapsible tap, the tapping attachment is not required because the tap automatically collapses at the required depth and returns without stopping or reversing the spindle.
2. *Single-spindle tapping machines.* Used for small to medium production lots. The simpler modes have no lead control devices, but depend on the screw action of the tap in the hole to control the feed (see Figure 4.12).

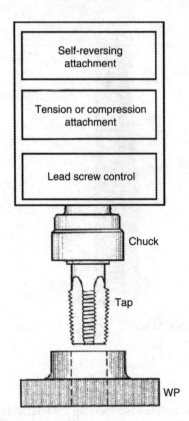

FIGURE 4.11 Tapping attachment.

FIGURE 4.12 Herbert flash tapping machine with automatic cycle. (Alfred Herbert Ltd., Coventry, UK.)

3. *Multiple-spindle tapping machines.* Used for high-volume production lots. They may have up to 25 spindles that are rotated by a common power source. Holes of different sizes can be tapped simultaneously. Spindles having axial float compensate for differences between the lead of the tap and the feed of the spindle. Thus, different thread pitches can be cut simultaneously on the same machine (see Figure 4.13).
4. *Gang machines.* Permit in-line drilling, reaming, and tapping operations and are generally used for low-volume production lots.
5. *Manual turret lathes.* Used for small production lots. Because the WP rotates, they are more accurate than machines that rotate the tap. The machine capability permits drilling, boring, and tapping on the same machine. A lead control device is used when tapping on the turret lathe.
6. *Automatic turret lathes.* Tapping may be included among the many other operations of an automatic turret lathe or in a single multiple-spindle bar or chucking-type machines. These machines require long setting times and are therefore used for large production lots. These machines use lead-control devices for regulating the feed.

The selection of a tapping machine depends on the following factors:

- Size and shape of the WP
- Production quantity
- Tolerance
- Surface finish
- Number of related operations
- Cost

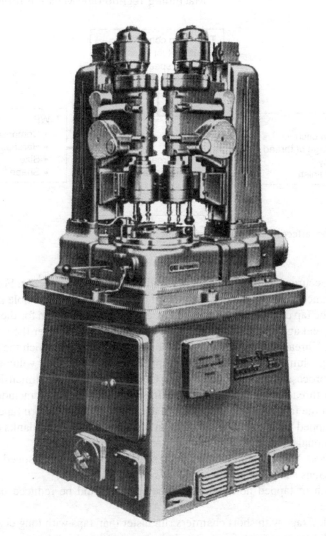

FIGURE 4.13 Jones and Shipman multiple-spindle automatic drilling and tapping machine.

Generally, small diameters and fine-pitch threads are cut on machines of relatively low power, and larger threads in harder materials require heavier machines with large power.

Thread Tapping Performance

Figure 4.14 summarizes the different factors that affect the performance measures of tapping in terms of quality, productivity, and cost. These include the following:

WP characteristics. The use of free-cutting metals is more recommended where better accuracy and surface finish at higher production rates and lower cost are achieved. General purpose high speed steel (HSS) taps are used when the WP hardness is about 30 or 32 HRC; otherwise, highly alloyed HSS is recommended. The work material composition may affect the preparation of the hole before tapping. In this regard, reaming the hole improves the accuracy and finish in aluminum although stainless and carbon steels do not require such reaming process (*Metals Handbook*, 1989). Tapping problems occur with WPs that are too weak to withstand tapping forces. Under such circumstances, a loss of

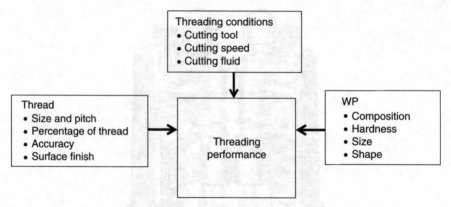

FIGURE 4.14 Factors affecting threading performance.

dimensional accuracy, bad surface quality, and WP damage may occur. For tapping blind holes, a clearance between the last full thread and the bottom of the hole should be compatible with the tap chamfer length. Such a clearance provides room for the produced chip to avoid tap breakage or hole damage by the compressed chip under the advancing tool.

Thread features. Thread size, pitch, and percentage of full depth to which the threads are cut determine the volume removed during the tapping operation. Larger volumes have a direct effect on the process efficiency and tool life. Conditions that cause dimensional variations in the tapped threads cause rough surface finish of threads. These include concentricity error between the tap holder and the spindle and the WP center. Worn tapes, chip entrapment in the tapped hole, and chip build-up on the cutting edges and flanks of the tool also cause dimensional variations and deterioration of the surface finish.

Tapping conditions. WP material has the greatest effect on the tapping speed. The following recommendations should be followed (*Metals Handbook*, 1989):
- As the depth of tapped hole increases, the speed should be reduced because of chip accumulation.
- In short holes, taps with short chamfers run faster than taps with long chamfers.
- As the pitch becomes finer, for a given hole, tapping speed can be increased.
- The amount of cutting fluid and effectiveness of its application greatly influences the cutting speed.

During tapping, the teeth of the tap are more susceptible to damage by heat and the chips that are more likely congested. Cutting fluids are, therefore, used in tapping all metals except CI. However, for tapping holes longer than twice the diameter or blind holes in CI, a cutting fluid or an air blast is recommended (*Metals Handbook*, 1989).

4.2.4 Die Threading

Die threading is a machining process that can be used for cutting external threads of 6.35–114 mm rapidly and economically in cylindrical or tapered surfaces using solid or self-opening dies. The process is faster than single-point threading on a lathe. Die threading of materials having hardness greater than 36 HRC causes excessive tool wear or breakage. Therefore, single-point threading or thread grinding is recommended for metals harder than 36 HRC. Die threading produces threads that are capable of producing fine and coarse threads. The quality and accuracy of such threads are acceptable for most mass-produced articles. For a small shop, thread chasing may be less expensive than stocking a complete set of taps and dies.

4.2.4.1 Die Threading Machines

1. *Drill press.* Easy to set and simple to operate. Threading can be cut manually or by using lead control devices that may require more rigid machine tools such as lathes.
2. *Manual turret lathes.* Used for threading small to medium quantities in parts that require other machining operations such as drilling, turning, reaming, and so on. Turret lathes can handle bar- and chucking-type work and can thread larger parts that are difficult to manipulate in a drill press. Many turret lathes are equipped with lead control devices.
3. *Automatic machines.* Include automatic turret lathes, single- or multiple-spindle automatic bar- or chucking-type machines, which are used for medium- or high-production lots because their setting time is long and their running cost is high. Threading is performed in addition to several other machining operations.
4. *Special threading machines.* Available only for die threading in either cylindrical- or irregular-shaped parts. WP loading and unloading can be manual, hopper-fed, or fully automatic. These machines usually incorporate lead control devices. Bar-type machines with collets can handle long parts to thread rods, shafts, and pipes.

4.2.4.1.1 Solid Dies

Solid threading dies may be of one-piece construction with integral cutting edges or may have replaceable chasers. Nonadjustable, one-piece, solid dies (Figure 4.15) have all cutting edges in a rigidly fixed relationship and are available in standard sizes to fit various types of holders. Adjustable, one-piece, solid dies (Figure 4.16) have a slotted body, a spring, a relief hole, and an adjusting screw for small adjustments that compensate for tool wear and retain greater accuracy than that is possible with nonadjusting dies. In Figure 4.17, the collet adjustability is provided by forcing the jaws inward as the outer nut is tightened.

An inserted-chaser solid die consists of a holder and three or more chasers, which can be compensated for wear and can be removed for resharpening operations. Like one-piece dies, it can be removed from the WP by being back-tracked over the cut thread. Circular chasers are used in solid dies in sets of five, where the die cuts better with less torque. Solid dies are not preferred for high production because the spindle must be reversed for the die removal after the thread is cut, which is time-consuming and increases wear.

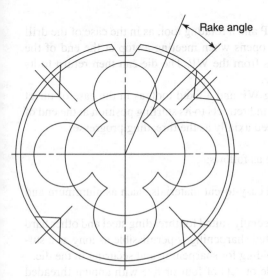
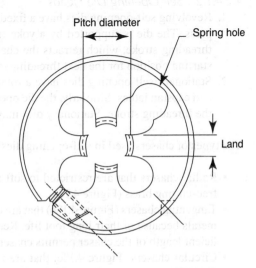

FIGURE 4.15 Solid nonadjustable die. **FIGURE 4.16** Solid screw-adjustable die.

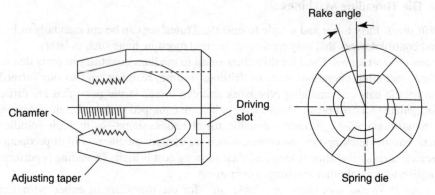

FIGURE 4.17 Spring type collet-adjustable die and holder.

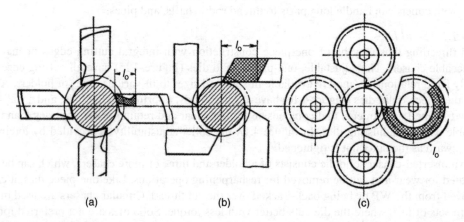

FIGURE 4.18 Threading die heads: l_o = Maximum layer of stock available for sharpening. (From Arshinov, V. and Alekseev, G., *Metal Cutting Theory and Cutting Tool Design*, Mir Publishers, Moscow, 1970. With permission.)

4.2.4.1.2 Self-Opening Die Heads
1. Revolving self-opening dies have a fixed WP and a rotating tool, as in the case of the drill press. The die is supported by a yoke that opens when meeting a stop at the end of the threading stroke, which retracts the chasers from the WP. The die can then return to its starting position for the next threading cycle.
2. Stationary self-opening dies have a rotating WP and a fixed tool, as in the case of turret and capstan lathes. Similarly, the die opens and retracts to its starting position at the end of the threading stroke. Stationary dies may feed axially as the threading progresses.

The types of chasers used in self-opening dies are as follows:

- Radial chasers that are restricted to soft and easy-to-cut materials, such as aluminum and free-cutting brass (Figure 4.18a).
- Tangential chasers (Figure 4.18b) that are especially suited for threading steel and other hard metals because of their long tool life. Repeated sharpening is permissible as long as a sufficient length of the chaser permits chaser holding for sharpening and securing in the die.
- Circular chasers (Figure 4.18c) that are made of sets of four or five with annual threaded form. They are normally used in high production for all metals that are threaded. They have a long lifespan because they can be resharpened many times.

Thread Cutting

Ease of the die removal from the WP is the greatest advantage of self-opening dies. Dies that return in the open mode do not require spindle stopping regardless of whether either the tool or the WP is rotating. This improves machining productivity by more than 50% and reduces the possibility of thread damage by trapped chips. As the chasers make one trip over the WP, their wear is greatly reduced, and hence their tool life is markedly increased. Figure 4.19 shows the general classification of threading die heads, and Figure 4.20 shows a self-opening die head with tangential chasers.

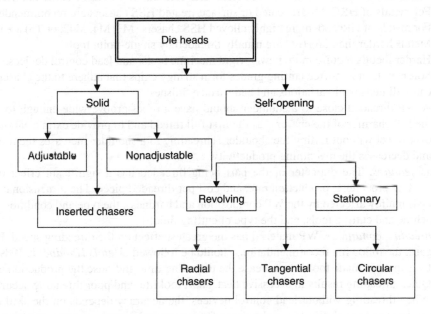

FIGURE 4.19 Classification of threading die heads.

FIGURE 4.20 Die head with tangential chasers. (Alfred Herbert Ltd., Coventry, UK.)

4.2.4.2 Die Threading Performance

Figure 4.14 summarizes the different factors that affect the performance measures of die threading in terms of quality, productivity, and cost. These include the following:

WP characteristic. The choice of free-cutting metals produces more accurate threads of better surface finish at higher production rate and low cost. Generally, the following recommendations are followed (*Metals Handbook*, 1989):
- For threading metals softer than 24 HRC, standard untreated HSS chasers are used.
- For metals of HRC 24–31, coated or surface-treated HSS chasers are recommended.
- For metals of HRC 36, more highly alloyed HSS chasers (M3, M4, M42, or T5) are used.
- Metals harder than 36 HRC are usually threaded by single-point tools.
- Harder metals require more power, rigid machine tools, and lead control devices.
- Soft metals of non–free cutting grades form stringy chips that adhere to the chasers and cause dimensional variations and bad surface finishes.
- A WP threaded close to a shoulder should have a relief groove wide enough to admit the full chamfer of the chaser plus the first full thread and to provide extra clearance for over-travel without hitting the shoulder. Threading to a shoulder increases the tool cost and decreases the machining productivity.

Thread features. The diameter of the part being threaded has a significant effect on the threading procedure, production rate, and cost per threaded piece. The dimensional accuracy is mainly affected by the WP composition and hardness, the type and condition of the machine and cutting tools, and the type of cutting fluid.

Die threading conditions. WP material has the greatest effect on the threading speed. In this regard, the following recommendations should be followed (*Metals Handbook*, 1989):
- Cuts that are made too slow increase the threading time and raise the production cost.
- Quick threading results in excessive heat, short tool life, and poor threading accuracy.
- When threading without lead control devices, the accuracy depends on the skill of the operator and the ability of the chasers to form a nut action.
- Manual control is satisfactory for threading diameters up to 6.4 mm.
- Sulfurized cutting oil is effective for most die-threading applications.

Threading using die heads has the following advantages:

1. Because the stopping and reversing of the spindle is eliminated, it is possible to save considerable time and increase the rate of production.
2. The use of die heads facilitates the withdrawal of any damaged chaser or the replacement of one chaser set by the other to suit the thread requirements so that several types of threads can be cut.
3. Thread manufacture with die heads is economical, because unskilled workers can operate the machines.
4. Thread accuracy is consistent.

However die heads have the following limitations:

1. Square threads are difficult to cut using die heads.
2. Screw threads running up to shoulder of the work cannot be cut.

4.2.5 Thread Milling

Thread milling is a machining process used for cutting screw threads with a single-form or multiple-form milling cutter (Figure 4.21). Threads having an accuracy of ±0.025 mm of pitch diameter and surface finish of 1.4 μm and spacing accuracy of multiple-start threads of ±0.01 mm can be cut.

Thread Cutting

Thread milling makes smoother and more accurate threads than a tap or a die. It is more efficient than using a single-cutting-point tool in a lathe. Thread milling is the most practical method for thread cutting near shoulders or other interfaces. Figure 4.22 shows thread-milled parts. This process is recommended for lot sizes greater than 20 units.

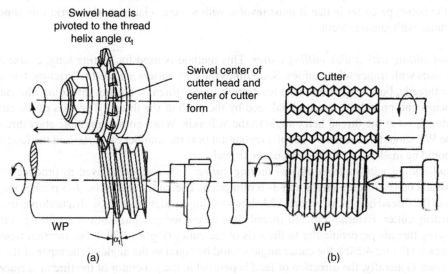

FIGURE 4.21 Thread milling operations: (a) disc cutter and (b) multiple-thread cutter.

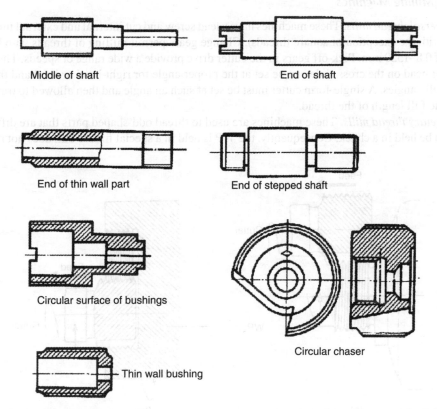

FIGURE 4.22 Thread milled parts. (From Barbashov, F., *Thread Milling Practice*, Mir Publishers, Moscow, 1984. With permission.)

Thread milling is used for cutting threads, usually of too large diameter for die heads. As the milling cutter is held on a stub arbor, the length of the thread is limited to short ones. The cutter rotates at a cutting speed of 0.6 m/s, and the work rotates at the correct feeding speed. As the work rotates, the cutter is fed outward under the action of a master lead screw. Right-hand and left-hand threads can be machined by controlling the direction of tool feed and WP rotation. The disadvantage of the hob-type cutter is that it must revolve with a fixed relation to the work; this is not true for the cutter with annular teeth.

Thread milling with a disk milling cutter. This method is used for cutting long, coarse and threads with trapezoidal profiles. Sometimes the disk cutters are used to machine triangular threads, but they are not used for cutting square threads. During threading, the cutter rotates and provides a longitudinal feed by the pitch of the thread. The axis of the cutter arbor is set at the thread helix angle to the WP axis. When cutting multiple-start threads, the WP should be turned by $1/n$ of a revolution (n is the number of starts) and the feed rate should be made equal to the lead of the thread.

Thread milling with multiple-thread milling cutter. This method is used to produce short threads of 15–75 mm length and 3–6 mm pitch. The cutter should be 2–3 pitches longer than the thread being cut. Figure 4.23 shows milling straight threads with a multiple-thread milling cutter. External tapered threads can be cut using multiple-thread milling cutters having threads perpendicular to the axis of the cutter (Figure 4.24a). For internal tapered threads (Figure 4.24b), the cutter angle should be equal to the angle of the taper of the cut thread. Generally, the direction of feed is parallel to the generator of the thread surface.

Thread Milling Machines

Universal thread mills. These machines have a lead screw and cut internal and external threads (with the exception of square threads). Change gears permit milling of threads with leads of 0.8–1520 mm. Pick-off gears in the cutter drive provide a wide range of speeds. The cutter head on the cross slide can be set at the proper angle for right-hand or left-hand thread helix angles. A single-form cutter must be set at such an angle and then allowed to traverse the full length of the thread.

Planetary thread mills. These machines are used to thread odd-shaped parts that are difficult to be held in a chuck. Consequently, the WP is held in a special fixture that does not rotate

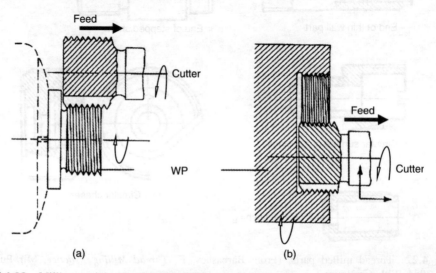

FIGURE 4.23 Milling straight (a) external and (b) internal threads.

Thread Cutting

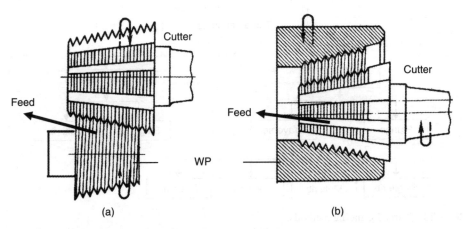

FIGURE 4.24 Milling (a) external and (b) internal tapered threads.

during thread cutting. The milling cutter rotates around its axis and revolves around the work. Double heads can be used to cut both ends of the part, and external and internal threads can also be machined at the same time.

NC machines. These machines are used for thread milling together with other operations in a single WP operation. Long cutter life and high-quality threads are some of the advantages of these machines.

4.2.6 THREAD BROACHING

Thread broaching is a newly developed thread cutting process that has been employed in the automotive field. Typical parts include internal threads on steering-gear ball nuts and ball-race nuts for various circulating ball-type assemblies. The WPs are given one or two passes (rough and finishing cuts), heat-treated, and then finish ground on an internal thread grinder. The broaches used for the application have a special form and are guided by lead screws. Threads are cut by drawing the part and fixture against the revolving tool. Threading broaches are available in sizes up to 50 mm diameter and 750 mm length.

4.3 THREAD GRINDING

Thread grinding is the preferred method of threading when the WP hardness is greater than 36 HRC or less than 17 HRC, and when a high degree of accuracy is required. Threads are ground by the contact between a rotating WP and a rotating GW that has been shaped to the desired thread form. In addition to the rotation, there is a relative axial motion between the wheel and WP to match the pitch of the thread being ground. The process can be used to produce either external or internal threads. Methods of thread grinding are classified in Figure 4.25.

4.3.1 CENTER-TYPE THREAD GRINDING

In this operation, the WP is held between centers or in the machine chuck. The material specifications and the form, length, and quality of the thread determine the number of passes required (from one to six passes). Depending on the design of the threading wheel, the following two basic methods can be identified:

1. *Single-rib wheel traverse grinding.* The most versatile method for which the highest accuracy can be obtained. The single rib wheel is adaptable, by truing, to many different profile configurations (Figure 4.26a).

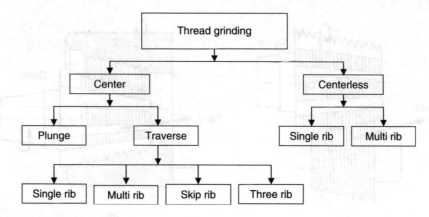

FIGURE 4.25 Thread grinding methods.

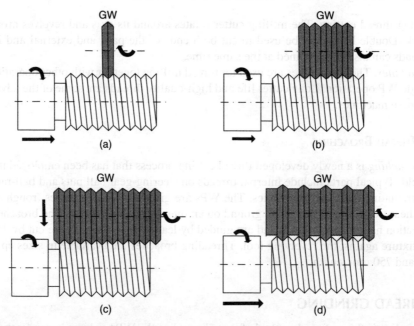

FIGURE 4.26 Thread grinding methods: (a) single-rib traverse grinding, (b) multi rib traverse grinding, (c) multi rib plunge grinding, and (d) skip-rib traverse grinding. (Adapted from *Metals Handbook*, Machining, Vol. 16, ASM International, Material Park, OH, 1984.)

2. *Multi rib-wheel grinding.* The wheels have two or more parallel grooves or ribs around the periphery of the wheel. Each rib is trued to the form of the thread to be ground. The thread form is imparted to the wheel by diamond or crush truing. Figures 4.26b and 4.26c show the different arrangements of multi rib-wheel thread grinding, which include the following:
   A. *Traverse grinding.* More productive than single-rib-wheel grinding, because of the higher material removal rate per pass. However, the pitch should not exceed 1/8 of the wheel width and threading against shoulders should be completely avoided (Figure 4.26b).
   B. *Plunge grinding.* The most productive thread grinding method; therefore used for the production of parts in substantial quantities. As shown in Figure 4.26c, the GW is advanced into the rotating WP.
   C. *Skip-rib traverse grinding.* This process uses a wheel, which has a spacing that is twice that of the thread pitch, basically used for threading accurate and fine pitches.

Thread Cutting

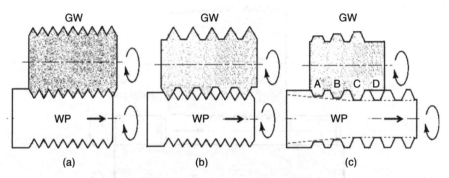

FIGURE 4.27 (a) Conventional grinding, (b) skip-rib grinding, and (c) three-rib grinding. (Adapted from *Metals Handbook*, Machining, Vol. 16, ASM International, Material Park, OH, 1989.)

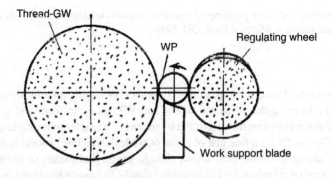

FIGURE 4.28 Centerless thread grinding. (Adapted from *Metals Handbook*, Machining, Vol. 16, ASM International, Material Park, OH, 1989.)

Threading is accomplished in two passes. In the first pass, the wheel grinds every other thread of the WP. In the second pass, the WP is advanced by a single pitch and the untouched threads are then ground (Figure 4.26d).

D. *Three-rib traverse grinding.* As shown in Figure 4.27, the GW has a roughing rib (A) that removes two-thirds of the material, and an intermediate rib (B) that takes the remainder of the material and leaves 0.13 mm for clean up by the finishing rib (C). A flattened area (D) is used to finish the crest of the thread. The process produces more accurate threads than single-rib-wheel thread grinding.

4.3.2 Centerless Thread Grinding

Centerless thread grinding is the most productive method of grinding screw threads that uses either single-rib or multi rib wheels. As shown in Figure 4.28, the regulating wheel rotates in the same direction as the GW. Screw threads are cut by feeding the blanks between the grinding and the regulating wheels in a continuous stream as shown in Figure 4.29.

Thread grinding machines are classified as external, internal, or universal; the universal machines are capable of threading external and internal threads (Chernov, 1984; Acherkan, 1968). The machine structure depends on the following:

- Type of GW used (single-rib or multi rib)
- The method of supporting the WP (centered or centerless)
- The method of restoring the contour of the GW (crushing or diamond dressing)

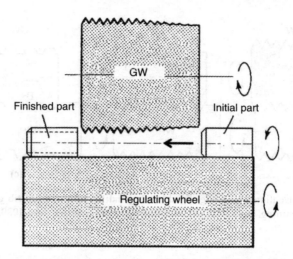

FIGURE 4.29 Traverse centerless grinding of headless screws. (Adapted from *Metals Handbook*, Machining, Vol. 16, ASM International, Material Park, OH, 1989.)

This process is similar to thread milling, in that it uses a GW having annular thread grooves formed around its periphery to cut a thread as the wheel and WP rotate to form and generate the thread. Internal or external threads can be finish-ground by means of a single- or multiple-edged GW. A vitrified bond is generally used with a fine grit of about 600 Mesh No. The process is carried on a special grinding machine having a master lead screw, change gears, and means of holding the work. The wheel rotates at a speed of 30 m/s and work is rotated slowly. In case of hardened stock, grinding is the only method of forming threads. The accuracy of thread grinding exceeds that of any other method, while the surface finish is exceeded only by a good thread rolling operation. Pitch diameters can be ground to an accuracy of ±0.002 mm/2.5 cm and accuracy of lead may be maintained within ±0.007 mm in 50 cm of thread length. Distortions due to heat treatment may be eliminated by grinding. Parts that would be distorted by milling threads can be satisfactorily ground. Parts that demand high accuracies and freedom from distortion and stress cracks are usually made by this method.

The GW has annular thread grooves around its periphery that can be produced either by crushing or by diamond dressing. The accuracy of thread profile is very important. In case of crushing, a roller of hardened steel having the required thread on it is fed under pressure into the wheel face, while a voluminous supply of lubricant is applied as the wheel slowly rotates. Two basic methods of centerless thread grinding that can be identified are as follows:

A. *Plunge grinding.* In this arrangement (Figure 4.28), the wheel is plunged into the WP to the full depth. The WP then makes one revolution, while the wheel traverses one pitch. This method gives a uniform wheel wear, but is used for short thread lengths.
B. *Traverse grinding.* The wheel is positioned at a full thread depth, then the work is traversed past the wheel. The first thread form on the wheel removes the majority of metal and therefore is subjected to the most wear; the following threads affect the finishing. A single-rib wheel may be used for large threads (Figure 4.29).

4.4 REVIEW QUESTIONS

1. Mark true (T) or false (F).
   [ ] Thread cutting on a lathe can be performed using multiple-point threading tools.
   [ ] Self-opening dies reduce tool wear as well as the threading time.

[ ] An acme thread is preferred over a square thread.
[ ] Thread rolling is not a machining process.
[ ] Change gears are not necessary when cutting threads on a lathe machine.
[ ] Free cutting materials produce threads that are accurate and of good surface finish.
[ ] Tapping blind holes is done easier and faster than through holes.
[ ] Thread grinding is recommended for WPs of 37 HRC.
[ ] Traverse centerless thread grinding is faster than plunge centerless thread grinding.
[ ] Plunge centerless thread grinding leads to more uniform wear than traverse grinding.

2. What are the main types of screw threads?
3. List the different methods of thread production.
4. List the different methods of thread cutting and grinding.
5. Show in a sketch how a thread is cut on a center lathe.
6. Calculate the suitable gear train when cutting the following threads on the lathe machine:
   - 3 mm pitch on 6 mm lead screw
   - 13 tpi on a 4 tpi lead screw
   - 6 threads in 12 mm on 6 mm lead screw
   - 2.5 mm pitch on 6 tpi lead screw
   - 10 tpi on a lathe having 6 mm pitch lead screw
7. Compare thread chasing and thread cutting on a lathe.
8. Compare thread milling using disk and multiple-thread cutters.
9. State the main advantages of self-opening die heads over solid dies.
10. Show the arrangement of thread chasers in threading die heads.
11. Show the arrangements of single-rib and multirib traverse grinding of threads.
12. Differentiate between skip-rib and three-rib thread grinding.
13. Show using line sketches how each of the following operations are performed:
    - Milling external tapered thread
    - Plunge centerless thread grinding
    - Traverse centerless thread grinding

REFERENCES

Acherkan, N. (1968) *Machine Tool Design*, 4 Volumes, Mir Publishers, Moscow.
Alfred Herbert Ltd., Coventry, UK.
Arshinov, V. and Alekseev, G. (1970) *Metal Cutting Theory and Cutting Tool Design*, Mir Publishers, Moscow.
Barbashov, F. (1984) *Thread Milling Practice*, Mir Publishers, Moscow.
Chapman, W. A. J. (1981) *Elementary Workshop Calculations*, Edward Arnold, London.
Chernov, N. (1984) *Machine Tools*, Mir Publishers, Moscow.
Metals Handbook (1989) Machining, Vol. 16, ASM International, Material Park, OH.
Rodin, P. (1968) *Design and Production of Metal Cutting Tools*, Mir Publishers, Moscow.

5 Gear Cutting Machines and Operations

5.1 INTRODUCTION

Gears are machine elements that transmit power and rotary motion from one shaft to another. An advantage they have over friction and belt drives is that they are positive in their action, a feature that most of the machine tools require, as exact speed ratios are sometimes essential. Thread cutting and indexing movements in gear cutting are typical examples, which require synchronized rotary and linear movements without any slip. As drive elements, gears are specifically used to

- Change the speed of rotation
- Change the direction of rotation
- Increase or reduce the magnitude of speed and torque
- Convert rotational movement into linear or vice versa (rack and pinion drive)
- Change angular orientation (bevel gears)
- Offset the location of rotating movement (helical gears and worm gear sets)

Depending on the specific application, gears can be selected from the following types:

Spur gears. These are the most common type, which transmit power or motion between parallel shafts or between a shaft and a rack. They are simple in design and measurement. If noise is not a serious problem, spur gears can be used. For aircraft gas turbines, spur gears of extra high quality can operate at pitch-line speed above 2000 m/min. In general applications, spur gears are not allowed to work at speeds over 1200 m/min.

Helical gears. These are used to transmit motion between parallel or crossed shafts, or between a shaft and a rack by meshing teeth that lie along a helix at an angle to the shaft. Because of this angle, teeth mating occurs in such a way that more than one tooth of each gear is always in mesh. This condition permits smoother action than with spur gears. However, some axial thrust is inevitable in helical gears, causing loss of power and requiring thrust bearings. External helical gears are generally used when both high speed and high power are involved.

Herringbone gears. These are sometimes called double helical gears. These gears transmit motion between parallel shafts. They combine the principal advantages of spur and helical gears, because two or more teeth share the load at the same time. Because they have equal right-hand and left-hand helixes, axial thrust is eliminated. Herringbone gears can be operated at higher velocities than spur gears.

Worm gear sets. These are used where the ratio of the speed of the driving member (worm) to the speed of the driven member (worm wheel) is large and for a compact right-angle drive. They are frequently used in indexing heads of milling machines and in hobbing machines.

Crossed-axes helical gears. These operate with shafts that are nonparallel and nonintersecting (Figure 5.1a). The action between mating teeth has a wedging effect, which results in sliding on tooth flanks. Therefore, these gears have low load carrying capacity, but are useful where shafts must rotate at an angle to each other.

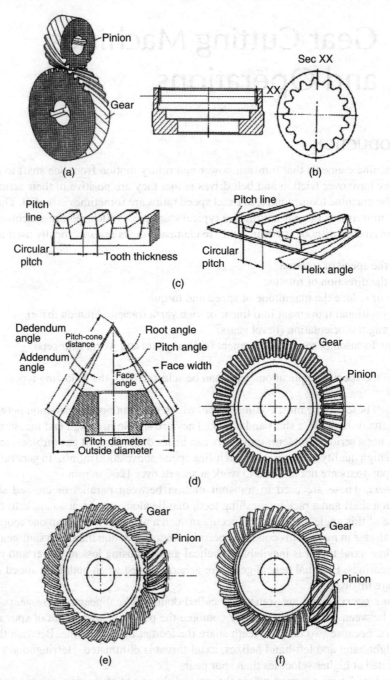

FIGURE 5.1 Common gear types: (a) crossed-axes helical gears, (b) spur internal gears, (c) spur and helical racks, (d) straight bevel gear terminology and a pair in mesh, (e) spiral bevel gears in mesh, and (f) hypoid bevel gears in mesh. (Adapted from *Metals Handbook*, Machining, ASM International, Material Park, OH, 1989.)

Internal gears (Figure 5.1b). They may be of spur or helical tooth form. Their main applications are as follows:
- Rear drives for heavy vehicles
- Planetary gears
- Toothed clutches

- Speed-reducing devices
- Compact design requirements

Racks (Figure 5.1c). A rack is a gear of infinite-pitch circular radius. The teeth may be at right angles to the edge of the rack and mesh with a spur gear, or at some other angle, and engage a helical gear.

Bevel gears. These gears transmit rotary motion between two nonparallel shafts. Bevel gears are of the following types.
- Straight bevel gears (Figure 5.1d). The figure indicates the terminology of the bevel gear. It also shows a pair of bevel gears in mesh. The use of straight bevel gears is generally limited to low-speed drives, and instances where noise is not important. These gears operate at high efficiencies of 98% or better and are used for nonparallel but intersecting shafts.
- Nonstraight bevel gears, which include spiral (Figure 5.1e), zerol, and hypoid gears. All these types are characterized by gradual and continuous engagement resulting in smooth-running. Hypoid gears (Figure 5.1f) do not have as good efficiency as straight bevel gears, but can transmit more power in the same space, provided that the speeds are not too high. They are used for nonparallel, nonintersecting shafts.

5.2 FORMING AND GENERATING METHODS IN GEAR CUTTING

Gears can be commercially produced by other methods like sand casting, die casting, stamping, extrusion, and powder metallurgy. All these processes are used for gears of low wear resistance, low power transmission, and relatively low accuracy of transmitted motion. When the application involves higher values for one or more of these characteristics, cut or machined gears are used.

Gear cutting is a highly complex and specialized art, that is why most of the gear cutting methods are single-purpose machines. Some of them are designed such that only a particular type of gear can be cut. Gear production by cutting involves two principal methods—forming and generating processes. Gear finishing involves four operations—shaving, grinding, lapping, and burnishing (Figure 5.2).

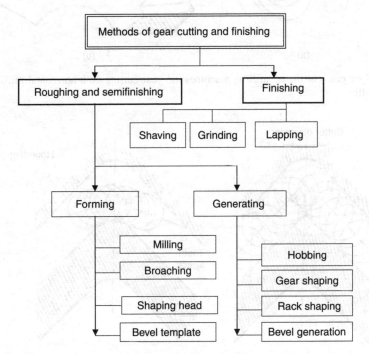

FIGURE 5.2 Methods of gear cutting and finishing.

5.2.1 Gear Cutting by Forming

The tooth profile is obtained by using a form cutting tool. This may be a multiple-toothed cutter used in milling, broaching machines, and shaping cutter head, or a single-point tool form for use in a shaper and a bevel gear planer.

5.2.1.1 Gear Milling

The usual practice in gear milling is to mill one tooth space at a time, after which the blank is indexed to the next cutting position. Figure 5.3 shows teeth in a spur gear cut by peripheral (horizontal) milling with a disk cutter. Similarly, end milling can also be used for cutting teeth in spur or helical gears and is often used for cutting coarse-pitch teeth in herringbone gears (Figure 5.4).

In practice, gear milling is usually confined to

- One-of-a-kind replacement gears
- Small-lot production
- Roughing and finishing of coarse-pitch gears
- Finish milling of gears with special tooth forms

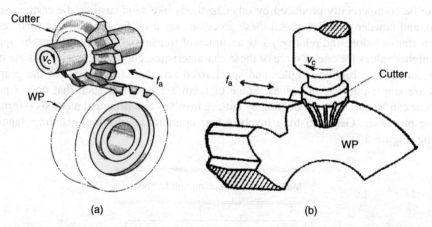

FIGURE 5.3 Spur gear cutting on milling machines: (a) gear cutting on a horizontal milling and (b) gear cutting by end mill.

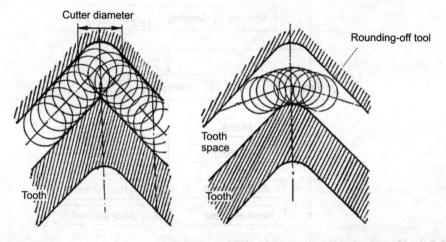

FIGURE 5.4 Herring gear cutting and rounding off the vertex.

TABLE 5.1
Gear Cutter Sets for Milling (According to ASA B.9-1959)

	8-Cutter Set for Spur Gears							
Cutter Number	1	2	3	4	5	6	7	8
Number of teeth	135-rack	$\frac{55}{134}$	$\frac{35}{54}$	$\frac{26}{34}$	$\frac{21}{25}$	$\frac{17}{20}$	$\frac{14}{16}$	$\frac{12}{13}$

	15-Cutter Set for Accurate Spur Gears							
Cutter Number	1	$1\frac{1}{2}$	2	$2\frac{1}{2}$	3	$3\frac{1}{2}$	4	$4\frac{1}{2}$
Number of teeth	135-rack	$\frac{80}{134}$	$\frac{55}{79}$	$\frac{42}{54}$	$\frac{35}{41}$	$\frac{30}{34}$	$\frac{26}{29}$	$\frac{23}{25}$
Cutter Number	5	$5\frac{1}{2}$	6	$6\frac{1}{2}$	7	$7\frac{1}{2}$	8	
Number of teeth	$\frac{21}{22}$	$\frac{19}{20}$	$\frac{17}{18}$	$\frac{15}{16}$	14	13	12	

Although high-quality gears can be produced by milling, the accuracy of tool spacing on older milling machines was limited by the inherent accuracy of the indexing heads. Most indexing techniques used on modern machines incorporate NC or CNC, and the accuracy can rival that of hobbing machines (*Metals Handbook*, 1989).

Moreover, as the tooth profile depends upon the module, pressure angle, and number of teeth, it is theoretically necessary to have a tool with a certain profile for each gear with a different number of teeth or module (module m_g = pitch diameter/number of teeth). In actual practice, however, sets of gear tooth milling cutters, according to ASA B.9-1959, are used (8 cutters per set, or for more accurate gears 15, and less frequently 26 cutters for each module of gear. Each cutter in the set is designed for cutting a limited range of numbers of teeth (Table 5.1).

Cutters for helical gears. When cutting helical gears, the size of the cutter (cutter number), as obtained from Table 5.1, has to be modified due to the helix angle β_g.

$$\text{Equivalent teeth number } Z' = \frac{\text{number of teeth of helical gear}}{(\cos \beta_g)^3}$$

Gear forming on milling machines (and shapers) has the following characteristics:
Advantages:

- General purpose equipment and machines are used.
- Comparatively simple setup is needed.
- Simple and cheap cutting tools are used.
- It is suitable for piece and small size production.

Drawbacks:

- It is an inaccurate process due to profile deviations and indexing errors.
- Low production capacity due to the idle time loss in indexing, approaching, and withdrawal of the tool. However, productivity can be enhanced by multi-WP setup.

ILLUSTRATIVE EXAMPLE 1

The following helical gears are to be produced on a milling machine. Determine the cutter number for each case using both sets listed earlier:

1. Helical gear of helix angle β_g = 10° and Z = 22 teeth
2. Helical gear of helix angle β_g = 20° and Z = 22 teeth

Solution

1. $\beta_g = 10°$ and $Z = 22$ teeth

   8-cutter set

   $$Z' = \frac{22}{(\cos 10)^3} = 23 \text{ teeth}$$

   Select cutter number 5.

   15-cutter set

   $$Z' = \frac{22}{(\cos 10)^3} = 23 \text{ teeth}$$

   Select cutter number $4\frac{1}{2}$.

2. $\beta_g = 20°$ and $Z = 22$ teeth

   8-cutter set

   Equivalent number of teeth Z'

   $$Z' = \frac{22}{(\cos 20)^3} = 26.5 \text{ teeth}$$

   Select cutter number 4.

   15-cutter set

   $$Z' = \frac{22}{(\cos 20)^3} = 26.5 \text{ teeth}$$

   Select cutter number 4.

ILLUSTRATIVE EXAMPLE 2

Calculate the machining particulars for milling the helical gear.

$$Z = 60 \text{ teeth}$$

$$\beta_g = 45°$$

$$m_g = 5 \text{ mm}$$

The lead screw pitch of the milling machine table is 6 mm and the indexing head is equipped by the change gears of 24, 24, 28, 32, 40, 44, 48, 56, 64, 72, 86, and 100 teeth.

Solution

$$\tan \beta_g = \frac{\pi \cdot d_p}{L} \quad (L = \text{lead of gear helix})$$

Normal module:

$$m_n = m_g \cdot \cos \beta_g$$
$$= 5 \times 0.707 = 3.5355 \text{ mm}$$

Gear Cutting Machines and Operations

Then

Addendum: $\quad h_a = 3.5355$ mm

Dedendum: $\quad h_d = 1.25 \times m_n$
$\quad\quad\quad = 1.25 \times 3.5355 = 4.4193$ mm

Tooth height: $\quad h_t = 2.25 \times m_n$
$\quad\quad\quad = 2.25 \times 3.5355 = 7.9548$ mm

Tooth thickness: $\quad s = 1.5708 \times m_n$
$\quad\quad\quad = 1.5708 \times 3.5355 = 5.5528$ mm

Fillet radius: $\quad r = 0.4\, m_n$
$\quad\quad\quad = 0.4 \times 3.5355 = 1.4142$ mm

Pitch diameter:
$$d_p = \frac{Z \cdot m_n}{\cos \beta_g} = Z \cdot m_g$$
$$= 60 \times 5 = 300 \text{ mm}$$

Outside diameter:
$$d_a = Z \cdot m_g + 2 m_n$$
$$= 60 \times 5 + 2 \times 3.5355$$
$$= 307.071 \text{ mm}$$

Indexing operation:

$$\text{Index crank movement} = \frac{40}{Z} = \frac{40}{60} = \frac{2}{3}$$

The index crank is moved 2/3 rev every tooth from the 60 teeth.

Cutter selection:

Using the 15-cutter accurate set, the equivalent cutter, Z', is selected by

$$Z' = \frac{Z}{(\cos \beta_g)^3} = \frac{60}{(\cos 45)^3} = 169.7 \approx 170 \text{ teeth}$$

The cutter number 1 (135 – rack) is selected.

Table tilting:

The table is tilted by 45°.

Lead of the machine L_m = pitch of machine lead screw $\times 40 = 6 \times 40 = 240$ mm.

Lead of gear L_w is calculated from the following relation:

$$L_w = \frac{\pi d_p}{\tan \beta_g} = \frac{\pi \cdot Z \cdot m_g}{\tan \beta_g} = \frac{\pi \times 60 \times 5}{\tan 45} = 942.5 \text{ mm}$$

Therefore,

$$\frac{\text{Driver}}{\text{Driven}} = \frac{L_m}{L_w} = \frac{240}{942.5} \approx \frac{240}{960} = \frac{1}{4} = \frac{24}{48} \times \frac{32}{64}$$

Taking 960 instead of 942.5 will not change the helix angle much, but enables standard change gears to be used.

Error in lead:

Lead produced = 960 mm
Lead desired = 942.5 mm
Error in lead = 960 − 942.5 = 17.5 mm

If this error is not acceptable, use other change gears.

5.2.1.2 Gear Broaching

Gear broaching is usually confined to cutting teeth in internal gears. However, not only internal but also external, spur, or helical gears can be broached. Figure 5.5a shows progressive broach steps in

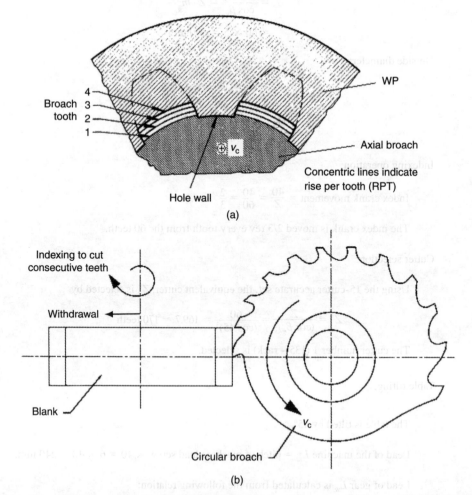

FIGURE 5.5 Gear broaching by forming: (a) broaching of an internal spur gear using an axial broach and (b) broaching of an internal spur gear using a rotating broach.

cutting an internal spur gear. The form of the space between gear teeth corresponds to the form of the broach teeth. The diameter of the broach increases progressively to major diameter that completes the tooth form on the WP. Figure 5.5b shows how an external spur gear is produced using a rotating broach. In such arrangements, the blank is withdrawn for indexing to cut another space between two teeth. Broaching is fast, accurate, and provides excellent surface quality. However, the cost of tooling is high; therefore, gear broaching is best suited to large production runs.

5.2.1.3 Gear Forming by a Multiple-Tool Shaping Head

This is a highly productive and accurate method of producing teeth in external and internal spur gears. This method is not applicable to helical gears. As in broaching of internal gears, all tool spaces are cut simultaneously and progressively (Figure 5.6). The cutter head has as many radially arranged form tools as the number of teeth on the gear being cut. The profile of the tool teeth has exactly the same shape as the gear tooth spaces. Prior to each cutting stroke, each tool is fed radially toward the blank by an amount equal to the prescribed infeed. All the tools are simultaneously retracted from the work on the return stroke to avoid rubbing the tool against the machined surfaces. The gear is finished when the tools reach the full depth of cut. Cutting speeds in this process are similar to those used for broaching the same work metal using the same tool material. Machines with shaping heads are available for cutting spur gears up to 500 mm in diameter, with a face width up to 150 mm.

For example, a machining time of not more than 1 min is required to produce a spur gear of 160 mm pitch diameter, face width of 30 mm, and a module of 4 mm; therefore, the process is best suited to large production runs. Drawbacks of the process are the comparatively complex shaping heads and the necessity of having a separate head for each gear size and module.

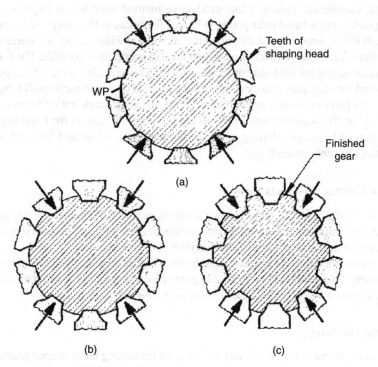

FIGURE 5.6 Cutting with progressive gear shaping head: (a) starting of cut (b) intermediate position, and (c) finished gear.

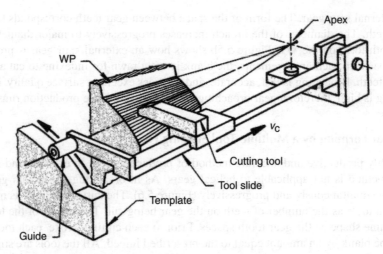

FIGURE 5.7 Template machining using a bevel gear planer.

5.2.1.4 Straight Bevel Gear Forming Methods

Two methods available are as follows:

a. *Straight bevel gear forming by milling.* This method is not widely used for two reasons.
   - It is of a very limited accuracy.
   - The operation is time-consuming.
   Sometimes straight bevel gears are rough-cut by milling and then finished by another method.
b. *Template machining.* This is a low productive method used to cut large bevel gears of coarse pitch using a bevel gear planer (Figure 5.7); because the setup can be made with a minimum effort, template machining is useful when a wide variety of coarse pitch gears are required. Under these conditions, a high level of accuracy is possible. The setup utilizes two templates, one for each side of the gear tooth. Theoretically, a pair of templates would be required for each gear ratio, but in practice a pair is designed for a small range of ratios. A set of 25 pairs of templates encompasses all 90° shaft angle ratios from 1:1 to 8:1 for either $14\frac{1}{2}°$ or 20° pressure angles (*Metals Handbook*, 1989). After the roughing operations are performed by simple slotting tools, the templates are set up and the teeth are finished by making two cuts on each side.

5.2.2 Gear Cutting by Generation

This technique is based on the fact that two involute gears of the same module and pitch mesh together—the WP blank and the cutter. So this method makes it possible to use one cutting gear for machining gears of the same module with a varying number of teeth.

Gear generation methods are characterized by their higher accuracy and machining productivity than gear forming. They comprise hobbing and gear or rack shaping for the manufacture of spur and helical gears, worm and worm wheels, and bevel gear generation.

5.2.2.1 Gear Hobbing

Hobbing is a gear generation method most widely used for cutting teeth in spur gears, helical gears, worms, worm wheels, and many special forms (Figure 5.8). Hobbing machines are not applicable to cutting bevel and internal gears. The tooling cost for hobbing is lower than for broaching and

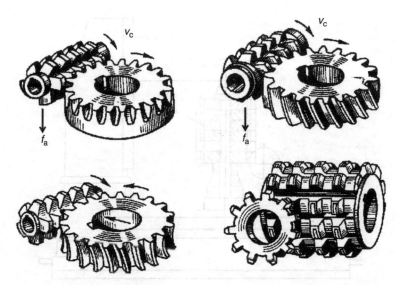

FIGURE 5.8 Various products that can be hobbed.

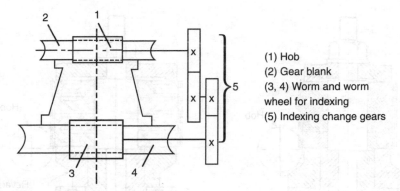

(1) Hob
(2) Gear blank
(3, 4) Worm and worm wheel for indexing
(5) Indexing change gears

FIGURE 5.9 Elementary hobbing machine setup.

multiple-tool shaping heads. For this reason, hobbing is used in low-quantity production or even for a few pieces. Compared with milling, hobbing is fast, accurate, and therefore suitable for medium and high quantity productions. The hob is a fluted worm of helix angle α with form-relieved teeth that cut into the gear blank in succession. A simplified gear train of a hobbing machine is shown in Figure 5.9.

The use of hobbing is sometimes limited by the shape of the WP; for example, if the teeth to be cut are close to a shoulder or a flange, the axial distance is not large enough to allow for the hob to over-travel at the end of the cut. This over-travel should be about one half of the hob diameter plus an additional clearance to allow for the hob thread angle.

The ability to cut teeth in two or more identical gears in one setup can encourage the use of this method (Figure 5.10). Inexpensive fixturing is often utilized for cutting two or more gears at one time when the ratio of the face width to pitch diameter is small. A typical hobbing fixture is illustrated in Figure 5.11, which is a common mandrel-type fixture for flat-face gears. Incorporated in the fixture is an interchangeable bottom plate to enable utilization of the same fixture for various sizes of gears. The clamp plate should be as large as possible and relieved to concentrate the

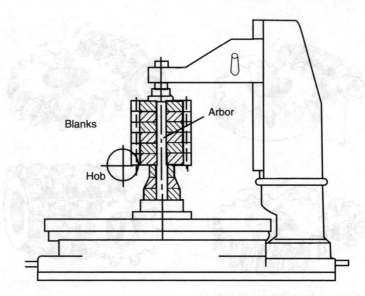

FIGURE 5.10 Clamping identical gears in one setup.

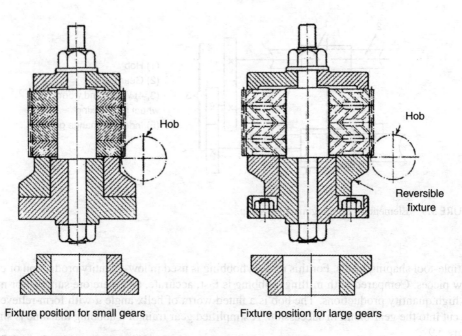

FIGURE 5.11 Interchangeable hobbing fixture to various size gears.

clamping action near the outer edge of the blank. Figure 5.12 illustrates the cutting action used for different types of gears. The rotary motions imparted to the blank and hob are the same as those of worm wheel and worm gearing.

Hobbing of Spur Gears

The hob is set up so that the thread of the hob on the side facing the gear blank is directed vertically along the axis. This is done by setting the hob axis at an angle α_h to the horizontal equal to the

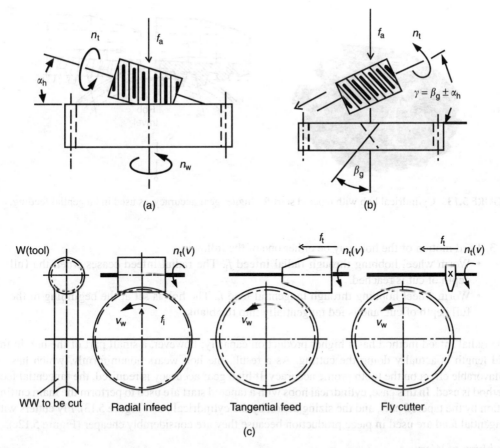

FIGURE 5.12 Cutting action for different types of gearing: (a) spur gear, (b) helical gear, and (c) worm wheel.

helix angle of the hob. The hob attains a continuous feed motion along the axis of the gear blank as shown in Figure 5.12a.

Hobbing of Helical Gears

To cut helical gears, the hob is set up so that the thread of the hob facing the gear blank is directed at the helix angle of the teeth. This is done by setting the hob at an angle $\gamma = \beta_g \pm \alpha_h$, where β_g is the helix angle of the helical gear being cut and α_h is the helix angle of the hob. If the hand of helical gear and that of the hob are different, the positive sign is considered; if the hand is the same, the negative sign should be used. Also, the hob attains a continuous feed motion along the axis of the gear blank (Figure 5.12b). In cutting helical gears, an incremental motion is imparted to the blank, with an angular velocity that would provide one full additional revolution of the blank during vertical feed of the hob through a distance equal to the lead of the helical teeth on the gear.

Hobbing of Worm Wheels

When cutting worm wheels, the axis of the hob is set perpendicular to the axis of rotation of the blank. The following principal motions are shown in Figure 5.12c:

1. Principal rotary cutting motion v of the hob.
2. Continuous indexing rotary motion v_w of the gear bank.

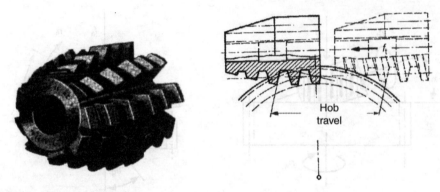

FIGURE 5.13 Cylindrical hob with tapered start for higher gear accuracy, as used in tangential feeding.

3. Feed motion of the hob may be either one of the following:
   - Worm wheel hobbing through radial infeed f_i. The radial infeed ceases when the full depth of cut is reached.
   - Worm wheel hobbing through tangential feed f_t. The hob is set at the beginning to the full depth of cut, and is fed tangentially into the blank.

The radial infeed method has a higher production capacity; however, a small part of the hob in the mid length is actually doing the cutting. As a result, the hob wears nonuniformly, which has an unfavorable effect on the tooth profile accuracy. If high gear accuracy is required, the tangential feed method is used. In this case, cylindrical hobs with a tapered start are used to perform the main cutting action by the tapered part, and the sizing action by the cylindrical part (Figure 5.13). Fly cutters with tangential feed are used in piece production because they are considerably cheaper (Figure 5.12c).

Hobbing of Worms
Hobbing produces the highest grade worm at the lowest machining cost, but can only be used when production quantities are large enough to justify the high tooling cost. The number of flutes in a worm hob is increased to improve surface finish.

Gear hobbing is characterized by the following:

1. High accuracy.
2. Flexibility for any production volume.
3. Low cost.
4. Adaptability to cut metals with higher than average hardness.
5. Any external tooth that is uniformly spaced about the center can be hobbed using a suitable hob.
6. One hob of a particular module can be used to cut teeth of all involute spur and helical gears of any number of teeth of the same module and pressure angle. It is thus a versatile process.
7. The accuracy of hobbed gears depends upon
   - Accuracy of the machine, blank, and tool
   - Care and accuracy of mounting work and hob
   - Feed method used
   - Machine rigidity

   A typical hobbing machine can produce gears of accumulated errors of tooth spacing not more than 20 μm.
8. The indexing is continuous, without an intermittent nature that can cause indexing errors.
9. Finish is dependent on the hob feed.

Gear Cutting Machines and Operations

10. Hobbing cannot be used to cut the following:
    - Bevel gears.
    - Internal gears.
    - Gears having adjacent shoulders larger than the root diameter of gear and that are close enough to restrict the approach or run out of the hob.

Kinematic Diagram and Gear Trains of Hobbing Machines

The hobbing machine is considered a model of versatile nature, capable of producing a wide spectrum of gear shapes. So it is intentionally selected to investigate its kinematic structure and gearing diagram. The whole kinematic structure of a hobbing machine is never simultaneously employed. The machine embraces different gear trains and change gears, which operate in accordance with the specific shapes of gears to be hobbed.

The hobbing machine is ordinarily furnished with a complete selection of change gears to provide flexibility in producing gear shapes. A representative kinematic structure of a gear hobbing machine is given in Figure 5.14a. Accordingly, the same structure is also represented by the chain shown in Figure 5.14b. Accordingly, the motion is transmitted from the drive motor (M_1) through point 3, and speed change gears i_v. At point 4, the motion is branched into 1–R_1 (hob rotation), a first input shaft (5) to differential Σ_1, point 6, indexing change gear i_x, point 2, point 9, feed change gears i_f to point 10. At point 10, the motion is branched to point 8, vertical feed screw t_1, hob slide that imparts elementary motion T_3, and likewise branched to differential change gears i_y, second input shaft 7 of the differential Σ_1, output shaft (6) of the differential, indexing change gears i_x and point 2, point 9, to the worktable to which the second elementary motion R_4 is imparted.

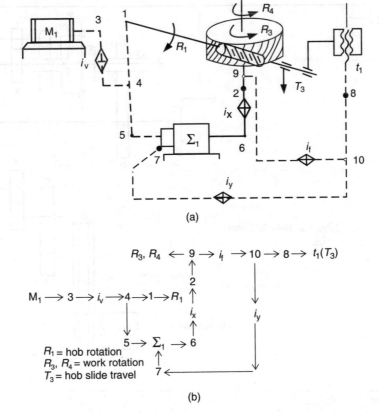

FIGURE 5.14 Kinematic and chain structures of a gear hobbing machine: (a) Kinematic structure and (b) chain structure. (From Acherkan, N., *Machine Tool Design*, Mir Publishers, Moscow, 1968. With permission.)

The representative structure shown in Figure 5.14 is applied in the most widely used gear hobbing machines produced in Europe and the United States. Figures 5.15a and 5.15b show in detail the different gear trains associated with a general purpose hobbing machine. In the same figure, the different change gears (i_v, i_x, i_f, i_y) are calculated based on a given example. The machine setting to hob a helical gear is considered to be the most complex type of gear machining. For calculating the cutting speed change gears (Figure 5.15a), the following is considered:

Assumed:
Hob diameter, d_{hob} = 100 mm (one start)
Cutting speed v = 60 m/min (HSS hob, mild steel blank)

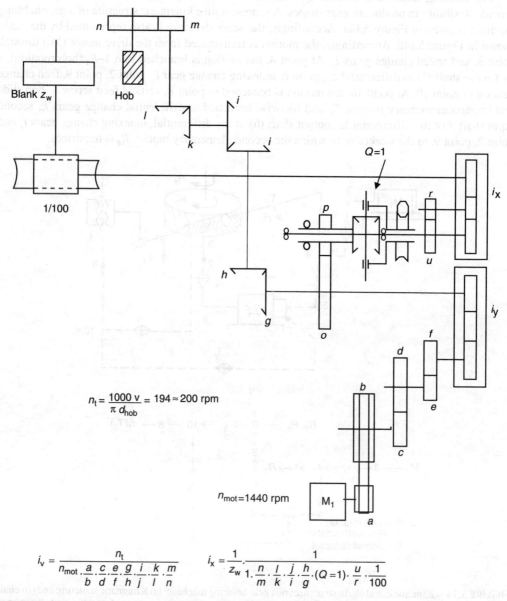

$$n_t = \frac{1000\,v}{\pi\,d_{hob}} = 194 \approx 200 \text{ rpm}$$

$n_{mot} = 1440$ rpm

$$i_v = \frac{n_t}{n_{mot}} \cdot \frac{a}{b} \cdot \frac{c}{d} \cdot \frac{e}{f} \cdot \frac{g}{h} \cdot \frac{i}{j} \cdot \frac{k}{l} \cdot \frac{m}{n}$$

$$i_x = \frac{1}{z_w} \cdot \frac{1}{1 \cdot \frac{n}{m} \cdot \frac{l}{k} \cdot \frac{j}{i} \cdot \frac{h}{g} \cdot (Q=1) \cdot \frac{u}{r} \cdot \frac{1}{100}}$$

FIGURE 5.15A Speed and indexing gear trains, i_r and i_x of general purpose hobbing machine.

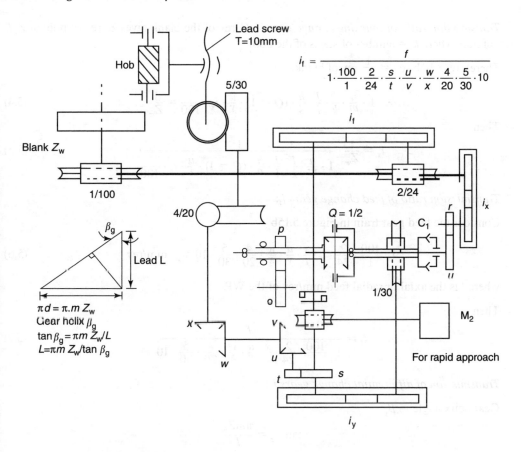

FIGURE 5.15B Feed and differential gear trains i_f and i_y of general purpose hobbing machine.

Therefore,

$$n_t = \frac{1000v}{\pi d_{hob}} = \frac{1000 \times 60}{\pi \times 100} = 194 \text{ rpm} \tag{5.1}$$

Assume $n_t = 200$ rpm.

Transmission ratio of cutting speed change gears is i_v

Consider the cutting speed gear train (Figure 5.15a):

$$n_{mot} \cdot \frac{a}{b} \cdot \frac{c}{d} \cdot i_v \cdot \frac{g}{h} \cdot \frac{i}{j} \cdot \frac{k}{l} \cdot \frac{m}{n} = n_t \tag{5.2}$$

Then,

$$i_v = \frac{n_t}{n_{mot} \cdot \frac{a}{b} \cdot \frac{c}{d} \cdot \frac{e}{f} \cdot \frac{g}{h} \cdot \frac{i}{j} \cdot \frac{k}{l} \cdot \frac{m}{n}} \tag{5.3}$$

Transmission ratio of indexing change gears i_x (1 rev of the blank gives Z_w rev of hob, or Z_w/k revs. of hob, where k = number of starts of the hob):

Therefore, 1 hob rev = $\frac{1}{Z_w}$ rev of blank.

$$1 \cdot \frac{n}{m} \cdot \frac{l}{k} \cdot \frac{j}{i} \cdot \frac{h}{g} \cdot (Q=1) \cdot \frac{u}{y} \cdot i_x \cdot \frac{1}{100} = \frac{1}{Z_w} \tag{5.4}$$

Then

$$i_x = \frac{1}{Z_w} \cdot \frac{1}{1 \cdot \frac{n}{m} \cdot \frac{l}{k} \cdot \frac{j}{i} \cdot \frac{h}{p} \cdot (Q=1) \cdot \frac{u}{r} \cdot i_x \cdot \frac{1}{100}} \tag{5.5}$$

Transmission ratio of feed change gears i_f:

Consider the feed gear train in Figure 5.15b

$$1 \cdot \frac{100}{1} \cdot \frac{2}{24} \cdot i_f \cdot \frac{s}{t} \cdot \frac{u}{v} \cdot \frac{w}{x} \cdot \frac{4}{20} \cdot \frac{5}{30} \cdot 10 = f \text{ mm/rev} \tag{5.6}$$

where f is the axial or radial feed mm/rev of the WP.

Then

$$i_f = \frac{f}{1 \cdot \frac{100}{1} \cdot \frac{2}{24} \cdot \frac{s}{t} \cdot \frac{u}{v} \cdot \frac{w}{x} \cdot \frac{4}{20} \cdot \frac{5}{30} \cdot 10} \tag{5.7}$$

Transmission of differential change gears i_y:

Gear helix angle = β_g

$$\tan \beta_g = \frac{\pi m Z_w}{L}$$

$$L = \frac{\pi m Z_w}{\tan \beta_g}$$

Consider the differential gear train [one rev of blank provides L (mm) longitudinal travel of hob]:

$$1 \cdot \frac{100}{1} \cdot \frac{1}{i_x} \cdot \frac{r}{u} \cdot \frac{1}{2} \cdot \frac{30}{1} \cdot \frac{1}{i_y} \cdot \frac{s}{t} \cdot \frac{u}{v} \cdot \frac{w}{x} \cdot \frac{4}{20} \cdot \frac{5}{30} \cdot 10 = L \tag{5.8}$$

C_1 engaged, then $Q = \frac{1}{2}$

Then,

$$i_y = \frac{\frac{100}{1} \cdot \frac{1}{i_x} \cdot \frac{r}{u} \cdot \frac{1}{2} \cdot \frac{30}{1} \cdot \frac{s}{t} \cdot \frac{u}{v} \cdot \frac{w}{x} \cdot \frac{4}{20} \cdot \frac{5}{30} \cdot 10}{\pi \cdot m \cdot Z_w / \tan \beta_g} \tag{5.9}$$

5.2.2.2 Gear Shaping with Pinion Cutter

This process is the most versatile of all gear cutting processes. Although shaping is most commonly used for cutting teeth in spur and helical gears, this process is also applicable to cutting herringbone teeth, internal gears (or splines), chain sprockets, elliptical gears, face gears, worm gears, and racks. Shaping cannot be used to cut bevel gears.

Figure 5.16 shows the principle of gear shaping with a pinion cutter. In this process, the cutter is mounted on a spindle that reciprocates axially as it rotates. The WP spindle is synchronized with the cutter spindle and rotates slowly as the tool meshes and cuts while it is being fed into the work at the end of each return (upward) stroke. The downward movement of the tool represents the principal cutting motion. To prevent the flanks of the cutter teeth from scoring the blank as the cutter is returned upward, the blank (or the cutter) is withdrawn radially in the direction of arrow X.

Because tooling cost is relatively low, gear shaping is practical for any production volume. WP design often prevents the use of milling cutters or hobs (e.g., cluster gears), and shaping is the most practical method for such cases (Figure 5.17a). Shaping can also be applied in cutting a worm

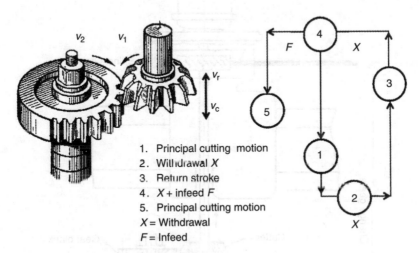

1. Principal cutting motion
2. Withdrawal X
3. Return stroke
4. X + infeed F
5. Principal cutting motion
X = Withdrawal
F = Infeed

FIGURE 5.16 Principles of gear shaping.

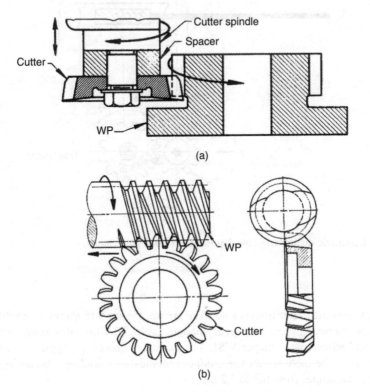

FIGURE 5.17 Shaping of (a) cluster gears and (b) a worm using gear shapers.

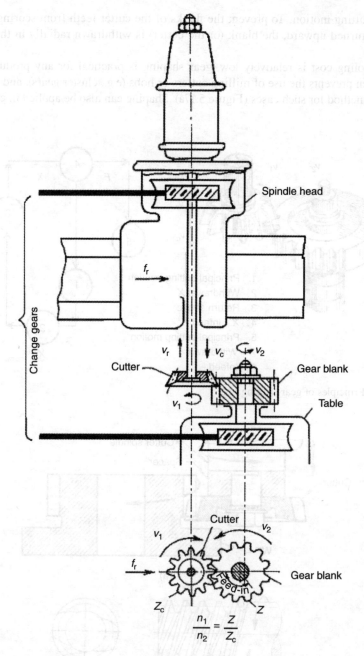

FIGURE 5.18 Kinematic diagram and mechanical drives of a gear shaper.

(Figure 5.17b) where the cutter involves no axial stroke. Figure 5.18 shows a simplified kinematic diagram and mechanical drives of a gear shaper. Table 5.2 illustrates some typical products produced on the Liebherr gear shaper WS1. The examples quoted are typical of the requirement of mass production. The table shows the product specifications, tooling, machining data, and the machining time that ranges from 0.3 to 1.2 min.

TABLE 5.2
Typical Products Machined on the Liebherr Gear Shaper Using HSS Cutters

Part Name	Product Specification	Machining Data	Cutter Teeth
Automotive gear	Material: SAE3120	$n = 70$ stroke/min	46
	$Z = 17$ teeth	$v = 52$ m/min	
	Module $m_g = 2.5$ mm	Number of cuts = 2	
	Helix $\beta_g = 31°$	Rotary feed = 0.64 mm/stroke	
		$t_m = 1.2$ min	
Lay shaft	Material: heat-treated steel	$n = 900$ stroke/min	44
	$Z = 16$ teeth	$v = 64$ m/min	
	Module $m_g = 2.85$ mm	Number of cuts = 3	
	Pressure angle = 20°	Rotary feed = 0.65 mm/stroke	
	Helix $\beta_g = 28°$	$t_m = 1.1$ min	
Cluster gear	Material: EC80	$n = 1000$ stroke/min	38
	$Z = 26$ teeth	$v = 56$ m/min	
	Module $m_g = 2$ mm	Number of cuts = 2	
	Pressure angle = 20°	Rotary feed = 0.58 mm/stroke	
		$t_m = 1.05$ min	
Starter pinion	Material: carbon steel	$n = 1000$ stroke/min	64
	$Z = 9$ teeth	$v = 50$ m/min	
	Module $m_g = 2.1$ mm	Number of cuts = 1	
	Pressure angle = 12°	Rotary feed = 0.54 mm/stroke	
		$t_m = 0.3$ min	
Clutch teeth	Material: 15CrNi6	$n = 2000$ stroke/min	26
	$Z = 27$ teeth	$v = 79$ m/min	
	Module $m_g = 5$ mm	Number of cuts = 1	
	Pressure angle = 20°	Rotary feed = 1 mm/ stroke	
		$t_m = 0.3$ min	
Internal gear	Material: Bakelite	$n = 1250$ stroke/min	25
	$Z = 60$ teeth	$v = 63$ m/min	
	Module $m_g = 1$ mm	Number of cuts = 1	
		Rotary feed = 0.34 mm/stroke	
		$t_m = 0.64$ min	

Source: High Production Gear Shaping Machine, WS1 Kaufbeurer Str. 141, Liebherr Verzahntechnik GmbH. D8960 Kempten, Germany. With permission.

Characteristics of gear shapers are as follows:

- They produce accurate gears.
- Both internal and external gears can be cut by this method.
- The production rate of gear shapers is lower than hobbers.
- Bevel and worm gears cannot be generated on gear shapers.

5.2.2.3 Gear Shaping with Rack Cutter

Gear shaping is performed by a rack cutter with 3–6 straight teeth (Figure 5.19). The cutters reciprocate parallel to the work axis when cutting spur gears, and parallel to the helix angle when cutting helical gears. In addition to the reciprocating action of the cutter, there is synchronized rotation of the gear blank with each stroke of the cutter, with a corresponding advance of the cutter in a feed movement. Rack cutters are less expensive than pinion cutters and hobs. A rack cutter is especially adapted for cutting of large gears of modules, typically of 5–10 mm.

5.2.2.4 Cutting Straight Bevel Gears by Generation

The generation principle of bevel gear cutting is based on reproducing the sides of the teeth on an imaginary crown gear in space by means of the cutting edges of rotating interlocking cutters or reciprocating two-tool generators. The profiles of the straight cutting edges coincide with the opposing sides of two teeth of the imaginary crown or generating gear that is in mesh with the gear being cut. The primary cutting motion, either rotation or reciprocation, is transmitted to these cutting edges.

5.2.2.4.1. Interlocking Cutters (Completing or Konvoid Generators)

In this method, two interlocking disk-type cutters rotate at the same speed on axes inclined to the face of the mounting cradle, and both cut in the same tooth space. The gear blank is held in a work spindle that rotates in timed relation with the cradle on which the cutters are mounted (Figure 5.20). A simplified kinematic diagram of Konvoid-type bevel gear generator is shown in Figure 5.21. A feed cam cycle begins with the work-head and blank moving into position for

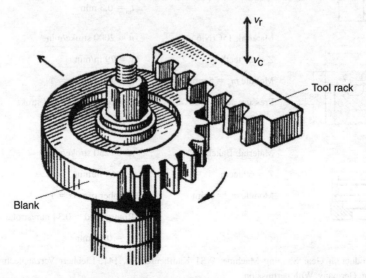

FIGURE 5.19 Principles of gear shaping using rack cutter.

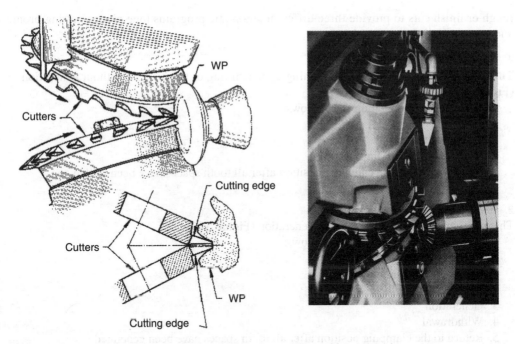

FIGURE 5.20 Bevel gear generating by interlocking cutters (Konvoid generators).

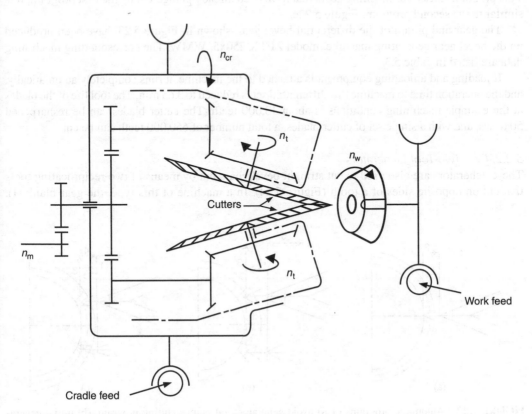

FIGURE 5.21 Simplified kinematic diagram of Konvoid generators.

rough or finish cuts to provide three different automatic programs (one plunge cutting program and two generating programs).

1. *Plunge Cutting Program*

This program is mainly used for roughing by machining of tooth space without generation (Figure 5.22a).

The sequence of operation is as follows:

1. Plunge cutting
2. Withdrawal of WP for indexing
3. Return of WP to the clamping position after all tooth spaces have been milled

2. *Initial Generation Program*

This program is used to cut spaces by generation (Figure 5.22b).

The sequence of operation is as follows:

1. Rolling return and indexing
2. Approaching the WP
3. Generation
4. Withdrawal
5. Return to the clamping position after all tooth spaces have been generated

3. *Infeed Generation Program*

This program is used for finishing gears that have been already plunge-cut by the first program. It is similar to the second program, Figure 5.22c.

The gear and pinion of the differential bevel gear, shown in Figure 5.23, have been produced on the bevel gear generating machine, model ZFTK 250x5, WMW. The corresponding machining data are listed in Table 5.3.

If loading and unloading equipment is attached to the machine, it runs completely automatically, and the operation time to machine this differential set is reduced to 3.65 min. The tool life of the blades at the example machining conditions is about 13,000 teeth. The cutter blades can be resharpened 50 times, and with a single set of cutter blades, a total number of 650,000 teeth can be cut.

5.2.2.4.2 Two-Tool Generators

These generators are also used to cut straight bevel gears but by means of two reciprocating tools that cut on opposite sides of a tooth (Figure 5.24). In a machine of this type, the gear blank (1),

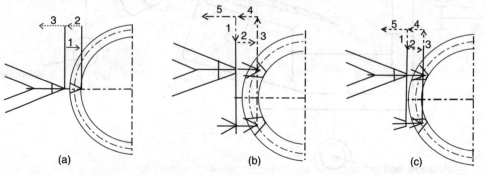

FIGURE 5.22 Automatic programs of Konvoid generators: (a) plunge cutting program, (b) initial generation program, and (c) infeed generation program. (From WMW, Bevel Gear hobbing machine ZFTX 250x5, Technical Information, 108 Berlin, Mohrenstr, 61 WMW-Export.)

FIGURE 5.23 Differential gear set, machined on bevel gear generating machine. (From WMW, Bevel Gear Hobbing Machine ZFTX 250x5, Technical Information, 108 Berlin, Mohrenstr, 61 WMW-Export. With permission.)

TABLE 5.3
Machining Data of the Differential Bevel Gear Set as Machined on the Bevel Gear Machine

Machining Data	Pinion	Gear
Material	16 MnCr5	
Cutting speed, v (m/min)	63	
Module, m_g (mm)	3.75	
Number of teeth Z	9	22
Machining time t_m (min)	1.5	3.45

Source: WMW, Bevel Gear Hobbing Machine ZFTX 250x5, Technical Information, 108 Berlin, Mohrenstr. 61 WMW-Export.

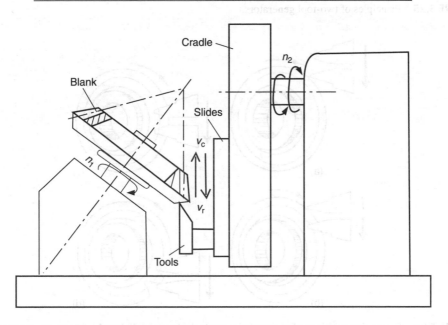

FIGURE 5.24 Operation of two-tool generators for the production of straight bevel gears.

is rotated at n_1; also the cradle (2) is rotated at n_2 with the reciprocating tools that represent kinematically the adjacent sides of a tooth in an imaginary crown gear. The slides (3) with tools (4) reciprocate at a speed v_c along ways arranged on the face of the cradle (2) (Figures 5.24 and 5.25). The tools cut in their motion toward the gear apex. They do not cut on their return stroke because they are withdrawn from the blank to avoid rubbing against the machined surfaces.

Figure 5.26 shows the successive positions of reciprocating tools and the gear blank during the generation process. For machining one of the side surfaces of the tooth, the tool starts to cut into the blank (position a). Then the second tool allocated to shape the other side of the tooth begins to cut (position b). At position c, both tools are in full engagement. Upon further rotation (roll) of the cradle, the tools run out of mesh with the gear blank (position d). At this stage the first tooth has been generated. Then the blank is automatically withdrawn from the engagement with the tools at the conclusion of each generating roll. The cradle work-spindle rolls back to the starting position, where the blank is indexed to the next tooth. This procedure is repeated until all teeth are finished.

The two tools are not subjected to the same load, as one of them cuts into the blank for each tooth and wears faster than the other tool. To eliminate the effect of nonuniform wear on the

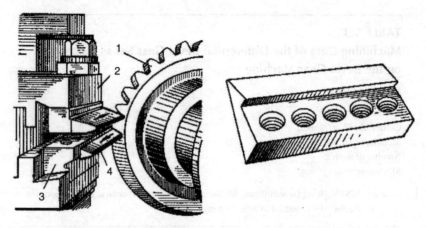

FIGURE 5.25 Principles of two-tool generators.

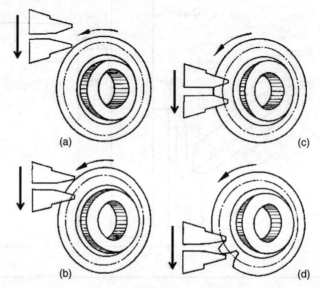

FIGURE 5.26 Successive positions of reciprocating tools during straight bevel gear generation.

profile accuracy, provision is made to make a finish cut after roughing, with most of the stock being removed in the roughing operation. The tooling cost of the two-tool generators is low, but production rates are lower than those of interlocking cutter generators, discussed previously.

Two-tool generators are usually used when

- The bevel gears are beyond a practical size range (larger than 250 mm pitch diameter).
- Gears have integral hubs or flanges that project above the root line, thus preventing the use of other generators.
- Small production quantity or variety of gear sizes cannot be accommodated by other types of straight bevel gear generators.

5.3 SELECTION OF GEAR CUTTING METHOD

Each gear cutting method discussed so far has a field of application to which it is best adapted. These fields overlap, however, so many gears can be produced satisfactorily by more than one method. In such cases, the equipment availability often determines which machining method will be used. The type of gears to be cut is usually the main factor in the selection. However, one or more of the following factors should be considered in the final choice of the method:

- Size of the gear and its module
- Configuration of the WP to be machined
- Batch size
- Gear ratio
- Accuracy
- Cost related to the tool and the machine
- Cycle time and productivity

Figure 5.27 summarizes the possibilities to produce a certain gear type by cutting. The outcome of this layout leads to the conclusions displayed in Table 5.4.

5.4 GEAR FINISHING OPERATIONS

Gear finishing operations are distinguished from gear cutting operations in that they are used for improving the accuracy, uniformity, and surface quality of the various gear tooth elements. The functional requirements of gears determine the degree of accuracy. Higher accuracy is necessary if the gears are required to operate quietly and at high speeds and to transmit heavy loads. Gear finishing methods include burnishing, shaving, lapping, and grinding. Unhardened teeth of gears are finished by shaving or burnishing, whereas hardened teeth are finished by grinding or lapping operations. Shaving is the main gear finishing process before hardening, whereas grinding is the main finishing process for hardened gears. A comparison between both processes with regard to machining time and machining allowance is presented in Figure 5.28 and Table 5.5.

The effect of the gear module (m_g) on both the machining time and the machining allowance is clear. Both increase with increasing gear module. Figure 5.28 depicts how time needed for grinding is about three times that needed for shaving. For the same module m_g, the machining time increases with the number of teeth for both shaving and grinding processes.

5.4.1 FINISHING GEARS PRIOR TO HARDENING

5.4.1.1 Gear Shaving

Gear shaving is a finishing process based on consecutively removing thin layers of chips (2–10 μm thick) from the profiles of the teeth by a tool called a gear shaving cutter. Shaving is currently the most widely used method of finishing spur and helical gear teeth following the gear cutting operation and

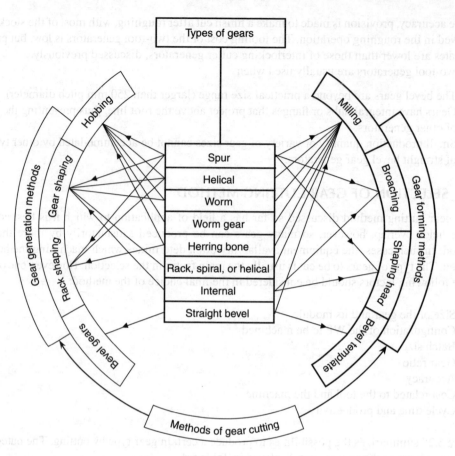

FIGURE 5.27 Selection of gear cutting method.

TABLE 5.4
Gear Cutting Methods and Their Capabilities to Produce Different Types of Gearing

	Gear Cutting Method						
	Forming				Generation		
	Milling	Broaching	Shaping Head	Bevel Template	Hobbing	Gear Shaping	Rack Shaping	Bevel Generators
Types of Gears	Spur	Spur external			Spur	Spur	Spur	
	Helical				Helical	Helical		
	Worm							
	Worm wheel			Straight bevel	Worm	Worm	Helical	Straight bevel
	Herringbone				Worm wheel	Rack	Herringbone	
	Rack	Spur internal				Internal		
	Straight bevel				Herringbone			

prior to hardening the gear. It is not intended to salvage gears that have been carelessly cut, although it can correct small errors in areas such as tooth spacing, helix angle, tooth profile, and concentricity. Shaving reduces noise level and tooth-end load concentration, and increases load-carrying capacity, surface quality, and accuracy.

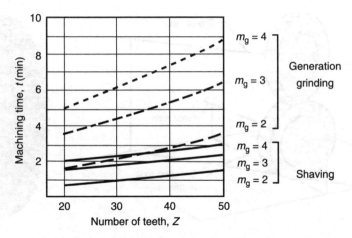

FIGURE 5.28 Machining time for shaving and grinding. (From WMW, Gear Cutting Practice, Technical Information, Special Edition 12, 108 Berlin, Mohrenstr 61 WMW-Export.)

TABLE 5.5
Machining Allowances for Gear Shaving and Grinding

Gear Module (mm)	1	2	3	4	5
Shaving (μm)	20	20	25	30	35
Grinding (μm)	50	60	80	100	120

Source: Düniβ, W., Neumann, M., and Schwartz, H., *Trennen Spanen and Abtragen*, VEB-Verlag Technik, Berlin, 1979.

1. *Principle of Operation*

Shaving is performed with a cutter and gear at crossed axes; the value of the crossed axes angle controls the finish produced to some extent. The smaller the angle, the finer the finish (Figure 5.29a). Angles ranging from 8° to 15° are generally ideal.

In the shaving process, helical cutters of a helix angle 10–15° are generally used for spur gears, and vice versa. In some cases, helical gears are shaved by helical cutters. The action between gears and cutter is therefore a combination of rolling and sliding. Vertical serration (0.6–1 mm deep) in the cutter teeth (Figure 5.29b) takes thin hair-like chips from the profile of the gear teeth. Actually, one member of the pair is driven and that makes the other to rotate. At the same time, a reciprocating axial feed movement is provided by the worktable. This movement ranges from 0.1 to 0.3 mm/rev of the work gear. After each stroke, the direction of cutter and work rotations is reversed to finish both sides of the teeth (Figure 5.30). Figure 5.31 shows the setup for machining spur and helical gears, respectively. The gear allowance increases from 10 to 130 μm for a corresponding increase of the gear module from 0.5 to 12.5 mm.

During shaving, the tip of the cutter must not contact the root fillet; otherwise, uncontrolled, inaccurate involute profiles will result. The serration depth governs the total cutter life in terms of the number of sharpening permitted. A shaving cutter is sharpened by regrinding the teeth profiles, thus reducing tooth thickness and consequently the general accuracy of shaved gears. Because the facilities necessary to produce high accuracy after resharpening of shaving cutters are not available in most gear manufacturing plants, cutters are ordinarily returned to the tool manufacturer for resharpening.

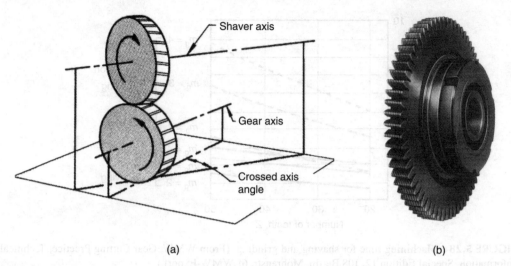

FIGURE 5.29 Gear shaving: (a) crossing of work gear with shaving cutter and (b) serrated shaving cutter.

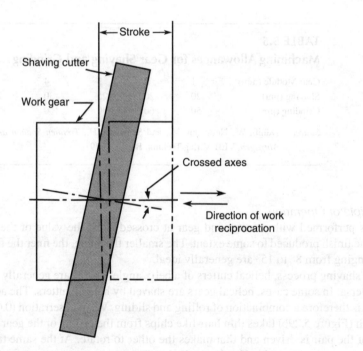

FIGURE 5.30 Axial traverse shaving.

A rack-type shaving cutter can be used instead of the gear shaving cutter. In rack shaving, the rack is reciprocated under the gear to be shaved, and infeed takes place at the end of each stroke. Because racks longer than 500 mm are impractical, 150 mm is the maximum diameter of gear that can be shaved by the rack method.

Regardless of the previously mentioned limitation of rack cutters, shaving has been successfully used in the finishing of spur and helical gears of a very wide spectrum of sizes and modules ranging from 6 to 5000 mm pitch diameter, and modules ranging from 0.15 to 12.5 mm. This process is ideal for finishing automotive and machine tool gearboxes after hobbing and before hardening.

Gear Cutting Machines and Operations

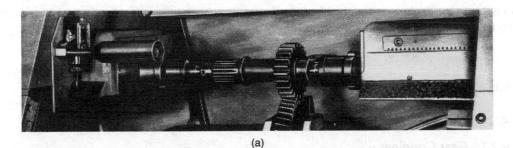

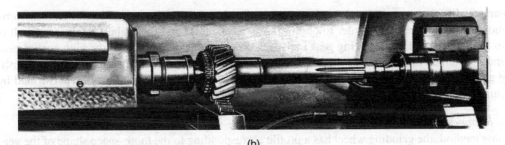

FIGURE 5.31 Gear shaving: (a) spur gear and (b) helical gear.

Comparison between Rotary and Rack Shaving

A rotary cutter is much less expensive than a rack-type cutter and its grinding cost is lower. Additional features include the following:

- Rotary cutters operate on simpler machines of smaller size.
- On rotary machines, internal gears can be shaved.
- A rotary cutter has a comparatively short tool life, and broken teeth cannot be replaced.
- Rotary cutters cannot remove excessive stock that impairs the final quality of shaved gears.

5.4.1.2 Gear Burnishing

Gear burnishing is another method of surface finishing for teeth of a gear, employed prior to heat treatment. It consists essentially of rolling the work gear with burnishing gears whose teeth are very hard, smooth, and accurate. The inaccuracies and asperities of the surface of the work gear are leveled by the kneading action of the material. Burnishing is of no use to gears that are to be subsequently heat-treated, as it may set up stresses that are released during heat treatment, hence leading to increased distortion, surface cracks, and peeling off the carburized and deformed surface layer.

Principle of Operation

Three burnishing gears (spur or helical, depending on the type of burnished gear) are meshed with and spaced at 120° positions around the work gear. One of the burnishing gears is the driver and the other two are idlers, which exert burnishing pressure against the work gear. The burnishing cycle starts by rotating the gears in one direction for the necessary period of time, then reversing the direction of rotation for an equal period of time (Figure 5.32). During burnishing, a lubricant is supplied to produce the desired surface quality and to prevent abrasion.

Burnishing gears are used until worn beyond the usable accuracy and are then reground to restore the original accuracy. It is possible to regrind burnishing gears several times before discarding. It is advisable to use the largest permissible gears, in order to obtain longest usable tool life. As seen in Figure 5.32, the maximum limit of burnishing gears addendum diameter, D_a, is given by

$$D_a = (2\sqrt{3} + 3)d_a$$

where d_a is the addendum diameter of work gear.

5.4.2 Finishing Gears After Hardening

5.4.2.1 Gear Grinding

Gear grinding is a specially adapted process to finish gears that have considerable stock to be removed after hardening in order to obtain the most accurate and the highest quality gears. It is also frequently used in producing gear tools. The low rate of production and the high cost of gear grinding exclude the use of this method for mass production. As a rule, it is used only for finishing gears of precise machinery. Similar to gear cutting, gear grinding also may be performed by forming or generation.

1. Formed Wheel Grinding

In this method, the grinding wheel has a profile corresponding to the tooth-space shape of the gear being ground and simultaneously machines the flanks of the two adjacent teeth.

The contour of the grinding wheel is profiled by a diamond dressing fixture (Figure 5.33a). The side diamonds are actuated by form templates and dress the tool profile on the grinding wheel. A variation of tooth forms can be produced by changing the contour of the templates by the dressing mechanism. Form grinding is performed by the wheel (1) that travels parallel to the axis of the work gear (2). After each full stroke of the wheel, the gear, mounted on an arbor, is automatically indexed by one or several teeth and the cycle is repeated (Figure 5.33b). Grinding is completed by three or four passes of the wheel in each tooth space. Grinding allowance from 50 to 120 μm on each side may be removed. This gear grinding method has a larger production capacity than generation grinding, but it is less accurate due to nonuniform wear of the wheel dressed to the tooth profile.

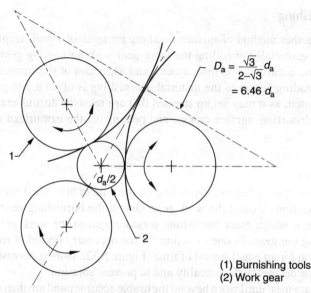

(1) Burnishing tools
(2) Work gear

FIGURE 5.32 Burnishing operation and maximum limit of addendum diameters of burnishing tools.

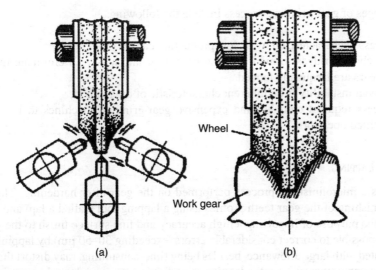

FIGURE 5.33 Gear grinding by forming: (a) diamond wheel dressing and (b) form grinding.

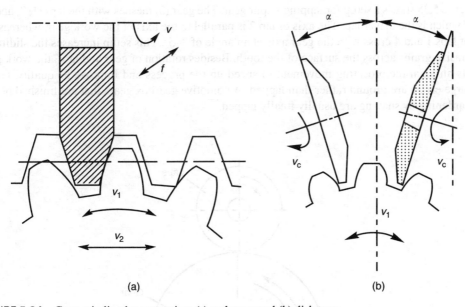

FIGURE 5.34 Gear grinding by generation: (a) rack type and (b) dish type.

2. *Generation Gear Grinding*

This method is based on reproducing the mesh of the gear being ground with a rack whose tooth is represented by a form grinding wheel or a pair of dish wheels. In Figure 5.34a, rotation (principal movement v) and reciprocating feed (movement in the direction of arrow f) are imparted to the grinding wheel. The gear is rotated about its axis at speed v_1 and is moved straight at speed v_2. These two movements (v_1 and v_2) are interrelated and form a complex generating roll movement. At this time, one tooth flank is ground. After reversal of the generating roll movement, the opposite flank in the same tooth space is machined. Upon the completion of the first tooth space, the grinding wheel is withdrawn, and the gear is indexed one tooth. Figure 5.34b illustrates grinding with two dish grinding wheels.

Disadvantages of gear grinding process include the following:

- The process is characterized by its low production capacity.
- The scratches or ridges formed increase both wear and noise. To eliminate this defect, ground gears are frequently lapped.
- Dimensional instability is an inherent characteristic of the method.
- The process requires complex and expensive gear-grinding machines to be tended by highly skilled operators.

5.4.3 Gear Lapping

Gear lapping is a microfinishing process performed on the gear after hardening. This method is based on the finishing of the gear teeth profiles using a lapping tool (called a lap) and fine-grained abrasive, with the purpose of imparting a high accuracy and fine surface finish to the gear teeth. It is, however, impossible to correct considerable errors (exceeding 30–50 μm) by lapping. Prolonged lapping associated with large allowance, besides being time-consuming, may distort the gear profile and impair the teeth accuracy. Usually, lapping is performed on special machines using three laps made of soft and fine-grained CI, where a lapping compound (oil and fine abrasives) is applied to the tools.

Figure 5.35 shows a setup for lapping a spur gear. The gear (3) meshes with the laps (1, 2, and 4), one of which is the driver lap. The axis of lap 2 is parallel to the axis of the work gear, whereas the axes of laps 1 and 4 cross with the gear axis at an angle of 3–5°. This setup increases the sliding of the abrasive grains across the surface of the tooth. Besides rotation of gear and laps, the work gear imparts an axial reciprocating movement to speed up the process and improve its quality. Gears with large errors are ground rather than lapped. Automotive gearbox gears that are finished before case hardening by shaving are usually finally lapped.

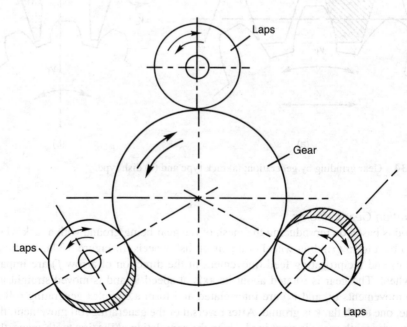

FIGURE 5.35 Gear lapping.

5.5 REVIEW QUESTIONS AND PROBLEMS

1. What are the disadvantages of milling a gear by a formed disk cutter?
2. Write down the relationships of the following in terms of module m_g and the number of teeth Z for a 20-gear tooth.
   - Pitch circle diameter d_p
   - Addendum h_a, dedendum h_d, clearance $(h_d - h_a)$
   - Working depth h_w, tooth height h_t
   - Outside diameter, d_a
   - Tooth thickness s, fillet radius r
3. What is the difference between hobbing and milling as gear cutting processes? Discuss their fields of application.
4. Discuss the inaccuracies that may result from gear cutting by hobbing.
5. What are the advantages of gear generation by shaping?
6. Mention three gear types that may be produced on the following gear cutting machines: hobbing, milling, gear shaping by rotary cutter.
7. Suggest only two types of gear cutting machines to produce the following gear types: helical gears, worms, straight bevel gears, worm wheels.
8. Make the necessary setups for milling the following helical gears on a horizontal universal milling machine:
   - 30° helix right-hand
   - 30° helix left-hand

   Show the direction of table feed and work rotation for each case.
9. Why a heat treatment process is not recommended after gear burnishing?
10. Draw a sketch to illustrate the principle of gear lapping operation.
11. Explain the main advantages and limitations when using a gear shaping head. Is it a forming or a generating gear production method?
12. What are the advantages of helical gears over spur gears?
13. What difficulty should be encountered in hobbing a herringbone gear? What modifications in design should be performed to permit them to be cut by hobbing?
14. Can a helical gear be machined on a universal milling machine?
15. Why is gear hobbing much more productive than gear shaping?
16. Under what conditions can shaving not be used for finishing gears?
17. Numerate methods of finishing gears before and after hardening.
18. An HSS-hob of pitch diameter 70 mm is used to cut a spur gear of 48 teeth. A cutting speed of 30 m/min is used, and the gear has a face width of 64 mm. The hob is fed axially at a rate of 2.1 mm/rev of the WP. What is the time required to achieve gear hobbing, provided that an approach and over-travel of 36 mm is assumed?

REFERENCES

Acherkan N. (1968) *Machine Tool Design* (four volumes). Mir Publishers, Moscow.

Düniβ, W., Neumann M., and Schwartz H. (1979) *Trennen Spanen and Abtragen*, VEB-Verlag Technik, Berlin.

High Production Gear Shaping Machine WS1 Kaufbeurer Str. 141, Liebherr Verzahntechnik GmbH. D8960 Kempten, Germany.

Metals Handbook (1989) Machining, Vol. 16, ASM International, Material Park, OH.

WMW, Bevel Gear Hobbing Machine ZFTX 250x5, Technical Information, 108 Berlin, Mohrenstr, 61 WMW-Export.

WMW, Gear Cutting Practice, Technical Information, Special Edition 12, 108 Berlin, Mohrenstr, 61 WMW-Export.

6 Turret and Capstan Lathes

6.1 INTRODUCTION

Turret and capstan lathes are the natural development of the engine lathe, where the tailstock is replaced by an indexable multistation tool head, called the capstan or the turret. This head carries a selection of standard tool holders and special attachments. A square turret is mounted on the cross slide in place of the usual compound rest in engine lathe. Sometimes a fixed tool holder is also mounted on the back end of the cross slide. Dimensional control is effected by means of longitudinal (for lengths) and traversal (for diameters) adjustable stops.

Therefore, capstan and turret lathes bridge the gap between manual engine lathes and automated lathes and are most practical for batch and short-run production. In comparison with manual lathes, the chief distinguishing feature of capstan and turret lathes is the multiple tool holders that enable the setting up of all the tools necessary to produce a certain job. Except for sharpening, the tools need no further handling.

Considerable skill is required to set and adjust the tools on such machines properly. But once the machines are set, they can be operated by semiskilled operators. Eliminating the setup time between operations reduces the production time considerably. The development of this group of lathes has been enhanced to provide the level of accuracy required for interchangeable production.

The main advantages of turret and capstan lathes include the following:

1. Less-skilled operators are needed, as compared with center lathes
2. No need to change tooling or move the work to another machine, as many operations can be performed without the need to change tooling layout

6.2 DIFFERENCE BETWEEN CAPSTAN AND TURRET LATHES

The essential components and operating principles of capstan and turret lathes are illustrated schematically in Figure 6.1. Capstan lathes are mainly used for bar work, whereas turret lathes are applicable for large work in the form of castings and forgings.

In a capstan or ram-type lathe, the hexagon turret is mounted on a slide that moves longitudinally in a *stationary saddle* (Figure 6.2a). During setup of the machine, the saddle is positioned along the bed to give the shortest possible stroke for the job. The advantage of the capstan lathe is that the operator has less mass to move, resulting in easier and faster handling. The disadvantage is that the hexagonal turret slide is fed forward such that the overhang is increased, resulting in the deflection of the ram slide, especially at the extreme of its position, which produces taper and reduces accuracy.

In the turret- or saddle-type lathe, the turret is mounted directly upon a *movable saddle*, furnished with both hand and power longitudinal feed (Figure 6.2b). This machine is designed for machining chuck work, in addition to bar work. Owing to the volume of the swarf produced, the guideways of the machine bed are flame-hardened and provided with covers that protect the sliding surfaces. The bed must be designed to allow free and rapid escape of swarf and coolant.

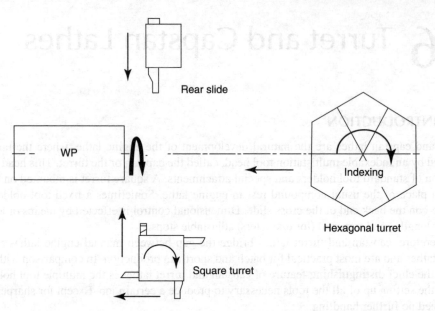

FIGURE 6.1 Essential components and operating principles of capstan and turret lathes. (Adapted from *Metals Handbook*, Machining, Vol. 16, ASM International, Materials Park, OH, 1989.)

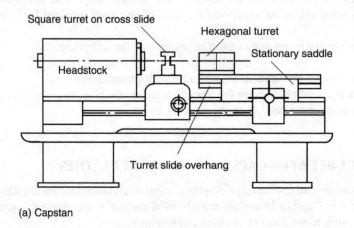

(a) Capstan

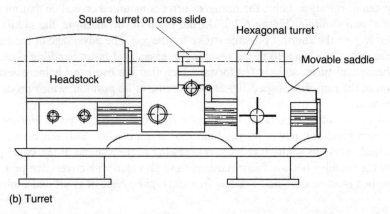

(b) Turret

FIGURE 6.2 Difference between capstan and turret lathes: (a) capstan and (b) turret. (From Browne, J.W., *The Theory of Machine Tools*, Book 1, Cassell and Co. Ltd., London, 1965.)

Advantages of the turret or saddle-type lathe include the following:

- It is more rigid and hence most suitable for heavier chucking work. Jobs up to 300 mm diameter can be machined on it.
- Its design eliminates the turret slide overhang problem inherent in the ram-type lathes.
- The power rapid traverse reduces the operator's handling effort.

Sometimes, the saddle-type machines are built with a cross turret feeding on the saddle to meet the requirement of specific jobs. The eight-sided turret, while offering two additional tooling stations, has the disadvantage of increasing the interference between turret and cross-slide tools and limits the size of the tools that can be mounted on the turret stations.

6.3 SELECTION AND APPLICATION OF CAPSTAN AND TURRET LATHES

Machine selection is based on two factors: lot size and complexity of operation. A lot size of 10–1000 pieces is usually considered suitable for capstan and turret lathe work. For lot sizes under 10 pieces, these machines can compete with engine lathes strictly on a time basis, but not on an economical basis. At the same time, it is impractical to use capstan lathes on very large lot sizes where the advantages of automatic equipment of turret-type machines can be economically utilized. A mathematical treatment should be developed for determination of unit cost in terms of the lot size, taking into consideration many factors, such as machine cost, labor cost, machine-setter cost, and also complexity of operations performed on the work.

Typically, turret and capstan lathe jobs contain multiple operations, such as turning, recessing, facing and boring, drilling, tapping, reaming, and so on. Jobs requiring simple operations should be done on simpler and less-expensive center lathes. Having decided that a turret or a capstan lathe is the best suitable machine for the work, the size of the machine must be selected.

To finalize the selection process, the following aspects are to be considered:

1. Select a machine with sufficient power and rigidity to remove the metal at the most economical rate.
2. Choose the smallest machine that has ample swing and bed length for the job to be performed.
3. Choose between a ram- and saddle-type machine. Long, accurate turning and boring operations dictate a saddle-type machine, while the ram-type is preferred for ease of handling.
4. Determine whether a power feed or a manual feed machine is required.
   - Determine whether a cross-feeding hexagonal unit makes sense for the job.
   - Consider whether spindle speeds and carriage feeds lend themselves to the job.
   - Consider whether an automatic headstock control would be worthwhile.

A word of caution should be inserted here on the use of capstan lathes equipped with extra-large-capacity spindles is that this machine is recommended only for light operations, in spite of its powerful spindle. If it is used for heavy work, excessive wear and ultimate breakdown results. Saddle-type lathes equipped with large-capacity spindles are recommended for heavy work and severe cuts.

6.4 PRINCIPAL ELEMENTS OF CAPSTAN AND TURRET LATHES

A ram or turret lathe has essentially the same elements as an engine lathe, with additional elements like hexagonal turrets and front and rear cross slides. However, the controls used are more complex. The motor is more powerful to enable the machine to perform overlapped cuts. The elements of a standard turret and capstan lathe are described in the following sections.

6.4.1 Headstock and Spindle Assembly

The headstock is heavier in construction than that of the engine lathe with a wider range of speeds. A typical layout from Heinemann Machine Tool Works-Schwarzwald is shown in Figure 6.3. Mounting of the free-running gears should be noted, in addition to the use of roller bearings with a taper bore for the spindle. The multidisk clutch drive is widely used in conjunction with constant mesh gearing. The use of these clutches provides rapid acceleration and the ability to sustain hightorque loads (Browne, 1965). In modern machines, pole-changing motors offer four speeds, which simplify the design of the gearbox and limit its size. One of the chief characteristics of the turret headstock is the provision for rapid stopping and starting, and for speed changing through speed preselectors. Through these measures, the minimum loss of time is realized.

When components are turned from bar stock fed through the hollow spindle of the machine, a collect chuck is used. The bar is generally of round or hexagonal shape. Collect chucks may be pneumatically or hand-operated. A sectional view of the hand-operated collet chuck is shown in Figure 6.4 (H. W. Ward and Co. Ltd.).

When the handle shown in Figure 6.4a is moved to the close position, the sliding sleeve (A), (Figure 6.4b) rotates and is therefore forced to move to the left, as the groove accommodating pads (B) are cut on a helix. Consequently, the sleeve forces the ball operating sleeve (C) to the left, which causes the right-hand (RH) ring of balls held in the ball cage (D) to move radially inward. This closes the sliding cone sleeve (E) and hence the collet. Moving the lever in the opposite direction reverses the action and the left-hand (LH) ring of balls (D) moves the sleeve (E) to release the collet. In the shown position, the collet is closed. The machine spindle (H) and the housing (K) are bolted to the headstock. The knurled cap (F) adjusts the collet for variations of the machined bar size. By a slight modification, the design can be altered such that the sliding sleeve can be actuated pneumatically to reduce the operator's fatigue and reduce the chucking time.

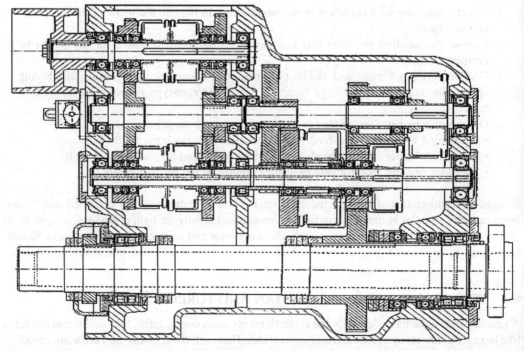

FIGURE 6.3 Typical headstock and spindle assembly of a turret lathe. (From Heinemann Machine Tool Works-Schwarzwald, Germany.)

Turret and Capstan Lathes

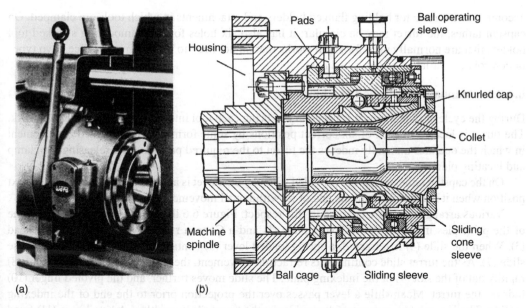

FIGURE 6.4 Hand-operated collet chuck: (a) general view and (b) sectional view.

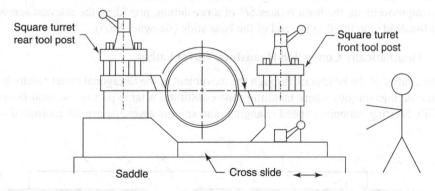

FIGURE 6.5 Cross slide and square turret tool posts.

6.4.2 Carriage/Cross-Slide Unit

The cross-slide unit on which the tools are mounted for facing, forming, recessing, knurling, and cutting off is made of four principal parts, namely, the cross slide, the square turret, the carriage, and the apron (Figure 6.5). The rear and front square turrets are mounted on the top of the cross slide. Each turret is capable of holding four tools ready for use. If additional tools are required, they are set up in sequence and can be quickly indexed and locked in correct chucking position.

The slide is provided with a positive stop to affect diametrical control of the depth of the cut. Dogs on the side of the cross slide engage these stops to regulate the cross-slide travel.

The carriage has two hand wheels for manual longitudinal and cross feed. In some machines, besides hand feed, a power feed (rapid or slow) can be engaged by a lever.

6.4.3 Hexagonal Turret

The hexagonal turret is carried on a saddle and is intended for holding and bringing the tools in a forward feed movement. On the turret-type, each face is provided with four tapped holes to accommodate screws for holding flanged holders and attachments in which tools are clamped. On capstan lathes, the turret may be circular; it has also six holes for accommodating shanked tool holders that are normally used for small works that do not need to be held in a flat face. Two types of control are available, as described next.

6.4.3.1 Manually Controlled Machines

During the cycle of operations, it is necessary to bring each tool into a position relative to the work. The turret is located in each of six correct positions by some form of hand-operated arrangement in which the operator manually indexes the turret to the required position after releasing the clamp and locating plungers.

On the capstan lathe, means are provided whereby the turret is automatically indexed to the next position when it reaches the extreme end of its withdrawal movement from the previous position.

Various arrangements are adapted in this respect. Figure 6.6 illustrates diagrammatically one of the principles involved. An indexing plate (1) and a Geneva ring (2) are secured to the head (3). When the slide (4) is retracted, a spring loaded lever (5) contacts a projection (4) on the base slide (7). As the turret slide continues its retracting movement, the lever moves the locking bar (8) rapidly out of the slot (9) of the indexing plate. The slide moves further, and the pivoted finger (10) indexes the turret. Meanwhile a lever passes over the projection prior to the end of the indexing motion. The locking bar moves rapidly and locks the turret (3) in position. A bevel gear (11) fixed in the underside of the turret meshes with a bevel pinion (12), the ratio being 5:1. The pinion shaft (13) carries a bush (14) in which six long screws (15) are filtered, one for each turret position. For one indexing movement, the bush rotates 5/4 of a revolution, providing the relevant screw, which moves to the dead stop fitted to the end of the base slide (Browne, 1965).

6.4.3.2 Automatically Controlled Headstock Turret Lathes

Automatic control of the headstock through the movement of the hexagonal turret results in considerable time savings on jobs where handling time constitutes a large part of the total floor-to-floor time (FFT). Starting, stopping, speed changing, and spindle reversing are all controlled by a unit

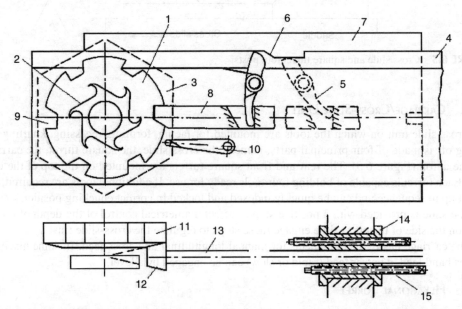

FIGURE 6.6 Turret indexing mechanism. (From Browne, J.W., *The Theory of Machine Tools*, Book 1, Cassell and Co. Ltd., London, 1965.)

actuated by the indexing of the turret head. The operator has to handle only the hexagonal turret, resulting in considerable time savings.

This control is best used on small machines where a high number of spindle changes take place in a short machining cycle. Plumbing fittings, aircraft fittings, small valve bodies, and other chucking work with short machining strokes are jobs ideally suited for the automatic controls. Bar work of the same cycle time and short strokes also shows time savings when the automatic spindle control is used instead of manually controlled machines.

The total percentage savings diminish as the machining time increases in relationship to handling time on longer-stroke jobs. For example, on a part requiring 0.5 min FFT of which 0.15 min is machining time, the handling time will be 0.35 min. If the use of automatic control would save 0.15 min/piece, the total time is therefore reduced by 30%. On a longer-stroke job requiring 4.0 min FFT in which 3.65 min is for machining and 0.35 min is elapsed for handling time, if again 0.15 min is reduced by the automatic control, the total time is reduced only by 3.7%. It can be readily seen through this comparison that a critical inspection should be made before deciding which type of control is to be used (*Tool Engineers Handbook*, 1959).

6.4.4 Cross-Sliding Hexagonal Turret

Sometimes, the hexagonal turret has a cross-sliding ability to feed in four directions. This characteristic adds greatly to the versatility of the turret lathe on certain difficult types of work. This unit is used only on the larger size turret lathe of the saddle-type construction. The mobility of this turret makes it especially adaptable to small-lot work where multiple inner surfaces can be machined using a minimum of quickly set up cutters (*Tool Engineers Handbook*, 1959).

Going beyond the small production lots, a cross-sliding turret offers other advantages on certain types of work. For example, it provides the possibility of machining large-diameter work, which prohibits the use of square turret cross slides. The graduated dial for the cross motion of the hexagonal turret enhances the accuracy and makes it the same as the square turret on the cross slides.

6.5 TURRET TOOLING SETUPS

6.5.1 Job Analysis

The jobs on turret lathe are simply a series of basic machining operations such as turning, facing, drilling, boring, reaming, threading, and so on. The setup for any job consists of arranging these machining operations in their proper sequence. The best tooling setup considers the tolerances, lot size, and the machining cost.

Once the turret lathe is properly tooled, an experienced operator is not required to operate the machine. However, skill is required in the proper selection and mounting of tools. The turret setup starts by the job analysis. The following guidelines should be considered in this analysis:

1. *Type of tooling*. In general, standard tools and holders should be used as much as possible, especially for small batches of work. Simple tool layout should be employed for small batches. Roller rest turning holders are used to support the bar work, whereas extended tooling is used for machining chucked parts such as castings and forgings.
2. *Machining operations*. Maximum productivity is achieved by a judicious combination of internal and external machining operations on both the square and the hexagonal turrets. As far as possible, cuts performed on square and hexagonal turrets should be combined in the tooling setup.
3. *Tool geometry and proper clamping*. Suitable cutting angles and edges should be ground on tools. Moreover, the tools should be set with minimum overhang and gripped firmly in their holders.
4. *Stops*. These should be set as accurately as part tolerances require. Hexagonal turret stops are usually set prior to cross-slide stops.

TABLE 6.1
Recommended Speeds and Feeds When Machining Different Materials on Capstan, Turret, and Automated Lathes Using HSS Tools

				Material			
					Structural, Case-Hardened, and Tempered Steels of σ_u (kg/mm²)		
Operation	Light Metals	Brass 70/30	Free-Cutting Steels	Up to 50	To 70	To 85	To 100
Cutting speeds[a] (m/min)							
Turning, forming, cutting off[b]	150–200	120–150	60–70	35–42	26–32	20–24	15–20
Drilling	80–120	70–120	40–50	30–35	20–26	16–20	12–15
Threading	30–50	30–60	5–9	5–7	4–6	2–4	1–3
Tapping[c]	10–20	8–16	5–8	3–7	3–5	2–3	1–2
Feeds (mm/rev)							
Turning	0.15–0.30	0.15–0.25	0.12–0.16	0.11–0.16	0.10–0.14	0.08–0.11	0.08–0.10
Forming	0.02–0.05	0.02–0.05	0.02–0.04	0.02–0.04	0.01–0.04	0.01–0.03	0.01–0.03
Cutting off	0.04–0.08	0.05–0.10	0.04–0.05	0.03–0.05	0.03–0.04	0.02–0.04	0.02–0.03
Drilling[d]	0.06–0.20	0.08–0.20	0.05–0.14	0.05–0.12	0.04–0.11	0.03–0.09	0.03–0.09
Core drilling	0.16–0.22	0.16–0.22	0.14–0.17	0.12–0.15	0.10–0.13	0.08–0.10	0.08–0.10

[a] Values are multiplied by a factor of 1.5–2.5 when carbide tools are used.
[b] Maximum values are used for turning; minimum values are applicable for forming and cutting off operations.
[c] Lower values for small taps of 0.5 mm pitch, higher values for larger taps of 1.5 mm pitch. When cutting external threads using dies, the values here are reduced by 50%.
[d] Lower values for small drill ($\varphi 2$ mm), higher values for larger drills ($\varphi 20$ mm).

Source: Reclassified and modified from Index-Werke, Esslingen, Germany Index 12-18-25 Berechnungsunterlagen.

---

5. *Speeds and feeds.* Each cut should be performed using the highest speed and feed as cutter and job permit. The speeds and feeds recommended for various operations, and work and tool materials, are available in machining handbooks. Extracted and modified speeds and feeds are given in Table 6.1. The use of maximum speeds and feeds depends on the machine power available as well as the rigidity of tooling, the holding mechanism, and the WP itself.

    Multiple or combined cuts on different diameters of the WP may call for the use of different grades of carbides or HSS to get the proper surface speeds. The cutters for the larger diameters will be carbide, while HSS will give the best results on smaller diameters with correspondingly lower machining speeds.

6. *Production time.* If the production time t_p consists of setup time t_s and FFT t_f (Figure 6.7), then

$$t_p = t_s + t_f \tag{6.1}$$

The setup time t_s is the time consumed in setting up the machine for a new job. It is evident that as the lot size decreases, the importance of setup time t_s increases. Adding cutters to take multiple cuts decreases the FFT t_f. However, it usually increases the setup time t_s and the tool cost. Simplifying the tooling setup by using fewer cutting tools reduces the setup time t_s, and tool cost, while increasing the FFT t_f. The FFT is the time that elapses between picking up a component to load for a machining operation and depositing it after machining.

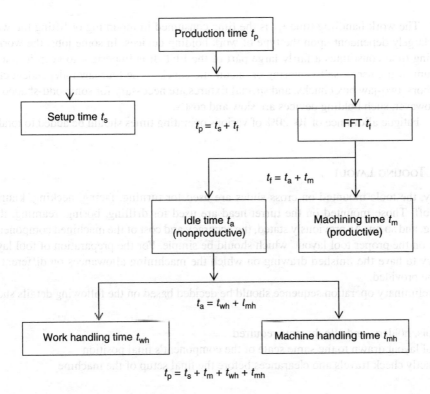

FIGURE 6.7 Production time on a turret lathe.

The FFT is a combination of the machining time t_m (productive time), and idle or auxiliary time t_a (nonproductive time); therefore,

$$t_f = t_m + t_a \tag{6.2}$$

The machining time t_m is the time consumed in the actual cutting operation and is controlled by the use of proper cutting tools, feeds, and speeds. It can be saved by performing multiple cuts or by increasing chip removal rate.

The idle time t_a is composed of the machine handling time t_{mh} and the work handling time t_{wh}; therefore,

$$t_a = t_{mh} + t_{wh} \tag{6.3}$$

The machine handling time t_{mh} is the time consumed in bringing the respective tools into the cutting positions. It can be reduced using multiple cuts, as several surfaces are machined with only one handling of the hexagonal turret unit, which is faster than a square turret.

Standard times (allowances) for machine handling time t_{mh} on turret and capstan lathes are as follows:

Feed to bar stop = 0.04 min
Hexagonal turret indexing = 0.08 min
Square turret indexing = 0.20 min
Speed changing = 0.05 min
Feed changing = 0.05 min

The work handling time t_{wh} is the time consumed in mounting or lifting the work. It is largely dependent upon the type of work holding devices. In some jobs, the work handling time constitutes a fairly large part of the FFT. It is important to keep it to a minimum by the use of self-centering or pneumatic chucks. The four-jaw independent chucks, arbors, two-jaw box chucks, and special fixtures are necessary for some odd-shaped parts. However, such holding devices are slow and costly.

Fatigue allowance of 10–20% of various operating times should be added to total FFT.

6.5.2 Tooling Layout

Basically, the tools mounted on cross slides are used for turning, facing, necking, knurling, and parting off. Those mounted on the turret head are used for drilling, boring, reaming, threading, recessing, and so on. As previously stated, the accuracy and cost of the machined component mainly depends on the proper tool layout, which should be simple. For the preparation of tool layout, it is necessary to have the finished drawing on which the machining allowances on different surfaces should be provided.

A preliminary operation sequence should be decided based on the following details such as:

- Tools, holders, and attachments required
- Tool layout drawn to the same scale of the component's final position
- Exactly check travels and clearances before the final setup of the machine

The tooling layout differs according to the nature of the WP. In this regard, checking- or bar-work are possible.

1. *Chucking work.* Figure 6.8 illustrates a standard tools setup for machining a CI ratchet wheel. In heavy cut stations (1 and 4), it is clear that the rigidity of the setup is enhanced by the use of pilot bar, fixed on the headstock and adapted in a socket in the turret head. Figure 6.9 represents the tooling setup for a cross-sliding hexagonal turret. Using such a basic setup with standard tools and tool holders, multiple cuts can be taken. The multiple turning heads (Figures 6.10 and 6.11) offer the possibility of multiple turning cuts at heavy

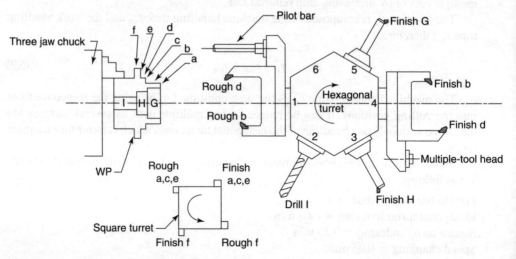

FIGURE 6.8 Tooling layout for chucking-type work.

Turret and Capstan Lathes

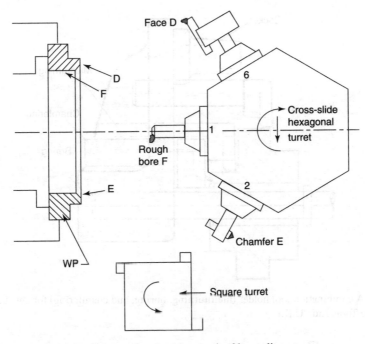

FIGURE 6.9 Cross-slide hexagon turrets for chucking work of large diameter.

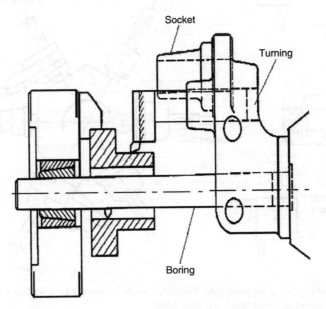

FIGURE 6.10 A knee-turning and boring attachment for chucking work. (From Herbert Machine Tools Ltd., U.K.)

metal removal (stations 1 and 4 in Figure 6.8) or for presetting tools for close tolerance finish turning cuts (*Metals Handbook*, 1989). Figure 6.10 shows a knee turning and boring attachment for chucking work; Figure 6.11 illustrates a combination tool holder for multi-turning, chamfering, and boring also for chucking-type work.

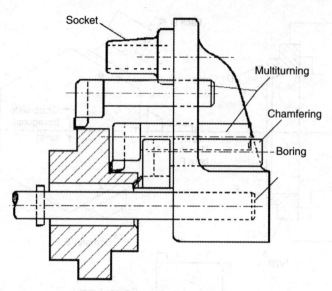

FIGURE 6.11 A combination tool holder (multiturning, boring, and chamfering) for chucking work. (From Herbert Machine Tools Ltd., U.K.)

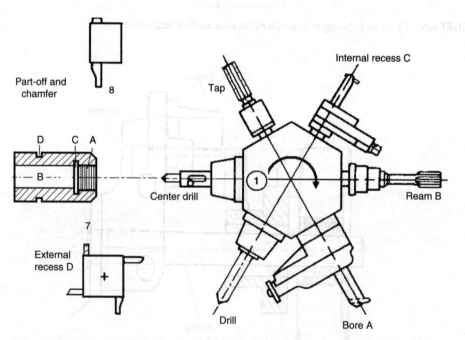

FIGURE 6.12 Tooling layout for producing a threaded adaptor. (Adapted from *Metals Handbook*, Machining, Vol. 16, ASM International, Materials Park, OH, 1989.)

2. *Bar work*. Figures 6.12 and 6.13 illustrate the tooling arrangement of a typical setup for bar-work. Such arrangement gives good production potential, realizing the minimum setup time. Figure 6.12 illustrates the multifunction capabilities of turret lathes in producing a thread adaptor shown in the same figure. Figure 6.13 illustrates the complex configuration of a steel shaft machined on a turret lathe. The single-cutter holder in position turn 6 removes metal at a maximum rate, as the rolls support the work at the point of cut. Behind the cutting tools, the support rolls burnish the work to a fine surface finish and accurate size (Figure 6.14).

Turret and Capstan Lathes 229

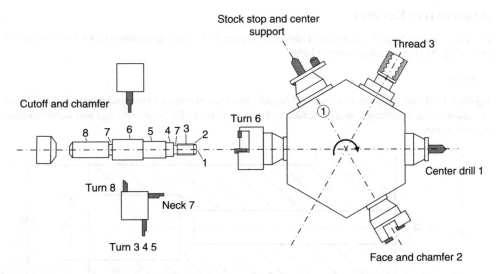

FIGURE 6.13 Tooling layout for producing a steel shaft of complex configuration. (Adapted from *Metals Handbook*, Machining, Vol. 16, ASM International, Materials Park, OH, 1989.)

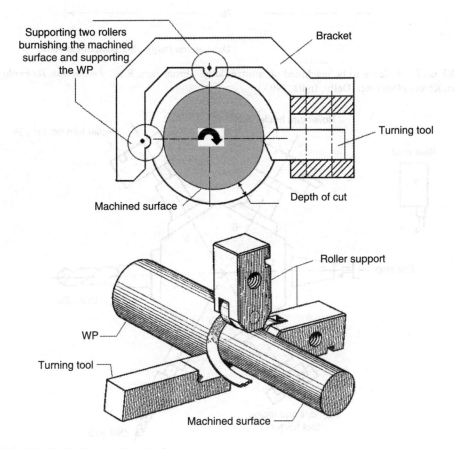

FIGURE 6.14 Roller box turning attachment.

ILLUSTRATIVE EXAMPLE

Draw a tool layout for the component shown in Figure 6.15. Also, determine the FFT necessary for producing the component on a turret lathe.

Solution

Figure 6.16 shows the tools and standard holders required to produce the component. Table 6.2 lists the sequence of operations, speeds, feeds, and operation times for productive (t_m) and nonproductive (t_i) movements.

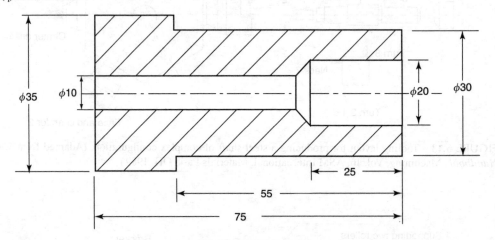

Dimensions (mm), Material: Brass 70/30

FIGURE 6.15 A sleeve to be machined on a turret lathe. (From Jain, R.K., *Production Technology*, 13th Edition, Khanna Publisher, Delhi, India, 1993.)

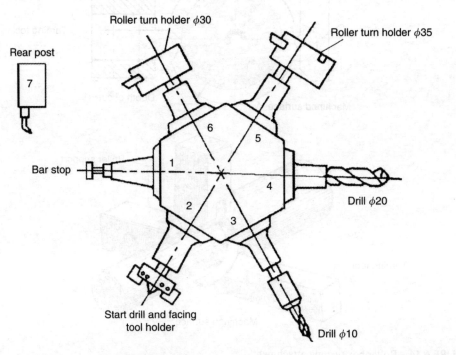

FIGURE 6.16 Tooling layout to produce the part in Figure 6.15. (From Jain, R.K., *Production Technology*, 13th Edition, Khanna Publisher, Delhi, 1993.)

TABLE 6.2
Turret Work Sheet for the WP Illustrated in Figure 6.15

Work material: Brass 70/30

Bar size: ⌀40 mm

Tooling: Turning; carbide K group; drilling, HSS

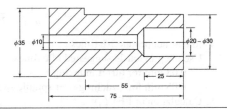

Operation	Tooling Station	Sequence of Operation	Spindle Speed (m/min)	Feeds (mm/rev)	t_f (min) t_m	t_i
1	—	Index turret to position 1	—	—	—	0.08
2	Turret 1	Feed to bar stop	—	—	—	0.08
3	—	Index turret to position 2	—	—	—	0.08
4	Turret 2	Center drill and face	1000	—	0.15 estimated	—
5	—	Select $f = 0.2$ mm/rev	—	—	—	0.05
6	—	Index turret to position 3	—	—	—	0.08
7	—	Select $n = 2000$ rpm	—	—	—	0.05
8	Turret 3	Drill $\phi 10 \times 75$ mm	2000	0.2	0.19	—
9	—	Select $n = 1000$ rpm	—	—	—	0.05
10	—	Index turret to position 4	—	—	—	0.08
11	Turret 4	Drill $\phi 20 \times 25$ mm	1000	0.2	0.13	—
12	—	Index turret to position 5	—	—	—	0.08
13	Turret 5	Turn $\phi 35 \times 77$ mm	1000	0.2	0.39	—
14	—	Index turret to position 6	—	—	—	0.08
15	Turret 4	Turn $\phi 30 \times 55$ mm	1000	0.2	0.28	—
16	—	Change to $f = 0.1$ mm/rev	—	—	—	0.05
17	Rear 7	Parting off past center (19 mm)	1000	0.1	0.19	—

Determination of FFT (t_f)   1.33   0.76
$t_f = t_m + t_i = 1.33 + 0.76 = 2.09$ min $t_f = 1.33 + 0.76$
Considering a fatigue allowance of 20%, then $t_f = 2.5$ min

1. Selection of cutting speeds and feeds:
   Material: brass 70/30
   Turning tools: all carbides, K-type
   Twist drills: HSS
   Referring to Table 6.1, the following speeds and feeds are depicted:

   Turning $v = 150–300$ m/min (select 150 m/min)
              $f = 0.15–0.25$ mm/rev (select 0.2 mm/rev and 0.1 mm/rev for parting off)

   Drilling $v = 70–120$ mm/min (select 70 m/min)
              $f = 0.08–0.2$ mm/rev, depending on diameter (select 0.2 mm/rev)

2. Determination of spindle speeds:

   Turning $n = \dfrac{1000v}{\pi D} = \dfrac{1000 \times 150}{\pi \times 40} = 1190$ rpm (select 1000 rpm)

   Drilling $n = \dfrac{1000v}{\pi D} = \dfrac{1000 \times 70}{\pi \times 20} = 1114$ rpm (select 1000 rpm)

   $= \dfrac{1000 \times 70}{\pi \times 10} = 2228$ rpm (select 2000 rpm)

   Then the spindle operates at two speeds, 1000 and 2000 rpm.

3. Sample calculation of the productive time t_m:
   Referring to Figure 6.16 and considering turret station 3, ⌀10 mm × 75 mm, the spindle speed 2000 rpm and the turret feed 0.2 mm/rev, then

   $$t_m = \frac{1}{n \cdot f} = \frac{75}{2000 \times 0.2} = 0.19 \text{ min}$$

4. Idle or nonproductive times t_a:
   Use the previously suggested idle times,
   0.08 min for turret indexing
   0.05 min for switching over spindle speeds and feeds

5. Calculation of FFT (t_f):
   All elementary times are added as shown in Table 6.2 and a fatigue allowance of 20% is considered:

   $$t_f = 1.2(t_m + t_a)$$

6.6 REVIEW QUESTIONS

1. Mark true or false.
   [ ] In turret lathes, the turret is mounted on a saddle.
   [ ] Capstan lathes are characterized by higher accuracy compared to turret lathes.
   [ ] Capstan lathes are ideal for heavy chucking work; therefore, they are equipped with powerful spindles.
   [ ] Heavier cuts can be taken by automatic turrets rather than the capstan machines.
2. What is the difference between a turret lathe and a capstan lathe?
3. What is the difference between ram-type and saddle-type turret lathes? What are their advantages and disadvantages?
4. Why is the saddle-type lathe suited to repetitive manufacture of complex cylindrical parts?
5. Draw a sketch to show indexing and locking of a turret of a capstan lathe.
6. For what purpose is the cross-sliding turret used?
7. Define FFT.

REFERENCES

Browne, J. W. (1965) *The Theory of Machine Tools*, Book 1, Cassell and Co. Ltd., London.
Heinemann Machine Tool Works-Schwarzwald, Germany.
Herbert Machine Tools Ltd., UK.
Index-Werke, Esslingen, Germany Index 12-18-25 Berechnungsunterlagen.
Jain R. K. (1993) *Production Technology*, 13th Edition, Khanna Publisher, Delhi, India.
Metals Handbook (1989) Machining, Vol. 16, ASM International, Materials Park, OH.
Tool Engineers Handbook (1959) ASTME, McGraw-Hill, New York.
H. W. Ward and Co. Ltd.

7 Automated Lathes

7.1 INTRODUCTION

Automated machine tools have played an important role in increasing production rates and enhancing product quality. Since they had been introduced in industry, they have contributed to mass production of spare parts and machine components. Especially, automated lathes are high-speed machines, and therefore, the application of safety rules in their operation is obligatory for all attendants and servicing personnel. Moreover, this type of machine tools and their setup can be entrusted only to persons with comprehensive knowledge of their design and principles of operation. If all servicing instructions are strictly observed, the operation of automated lathes presents no hazard at all to the operator.

Fully automatic lathes are those machines on which WP handling and the cutting activities are performed automatically. Once the machine is set up, all movements related to the machining cycle, loading of blanks, and unloading of the machined parts are performed without the operator. In semiautomatic machines, the loading of blanks and unloading of the machined components are accomplished by an operator.

From early times, machine tools—especially lathes—have been fitted with devices to reduce manual labor. A considerable range of mechanical, hydraulic, and electrical devices, or a combination of these have contributed to the development of automatic operation and control.

Lathes in which automation is achieved by mechanical means are productive and reliable in operation. Much time, however, is lost in switching over from one job to another. Therefore, automatics are used in mass production, while semiautomatics are used in lot and large-lot production. Machine tools and lathes that are automated by other than mechanical means (NC and CNC, using numerical data to control its operating cycle) can be set up for new jobs much more rapidly, and are therefore efficiently employed in lot, batch, and even single-piece production (Chapter 8). Automatic and semiautomatic lathes are designed to produce parts of complex shapes by machining the blank (or bar stock). They are designed to perform the following machines operations:

- Turning
- Centering
- Chamfering
- Tapering and form turning
- Drilling, reaming, spot facing, and counter boring
- Threading
- Boring
- Recessing
- Knurling
- Cutting off

Special attachments provide additional operations, such as slotting, milling, and cross drilling.

7.2 DEGREE OF AUTOMATION AND PRODUCTION CAPACITY

Generally, metal cutting operations are classified into one of the following categories:

1. Processing or main operations in which actual cutting—that is, chip removal—takes place.
2. Handling or auxiliary operations, which include loading and clamping of the work, releasing and unloading of the work, changing or indexing the tool, checking the size of the work, changing speeds and feeds, and switching on and off the machine tool.

The operator of a nonautomated machine performs the handling or auxiliary operations. With automated machines, some or all of the auxiliary operations are performed by corresponding mechanisms of the machine. The faster the auxiliary operations are performed in the machine, the more WPs can be produced in the same period of time; that is, a higher production or automation rate is realized (Figure 7.1). The selection of a suitable degree of automation should be based on the feasibility of machining parts at the specified quality and desired rate of production. It should be emphasized that an increase in the number of spindles and the degree of automation leads to an increase in the time required for setting up the machine for a new lot of WPs, which calls for an increased lot size.

Each type of automatic lathe has an optimum range of lot size in which the cost per piece is minimal. Figure 7.2 shows that the physical and psychological efforts exerted by the operator decrease with increasing degree of automation and consequently, the lot size also increases. Greater psychological effort is required with a lesser degree of automation, whereas the physical effort predominates in higher range of automation. A higher degree of automation realizes the following advantages:

- Increases the production capacity of the machine.
- Insures stable quality of WPs.
- Veressitates less number of machines in the workshop thus achieving higher output per unit shop floor area.
- Reduces the physical effort required from the operator and releases him from tediously repeated movements and from monotonous nervous and physical stresses.
- Avoids direct participation of the operator and therefore enables him to operate several automatic machines at the same time.

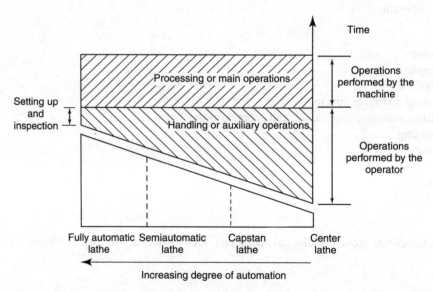

FIGURE 7.1 Degree of automation as affected by auxiliary operations.

Automated Lathes

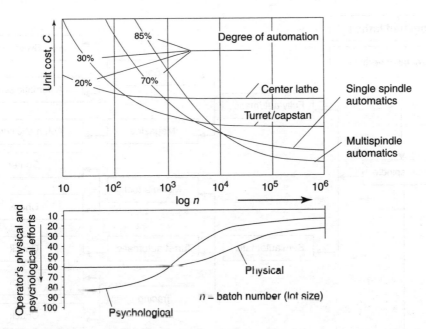

FIGURE 7.2 Unit cost, as well as physical and psychological efforts against batch number, for different degrees of automation. (From Pittler Machinenfabrik AG, Langen bei Frankfurt/M, Germany.)

To increase the production capacity of automated lathes, it is necessary to reduce the time required by the operating cycle of the machine through the following measures:

1. Concentrating cutting tools at each position or station. The concentration factor, q, represents the ratio of the total number of tools/number of stations. As a rule, automated lathes are multiple-tool machines
2. Overlapping working travel motions
3. Providing independent spindle speeds and feeds at each position or station
4. Machining several WPs in parallel
5. Employing throwaway tipped carbide cutting tools

7.3 CLASSIFICATION OF AUTOMATED LATHES

The principal types of general-purpose automated lathes are visualized in Figure 7.3. They may be classified according to the following features:

1. *Spindle location (horizontal or vertical)*. Vertical machines are heavier, more rigid, more powerful, and occupy less floor space. They are especially designed for machining large-diameter work of comparatively short length.
2. *Degree of automation (fully or semiautomatic)*. As mentioned earlier, the decision to choose between fully and semiautomatic depends mainly on the lot size.
3. *Number of spindles (single- or multispindle)*.
    A. *Single-spindle automated lathe are classified as*
        - Fully automatic (Swiss-type and turret-screw automatics)
        - Semiautomatic

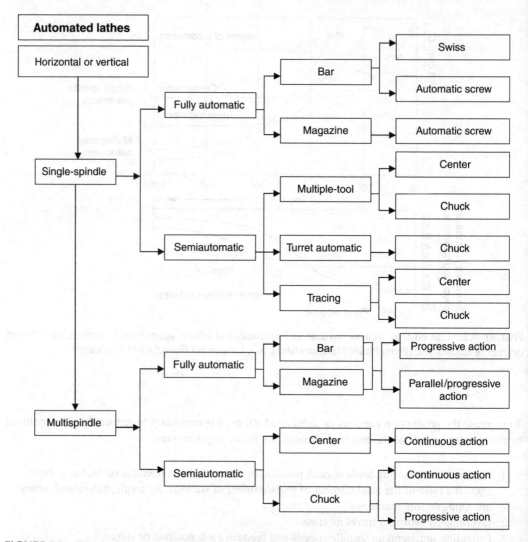

FIGURE 7.3 Principal types of general-purpose automated lathes.

   B. *Multispindle automated lathes.* These machines have 2–8 horizontal or vertical spindles. Their production capacity is higher than that of single-spindle machines, but their machining accuracy is somewhat lower. They are further classified as follows:
   – *Fully automatic.* These machines are suitable for both bar and magazine work. They are widely used for mass production and need a lot of setup work. Large multispindle automatics are equipped with an auxiliary small power motor, which serves to drive the camshaft when the machine is being set up. The rate of production of multispindle automatic is less than that of a corresponding sized single-spindle automatic. The production capacity of a four-spindle automatic, for example, is only 2.5–3 times (not 4 times) as large as that of a single-spindle automatic, assuming the same product size, shape, and material.
   – *Semiautomatic.* Semiautomatic multispindle machines are mostly of vertical type.
4. *Nature of workpiece stock (bar or magazine).* Automated lathes use either coiled wire stock (upto 6 mm in diameter), bar, pipe, or separate blanks. Bar stock is available in great variety of shapes and sizes; however, it is considered poor practice to use bar stock over 50 mm

in diameter, as the waste metal in the form of chips will be excessive. Separate blanks are frequently used in semiautomatics. The blanks should approach the shape and size of the finished product; otherwise, the cycle time increases, thus increasing the production cost.

Automated lathes are broadly classified according to the stock nature into the following main categories:

A. *Automatic bar machine.* These are used for machining WPs from bar or pipe stock.
B. *Magazine loaded machine.* These are used to machine WPs in the form of blanks, which have been properly machined to appropriate dimensions, prior to feeding them into the machine.

The introduction of any form of automatic feeding results in higher degree of automation and economy, and makes it possible for one operator to observe a number of machines instead of being confined to one.

5. *WP size and geometry.* The size and geometry of the WP determine the suitable machine to be used. In this regard, long accurate parts of small diameters are produced on the Swiss-type automatics, whereas parts of complex external and internal surfaces are machined using turret-type automatic screw machines.
6. *Machining accuracy.* Generally, bar automatics are employed for machining high-quality fastenings (screws, nuts) bushings, shafts, rings, etc. The design configuration of the Swiss-type automatics makes them superior with respect to the production accuracy, especially when producing long slender parts

The machining accuracy of multispindle automatics is generally lower than single spindle automatics due to the errors in indexing of spindles and large number of spindle-head fittings.

7.4 SEMIAUTOMATIC LATHES

7.4.1 SINGLE-SPINDLE SEMIAUTOMATICS

The three main types of single-spindle semiautomatic lathes are as follows:

- Multiple-tool semiautomatic lathes
- Turret semiautomatic lathes
- Hydraulic tracer–controlled semiautomatic lathes

All of these are equipped either by centers or chucks. WPs several times longer than their diameters are normally machined between centers, while short WPs with large diameters should be chucked.

1. *Multiple-tool semiautomatic lathes.* These machines operate on a semiautomatic cycle. The operator only sets up the work, starts the lathe, and removes the finished work. This feature allows one operator to handle several machine tools simultaneously (multiple machine tool handling). Figure 7.4 shows the multiple-tool machining of a stepped shaft, mounted between centers, using several tools mounted on the main and cross slides (cross and longitudinal feeds are designated by arrows). Tailstock centers are most often ball- or roller bearing–type to withstand heavy static and dynamic forces. These machines have found extensive applications in large-lot and mass production.
2. *Turret semiautomatic lathes.* Turret semiautomatics, commonly referred to as single-spindle chucking machines, are used basically for the same type of work carried out by the turret lathe. They generally require hand loading and unloading and complete the machining cycle automatically. These machines are used when production requirements are too high for hand turret lathes and too low for multispindle automatics to produce economically. The setup time is much lower than that of multispindle automatics.

It is important to realize that during the automatic machining operation, the operator is free to operate another machine or to inspect the finished part without loss of time.

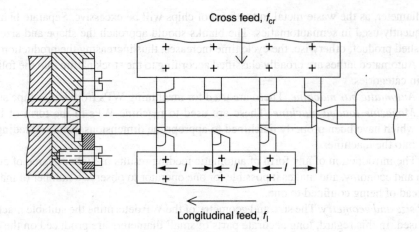

FIGURE 7.4 Multiple-tool machining of a stepped shaft.

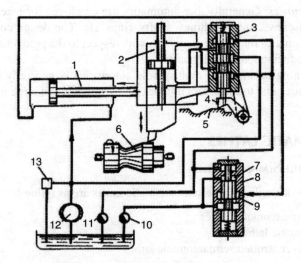

FIGURE 7.5 Hydraulic circuit diagram of a tracer control system.

The turrets normally consist of four or six tooling stations. Cross-slide tooling stations are also available in the front and the rear slides.

The machine has a control unit that automatically selects speeds, feeds, length of cuts, and machine functions such as dwell, cycle stop, index, reverse, cross-slide actuation, and many others.

3. *Hydraulic tracer–controlled semiautomatic lathes.* These are intended to turn complex shaped and stepped shafts between centers by copying from a template or master WPs. Figure 7.5 represents the longitudinal and cross-feed movements to produce the part (6). The casing (3) of the tracer valve is rigidly attached to the tracer slide, the valve spool being pressed by a spring to template (5) through a tracer stylus (4). Additionally, the feed pump (12) delivers oil into the right chamber of the cylinder (1). If a part of the template profile is parallel to the axis of the machine, the cylinder surface of the WP (6) is turned. Oil is exhausted from the left chamber of the cylinder (1) into the tank through the groove of automatic regulator (8) and a throttle (10). As the stylus (4) moves downward or upward to the template profile, the speed and direction of the tracer control slide change. When the valve spool move downward, the pump (12) delivers oil in the bottom chamber of the

cylinder (2), the tracer-controlled slide also moves downward, the oil from the upper chamber of the cylinder (2) is exhausted into the tank through the throttle (11). The valve (13) is a safety element for the system.

7.4.2 Multispindle Semiautomatics

Multispindle semiautomatics may be of continuous or progressive action (Figure 7.6).

1. *Machines of continuous action.* This type is designed for holding the work either between centers or in a chuck. Its operation is shown diagrammatically in Figure 7.7.
   - The outer column (1), connected to the spindle (2), rotates continuously and slowly.
   - The work is clamped in six chucks (Figure 7.7a) or between centers in six spindles (Figure 7.7b), and the longitudinal and cross-tool slides (4) are located on the outer column (1).
   - The same machining operation is performed at each tooling station, except at the loading zone. Each slide is set up with the same tooling.
   - In a definite zone, the work spindles cease to rotate, the finished work is removed from the chuck (3), and a new block (or bar) is loaded.
2. *Machines of progressive action.* This type is designed for chucking operations only. They are available with either six or eight spindles. Its operation is illustrated in Figure 7.8. Referring to Figure 7.8a:
   - The carrier (1) is periodically indexed through 60°. Each spindle rotates at its own setup speed, independent of other spindles.
   - A hexagonal column (5) carries only five tool slides (3 and 4). The WPs are clamped in chucks (2).
   - Work is loaded periodically in the loading station after the carrier (1) indexes through 60°, while the finished work is removed from the loading station.
   - At the other five stations, the WPs are machined simultaneously; at each consecutive station, the work is machined by the tools set up at that station.

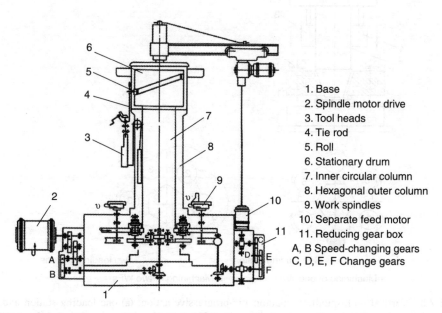

FIGURE 7.6 Gearing diagram of a vertical multispindle semiautomatic. (From Acherkan, N., *Machine Tool Design*, Mir Publishers, Moscow, 1969. With permission.)

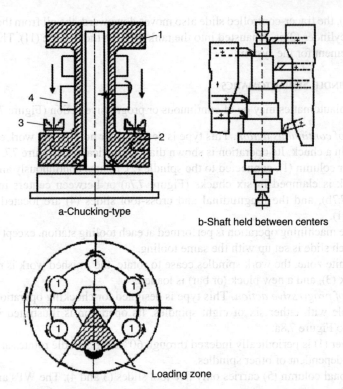

FIGURE 7.7 Vertical multispindle semiautomatic machines of continuous action. (From Maslov, D. et al., *Engineering Manufacturing Processes*, Mir Publishers, Moscow, 1970. With permission.)

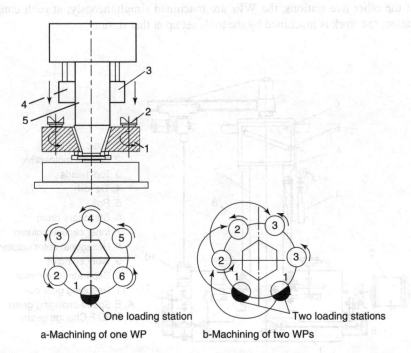

FIGURE 7.8 Vertical multispindle semiautomatic-progressive action: (a) one loading station and (b) two loading stations. (From Maslov, D. et al., *Engineering Manufacturing Processes*, Mir Publishers, Moscow, 1970. With permission.)

- The spindle speed is automatically changed to the setup value for each station and each particular spindle stops rotating when reaching the loading station.
- Referring to Figure 7.8b, the machine may be adjusted to perform the following machining duties:
  - Turning two different WPs in one operation cycle. This duty is applicable only if two tooling stations are sufficient to machine each WP on a six-spindle machine.
  - Turning both sides of the work consecutively.

For the last two machining duties, there should be two loading stations and the spindle carrier should be indexed through 120° each time.

7.5 FULLY AUTOMATIC LATHES

These are mainly based on mechanical control systems and are characterized by a rigid linkage between the working and auxiliary operations. The two following mechanical systems are frequently employed:

1. A control system composed of a single camshaft, which provides all the working and auxiliary motions. This is the simplest arrangement and is further classified into two types:
   - *Systems in which the camshaft speed is set up and remains constant during the complete cycle*: These systems may be applied only for automatics with a short machining cycle (up to 20 s), such as Swiss-type automatics.
   - *Systems in which the camshaft speed is set up for the working movements*: Auxiliary movements are performed at a high constant speed of the camshaft that is independent of the setup. The mass of the camshaft with the cam drums and disks is comparatively large and has large inertia torques, which lead to impact loads at the moment of the camshaft switching over from high to low speeds or vice versa. This system is most extensively used for multispindle automatics and semiautomatics.
2. A control system composed of a main and an auxiliary camshaft. The machining cycle is completed in one revolution of the main camshaft. Trip dogs on the main camshaft perform the auxiliary operations through signals that link the operative devices for auxiliary movements to the auxiliary camshaft. This system is more complex than the preceding one due to the very large number of transmission elements and levers. It is used in automatic screw machines and vertical multispindle semiautomatic chucking machines.

7.5.1 SINGLE-SPINDLE AUTOMATIC

The range of work produced by these machines extends from pieces so small that thousands can be put in a household thimble to complex parts weighing several kilograms. Two distinct basic machining techniques involved are turret automatic screw machines and Swiss-type automatics.

7.5.1.1 Turret Automatic Screw Machine

This type is regarded as the final stage in the development of the capstan and turret lathes. Its main objective is to eliminate (as much as is possible) the operator's interference by extensive use of levers and cams. Although this machine was originally designed for producing screws, currently it is used extensively for producing other complex external and internal surfaces on WPs by using several parallel working tools. Figure 7.9 shows typical parts produced on turret automatic screw

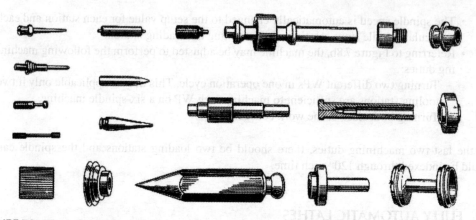

FIGURE 7.9 Typical parts produced on turret automatic screw machine. (From Acherkan, N., *Machine Tool Design*, Mir Publishers, Moscow, 1969. With permission.)

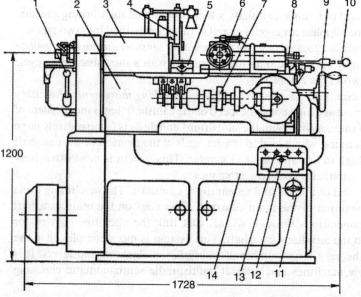

1. Lever to engage auxiliary shaft
2. Bed
3. Headstock
4. Tool slide (vertical)
5. Turret-tool slide (horizontal)
6. Turret slide
7. Main cam shaft
8. Adjustable rod for positioning turret slide with respect to spindle nose
9. Hand wheel to rotate auxiliary shaft
10. Lever to traverse turret slide
11. Rotary switches
12. Console panel for setting up spindle speeds
13. Push button controls of spindle drive
14. Base

FIGURE 7.10 General view of the automatic screw machine. (From Acherkan, N., *Machine Tool Design*, Mir Publishers, Moscow, 1969. With permission.)

machines. A general view of a classical automatic screw machine is shown in Figure 7.10, along with its basic elements.

The main specifications of turret automatic screw machine include the following:

a. Bar capacity
b. Maximum diameter of thread to be cut (in steel or in brass)
c. Maximum travel of turret
d. Maximum radial travel of cross slides
e. Maximum and minimum production times (maximum and minimum cycle times)
f. Range of spindle speeds (left and right)
g. Main motor power

Automated Lathes

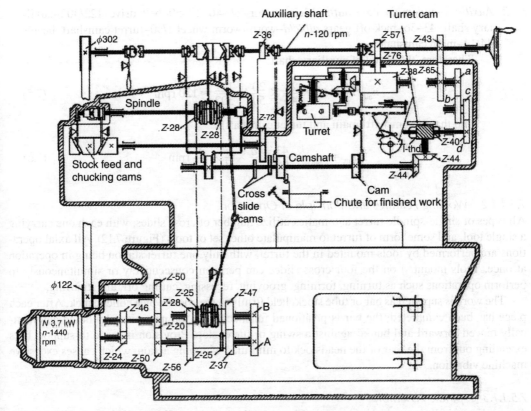

FIGURE 7.11 Gearing diagram of automatic screw machine. (From Boguslavsky, B. L., *Automatic and Semi-automatic Lathes*, Mir Publishers, Moscow, 1970. With permission.)

h. Auxiliary motor power
i. Overall dimensions

An important feature of the turret automatic screw machine is the auxiliary shaft system, which will be described together with the main characteristics of this automatic.

7.5.1.1.1 Kinematic Diagram

A simplified kinematic diagram of a typical automatic screw machine is given in Figure 7.11. The main motor of 3.7 kW and 1440 rpm imparts the required motion to the following components:

1. *Main spindle.* For one pair (seven pairs existing) of pick-off gears, the motion is transmitted through the following gear train (Figure 7.11): Main motor–24–46–20–50–back gear 56/37–pick-off gears A/B–sprockets 25/28 and 25/28. Accordingly, four spindle speeds are obtained (two forward n and two reverse n_r).

$$n = 1440 \times \frac{24}{46} \times \frac{20}{50} \times \frac{A}{B} \times \frac{28}{56} \times \frac{25}{28} \text{ (slow forward)} \quad (7.1)$$

$$n_r = 1440 \times \frac{24}{46} \times \frac{20}{50} \times \frac{A}{B} \times \frac{25}{50} \text{ (slow reverse)} \quad (7.2)$$

$$n = 1440 \times \frac{24}{46} \times \frac{20}{50} \times \frac{A}{B} \times \frac{47}{37} \times \frac{25}{28} \text{ (fast forward)} \quad (7.3)$$

$$n_r = 1440 \times \frac{24}{46} \times \frac{20}{50} \times \frac{A}{B} \times \frac{47}{37} \times \frac{56}{28} \times \frac{25}{28} \text{ (fast reverse)} \quad (7.4)$$

2. *Auxiliary and main camshafts.* Main motor–24–46–20–50–belt drive 122/302–auxiliary shaft–43–56–pick-off gears *a/b/c/d*–worm/worm wheel 1/40–turret camshaft–bevels 44/44–main camshaft.

Therefore, the speed of auxiliary shaft:

$$n_{aux} = 1440 \times \frac{24}{46} \times \frac{20}{50} \times \frac{122}{302} \approx 120 \text{ rpm} \quad (7.5)$$

and that of camshaft (main and turret camshaft):

$$n_{cam} = n_{aux} \times \frac{43}{65} \times \frac{a}{b} \times \frac{c}{d} \times \frac{1}{40} \text{ rpm} \quad (7.6)$$

7.5.1.1.2 Working Features and Principle of Operation

All types of single-spindle turret automatics utilize number of cross slides, with each one carrying a single tool, and some form of turret to manipulate other set of tools (Figure 7.12). All axial operations are performed by tools mounted in the turret, with only one turret station being in operation at once. Tools mounted on the four cross slides can perform consecutively or simultaneously to perform operations such as turning, forming, grooving, recessing, cutting off, and knurling.

The work is supplied as bar or tube stock, held firmly in the spindle by a collet chuck. After each piece has been completed, the bar is positioned for machining the next piece by being automatically moved forward and butted against a swing or turret stop. Provision is made to support bars extending out from the rear of the headstock to minimize whipping action, which causes excessive machine vibration.

7.5.1.1.3 Spindle Assembly

A complete spindle assembly is shown in Figure 7.13b. A spring collet chuck arranged in spindles is commonly used in automatic bar lathes. Three widely used types are illustrated in Figure 7.13b through 7.13d:

a. Push-out type (Figure 7.13b) in which the collet (a_1) is pushed by the collet tube (g) to the right into the tapered seat of the spindle nose (b) for clamping.
b. Pull-in type (Figure 7.13c) in which the collet (a_1) is pulled by the collet tube (g) to the left into the tapered spindle nose (b) for clamping.
c. Immovable or dead-length type (Figure 7.13d) in which the shoulder of the chuck (b_1) bears against a nut (b), screwed on the spindle nose (Figure 7.13a). Hence, the axial movement is not exerted when the clamping sleeve (c) is actuated. The spring shown in Figure 7.13a shifts back tubes (c and g) to the left, while releasing the bar. Figure 7.13a shows also the

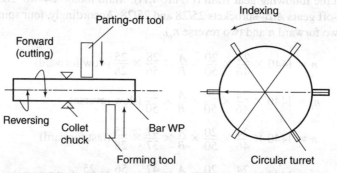

FIGURE 7.12 Essential components and operating principles of single-spindle automatic screw machine.

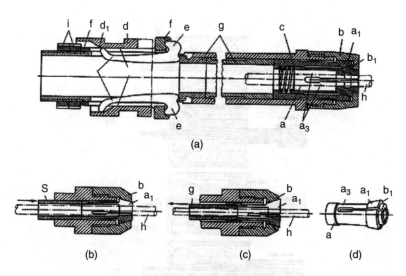

FIGURE 7.13 Spindle assembly and types of spring collets of automatic screw machines.

clamping levers (e) and clamping sleeve (d) actuated by the chucking cam drum. Nuts (i) are used for the fine adjustment of the ring (f) and the supporting levers (e) to adapt the collet to limited changes of the bar stock diameter.

The push-out and pull-in types have the disadvantage that during clamping, the bar (h) has an axial movement that affects the accuracy of axial bar positioning. Moreover, in the push-out type, the collet tube may be subjected to buckling if the tube is long and high clamping forces are exerted.

Spring collets locate the bar stock with high accuracy. Bars up to 12 mm in diameter will turn true on a length of 30 or 35 mm to within 20 or 30 µm, whereas bar stocks of 40 mm in diameter will run true within 50 µm on a length of 100 mm.

In the chucking arrangement, shown in Figure 7.14, the bar is clamped by the dead-length spring collet (2), on which the sleeve (3) is pushed by the collet tube, which closes the collet. The feeding tube (4) is arranged inside the collet tube and carries a spring feeding finger (5). The feeding finger (also called the bar stock pusher) is screwed in the frontal end of the feeding tube. The finger is a slitted spring bushing in which jaws are closed before hardening and tempering (Figure 7.15). The pressure of the slitted feeding finger is sufficient to move the bar through the open collet and slide over the bar when the collet closes. The bar stock feeding and chucking occurs in the following order shown in Figure 7.14.

The stock feeding is actuated by the stock feeding cam, located under the spindle shown in Figure 7.11, which actuates the bar feeding mechanism. The amount of the stock feed movement can be adjusted by a screw (1 in Figure 7.16) through setting the sliding block (4) up or down in the slot of the rocker arm (3).

7.5.1.1.4 Control System

Figure 7.17 visualizes how the auxiliary shaft and the camshaft control the operation of turret automatic screw machine. The auxiliary shaft rotates at a relatively high speed of 100–200 rpm. It helps mainly in bridging the idle (auxiliary) movements of the machining cycles by reducing their actuation times.

In contrast, the cutting movements of the machining cycle are controlled by the camshaft, the speed of which is exactly equal to the production rate; that is, one WP produced per one revolution of the cam shaft. The turret head has two types of movements: cutting and indexing movements. The cutting movement is actuated by the multicurve disk cam mounted on the camshaft, whereas the indexing movement is performed by the auxiliary shaft.

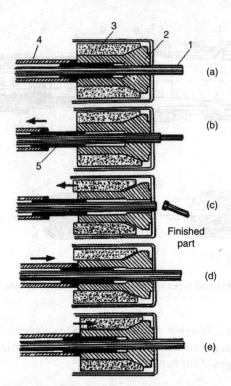

FIGURE 7.14 Operation of the bar feeding and chucking mechanisms using a dead-length chuck: (a) bar feed, stock clamping, (b) retraction of the feeding finger which slides over the clamped WP, (c) work cutting-off and releasing the collet, (d) feed of bar stock by feeding finger to stop of the first turret station, and (e) bar stock clamping. (From Boguslavsky, B. L., *Automatic and Semi-automatic Lathes*, Mir Publishers, Moscow, 1970. With permission.)

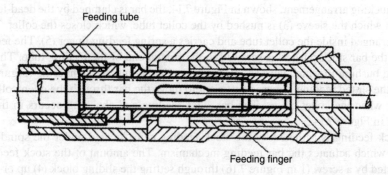

FIGURE 7.15 Feeding finger (stock pusher) in a pull-in collet chuck. (From Pittler Machinenfabrik AG, Langen bei Frankfurt/M, Germany.)

It should be emphasized that the control system of the automatic screw machine is based on the following:

1. All cutting movements performed by tools mounted on the cross slides and the turret head are controlled by the camshaft.
2. The idle or auxiliary movements are rapidly performed by an auxiliary shaft rotating at higher speeds.

The exact functions of both auxiliary shaft and camshaft are presented in Figure 7.17.

Automated Lathes

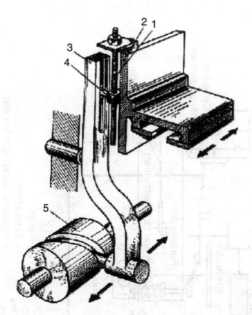

FIGURE 7.16 Mechanism of adjusting the travel of the feeding finger.

Auxiliary Shaft

An assembly of the auxiliary shaft is shown in Figure 7.18. The shaft carries three dog clutches (single-revolution clutches), which are operated through a lever system by trip dogs on drum cams that are mounted on the camshaft in the correct angular position (Figures 7.19 and 7.20). The dog clutches can be made to operate when required during the cycle of operations.

These three dog clutches, as arranged from left to right (Figure 7.18), when operated cause the following to occur:

1. Opening of the collect chuck, the bar feeding to the bar stop in turret, and closing the collet again for gripping the bar
2. Changing over of the spindle speed from fast to low or vice versa
3. Indexing of the turret head

Figure 7.21 shows a detailed drawing of the dog clutch (single-revolution). This type of clutch is not recommended for a rotational speed that exceeds 200 rpm. It operates in the following manner:

- The gear (1) is rotated only one revolution by the auxiliary shaft (2) and is then automatically disengaged.
- Long jaws are provided at the end of the gear (1) and clutch members (B), so that they do not disengage when a member is shifted to engage the jaws at the right end.
- In the disengaged condition, the clutch member is retained by the lock pin (D).
- If the lock pin is withdrawn from its slot, the spring forces the clutch member to the right, and engages the disk (3). The gear starts rotating through the clutch members.
- The pin slides along the external surface of a clutch member until it drops into the recess of a beveled surface, which forces the clutch member, upon further rotation, toward the left to disengage the gear from the rotating disk after one complete revolution.
- The lock pins (D and A) are backed by their springs.

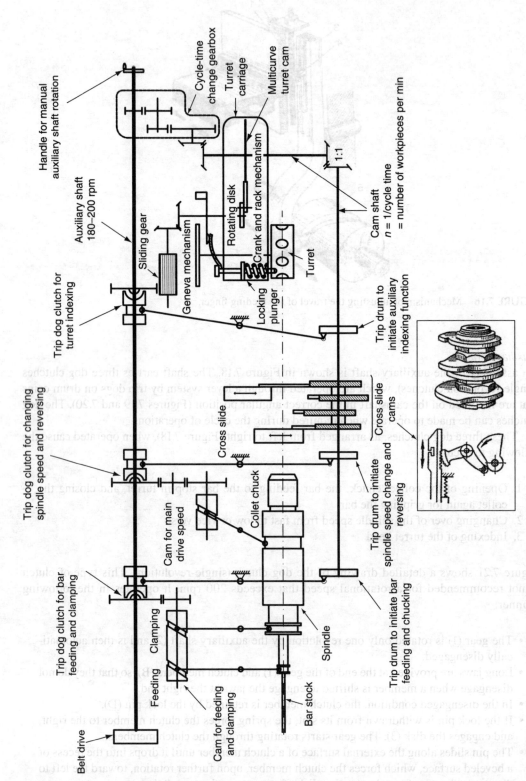

FIGURE 7.17 Control of automatic screw machine.

Automated Lathes

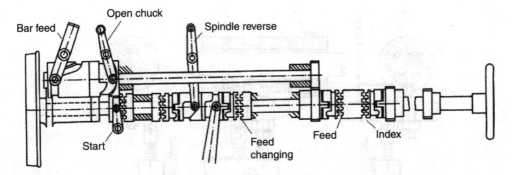

FIGURE 7.18 The auxiliary shaft assembly.

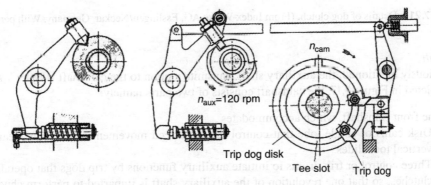

FIGURE 7.19 Trip levers controlling chucking and speed change over. (From Index-Werke AG, Esslingen/Neckar, Germany. With permission.)

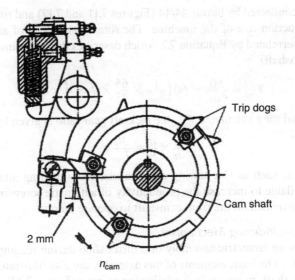

FIGURE 7.20 Turret indexing trips. (From Index-Werke AG, Esslingen/Neckar, Germany. With permission.)

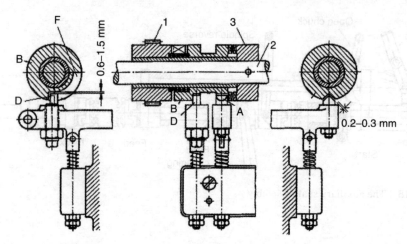

FIGURE 7.21 Details of dog clutch. (From Index-Werke AG, Esslingen/Neckar, Germany. With permission.)

Camshaft

As previously mentioned, the auxiliary shaft transmits motion to the camshaft (Figure 7.11). As is also depicted in Figure 7.17, the camshaft consists of two parts namely:

1. The front camshaft, which accommodates
   - Disk cams of single lobe that control the cross feed movement of the front, rear, and vertical tool slides.
   - Three control or trip drums to initiate auxiliary functions by trip dogs that operate dog clutches, so that one revolution of the auxiliary shaft is imparted to perform chucking, speed changing or reversing, and indexing in the required times.
2. A cross camshaft, which carries a multilobe cam to control the main movement of the turret head.

Both camshafts are connected by bevels 44/44 (Figures 7.11 and 7.17) and rotate at rotational speed that equals the production rate of the machine. The rotational speeds of auxiliary shaft n_{aux} and camshaft n_{cam} are interrelated by Equation 7.7, which describes the cycle time T_{cyc} in seconds (time/revolution of the camshaft):

$$T_{cyc} = \frac{60}{n_{cam}} = 60\left(\frac{1}{n_{aux}} \times \frac{65}{43} \times \frac{a}{b} \times \frac{d}{c} \times 40\right) \text{s} \qquad (7.7)$$

If $n_{aux} = 120$ rpm, and for a calculated T_{cyc}, the pick-off gear ratio is given by

$$\frac{a}{b} \times \frac{d}{c} = \frac{30}{T_{cyc}} \text{s} \qquad (7.8)$$

Many special devices, such as slot sawing attachments, cross drilling attachments, and milling attachments, are available to increase the productivity of automatic screw machine. These attachments require cams provided on the front camshaft to operate them.

Turret Slide and Turret Indexing Mechanism

Figure 7.22 illustrates an isometric assembly of a turret slide during feeding travel as actuated by the multicurve cam (6). The basic elements of this assembly are also illustrated. The indexing cycle, actuated by auxiliary shaft, proceeds in the following manner (Figure 7.23):

a. *Beginning of indexing.* The roll of segment gear runs down along the drop curve of the turret cam. Under the action of the spring, the slide travels to the right. The single-revolution clutch on the auxiliary shaft is engaged. The crank begins to rotate.

Automated Lathes

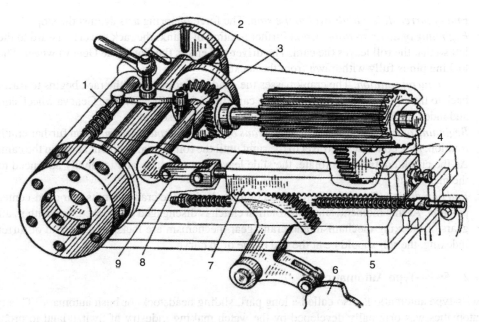

FIGURE 7.22 Isometric assembly of a turret slide of automatic screw machine. (From Acherkan, N., *Machine Tool Design*, Mir Publishers, Moscow, 1969. With permission.)

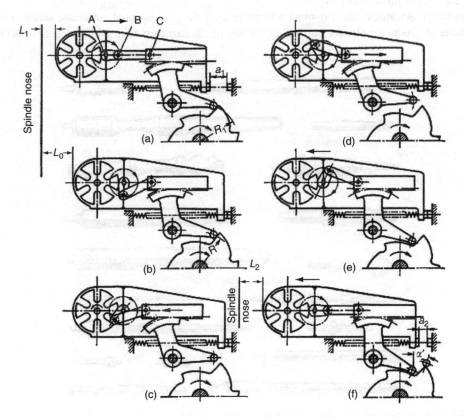

FIGURE 7.23 Steps of turret indexing in an automatic screw machine. (From Acherkan, N., *Machine Tool Design*, Mir Publishers, Moscow, 1969. With permission.)

b. *End of turret slide withdrawal to the stop.* The force of spring acts against the stop.
c. *Beginning of turret rotation.* Upon further rotation of crank, the rack travels forward to the left so that the roll leaves the cam. The driver roll enters the slot of the Geneva wheel. The locking pin is fully withdrawn from the socket.
d. *End of turret rotation.* The crank passes the dead-center position, the rack begins to travel back to the right, the roll approaches the cam, the driver roll leaves the Geneva wheel slot, and indexing is completed. The locking pin reenters the socket.
e. *Beginning of turret slide approach by means of crank-gear mechanism.* Upon further crank rotation, rack continues to travel to the right until the roll of gear segment lands on the cam. As the crank continues to rotate, the slide leaves the end stop, and is rapidly advanced to the machining zone.
f. *End of indexing.* The slide approach is completed when the crank is in its rear dead-center position. At this moment, single-revolution clutch is disengaged and locked, and with it, all gears of indexing mechanism and crank gear mechanism are locked. At the end of turret indexing, the roll should be at the end of the drop curve.

7.5.1.2 Swiss-Type Automatic

A Swiss-type automatic is also called a long part, sliding headstock, or bush automatic. This type of automatics was originally developed by the watch-making industry of Switzerland to produce small parts of watches. It is now extensively used for the manufacture of long and slender precise and complex parts, as shown in Figure 7.24.

7.5.1.2.1 Operation Features
A Swiss-type automatic has a distinct advantage over the conventional automatic screw in that it is capable of producing slender parts of extremely small diameters with a high degree of accuracy,

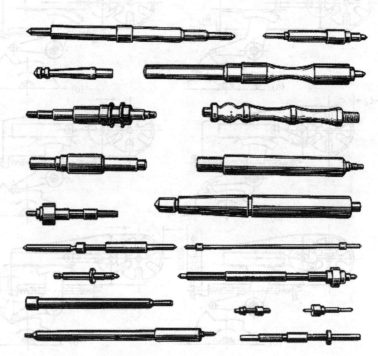

FIGURE 7.24 Typical parts produced by Swiss-type automatics.

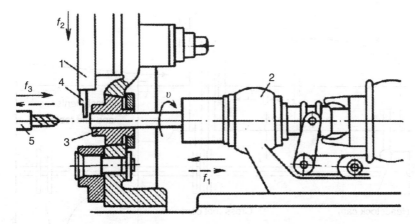

FIGURE 7.25 Operation of a Swiss-type automatic. (From Acherkan, N., *Machine Tool Design*, Mir Publishers, Moscow, 1969. With permission.)

concentricity, and surface finish. This is possible due to its different machining technique, which is based on the following exclusive features (Figure 7.25):

- The machining is performed by stationary or cross-fed single point tools (at f_2) in conjunction with longitudinal working feed f_1 of the bar stock.
- Longitudinal feed is obtained by the movement of the headstock or of a quill carrying the rotating work spindle.
- The end of the bar stock, projecting from the chuck, passes through a guide bushing, directly beyond which the cross-feeding tool (4) slides are arranged.
- Turning takes place directly at the guide bushing supporting the bar stock. The bushing then relieves the turned portion from tool load, which is almost entirely absorbed by the guide bushing. It is possible to turn a diameter as small as 60 μm.
- A wide variety of formed WP surfaces are obtained by coordinated, alternating, or simultaneous travel of headstock and the cross slides f_2.
- Holes and threads are machined by a multispindle end attachment, carrying stationary or rotating tools performing axial feed f_3.

The clearance between bar stock and the guide bush is controlled to practically eliminate all radial movements. Best results are obtained by using centerless-ground bar stock as round as possible, and of uniform diameter throughout the bar length. High machining accuracy is an important feature of the Swiss-type automatic. A tolerance of ±10 μm may be attained for diameters and ±20 to ±30 μm for lengths. When a wide tolerance is permitted and when the parts are not too long, the automatic screw machine is preferred for its higher productivity compared to the Swiss-type automatic. This is due to the reduced idle time of the automatic screw, whose control is based on the auxiliary shaft system.

7.5.1.2.2 Machine Layout and Typical Transmission

The Swiss-type automatic bears slight resemblance to a center lathe. Figure 7.26 shows a general layout of this machine. The bar is fed by a sliding headstock and held in a collet chuck. The movement of the headstock is controlled by a bell or disk cam designed to suit each component. The tool slide block carries four or five radial tool slides; the radial movement of each slide is controlled by a cam (Figure 7.27). A precise stationary bush is inserted in the tool block to guide round bars (Figure 7.25).

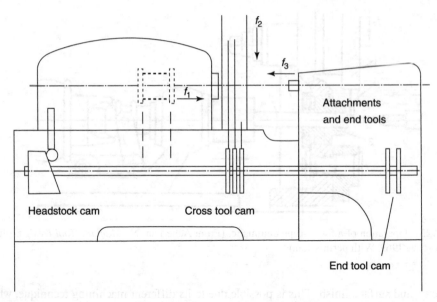

FIGURE 7.26 General layout of a Swiss-type automatic. (From Browne, J. W., *The Theory of Machine Tools*, Cassel and Co. Ltd., 1965.)

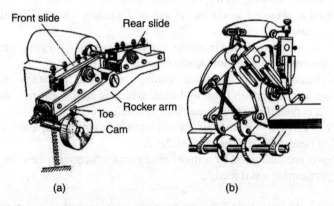

FIGURE 7.27 Radial feed of slides in Swiss-type automatics: (a) rocker arm and (b) overhead tool slides. (From Boguslavsky, B. L., *Automatic and Semi-automatic Lathes*, Mir Publishers, Moscow, 1970. With permission.)

A running bush must be used for a hexagonal bar. The bar is moved past the radially acting tools by the headstock to provide the longitudinal feed. In addition, it can move backward or pause during cutting operation as dictated by the operational layout.

A simplified line diagram of a typical transmission is shown in Figure 7.28. The machine is equipped with two motors. The main motor drives a back shaft at relatively high speed and imparts rotation to the spindle. The motion is then transmitted from the backshaft to the main camshaft through: worm and worm wheel–change gears belt drive-worm and worm wheel-main camshaft. On the main camshaft are various cams to control machine movements. The end attachment has an independent motor drive. The three spindles carrying end tools can be either stationary, all running, or a combination of both. The attachment can be shifted laterally according to the required sequence by a cam mounted on the camshaft. One revolution of the camshaft presents the cycle time and produces one component. The change gears in Figure 7.28 are selected according to the required cycle time.

Automated Lathes

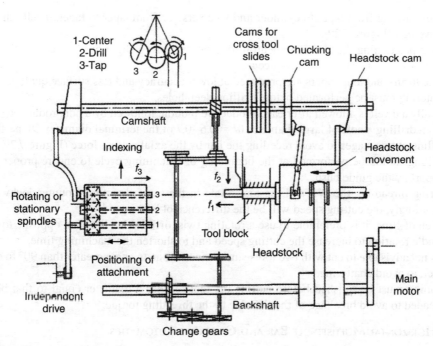

FIGURE 7.28 A simplified transmission diagram of a typical Swiss-type automatics. (From Browne, J. W., *The Theory of Machine Tools*, Cassell and Co. Ltd., 1965.)

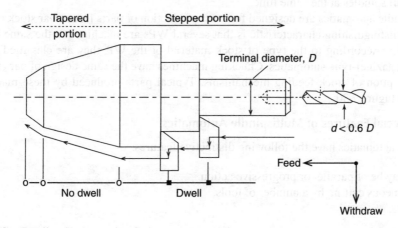

FIGURE 7.29 Dwells when operating Swiss-type automatics.

7.5.1.2.3 General Guidelines When Operating Swiss-Type Automatics

The guidelines to be followed when operating Swiss-type automatics:

1. A parting-off tool is used as a bar stop at the end of the machining cycle.
2. Recessing is accomplished by cross-feeding tools, as the WP is stationary.
3. Low feeds should be used when turning far from the guide push.
4. Using a wide parting-off tool initiates vibrations and inaccurate and rough WPs.
5. When turning stepped WPs, one tool is used, starting with the smallest diameter; then there is a stop in feeding (or a *dwell*), before the tool is moved to the next larger diameter, then it is fed for turning, and so on (Figure 7.29).

6. When turning from taper to cylinder and vice versa or from taper to taper, dwells should be avoided (Figure 7.29).
7. When positioning the tool from neutral for taper turning, a slight dwell must be allowed; otherwise, incorrect taper is produced.
8. Wide forms and long tapers are produced at lower accuracy and less surface quality.
9. Centering must be performed before drilling deep holes.
10. Usually a dwell is allowed after each productive motion, followed by a nonproductive one.
11. When drilling a hole of large diameter ($d = 0.6–0.7$ of the terminal diameter D), perform drilling at two stages to avoid receding the bar by the axial cutting force (Figure 7.29).
12. Drilling should be performed at the beginning of machining cycle to ensure proper WP support by the guide bush.
13. During threading, the WP and the tool rotate in the same direction (counterclockwise); accordingly, the cutting speed will be the difference of both speeds.
14. When drilling, it is preferable to use a rotating twist drill in the direction opposite to WP spindle rotation to increase the cutting speed and to shorten the machining time.
15. It is not advisable to cut with two tools simultaneously, at an angle greater than 90° to each other, to avoid chattering.
16. Tapping must be finished before cutting-off tool reaches a diameter equal to that being threaded to avoid breakage of the WP due to the threading torque.

7.5.2 Horizontal Multispindle Bar and Chucking Automatics

The principal advantage of the multispindle automatics over the single-spindle automatics is the reduction of cycle time. In contrast, with single-spindle automatics, where the turret face is working on one spindle at a time, in the multispindle automatics, all turret faces of the main tool slide are working on all spindles at the same time.

Multispindle automatics are designed for mass production of parts from a bar stock or separate blanks. The distinguishing characteristic is that several WPs are machined at the same time from bars or blanks. According to the type of stock material of the WP, they are classified as bar- or chucking (magazine)-type automatics. Chucking machines have the same design of bar automatics, with the exception of stock feeding mechanisms. Typical parts produced by these machines are illustrated in Figure 7.30.

7.5.2.1 Special Features of Multispindle Automatics

Multispindle automatics have the following distinctive features:

- They may be of parallel or progressive action.
- Simultaneous cutting by a number of tools.

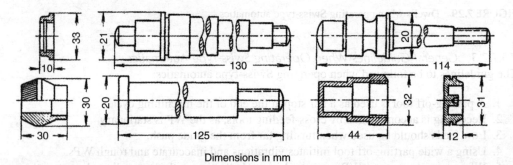

FIGURE 7.30 Typical parts produced on multispindle automatics. (From Acherkan, N., *Machine Tool Design*, Mir Publishers, Moscow, 1969. With permission.)

Automated Lathes

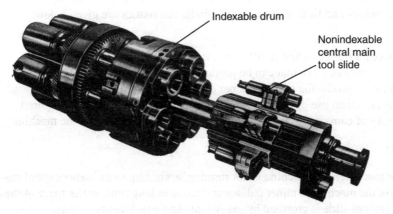

FIGURE 7.31 Nonindexable central main tool slide of a multispindle automatic. (From Pittler Machinenfabrik AG, Langen bei Frankfurt/M, Germany.)

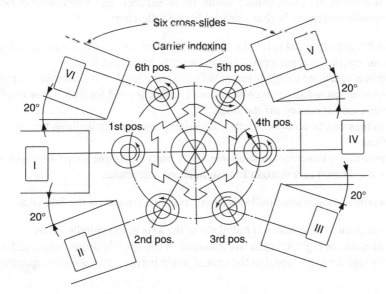

FIGURE 7.32 Cross slides of a progressive-action multispindle automatic. (From Acherkan, N., *Machine Tool Design*, Mir Publishers, Moscow, 1969. With permission.)

- A nonindexable central main tool slide has a tooling position for each spindle serving the same function of the turret of single screw automatic. It provides one or more cutting tools for each spindle and imparts axial feed to these tools (Figure 7.31).
- Progressive-action automatics are available in four, five, six, or eight spindles. Six-spindle automatics are the most common. The equispaced work spindles are carried by a rotating drum (headstock) that indexes consecutively to bring each spindle into a different working position (Figure 7.31).
- The parallel-action automatics are simpler than progressive action automatics in construction, as no indexing is required.
- Multispindle automatics (parallel or progressive) have a nonindexable cross slide at each position, so that an additional tool can be fed crosswise (Figure 7.32).
- Cams are used to control the motions of the cross slides and the main tool slide. They are either specially designed or selected from a standard range at some sacrifice of optimum output.

The advantages and limitations of multispindle automatics are given below:

Advantages:

- More tooling positions and resultant higher productivity
- Greater variety of work that can be produced
- Possibility of producing two pieces per cycle
- More economical use of floor space when continuous high output is required
- Simplicity of cams controlling the movement of different parts of the machine

Limitations:

- Higher loss when the machine is not running, according to its higher capital cost
- Set up of the machine is rather tedious and requires long time, as the space of the head stock and main tool slide is crowded by many tools and attachments

7.5.2.2 Characteristics of Parallel- and Progressive-Action Multispindle Automatic

Figure 7.33 illustrates the multispindle automatic of parallel- and progressive-action. A parallel-action multispindle automatic is characterized by the following:

1. Its spindles are arranged vertically (Figure 7.33a) and is usually a four-spindle machine.
2. The same operation is performed simultaneously in all spindles.
3. During one operating cycle, as many WPs are completed as the number of spindles.
4. Each spindle has usually two cross slides. The first is used for forming or chamfering and the other used for cutting off the stock.
5. The machine can be equipped with multiple-tool spindles for drilling, boring, or threading operations.
6. Such machine produces comparatively short parts of simple shape from bar stock. This type is also known as a straight four-spindle bar automatic.

And a progressive-action multispindle automatic is characterized by the following:

1. The arrangement of spindles is radial about the axis of the spindle drum.
2. Four, five, six, or eight spindles are mounted in the spindle drum, which indexes periodically through an angle equal to the central angle between two adjacent spindles.

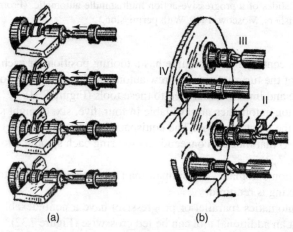

FIGURE 7.33 Multispindle bar automatics: (a) parallel-action and (b) progressive-action. (From Chernov, N., *Machine Tools*, Mir Publishers, Moscow, 1975. With permission.)

3. Only one machining stage is performed at each spindle position, and each WP passes consecutively through all positions according to the sequence of operations established in the set up (Figure 7.33b).
4. The setup is designed so that the WP is completely machined in one full revolution of the spindle drum, one part being completed at each indexing.
5. One of the positions is the loading or feeding position. In the bar-type automatic, the finished WP is cut off in this position, the bar is fed out to the stop and then clamped by the collet chuck. In the chucking type, the finished part is released by the chuck; in the loading position, a new blank is loaded into the chuck from the magazine and clamped.

A parallel- or progressive-action multispindle automatic is characterized by the following:

1. Sometimes provision may be made in the design of six- or eight-spindle machines for two loading (or feeding) positions, usually diametrically opposed in case of bar automatics, and adjacent in case of chucking automatics. Table 7.1 illustrates the switching sequence for six- and eight-spindle bar and chucking automatics.
2. Single indexing is required in case of bar automatic; double indexing is required in case of chucking automatic (Table 7.1).

TABLE 7.1
Switching Sequence for Six- and Eight-Spindle Bar and Chucking Automatics

Automatic	Progressive Bar/Chucking	Parallel/Progressive	
		Bar	Chucking
Loading and Feeding Stations	(6)	(3)–(6)	(1)–(2)
	(8)	(4)–(8)	(1)–(2)
Six-spindle	1–2–3–4–5–(6)	1–2–(3) 4–5–(6)	3–5–(1) 4–6–(2)
Eight-spindle	1–2–3–4–5–6–7–(8)	1–2–3–(4) 5–6–7–(8)	3–5–7–(1) 4–6–8–(2)

Source: Technical Data, Mehrspindel-Drehautomaten, 538-70083. GVD Pittler Maschinenfabrik AG, Langen bei Frankfurt/M, Germany.

3. Two WPs are completely machined during one full revolution of the spindle drum in a bar automatic, whereas two revolutions of the spindle drum are needed to produce two WPs in case of a chucking automatic.
4. Parallel/progressive-action is applicable for machining parts of simple shape at a high rate.

7.5.2.3 Operation Principles and Constructional Features of a Progressive Multispindle Automatic

The multispindle automatic has a rigid frame base construction, in which the top brace connects the headstock and the gearbox mounted at right side of the heavy base. The base also serves as a reservoir for cutting fluid and lubricating oil. The headstock has a central bore for the spindle drum with the work spindles.

The gearing diagram of the spindles of a horizontal four-spindle automatic is shown in Figure 7.34. The power is transmitted from an electric motor (7 kW, 1470 rpm) through a belt drive, change gears Z_1/Z_2, continuously meshing gears Z_3 and Z_4, a long central shaft, central gear Z_5, and a gear (Z_6) to impart rotational motion to the spindles. The long central shaft should be hollow and strong to have sufficient torsional rigidity. It is evident that all spindles rotate in the same direction at the same speed. Both bar and chucking multispindle automatics are made in a considerable range of sizes. The sizes are mainly determined by the diameter of stock that can be accommodated in the spindles. The following are the main specifications of multispindle bar automatic DAM 6 × 40:

Number of work spindles	6
Maximum bar diameter (mm)	⌀42, Hex. 36, Sq. 30
Maximum bar length (mm)	4000
Maximum length of stock feed (mm)	200
Maximum turning length (mm)	180
Maximum traverses	
Bottom and top slides (mm)	80
Side slide (mm)	80
Height of centers over main slide (mm)	63
Speed range, normal (rpm)	100–560
Speed range, rapid (rpm)	400–2240
Progressive ratio (rpm)	1.12
Range of machining time per piece, normal (s)	8.9–821
Range of machining time per piece, rapid (s)	5.5–206
Rated power of drive motor (kW)	17
Overall dimensions (L × W × H) (mm)	6000 × 1400 × 2280
Weight (kg)	11,000

A brief description of the machine elements is as follows:

Spindle-drum (carrier) and indexing mechanism. The spindle drum (2) is supported by and indexes in the frame of the headstock (I). It is indexed by the Geneva mechanism (3) through index arm (4, Figure 7.35a), which revolves on the main camshaft (5). The indexing motion is geared to the drum. During the working position of the machine cycle, the spindle drum is locked rigidly in position by a locking pin (6), which is withdrawn only for indexing (Figure 7.35b). A Geneva cross of five parts is preferred to index the drum of four-, six-, and eight-spindle automatics. The division

Automated Lathes

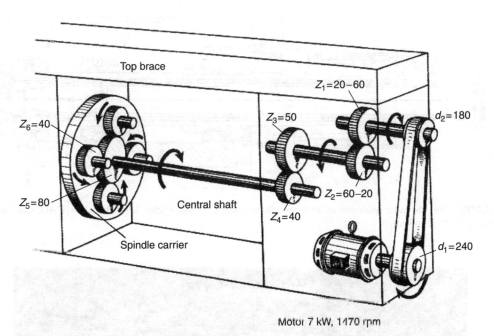

FIGURE 7.34 Gearing diagram of a four-spindle automatic. (From Boguslavsky, B. L., *Automatic and Semi-automatic Lathes*, Mir Publishers, Moscow, 1970. With permission.)

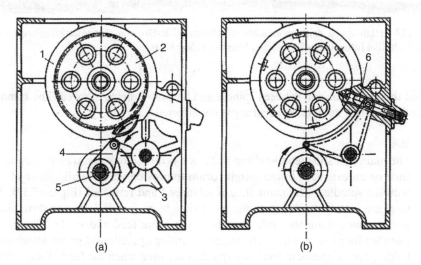

FIGURE 7.35 Drum indexing and locking of a six-spindle automatic: (a) indexing and (b) locking.

into five parts renders a favorable transmission of acceleration and power, thus granting a light and smooth indexing of the spindle drum.

Spindle assembly. Figure 7.36 shows a section through a typical assembly of one of the machine spindles. The collet opening and closing unit is similar to that of the single-spindle automatic. The spindle is mounted on fixed front bearings and a floating rear double raw tapered roller bearing; thus

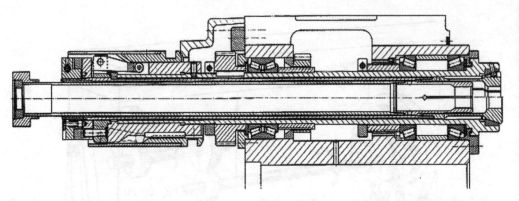

FIGURE 7.36 Spindle assembly of a six-spindle automatic. (From Pittler Maschinenfabrik AG, Langen bei Frankfurt/M, Germany.)

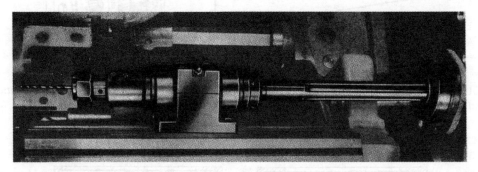

FIGURE 7.37 High-speed drilling attachment. (From VEB-Drehmaschinenwerk/Leibzig, Pittlerstr, 26, Germany, Technical Information Prospectus Number 1556/e/67.)

differential thermal expansion between spindle and housing is being allowed. The spindle expands only backward, so that its running accuracy is not affected.

Tool slides
1. The main tool slide (end working slide) is a central block that traverses upon a round slide on an extension to the spindle drum to provide accurate alignment of the slide with the spindles. The main slide is advanced and retracted (Figure 7.37). The end tools are mounted directly on the main slide by means of T-slots or dovetails. Every tool mounted upon the slide must have the same feed and stroke. These tools are intended for plain turning, drilling, and reaming operations. Special attachments and holders for independent feed tool spindles are used when the feed of any cutting tool must differ from that of the main slide. These attachments and holders are actuated by drum cams. Figure 7.37 shows a holder carrying a high-speed drilling attachment, whereas Figure 7.38 shows an independent feed, high-spindle speed attachment. The drive mechanism of the end tool slide is shown in Figure 7.39 and performs the following steps:
   – Rapid approach of the tool slide may be either 75 or 120 mm, while the working feed may be adjusted in a range from 20 to 80 mm. The rapid approach is effected by the advance of the carriage (1) with the feed lever (5) held stationary. The carriage is traversed by a corresponding cam of the main slide through the roll (2, Figure 7.39).

FIGURE 7.38 Independent feed/high-speed drilling attachment. (From VEB-Drehmaschinenwerk/Leibzig, Pittlerstr, 26, Germany, Technical Information Prospectus Number 1556/e/67.)

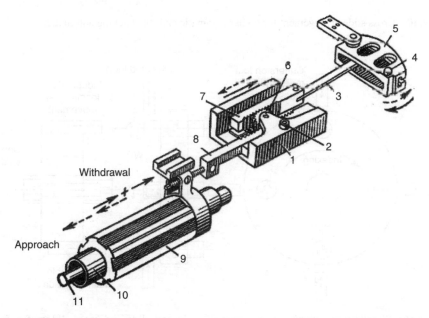

FIGURE 7.39 Drive mechanism of the end tool slide. (From Chernov, N., *Machine Tools*, Mir Publishers, Moscow, 1975. With permission)

- Rapid approach proceeds until the carriage runs against a stop screw (not shown in figure). The gear (6) travels together with the carriage. This gear meshes simultaneously with the rack (7) of the tie rod (3) and the rack of the main slide (8). As rack (3) is stationary, the rack (8) and correspondingly the main tool slide travels a distance twice that of the carriage.
- At the end of rapid approach, the carriage stops and is held stationary by the stop screw and the carriage driving cam (16).
- Immediately after this, another cam mounted on the camshaft actuates the feed lever (5, Figure 7.39) through the roll (4). The rack (7) moves the gear (6), which imparts the movement to the rack (8) and to the end tool slide.
- The length of the main slide working travel is set up by positioning a link (3) in the slot of the lever (5) with the aid of the scale located on the lever.
- Rapid withdrawal is engaged at the end of working feed. In this case, both the carriage (1) and lever (5) return to their initial positions at the same time.

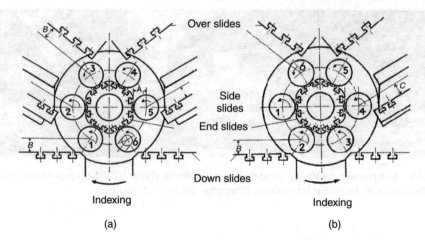

FIGURE 7.40 Cross slide arrangement for (a) bar automatic and (b) chucking automatic.

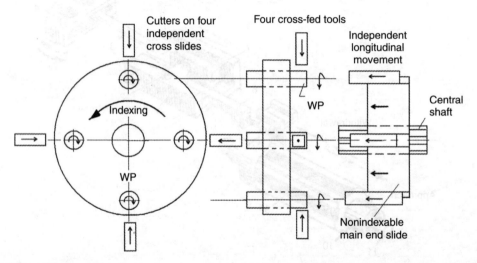

FIGURE 7.41 Simultaneous working movements of main end slide and cross slides of a four-spindle automatic.

2. Cross slides are intended for the plunge-type cutting operations such as facing, grooving, recessing, knurling, chamfering, and cutting. They are directly mounted on the headstock of the machine, and move radially to the center line of the work. Figures 7.40a and 7.40b show the cross slide arrangement for both bar and chucking six-spindle automatics, respectively.

The cross slides are cammed individually; each is driven by its own cam drum. Therefore, the feed rate can be different for each side tool. The side tools feed slowly into the work to perform their cutting operations and then return to clear out spindles for indexing. In general, two slides are allocated for making heavy roughing and forming cuts. The other slides are used to complete subsequent finishing operations to the required accuracy. Except for a stock feed stop at one position, the tools on the main tool slide move forward and make the cut essentially simultaneously. At the same time, the tools in cross slides move inward and make their plunge cuts (Figure 7.41).

FIGURE 7.42 Standard cams of cross overslides. (From Pittler Maschinenfabrik AG, Langen bei Frankfurt/M, Germany.)

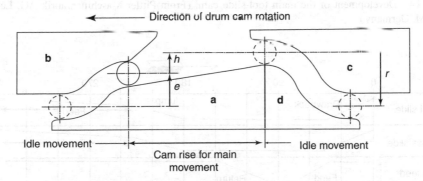

FIGURE 7.43 Development of the cross overslide in the direction of drum cam rotation (From Pittler Maschinenfabrik AG, Langen bei Frankfurt/M, Germany.)

Camming and cyclogram. The main camshaft, either directly or indirectly controls the cam movements. Hence, cams of various machining operations must be selected from a range of standard cams according to rise and feeds required. The cams for the idle motions, such as stock feeding, chucking, indexing, and so on, are standard cams and are not changed. Multispindle cams are generally composed of specially shaped segments that are bolted onto a drum to control motions. Figure 7.42 shows the cams of the cross overslides, which reduce the need for special cams.

Figures 7.43 and 7.44 show the developments of the cross overslide and the main tool slide cam drums of a six-spindle automatic DAM 6 × 40. The working feeds of both drums occupy about 105° and the auxiliary activities occupy 255° of the total cycle time 360°. Figure 7.45 shows the complete cyclogram of the working and auxiliary cams of a four-spindle bar automatic (on the basis of camshaft rotation angle). The cyclogram shows the sequence of events in the production of a single piece during one complete revolution of the main tool-slide cam, or cross-slide cams. Prior to stock feeding, there is a rapid rise or jump toward the work, and at the end of cut, an equal and rapid withdrawal or drawback is followed by a dwell while indexing and stock feeding. The dwell is denoted by a horizontal line, while a rising or a falling line denotes movement. Cyclograms may be of circular or developed types. The developed cyclograms are more easily read. Chucking events occur during the rapid drawback of the slide (Browne, 1965).

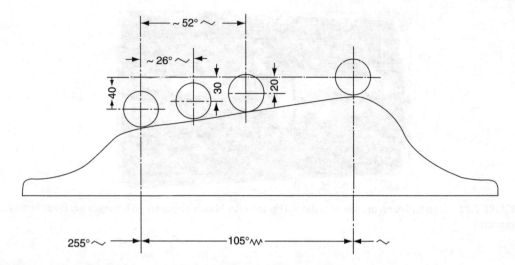

FIGURE 7.44 Development of the main tool-slide cam. (From Pittler Maschinenfabrik AG, Langen bei Frankfurt/M, Germany.)

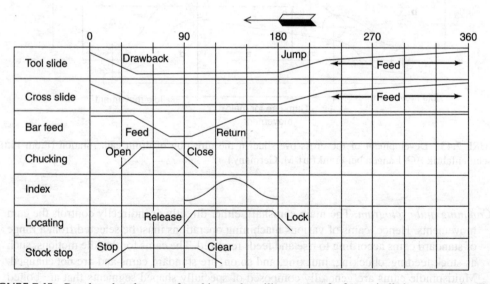

FIGURE 7.45 Developed cyclogram of working and auxiliary cams of a four-spindle bar automatic. (From Browne, J. W., *The Theory of Machine Tools*, Cassell and Co. Ltd., 1965.)

Setting time and accuracy of multispindle automatics. Setting the multispindle automatic for a given job requires 2–20 h while a piece can be often completed every 10 s. The precision of multispindle chucking or bar automatics is good, but seldom as good as that of single-spindle automatic. Tolerances of ±13 to ±25 μm on the diameter are common (*Metals Handbook*, 1989), and the maximum out-of-roundness may reach 15 μm.

7.6 DESIGN AND LAYOUT OF CAMS FOR FULLY AUTOMATICS

The production of a WP on an automatic machine represents a symphonic master work in which different instruments (cams and tools) contribute in harmony to compose or produce the work in a predetermined playing or cycle time. The machine is the orchestra; the contributors (owners) of the work are the process engineer, cam designer, and the machine setter. Setting up an automatic involves all the preparatory work required to manufacture a WP in accordance with the part drawing and specifications.

Setting up includes the following steps:

1. Planning the sequence of operation
2. Working out the calculation sheet for the set up
3. Manufacturing the necessary cams and tooling
4. Setting up the kinematic trains to obtain the required speeds and feeds
5. Installing and adjusting cams and tools on the machine

7.6.1 Planning a Sequence of Operation and a Tooling Layout

The sequence of operations is worked out on the basis of the specifications of the automatic as given in the machine manuals and the specifications of the WP as obtained from the working drawing.

There are general rules for developing the tool layout of a general purpose automatic. These are based on the following machine processing features:

1. Determine the quickest and best operation sequence before designing the cams.
2. Use the highest spindle speeds recommended for the material being machined, provided the tools will stand such conditions.
3. Begin finishing only after rough cutting is completed.
4. Wherever possible, overlap the working operations and try to increase the number of the tools operating simultaneously at each position.
5. Overlap idle operations with one another and with the working operations.
6. Do not permit substantial reduction of the WP rigidity before rough cuts have been completed.
7. Accurate dimensions along the WP length should be obtained by cross slide tools, and not turret tools.
8. Speed up cutting-off operations, which require much time, especially in solid stock. The feed may be decreased near the end of the cut, where the piece is separated from the bar.
9. Wherever possible, break down form turning operations into a rough cut and a finish cut.
10. Provide a dwell at the end of the cross slide travel for clearing up the surface, removing the out-of-roundness, and improving the surface finish.
11. When deep holes are drilled, it is sometimes necessary to withdraw the drill a number of times. This facilitates chip removal and permits drill cooling.
12. When drilling a hole, first spot (center) drill the work using a short drill of a larger diameter.
13. When drilling a stepped hole, first drill the largest diameter and then smaller diameters in succession. This approach reduces the total working travel by all drills.
14. If strict concentricity and alignment are required between external and internal surfaces, or stepped cylindrical surfaces, finish such surfaces in a single turret position.
15. Do not combine thread cutting with other operations. The calculated length of working travel should be increased by two or three pitches in comparison with thread length specified in the part drawing. Moreover, the actual length of travel is reduced by 10–15% of the calculated length by correspondingly reducing the radius of the cam at the end of working travel movement. Thus the slide feed lags behind the tap or die movement along the thread being cut, excluding any possibility of stripping the thread due to incorrect feed of the slide by the cam. The tap or the die should have a certain amount of axial freedom in its holder.

16. If only two or three positions of the turret are occupied:
    - Index the turret through every other position and use a swing stop.
    - Machine two WPs every cycle.
17. To obtain equal machining times at all positions of multispindle automatics, divide the length to be turned into equal parts, or increase the feed or cutting speeds at positions where a surface of longer length is to be turned.
18. In all cases, use standard tools and attachments whenever possible.

Tooling layout. The tooling layout for all operations consists of sketches drawn to a convenient scale of WP, tools, and holders in the relative position that they occupy at the end of the working travel movement. The lengths of travel should be indicated. These sketches are checked against the setup characteristics of the working members, and also used to check whether the tools and slides interfere with one another during operation. The tooling layout serves as initial data for working out the operation sheet and the cam design sheet.

7.6.2 Cam Design

The cam design for automatics is a tedious work including definite steps depending on the type of automatic machine. However, the following main guidelines are to be generally observed in the design process:

1. Determine the number of spindle revolutions required for each operation and idle movements.
2. Overlap those operations and idle movements that can take place simultaneously.
3. Proportion the balance of spindle revolutions on the surface of the cam so that the total of these revolutions equals the full circumference of the cam.

Within the spaces reserved for turret operations, in case of automatic screw machines, lobes are developed to feed the tools on the work. The radial height (throw) of these lobes equaling the length a tool will travel on the work, and the gradient of cam lobe governs the rate of tool feed. The lobes are connected by drops or rises, and in these spaces, idle movements take place. The cross-slide cams revolve at the same rate as the turret cam and operations performed by cross-slide tools are laid out on these cams.

The cam blank surface is divided into 100 equal divisions (Figure 7.46). The radius R is equal to the distance from the cam follower center to the fulcrum of the lever carrying the roll; the arc centers are located on the lever fulcrum circle of the radius R_1, given in the machine manual. The cams are drawn on a full scale. Marking out the cam begins from a zero arc and proceeds clockwise, provided that the turret slide cam is watched from the rear side of the machine and the cross-slide cams from the turret side. Feeding and clamping the bar begins from zero arc. Whenever a tool has not been moved, the corresponding cam outline is formed by a circle arc drawn from the cam center. To construct the withdrawal and approach curves of the turret-slide cam and cross-slide cams, a special drawing template is provided as a supplement to the machine's service manual (Figure 7.47).

The detailed procedure of cam layout for different automatics can be carried out in the following sequences.

1. Single-spindle screw automatic:
   - Determine the machine size and specifications.
   - Determine the operational sequence.

Automated Lathes

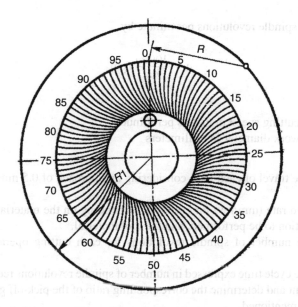

FIGURE 7.46 Blank of disk cam. (From Chernov, N., *Machine Tools*, Mir Publishers, Moscow, 1975. With permission.)

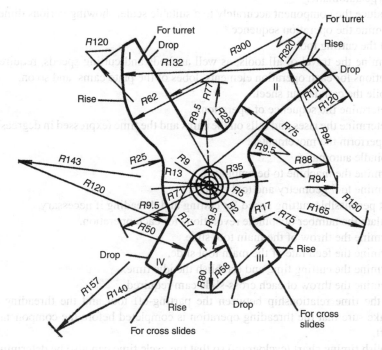

FIGURE 7.47 Cam template for drawing curves for idle travel movements of an automatic screw machine.

- Determine the tool geometry and tool material to be employed.
- Select permissible cutting speeds for the material to be machined according to the operations to be preformed.

- Calculate the spindle revolutions per minute by

$$n = \frac{1000v}{\pi D} \text{ rpm}$$

where,
v = cutting speed in meters per minute
D = work diameter in millimeters

- Determine the travel of each tool, considering an approach of 0.5 mm to avoid damage of tools.
- Select the feed rate (mm/rev) for each tool, depending on the material to be machined, type of operation to be performed, and the tool material.
- Calculate the number of spindle revolutions for each cutting operation and the idle movements.
- Determine the cycle time expressed in number of spindle revolutions required to complete one component and determine the corresponding ratio of the pick-off gears $(a/b) \times (c/d)$, as previously mentioned.
- Calculate the hundredths of cam surface needed for both cutting and idle operation by converting revolutions into hundredths.
- Establish the operation and cam design sheets.

2. Swiss-type automatic:
   - Reproduce the component accurately to a suitable scale, showing various dimensions.
   - Determine the operation sequence.
   - Select the cutting tools.
   - Determine the travel of all tools, as well as feeds and cutting speeds, required spindle revolutions for each operation elements, lobes of the plate cams, and so on.
   - Compile the cam layout sheet.
     - Determine the sequence of operations.
     - Determine the rises and falls on the cams and the time (expressed in degrees) required to perform the movements.

3. Multispindle automatic:
   - Determine the machine to be employed.
   - Determine tool geometry and tool material.
   - Select permissible cutting speed for cutting and threading if necessary.
   - Calculate the number of spindle revolutions for each operation.
   - Determine the throw of the main tool slide.
   - Determine the feed rate of the main tool slide.
   - Determine the cutting time and establish the idle time.
   - Determine the throw of each cross-slide cam required.
   - Find the time relationship between the parting-off tool and the threading operation to make sure that the threading operation is completed before the component is finally cut off.
   - Establish timing chart (cyclogram) so that the cycle time can also be determined.
   - Draw the tool layout and record data.

Illustrative Examples of Cam Layout

The following examples illustrate the cam layout procedure as applied on different types of automatics.

Example 1

The component shown in Figure 7.48 is to be produced in mass production on a turret automatic screw. It is made of brass 85, and the bar stock is of 28 mm diameter. To increase productivity, the machine is supplemented by a milling attachment to take care of milling the product after the cutting off operation, by gripping and moving it far from the machining area while the next product is in operation.

Design and draw a set of plate cams required for the production of this component. Illustrate the tool layout without considering the cam layout necessary for the milling, attached for simplicity.

Solution

After specifying the machine, the operational sequence is determined as indicated in Figure 7.49.

1. The cutting speed for brass in case of using HSS tools:

Turning: $v = 132$ m/min
Threading: $v = 42.5$ m/min

Determination of spindle rotational speed:

$$\text{Turning: } n_s = \frac{1000v}{\pi D} = \frac{1000 \times 132}{\pi \times 28} = 1500 \text{ rpm}$$

$$\text{Threading: } n_s = \frac{1000 \times 42.5}{\pi \times 18} = 750 \text{ rpm}$$

2. Sequence of operation is illustrated in Table 7.2 and the tool layout is given in Figure 7.49.
3. Throw or travel of each tool is determined as in Table 7.2, column (b). Add up to 0.4 mm approach to avoid tool damage.
4. Selected feeds per revolution are illustrated in Table 7.2; column (c). The following notes are especially important:
   - Feeds for forming and cutting off are much smaller than those for drilling and turning.
   - In high-speed drilling (turret station 5), the feed for a stationary drill equals 0.11 mm/spindle revolution. If the drill is rotating in the opposite direction from the spindle at 1600 rpm, then the equivalent feed should equal $0.11 \times (1500 + 1600)/1600 = 0.23$ mm/rev.
   - In threading, the feed equals the pitch of the thread. To avoid the possibility of stripping the thread, the slide feed should lag behind the die movement by about 10% of its throw (Figure 7.50); see turret station 4. Also it is evident from Figure 7.49, at turret station 5, that threading is performed alone and not combined with other operations.

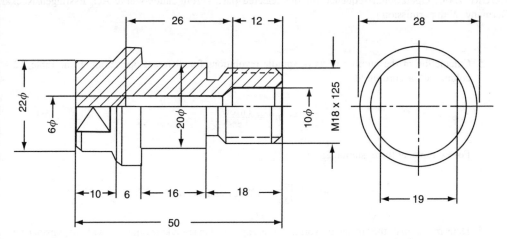

FIGURE 7.48 Product to be produced on an automatic screw machine. (From Index-Werke AG, Esslingen/Neckar, Germany. With permission.)

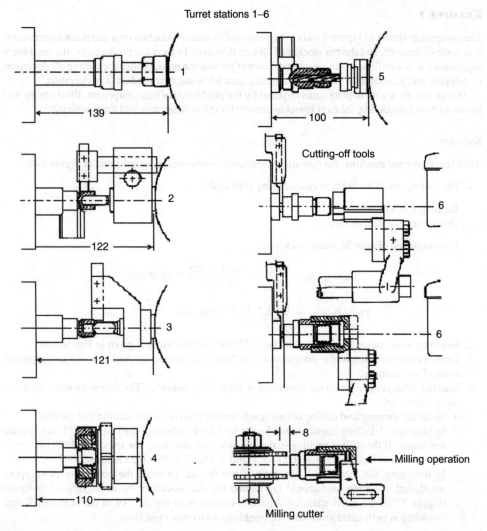

FIGURE 7.49 Operational sequence for the selected part. (From Index-Werke AG, Esslingen/Neckar, Germany. With permission.)

5. Calculation of the number of work spindle revolutions for different operations, from Table 7.2 column (d), according to

$$n^* = \frac{\text{Throw}}{\text{Feed}}$$

For example, at turret station 2:

$$n^* = \frac{17}{0.133} = 128 \text{ rev and so on.}$$

6. Determination of the cutting or working time expressed in spindle revolutions without considering time of overlapped operations (Table 7.2), column (h).
   Sum of spindle revolutions: $n = 319$ rev

TABLE 7.2
Operation Sheet of the Part Produced on Automatic Screw Machine Index 24

Part Drawing (see Figure 7.48)
Machine: Index 24
Work Material: Brass Ms 58, 28 ∅

Spindle Speed (rpm): 1500 / 750
Cutting Speed (m/min): 132 / 42.5

Tool Station		Operation Sequence	Throw (mm)	Feed/Rev (mm)	Revolutions Per Operation	% of Cam Circumference — Number	Turning Cam From	Thread Cutting Cam To	Turning Main (In rev)	Thread Cutting Main (In rev)	Auxiliary (%)	Cam Drawing From	Cam Drawing To
		a	b	c	d	e	f	g	h		i	k	l
Turret Slide	1	Feed stock to stop (1 s)									5	0	5
		Index turret (2/3 s)									3	5	8
	2	Turn for thread/drill 10∅	17	0.133	128	23	—	—	128			8	31
		Index turret									3	31	34
	3	Chamfer for thread or break hole edge	1.5	0.14	11	2			11		—	34	36
		Index turret									4	36	40
	4	Thread cutting on 1:2	16th	1.25	32	6				32	—	40	46
		off 1:1	16th		16	3				16	—	46	49
	5	Drilling 6∅ with high speed drill 1600 rpm	25.5	0.11 / 0.23	110	20			110		4	49	53
		Index turret (half) 1/3 s				4	73	77			2	53	73
	6	Vacant hole to clear grip arm					77	96				73	75
		Index turret (half) 1/3 s				4	96	0			2	98	0

(continued)

TABLE 7.2 Continued
Operation Sheet of the Part Produced on Automatic Screw Machine Index 24

| Tool Station | Operation Sequence | Throw (mm) | Feed/Rev (mm) | Revolutions Per Operation | % of Cam Circumference | | | Figures for Calculating Cycle Time | | | |
					Number	Cam From	To	Main (In rev)	Auxiliary (%)	Cam Drawing From	To
	a	b	c	d	e	f	g	h	i	k	l
Grip Arm	Swing down (half)				6	72	78		3	75	78
	Dwell					—	—		1	78	79
	Advance for picking up					—	—		6	79	85
	Dwell while cutting off				4	85	89		—	—	—
	Dwell after cutting off					—	—		1	89	90
	Relief arm from stop				2	90	92		—	—	—
	Withdrawn					—	—		3	90	93
	Swing up (half)				10	92	102		5	93	98
Cross Slides	Front: 28–16⌀	6	0.04	150	27	7	34				
	Back: 28–20⌀	4	0.04	100	18	55	73				
	Cutting off 28–3⌀	12.5	0.065	192	35	50	85			85	89
	3-center	1.5	0.065	22	4	—	—	22			
	Past center	1	0.1	10	2	89	91				
								319	42		

Revolutions/four pieces = $\frac{319}{58} \times 100 = 550$ rev

Cycle time = $\frac{550}{1500} \times 60 = 22$ s

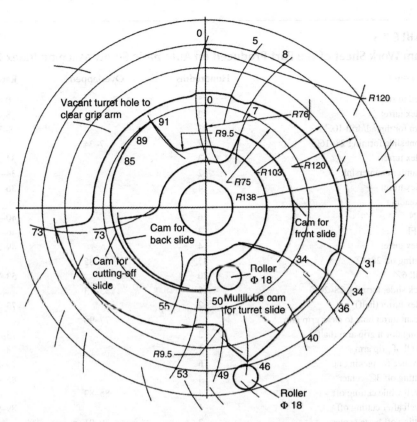

FIGURE 7.50 Cam layouts for the selected part. (From Index-Werke AG, Esslingen/Neckar, Germany. With permission.)

7. Determination of idle movements in hundredths of cam circumference (Table 7.2), column (i), assuming the following allowances.

Feeding stock to stop and clamping (1 s)	= 5%
Turret indexing (2/3 s)	= 3–4%
Turret half indexing (1/3 s)	= 2%
Grip arm allowances	
Swing down (half)	= 3%
Dwell	= 1%
Advance for picking up	= 6%
Dwell after cutting off	= 1%
Withdraw	= 3%
Swing up (half)	= 5%

Columns (e)–(f)–(g)–(k) and (l), (Table 7.2) could now be completed.

8. Calculation of cycle time in seconds, referring to Table 7.2.

$$\text{Total revolutions/piece} = \frac{319}{58} \times 100 = 550 \text{ rev}$$

$$\text{Cycle time, } T_{\text{cyc}} = \frac{550}{1500} \times 60 = 22 \text{ s}$$

Therefore, the cam work sheet (Table 7.3), for the required component can be constructed and the cam layout is shown in Figure 7.50.

TABLE 7.3
Cam Work Sheet of the Part Produced on Automatic Screw Machine Index 24

Operation	Hundredths	Overlapped	Range
Feed to stop	5		0–5
Index turret	3		3–8
Turn for thread/drill 10∅	23		8–31
Front slide forming 28–16∅	27	7–34	
Index turret	3		31–34
Chamfer/countersink	2		34–36
Index turret	4		36–40
Threading			
ON	6		40–46
OFF	3		46–49
Index turret	4		49–53
Cutting off 28/3∅	35	50–85	
Drill 6∅	20		53–73
Back slide, forming 28–20∅	18	55–73	
Index turret (half)	2		73–75
Vacant turret hole to clear grip arm		77–96	
Swing down grip arm (half)	3		75–78
Dwell of grip arm	1		78–79
Advance for picking up	6		79–85
Cutting off 3∅-center	4		85–89
Dwell while cutting off	4	85–89	
Dwell after cutting off	1		89–90
Cutting off paste center	2	89–91	
Relief arm from stop	2	90–92	
Withdrawn	3		90–93
Swing up (half)	5		93–98
Index turret (half)	2		98–100

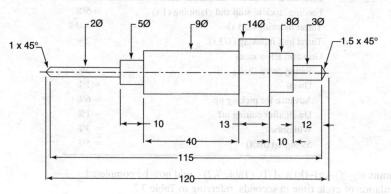

FIGURE 7.51 Part to be produced on a typical Swiss-type automatic.

ILLUSTRATIVE EXAMPLE 2

It is required to produce in mass production the long part (brass) shown in Figure 7.51 on the Swiss-type automatic, Model 1 π16 Stankoimport (spindle speed: 400–5600 rpm, bar capacity: 16 mm, and rated power: 3 kW).

Automated Lathes

Suggest an operational sequence and tooling layout. Establish a cam design sheet and calculate the product cycle time.

Solution

Figure 7.52 illustrates the proposed tooling layout. The operational sequence is shown in Table 7.4. Three tools are sufficient to perform the work. Tools I and II are mounted on the rocker arm, and tool III is mounted on the overhead slide.

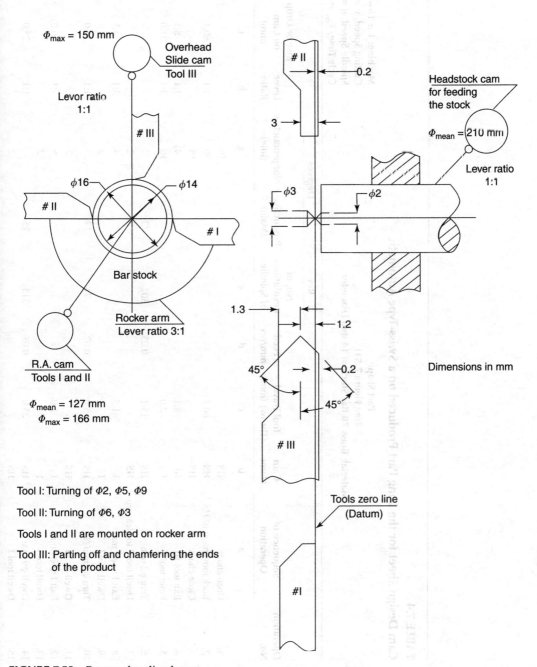

Tool I: Turning of $\Phi 2$, $\Phi 5$, $\Phi 9$

Tool II: Turning of $\Phi 6$, $\Phi 3$

Tools I and II are mounted on rocker arm

Tool III: Parting off and chamfering the ends of the product

FIGURE 7.52 Proposed tooling layout.

TABLE 7.4
Cam Design Sheet for the Long Part Produced on a Swiss-Type Automatic

Part Shape (see Figure 7.51)
Material: Brass 70 Bar Stock of 14 mm Diameter

Machine: 1 π 16—Swiss-type, Stankoimport
Cutting Speed, v = 100 m/min
Spindle Speed, n = 2240 rpm
Cycle Time, T_{cyc} = 40.5 s

Operation No.	Sequence of Operation	Cam Name	Tool Travel (Throw) (mm)	Feed mm/rev	Rev No. of Revolutions of Spindle	Degrees Productive (Main)	Degrees Nonproductive (Idle)	Lever Ratio	Rise or Drop on Cam (mm)	Cam Layout Data Degrees (Range)	Cam Layout Data Lobe Radius (mm) (Range)
	a	b	c	d	e	f	g	h	i	j	k
1.	Open chuck	HS					10			0–10	105–105
2.	Back movement of HS	HS	60.1				30	1:1	60.1	10–40	105–39.9
3.	Close chuck	HS					15			40–55	39.9–39.9
4.	Exit tool III	III	8.1				(4)	1:1	8.1	(55)–(59)	75–66.9
5.	Enter tool I	I	7.0				10	3:1	21.0	55–65	63.5–42.5
6.	Turn ϕ2 mm	HS	15.1	0.05	302	72				65–137	39.9–55
7.	Dwell tool I to clean up	HS					2			137–139	55–55
8.	Exit I to ϕ5 mm	I	1.5				4	3:1	4.5	139–143	42.5–47
9.	Dwell tool I	I					2			143–145	47–47
10.	Turn ϕ5 mm (I)	HS	5.0	0.07	71	17		1:1	5.0	145–162	55–60
11.	Dwell tool I	HS					2			162–164	60–60
12.	Exit I to ϕ9 mm	I	2.0				6	3:1	6.0	164–170	47–53
13.	Dwell tool I	I					2			170–172	53–53
14.	Dwell ϕ9 mm (I)	HS	25.0	0.08	313	75		1:1	25.0	172–247	60–85
15.	Dwell tool I	HS					2			247–249	85–85

Automated Lathes

16.	Exit I to φ16 mm	I	3.5		10	3:1	249–259	53–63.5
17.	Stock feeding	HS	9.3		10	1:1	259–269	85–94.4
18.	Enter II to φ14.2 mm	II	0.9		(3)	3:1	(269)–(272)	63.5–66.2
19.	Feeding II to φ8 mm	II	3.1	0.05	14	3:1	269–283	66.2–75.5
20.	Dwell II	II			2		283–285	75.5–75.5
21.	Turn φ8 mm (II)	HS	5.0	0.08	15	1:1	285–300	94.3–99.3
22.	Dwell tool II	II	9.3		2		300–302	75.5–75.5
23.	Enter II to φ3 mm	II	2.5	0.04	15	3:1	302–317	75.5–83
24.	Dwell tool II	II			2		317–319	83–83
25.	Turn φ3 mm (II)	HS	5.7	0.06	19	1:1	319–338	99.3–105
26.	Dwell tool II	HS			2		388–340	105–105
27.	Exit II to φ16 mm	II	6.6		(8)	3:1	(340)–(348)	(83)–63.5
28.	Enter III to φ3.2 mm	III	6.4		8	1:1	340–348	66.9–73.3
29.	Feed III for parting-off	III	1.7	0.04	10	1:1	348–358	73.3–75
30.	Dwell	III			2		358–360	75–75
				44	996	123	237	Cycle time in revolutions

$$= \frac{996}{237} \times 360 = 1513 \text{ rev}$$

$$T_{\text{cyc}} = \frac{1513}{2240} \times 60 = 40.5 \text{ s}$$

The procedure is carried out according to the following sequence:

1. Determination of the spindle speed:

$$v = 100 \text{ m/min (WP: brass, Tool: HSS)}$$

$$n = \frac{1000v}{\pi D} = \frac{1000 \times 100}{\pi \times 14} = 2274 \text{ rpm}$$

The spindle speed n is selected to be 2240 rpm.

2. The tool travel (throw), and the selected feeds are listed for each operation in columns (c) and (d) of Table 7.4. Accordingly, the number of spindle revolutions column (e) can be calculated. For example, for operation 6 (Table 7.4):

Tool travel	= 15.1 mm
Feed	= 0.05 mm/rev
Number of revolutions	= 15.1/0.05 = 302 rev

From Table 7.4, the total number of revolutions to perform the main (productive) operations = 996.

3. Allowances for idle (nonproductive) activities are assumed in degrees, column (g)
   Total of idle activities = 123°.
   Therefore, total of main activities = 237°.
4. Determination of the time T_{cyc}:
   - Expressed in revolutions = $996 \times \frac{360°}{237°} = 1513$ rev
   - Expressed in seconds = $\frac{1513}{2240} \times 60 = 40.5$ s
5. The main (productive) activities, as expressed in degrees instead of revolutions, are listed in column (f) of Table 7.4.
6. Rises and drops on different cams are calculated, column (i), by considering the lever ratio of each slide, column (h).
7. The cam layout data are completed:
   - Degrees on cam circumference, column (j).
   - Lobe radii (mm), column (k).
8. Use data in columns (j) and (k) to draw the cams.

ILLUSTRATIVE EXAMPLE 3

A batch size of 50,000 pieces is to be produced on a six-spindle bar-type automatic. The part (Figure 7.53) is made of steel 20 (σ_u = 40–50 kg/mm²). The bar size is 27 mm diameter. Provide a tooling layout and calculate the cycle time.

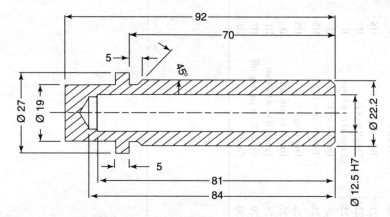

FIGURE 7.53 Part produced on a six-spindle automatic.

Automated Lathes

Solution

Tooling layout. The sequence of operation and tooling layout should be written in advance, and the best one should be chosen with regard to lower tooling cost. In multispindle automatics, a proper sequence of operation necessitates distributing the machining operation so that all operations have approximately the same machining time. Elements of operation that require much time are sometimes divided between two or even three positions, in order to increase the production capacity.

The tooling setup is shown in Figure 7.54. It may be written as follows:

1st station: Turn length 40 mm and spot drill before drilling by the central main slide. Rough front forming by cross slide.
2nd station: (Drilling needs too much time so it is divided in to three parts performed in 2nd, 3rd, and 4th stations). Turn remainder (40 mm) and drill 1st part (30 mm) by the central main slide.
3rd station: Rough face, and form by cross slide, support, and drill 2nd part (30 mm) by the central main slide.
4th station: Finish rear forming by cross slide, support and drill the 3rd part by the central main slide.
5th station: (Reaming is not recommended with cutting-off station). Fine face, form, chamfer, and sizing the outer flange by cross slide, support, and ream by the central main slide.
6th station: Cutting-off by cross slide.

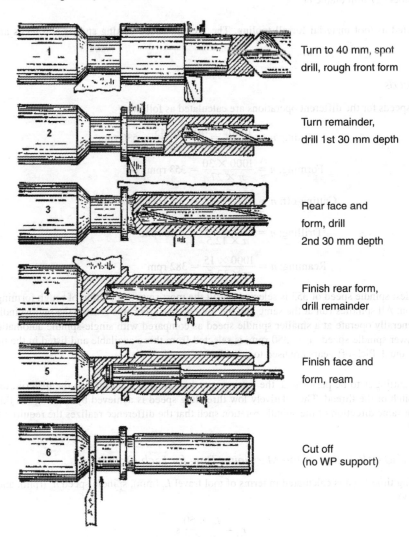

FIGURE 7.54 Tooling setup for the part in Figure 7.53.

TABLE 7.5
Recommended Speeds and Feeds for Different Operations, Tool HSS and WP = Steel 20

Operation	Speed (m/min)	Feed (mm/rev)
Turning	45–55	0.05–0.18
Forming and cutting-off	30–40	0.02–0.05
Drilling	40–50	0.04–0.12
Reaming	10–15	0.10–0.18

DETERMINATION OF SPINDLE SPEEDS AND TOOL FEEDS

The material of the WP: steel 20
The bar stock: 27 mm diameter

HSS is selected as tool material for all tooling. The recommended cutting speeds and feeds are listed in Table 7.5.

Spindle Speeds

The cutting speeds for the different operations are calculated as follows.

$$\text{Turning: } n = \frac{1000v}{\pi \times D} = \frac{1000 \times 45}{\pi \times 27} = 530 \text{ rpm}$$

$$\text{Forming: } n = \frac{1000 \times 30}{\pi \times 27} = 353 \text{ rpm}$$

$$\text{Cutting-off: } n = \frac{1000 \times 30}{\pi \times 19} = 502 \text{ rpm}$$

$$\text{Drilling: } n = \frac{1000 \times 40}{\pi \times 12.5} = 1018 \text{ rpm}$$

$$\text{Reaming: } n = \frac{1000 \times 15}{\pi \times 12.5} = 382 \text{ rpm}$$

The smallest spindle speed of 353 is selected to suit the most severe operation of rough forming at the first station. All spindles run at the same and the lowest speed. For this reason, the multispindle automatics generally operate at a smaller spindle speed as compared with single-spindle automatics. The nearest lower spindle speed, $n_s = 350$ rpm, is selected from those available and listed in the machine service manual. Pick-off gears are used to provide this speed.

If thread cutting is to be performed, the cutting speed is selected according to the material to be cut and the pitch of the thread. The relatively low threading speed is achieved by rotating the threading tool in the same direction of the spindle rotation such that the difference realizes the required threading speed.

Tool Feeds and Calculation of the Machining Time for Each Operation

The machining time t_m (s) is calculated in terms of tool travel L_t (mm), spindle speed n_s (rpm), and tool feed rate f (mm/rev)

$$t_m = \frac{L_t \times 60}{n_s \times f} \text{ s} \tag{7.9}$$

The operational machining time (t_m) is calculated for the different stations. Consider again the sequence of operation and tool layout.

1st station:
  Turn, $L = 40$ mm, $f = 0.1$ mm/rev, $n = 350$ rpm, $t_m = 69$ s
  Rough form, $L = 4.5$ mm, $f = 0.1$ mm/rev, $n = 350$ rpm, $t_m = 8$ s, overlap
2nd station:
  Turn, $L = 40$ mm, $f = 0.1$ mm/rev, $n = 350$ rpm, $t_m = 69$ s
  Drill, $L = 30$ mm, $f = 0.1$ mm/rev, $n = 350$ rpm, $t_m = 52$ s, overlap
3rd station:
  Drill, $L = 30$ mm, $f = 0.1$ mm/rev, $n = 350$ rpm, $t_m = 52$ s
  Rough form, $L = 4.5$ mm, $f = 0.1$ mm/rev, $n = 350$ rpm, $t_m = 8$ s, overlap
4th station:
  Drill, $L = 30$ mm, $f = 0.1$ mm/rev, $n = 350$ rpm, $t_m = 52$ s
  Finish form, $L = 4.5$ mm, $f = 0.1$ mm/rev, $n = 350$ rpm, $t_m = 8$ s, overlap
5th station:
  Ream, $L = 81$ mm, $f = 0.18$ mm/rev, $n = 350$ rpm, $t_m = 77$ s
  Separate reaming attachment is required to allow additional reamer
  Feed of 0.08 mm/rev relative to that of the central slide
  Fine face, $L = 4.5$ mm, $f = 0.1$ mm/rev, $n = 350$ rpm, $t_m = 8$ s, overlap
6th station:
  Cut-off, $L = 15$ mm, $f = 0.05$ mm/rev, $n = 350$ rpm, $t_m = 52$ s

The time of reaming is found to be the largest, and therefore it determines the machine productivity. Assuming idle time $t_a = 3$ s, then the cycle time T_{cyc} (FFT) can be calculated as follows:

$$T_{cyc} = (t_m)_{max} + t_a$$
$$= 77 + 3 = 80 \text{ s}$$
$$= 1.33 \text{ min}$$

And the hour production rate $= 60/1.33 = 45$ pieces/h.

7.7 REVIEW QUESTIONS AND PROBLEMS

1. Mark true or false:
   [ ] Swiss-type automatics are best suited for turning long slender parts in mass production.
   [ ] In automatics, the main camshaft rotates one revolution per machining cycle.
   [ ] In a turret automatic screw machine, the auxiliary shaft rotates slower than the main camshaft.
   [ ] In Swiss-type automatics, turning occurs near to the guide bush.
   [ ] Threading and parting-off can be performed simultaneously on Swiss-type automatics.
   [ ] Draw-in collet chucks produce the most accurate parts on automatics.
   [ ] The spindle rotational speed in automatics considerably affects the cycle time of a product.
   [ ] The productivity of six single-spindle automatics is exactly the same as that of a six-spindle automatic of the same size.
2. What are the necessary measures to reduce the cycle time in automatics?
3. List some important rules to be considered when operating a Swiss-type automatic.
4. What are the main operation features of a Swiss automatic?
5. What is the distinct advantage of a Swiss automatic over a single-spindle automatic screw machine?

6. "The material feeding in a Swiss-type automatic is not an auxiliary movement." Discuss this statement briefly.
7. What are spring collets available for bar automatics? What type do you recommend for:
    a. Single-spindle automatic
    b. Multispindle automatic
8. What is a long part? On which machine can it be produced, and why?
9. A single-spindle bar automatic is set as shown in the following table to produce the same component at each setting.

Setting	Spindle Speed (rpm)	Camshaft Speed (rpm)
First setting	2000	2
Second setting	1500	3

Compare the two settings from the following points of view:
   a. Productivity
   b. Accuracy and surface finish
10. What are the two main types of multispindle automatics?
11. In automatic screw machine, at what speed does the auxiliary control shaft revolve? List the functions it performs. How is the speed of the camshaft set up? List the functions of the main camshaft.
12. Mention one of the special attachments that can be provided on automatic screw machine. Why these special attachments are sometimes necessary?

REFERENCES

Acherkan, N. (1969) *Machine Tool Design*, Vols. 1–4, Mir Publishers, Moscow.
Boguslavsky, B. L. (1970) *Automatic and Semi-automatic Lathes*, Mir Publishers, Moscow.
Browne, J. W. (1965) *The Theory of Machine Tools*, Vols. 1 and 2, 1st Edition, Cassel and Co. Ltd.
Chernov, N. (1975) *Machine Tools*, Mir Publishers, Moscow.
VEB-Drehmaschinenwerk/Leibzig, Pittlerstr, 1967, 26, Germany, Technical Information Prospectus Number 1556/e/67.
Index-Werke AG, Esslingen/Neckar, Germany.
Maslov, D., Danilevesky, V., and Sasov, V. (1970) *Engineering Manufacturing Processes*, Mir Publishers, Moscow.
Metals Handbook (1989) Machining, Vol. 16, ASM International, Materials Park, OH.
Technical Data, Mehrspindel-Drehautomaten, 538-70083. GVD Pittler Maschinenfabrik AG, Langen bei Frankfurt/M, Germany.

8 Numerical Control and Computer Numerical Control Technology

8.1 INTRODUCTION

In conventional or manually operated machine tools, the process starts from the part drawing, and the machinist is responsible for the entire job. The machinist determines the machining strategy, sets up the machine, selects proper tooling, chooses machining feeds and speeds, and manipulates machine controls to cut a part that will pass inspection. It is clear that using this method of machining involves a considerable number of decisions that influence the accuracy and surface finish of the machined part.

Numerical control (NC) is a system that uses prerecorded information prepared from numerical data to control a machine tool or the machining process. NC describes the control of machine movements and various other functions by instructions expressed as a series of numbers and initiated via electronic control system. Figure 8.1 shows the operator-controlled and numerically controlled machine tools.

Computer numerical control (CNC) is the term used when the control system includes a computer. Figure 8.2 shows the difference between NC and CNC of machine tools. Manufacturing areas of NC, CNC, and DNC include flame cutting, riveting, punching, piercing, tube bending, and inspection. NC and CNC are particularly suitable for the manufacture of a small number of components needing a wide range of work, such as those with complex profiles or a large number of holes. They are also suitable for batch work. In NC machining, the part programmer analyzes the drawing, decides the sequence of operations, and prepares the manuscript in a language that the NC system can understand. As shown in Figure 8.3, the NC system consists of data input devices, a machine control unit (MCU), servo drive for each axis of motion, a machine tool operative unit, and feedback devices. The program written and stored on the tape is read by the tape reader, which is a part of the MCU. The MCU translates the program and converts the instructions into the appropriate machine tool movements. The movement of the operative unit is sensed and fed back to the MCU. The actual movement is compared with the input command and the servo motor operates until the error signal is zero.

The history and development of NC dates back to 1952, when the first NC conventional milling machine was demonstrated at the Massachusetts Institute of Technology (MIT). In 1957, aircraft manufacturers installed a milling machine—the beginning of NC technology—that was used for machining complex profiles for the aircraft and aerospace industries. Drilling machines, jig borers, lathes, and other NC machine tools were soon developed with less tooling, more operations performed in the same setup, and involvement of the operator in controlling the machine was avoided. NC machining centers and turning centers then appeared and gave the machine designers and builders a chance to improve NC-machined products in terms of accuracy and surface quality.

The development in electronics industry played a key role in the growth and acceptance of NC machine tools. Since the 1960s, smaller electronic components such as transistors, resistors, and diodes have increased the reliability and reduced the size and cost of machine tools. The development of integrated circuits in 1965 led to a further reduction of the size and cost of the

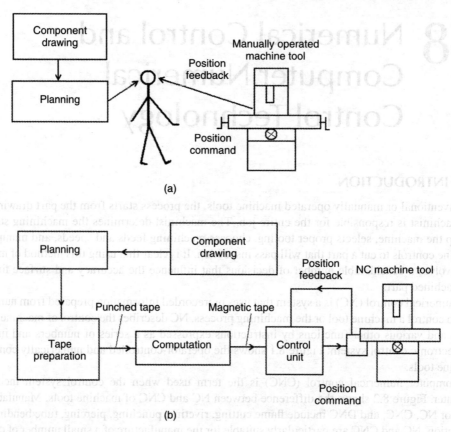

FIGURE 8.1 Operator-controlled and numerically controlled machine tools: (a) manual machine tool and (b) NC machine tool.

control units and provided the basis for the use of minicomputers in CNC and direct numerical control (DNC) machining.

Earlier systems of NC machines consisted of a specially built control unit permanently connected to the machine tool. They are relatively inflexible, as they are special-purpose machine tools. Developments in the area of miniaturization and integration of circuits has led to the introduction of new, small, and powerful computers that are used to control the machine tools (CNC) instead of a conventional controller. The advantages of CNC are related to the control system, which allows a great deal of flexibility unobtainable with NC. DNC involves controlling more than one machine using the same computer and data transmission lines. The major advantage of CNC and DNC over NC is that punched tapes are not used directly to control the machine tool. Instead, all information flows from a computer that interfaces with each MCU (see Figure 8.2).

NC machines cost approximately five to ten times as much as the cost of conventional machines of the same size depending on the capacity of the control system and accessories. Figure 8.4 shows the total cost against the total quantity of parts being produced using different machining methods. At a volume of zero, the fixed cost of machining by NC includes tape preparation and setup in addition to the costs related to the design and fabrication of holding fixtures whenever required. When using conventional machine tools, this cost includes the design and fabrication of tooling, fixtures (when required), and setup. Manual preparation and machine adjustments require more time than tape preparation. For special-purpose and automatic machines, the design and fabrication of special tooling, manual setup, and adjustment of the machine are expensive.

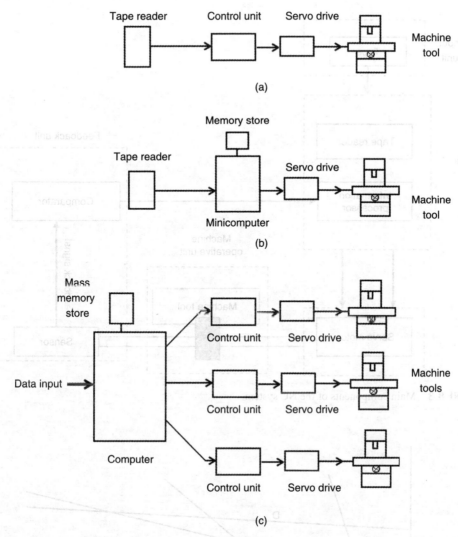

FIGURE 8.2 NC, CNC, and DNC concepts: (a) conventional NC, (b) CNC, and (c) DNC.

With NC flexibility, the setup costs are often less than conventional machines, smaller lot sizes are economical, and less floor space is needed for materials in process and storage. Referring to Figure 8.4, it is obvious that NC cannot compete with fixed-program special-purpose machines and tools when producing large quantities of pieces. In this regard, cam-operated automatic machines are simple, direct, and fast for turning operations, while transfer machines are specialized for machining certain products more effectively than NC machines for large quantities. NC cannot compete in terms of the machining cost with the special-purpose machines used for mass production. Their ultimate benefit is achieved when machining small and medium-size runs. Generally, NC can be used when

- The tooling cost is high compared to the machining cost by conventional method
- The setup time is large in conventional machining
- Frequent changes in tooling and machine setting are required
- Parts are produced intermittently
- Complex-shaped components are needed

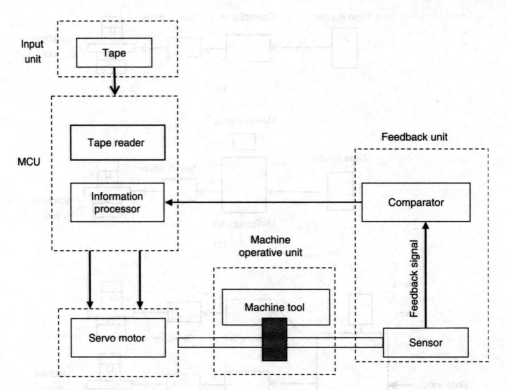

FIGURE 8.3 Main components of the NC system.

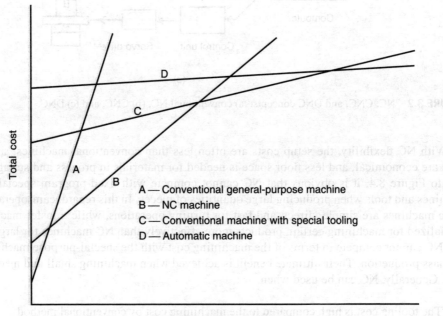

FIGURE 8.4 NC cost compared to other methods.

- Expensive parts where human errors are costly
- Design changes are frequent
- 100% inspection is required

Advantages of NC include the following:

1. *Greater flexibility.* With NC, a wide variety of operations can be performed, changeovers from one run to another through tape or program changes can be made rapidly, and design changes to parts can be made rapidly through minor changes to the part program.
2. *Elimination of templates, models, jigs, and fixtures.* The NC control tape takes over the job of locating the cutting tools, which eliminates the design, manufacture, and the use of templates, jigs, and fixtures.
3. *Easier setups.* By using more simple work holding and locating devices, the operator does not have to set table limit stops or dogs, or depend on the feed screw dials when setting up for machining.
4. *Reduced machining time.* Machining with NC allows the use of a wider range of speeds and feeds than conventional machine tools. Optimum selection of feed rates and cutting speeds is ensured. The NC equipment can also move from one cutting operation to the next faster than the operator, which significantly reduces the total machining time.
5. *Greater accuracy and uniformity.* During NC machining, no human errors are possible and machining of the same part is performed in the same way through the stored tape or program, which improves the uniformity and interchangeability of the machined parts. Therefore, inspection time is greatly reduced, and is necessary for the first piece only, in addition to random checks for critical dimensions. Hence, scrape and rework are greatly reduced or completely eliminated by using NC.
6. *Greater safety.* The operator is not as closely involved with the actual machining operations as with conventional machine tools. As the tape is checked out before actual production runs, there is less chance of machine damage that may cause human injuries.
7. *Conversion to the metric system.* An NC system can be converted to accept either inch or metric inputs.

Disadvantages of NC include the following:

1. NC follows programmed instructions that can lead to machine destruction if not properly prepared.
2. NC cannot add any extra machining capability to the machine tool, as no more power from the original drive motor and no more table travel than originally built into the machine tool can be added.
3. NC machines cost five to ten times more than conventional machines of the same working capacity. The machine, therefore, cannot remain idle and needs special maintenance.
4. The skills required to operate an NC are usually high, because of the sophisticated technology involved, which requires part programmers, tool setters, punch operators, and maintenance staff who are more educated and well-trained than conventional machine operators.
5. Special training for personnel in software and hardware is very important for successful adoption and growth of the NC technology.
6. NC requires high investments in terms of wages of highly skilled personnel and expensive spare parts.

8.2 COORDINATE SYSTEM

8.2.1 MACHINE TOOL AXES FOR NC

In NC, the standard axis system is used to plan the sequence of positions and movements of the cutting tool. A drilling machine can be described using two or three axes X, Y, and Z. There are three rotational axes a, b, and c around X, Y, and Z, respectively. The right-hand rule, shown in Figure 8.5, defines the relative positions of X, Y, and Z; Figure 8.6 shows the direction of positive rotation a, b, and c around X-, Y-, and Z-axes. The machine axis and motion nomenclature are published according to the Electronics Industry Association (EIA) standard. Figure 8.7 shows the designation of some are typical machine tools. Accordingly, in NC turning and cylindrical grinding machines X- and Z-axes are only required, where X is in the radial direction and Z is in the axial direction of the WP.

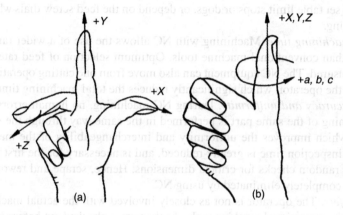

FIGURE 8.5 Right-hand rule.

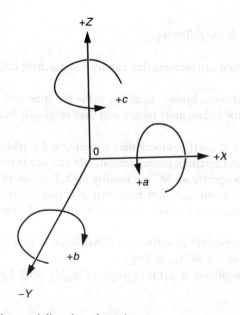

FIGURE 8.6 Relative positions and direction of rotation.

Numerical Control and Computer Numerical Control Technology

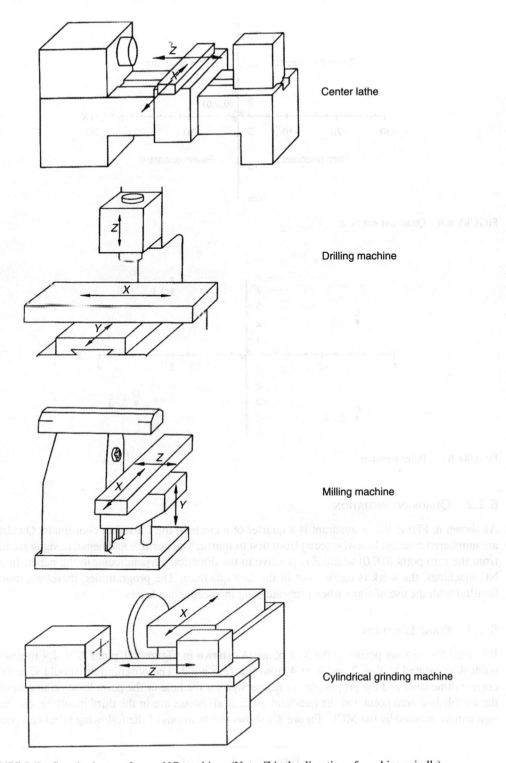

FIGURE 8.7 Standard axes of some NC machines (Note: Z is the direction of machine spindle).

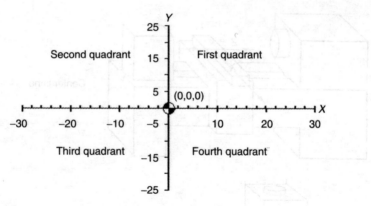

FIGURE 8.8 Quadrant notation.

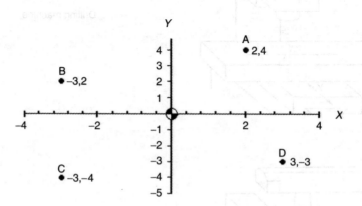

FIGURE 8.9 Point location.

8.2.2 QUADRANT NOTATION

As shown in Figure 8.8, a quadrant is a quarter of a circle in the Cartesian coordinate. Quadrants are numbered counterclockwise (ccw) from first to fourth. The positive and negative signs are taken from the zero point (0,0,0) where Z is positive in the direction perpendicular to the paper. In most NC machines, the work is carried out in the first quadrant. The programmer, therefore, must be familiar with the use of signs when programming in specific quadrants.

8.2.3 POINT LOCATION

It is used for locating points in the X–Y plane. As shown in Figure 8.9, point A = 2,4 means that point A is located at $X = 2$ and $Y = 4$ from the zero point. The programmer should specify the correct dimension and the proper plus or minus sign for the hole or the point location in relation to the established zero point and the quadrant used. If all points are in the third quadrant, the minus sign can be avoided by the MCU. Figure 8.9 shows the locations of the following tabulated points:

Point	X	Y
A	2	4
B	-3	2
C	-3	-4
D	3	-3

Numerical Control and Computer Numerical Control Technology

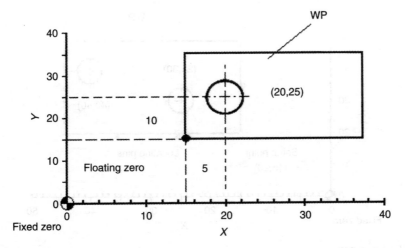

FIGURE 8.10 Fixed zero and floating zero.

The centerline of the machine spindle is usually taken as the Z-axis, which is positive in the direction from the WP toward the tool. The plane formed by the X- and Y-axes, as in algebra, is perpendicular to the Z-axis. Although machine tools of two or three axes can be easily programmed, machine tools of four or five axes require computer assistance in writing the NC programs.

8.2.4 Zero Point Location

The zero point location is where X, Y, and Z and the point from which all coordinate dimensions are measured. This point can be either fixed by the manufacturer (fixed zero) or determined by the programmer (floating zero). In the fixed zero location, the point of $X = 0$ and $Y = 0$ is located at specific point on the machine table and cannot be changed. Accordingly, the coordinates of the center of the hole in Figure 8.10 are (20,25). Floating zero is found in some NC machine tools where the programmer can select the location of the zero point at any convenient spot on the machine table. Accordingly, the center of the hole location is (5,10). Figure 8.10 shows fixed and floating zeros.

8.2.5 Setup Point

The setup point is actually on the WP or the fixture holding the WP, which tells the setup person where to place the part or the fixture (holding the part) on the machine table. Hence, holes and other machining operations are performed in the correct locations on the part when the tape or program is used. As can be seen in Figure 8.11, the setup point may be the intersection of two previously machined edges. In other situations, it may be the center of a previously machined hole, or a dowel or a hole on a given location on the fixture. The setup point must be accurately located in relation to the zero point. In some cases, the zero and the setup points may coincide with each other in one point. The setup point should keep the part in a convenient place on the machine tool that ensures the ease of loading and unloading of the machined part.

8.2.6 Absolute and Incremental Positioning

In absolute positioning, the tool locations are always defined in relation to the zero point. This setup is easy to check and correct and programming mistakes affect only one line of the NC program.

In incremental positioning, the next tool position or location is defined with reference to the previous tool location, which is usually considered to be (0,0). In such a system, any

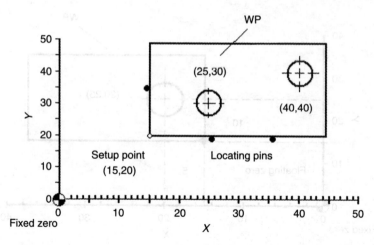

FIGURE 8.11 Setup point.

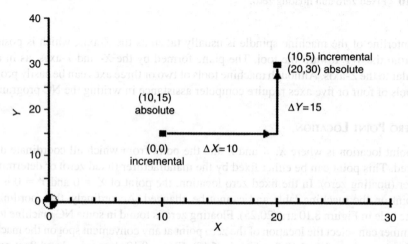

FIGURE 8.12 Absolute and incremental positioning.

mistakes will affect all subsequent programmed positions. To check the incremental positioning, the tool must return to the original position of the program. Figure 8.12 shows an example of the absolute and incremental positioning methods.

8.3 MACHINE MOVEMENTS IN NUMERICAL CONTROL SYSTEMS

NC control systems are built to provide specific movements such as simple movements used in drilling holes or complex ones used in the milling of dies and mold cavities. These movements include the following:

Point-to-point (PTP) NC: As shown in Figure 8.13, when drilling holes in a WP, the following steps are performed:
1. The spindle goes to the specific hole location on the WP (X,Y) position.
2. The tool then stops to perform drilling, reaming, boring, taping, counterboring, and countersinking at the programmed feed rate.
3. When the operation is finished, the tool goes to the next location, stops, and performs another operation.
4. The tool does not contact the WP as it moves from one point to another at a traverse speed of 2.54 m/min in a tool path that is not important.

Numerical Control and Computer Numerical Control Technology

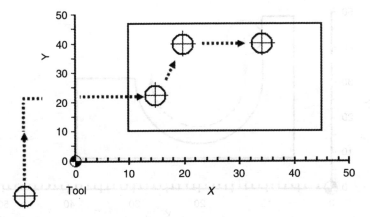

FIGURE 8.13 PTP control.

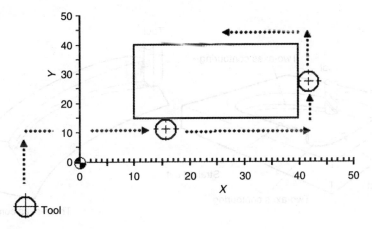

FIGURE 8.14 Straight-cut NC.

Straight-cut NC: This movement is used during the machining of successive shoulders in a WP or cutting rectangular shapes on the milling machine, as shown in Figure 8.14. Such a system is equipped to control the feed rate as the tool travels from one point to another. Tool movements are restricted to lines parallel to the coordinate axes of the machine or at 45° to the axes. NC machine tools are often equipped with PTP systems in addition to the straight-cut movement that can be used for hole drilling and simple milling operations.

Contouring (continuous-path) NC: This movement is used for machining contours and other complex shapes. According to Figure 8.15, the tool moves at a controlled feed rate in any direction in the plane described by two axes. The cutting tool motion is limited only by the number and range of axes under control (three to five axes). The method by which a continuous-path system moves the tool from one programmed point to the next is called *interpolation*; *resolution* is the minimum movement that can be commanded by the NC control unit. It is the table or the slide movement resulting from a single pulsed command. If a single pulse produces 0.025 mm of slide (table) movement, the resolution is then 0.025 mm. Machine tools are available at different resolutions depending upon the degree of precision required by the user of the machine. The smaller the resolution, the higher the possible accuracy. PTP systems are normally built with 0.026 mm, and contouring NC systems are built with 0.0025 mm. Figure 8.16 shows typical profiles cut by straight-cut and contouring NC.

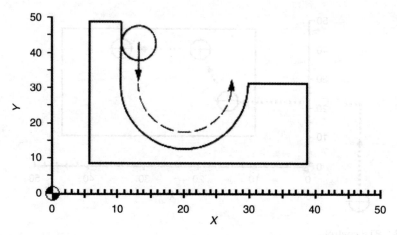

FIGURE 8.15 Contouring NC.

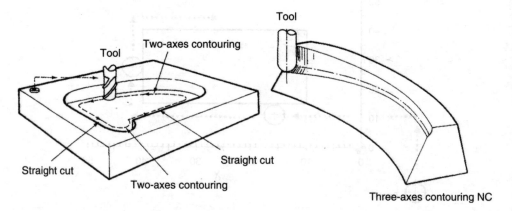

FIGURE 8.16 Examples of straight and contouring NC.

Combination systems: Although a PTP system is effective for drilling, taping, and boring operations, straight-cut NC is effective for face milling. Combination systems may include PTP and straight-cut systems, which are common in machine tools used for milling, drilling, and boring. Additionally, PTP and continuous-path systems are used for machining profiles in addition to the work done by PTP system.

8.4 INTERPOLATION

Interpolation is the method of getting from one programmed point to the next so that the final WP shape is a satisfactory approximation of the programmed design. Today, continuous-path NC control systems can be supplied with four general types of interpolation: linear, circular, parabolic, and cubic interpolation.

Linear interpolation. The cutting tool motion between two points is controlled in a straight line. Curves are then broken in to a series (number) of straight-line tool movements that is sufficient enough to produce an approximation of the desired WP shape within the given tolerance. The smaller the tolerance, the more the points required to define the desired shape, as shown in Figure 8.17a.

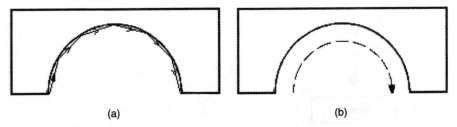

FIGURE 8.17 (a) Linear and (b) circular interpolations.

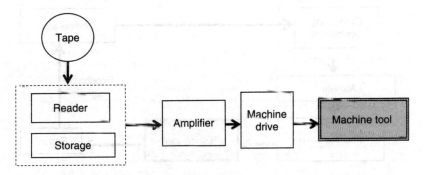

FIGURE 8.18 Open-loop control system.

Circular interpolation. For machining an arc, the points needed are the coordinates of the center point and the start and finish points. A code is required to specify the direction of the cut in addition to the desired feed rate (Figure 8.17b).

Parabolic interpolation. Produces parabolic tool paths with the minimum inputs and uses fewer blocks than circular interpolation.

Cubic interpolation. Developed with the use of computers where sophisticated cutter paths can be produced with few input data points.

8.5 CONTROL OF NUMERICAL CONTROL MACHINE TOOLS

The movement of NC machine tools is controlled using automatic control systems, which include the MCU, the drive motors, and other equipment. The main function of the control unit is to read and interpret instructions, store information until the time comes to use it, and send signals to the machine tool to get the appropriate movement for creating the finished WP. The two types of control systems are as follows:

1. *Open-loop control system.* As shown in Figure 8.18, this system is used in machine tools to perform specific movements without any check on whether the desired movements actually take place. Such a control system is simple and inexpensive. Because there is no provision for checking the actual movements, they must have extremely accurate, reliable, and responsive drive mechanisms, such as stepping motors. In this case, the frequency and the total number of pulses determine the direction (+ or − charge), speed, and distance of travel resulting from the stepping motor.
2. *Closed-loop control system.* In such a system (Figure 8.19) the actual movement is checked by the feedback system. The measured slide or tool movement is then compared with the original input instruction, and the difference produces an error signal that is used to drive the system. In this system, the servo mechanisms are used to control the machine tool movements to move the slide or the table to the desired position. In a closed-loop control

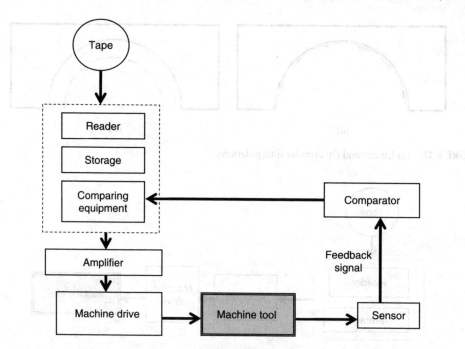

FIGURE 8.19 Closed-loop control system.

system, linear transducers, fitted to the machine table, provide the necessary feedback required for the servomotors to position the worktable accurately in accordance with the requirements of the program. Rotary transducers measure the angular displacement of a machine rotary element such as a lead screw or the spindle of the lathe machine. This measurement is essential for synchronizing the rotation of the WP and the axial movement of the tool during screw cutting on the lathe.

The control of NC machine tools includes spindle rotation, slide movements, tooling, work holding, and supplementary functions:

1. *Slide movements*. The accuracy of machined parts is affected by operator's skill, especially when positioning the WP and tool in the correct position to each other. Machine tools may have more than one slide and therefore the slide required to move must be identified. The plane of the slide movement may be horizontal, transverse, or vertical. These planes are referred to as axes and are designated by letters X, Y, and Z. The Z-axis always relates to a sliding motion parallel to the spindle axis. The rate of travel in millimeters per revolution or in millimeters per minute is proportional to the revolutions per minute of the servomotor. Therefore, controlling that motor will in turn control the slide movement. The motor is controlled electronically via the MCU in a variety of ways such as paper tape, magnetic tape, and computer link.

2. *Control of spindles*. Rotary movements are controllable via the machining program. They are identified by letters a, b, or c (see Figure 8.6). Machine tool spindles are driven directly or indirectly by electric motors. The degree of automatic control over these movements includes stopping and starting and speed and direction of the rotation. The speed of the spindle is often infinitely variable and will automatically change as cutting takes place to maintain a programmed surface speed.

3. *Control of tooling*. NC machine tools may incorporate in their design turrets or magazines that hold a number of cutting tools. The machine controller can be programmed to cause

indexing of the turret or the magazine to present a new tool to the machining operation or to facilitate tool removal and replacement when automatic tool changing devices are involved.
4. *Control of work holding.* NC machine tool control can extend to loading the WP by the use of robots and securely clamping it by activating hydraulic or pneumatic clamping systems.

8.6 COMPONENTS OF NUMERICAL CONTROL MACHINE TOOLS

NC machine tools are made from specially designed parts that ensures the production of accurate machined WPs. These parts include the following:

1. *Machine tool structures.* CI is widely used for building NC machine tool structures, as it possesses an adequate strength and rigidity, in addition to a high tendency to absorb vibrations. Complex shapes can also be produced by casting of one-piece box construction, which is heavily ribbed to promote rigidity and stabilized by an appropriate heat treatment. For large machines, fabricated steel structures are used that ensured a reduction in weight while ensuring adequate strength and rigidity. The general use of steel structures is, however, limited by the problems of making complex structures and the low tendency of vibration damping. Concrete is used as a machine tool foundation; it has low cost and good damping characteristics.
2. *Machine spindles.* The machine tool spindles are subjected to a radial load that may cause deflection. Additionally, spindle assembly is subjected to an axial load acting along its axis. Inadequate spindle support leads to dimensional inaccuracies, poor surface finish, and chatter. Spindle overhang of the turning and other horizontal machines must be kept to a minimum, as shown in Figure 8.20. Vertical machining spindles may slide up and down, which makes them extended, thus raising the risk of deflection. To overcome such a problem, the spindle assembly is made to move up and down (Figure 8.21a). To avoid the possible twist of the spindle housing that is located between two substantial slideways a bifurcated or two-pillar structure is used (Figure 8.21b).

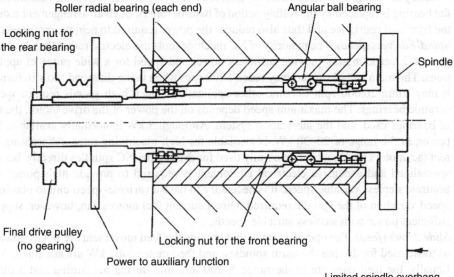

FIGURE 8.20 Spindle assembly for NC machine tools.

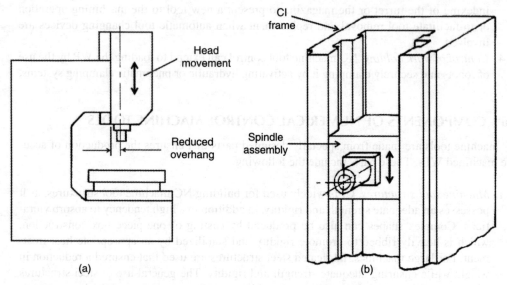

FIGURE 8.21 Minimizing spindle deflection.

3. *Lead screws*. NC machines are fitted with recirculating ball screws that replace the normal sliding motion with the rolling motion, resulting in the reduction of the frictional resistance. As shown in Figure 8.22, the balls make opposing point contact, which eliminates backlash. Ball screws, in comparision to Acme screws, offer longer life, less wear, less frictional resistance, less necessary drive power, higher traversing speeds, no stick/slip effect, and more precise positioning over the total life of the machine.
4. *Machine slides*. Machine tool slides must be smooth and have minimum frictional resistance and low wear to ensure dimensional inaccuracies. Slides of flat bearing surfaces are widely used for NC machine tools. Such surfaces are usually hardened, ground, and coated with polytetrafluoroethylene (PTFE). The coated material has low coefficient of friction plus the tendency of retaining lubricant with superior load-carrying capacity. In some machines, the flat bearing is replaced by the rolling action of balls or rollers. Such an arrangement reduces the frictional resistance and thus also reduces the power required to achieve movement.
5. *Spindle drives (speed)*. The majority of NC machine tools use electric rather than hydraulic motors. Electric motors provide sufficient power and speed for a wide range of applications. The main spindle speed may reach 5000 rpm when using diamond tools in turning. It may attain 20,000 rpm in some other applications. Such high speeds require special ceramic bearings. The maximum speed depends on the power of the drive motor, the type of bearings used, and the lubrication system. Although 5 kW is normally available, high power, in the range of 20–30 kW, is available for high machining rates. Alternating current (ac) motors have not been generally used for driving the NC spindles directly, because specialized and expensive electrical equipment is required to provide high power with accurate stepless variable speed. It is necessary to have a variable-speed unit to obtain the speed variation of the spindle required. Direct current (dc) motors can, however, supply a sufficient power with stepless variable speeds.
6. *Slide drives (feed)*. The operative units that provide the feed movement are not as powerful as those used for driving the main spindles, and feed motors of 1 kW are adequate. Additionally, the feed rates are in the range 5–200 mm/min during machining and 5 m/min during rapid positioning. Such feed rates can be obtained using a screw and nut driving system where a screw of 5 mm pitch rotates at 1–40 rpm during machining and 1000 rpm

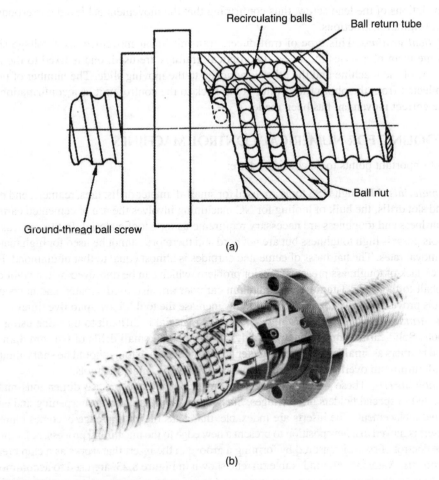

FIGURE 8.22 External recirculating ball screw.

during rapid positioning. It is essential that the movement provided by the feed motors is controlled very precisely and accurately. Generally, dc motors are used in closed-loop control systems for moving the tools or WP under precise control. Open-loop systems use stepper motors where the drive unit receives a direction input (clockwise [cw] or counter clockwise [ccw]) and pulse inputs. For each pulse received, the drive unit manipulates the motor voltage and current, causing the motor shaft to rotate by a fixed angle (one step). The lead screw converts the rotary motion of the motor shaft into linear motion of the WP or tool feed.

7. *Power units for ancillary services.* Alternating current (ac) induction motors are generally used for coolant pumps, chip removal equipment, and driving hydraulic motors, where the only control required is on/off switching.

8. *Positional feed back.* Rotary-type synchronic systems transmit angular displacement to voltage signals. Such systems are composed mainly of the rotor and the stator. The rotor rotates with the lead screw, while the stator is fixed around its periphery. The stator winding is fed with the electrical power at a rate that is determined by the MCU in response to the digital information related to the required slide movement, received via the part program. As the lead screw rotates, a voltage is induced in the rotor that will vary according to the angular position of the lead screw in relation to the stator windings. Information related to the induced voltage is fed back to the control unit, which counts the number of

revolutions of the lead screw, thus confirming that the movement achieved corresponds to the original instructions.
9. *Optical gratings.* This type of transducers transmit linear movement as a voltage signal in the form of a series of pulses. Two optical gratings are used; one is fixed to the main frame of the machine and the other is attached to the moving slide. The number of pulses collected from the photo transmitter is fed back to the control unit as a confirmation that the correct movement has been made.

8.7 TOOLING FOR NUMERICAL CONTROL MACHINES

The most important points to be considered are:

Tool materials. Although HSS tools are used for small-diameter drills, taps, reamers, end mills, and slot drills, the bulk of tooling for NC machining involves the use of cemented carbides. Hardness and toughness are necessary requirements for a tool material. In this regard, HSS tools possess high toughness but are not hard and therefore cannot be used for high material removal rates. The hardness of cemented carbides is almost equal to that of diamond. However, lack of toughness presents a major problem, which can be improved by the addition of cobalt to the WCs. Titanium and tantalum carbides are also used. Coated and nanocoated tools provide high wear resistance and thus increase the tool life by up to five times.

Solid carbide tools. These are used when the WP material is difficult to machine using HSS tools. Solid carbide milling cutters of 1.5 mm diameter, small drills of 0.4 mm diameter, and reamers as small as 2.4 mm diameter are available. Such tools should be short, mounted with minimum overhang, and used on vibration-free NC machine tools.

Indexable inserts. These have the correct cutting geometry and precise dimensions and are located in special holders or cartridges. Such inserts do not require resharpening and ensure rapid replacement. The inserts are indexable; that is, as the cutting edge becomes blunt, the insert is moved to a new position to present a new edge to the machining process. A facility for the control of swarf is ensured by forming a groove in the insert that works as a chip breaker.

Tool turrets. Automatically indexable turrets, shown in Figure 8.23, are used to accommodate cutting tools. These turrets are programmed to rotate to a new position so that a different tool can be presented at work. Indexable turrets are used in the majority of turning centers

FIGURE 8.23 Indexable tool turrets.

as well as some NC milling and drilling machines. Turrets are now available that can accommodate eight to ten tools. Some machines have two turrets; one is in use while the other is loaded with tools for a particular job and attached to the machine when required. Turrets fitted to NC drilling and milling machines have to rotate their tools at a predetermined speed, as it acts as a spindle. Tool stations are numbered according to the tooling stations available. When writing the part programs, the programmer provides each tool with a corresponding number in the form of a letter T followed by the corresponding numerical identity in two digits T01, T02, and so on.

Tool magazines. A tool magazine, shown in Figures 8.24 and 8.25, is indexable storage used on a machining center to store tools not in use. They are available as rotary drum and chain types. When the tool is called into use, the magazine indexes by the shortest route to bring the tool to a position where it is accessible to a mechanical handling device. At the end of use, the tool is returned to its slotted position in the magazine before calling the next tool. Rotary drums with 12–24 stations are available and 24–180 stations are available for the chain type.

Tool replacement. Cutting tools should be replaced when affected by wear or breakage. Tool changes must be made rapidly. The replaced tool must be of identical dimensions to the original one, which is achieved by using a qualified or preset tooling. Temporary modifications could also be achieved by offsetting the tool from its original datum. The preset tooling concept, shown in Figure 8.26a, is used for both turret and spindle-type machines. For NC machines, the cutting tool is preset to a specific length and diameter while it is off the machine using special fixtures and gauges. The tool length is used by the part programmer to develop the Z-axis coordinate. Preset tool holders and boring bars are available for many NC turning machines. Once these are preset to the appropriate dimensions, inserts can be changed and WP tolerances are maintained by minor adjustment of the tool offset switches.

Qualified tool holders for NC lathes are ground to standardized dimensions at close tolerances and no presetting is required (Figure 8.26b). The qualified tool holder is usually inserted into the turret tool block and tightened in position. The dimensions provided by the manufacturers of the qualified tool holder are used by the part programmer and minor adjustments are easily made by tool offset switches.

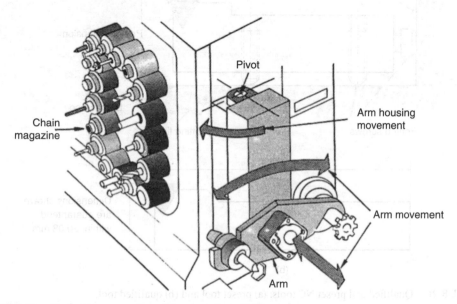

FIGURE 8.24 Chain-type magazine with automatic tool changer.

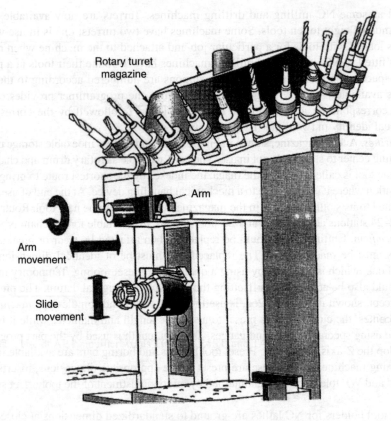

FIGURE 8.25 Rotary-type magazine with automatic tool changer.

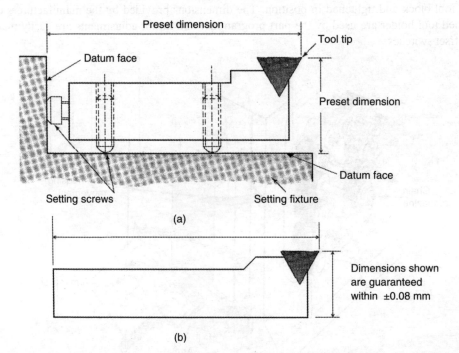

FIGURE 8.26 Qualified and preset NC tools: (a) preset tool and (b) qualified tool.

8.8 NUMERICAL CONTROL MACHINE TOOLS

The most important types of NC machine tools are:

1. *NC drilling machines.* An NC drilling machine holds, rotates, and feeds the drilling tool into the WP. They are available in a wide range of types and sizes that are built with single spindle or multiple spindles. Some machines are equipped with turrets and others with tool-changing mechanisms. Either two or three axes is available, and some drilling machines are even capable of performing milling operations.
2. *NC milling machines.* These machines (Figure 8.27) are used to machine flat surfaces and produce contours and curved surfaces. The orientation of the spindle may be horizontal or vertical and provided with single spindle or multiple spindles. Milling machines with two perpendicular spindles provide machining a hole and a vertical surface simultaneously. Such a facility is useful when machining large components, as shown in Figure 8.28. On the other hand, machining of vertical and horizontal plains, simultaneously could be secured in this setup by replacing the boring tool in the vertical spindle by a face milling cutter. NC milling machines may have from two to five axes under tape control. NC milling machines can do some of the work normally performed on NC drilling machines such as drilling, boring, and tapping.
3. *NC turning machines.* Lathes are primarily used for producing cylindrical shapes in addition to cutting tapers, boring, drilling, and thread cutting. NC lathes are equipped with either straight-cut or continuous-path control systems. Most of NC lathes produced today are equipped with continuous-path control and circular interpolation. They are capable of tool offset so that the machine operator can make fine adjustments in the cutting tool location to achieve the required part size. Figure 8.29 shows a typical NC turning machine.

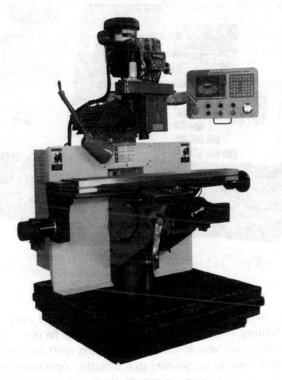

FIGURE 8.27 Typical CNC milling machine. (From Harding Inc.)

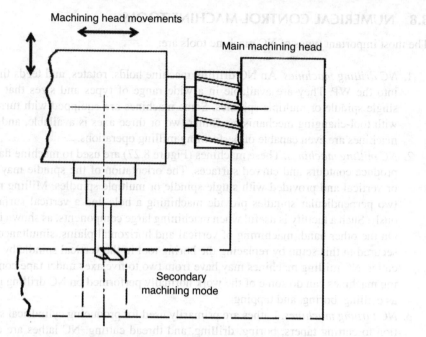

FIGURE 8.28 Machining a hole and a vertical surface simultaneously. (From Gibs, D., *An Introduction to CNC Machining*, ELBS Cassell Publishers Ltd., London, 1988. With permission.)

FIGURE 8.29 CNC lathe QUEST 10/56 of Harding Incorporation. (From Harding Inc. With permission.)

4. *NC machining centers.* Machining centers perform a wide range of operations that include milling, drilling, boring, tapping, countersinking, facing, spot facing, and profiling. Machining centers are able to change the cutting tools automatically, which allow most of the machine time to be devoted to the cutting operation. Most NC machining centers have three axes. In a four-axes system, the fourth axis is used to rotate the table, which allows for the machining of four sides of a part. In many cases, it is possible to

Numerical Control and Computer Numerical Control Technology

FIGURE 8.30 Five-axes vertical machining center (5ax400) of Harding Incorporation. (From Harding Inc.)

FIGURE 8.31 CNC (Super Quadrex 250M) turning center. (From MAZAK Corporation.)

machine a part completely without removing it from the machine. Figure 8.30 shows a typical machining center.

5. *NC turning centers.* These machines combine the features of bar-type, chucking-type, and turret lathes. They are built with four axes of control and are also equipped with continuous-path NC systems with circular interpolation. Their design may include a slanted or vertical bed rather than the horizontal one normally used with conventional center lathes. The capabilities of turning centers can be extended by providing two turrets such that two tools can cut simultaneously. Power-driven tool holders (that rotate when the WP is stationary) permit milling of flats, keyways, and slots in addition to the drilling of holes offset from the machine axis. Figure 8.31 shows a typical turning center. The use of tooling magazines extends the range of tooling that may be used as shown in Figure 8.32.

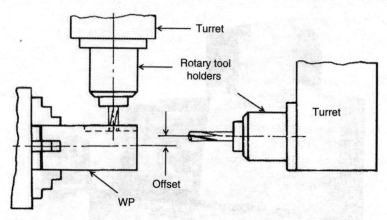

FIGURE 8.32 Additional tooling facilities. (From Gibs, D., *An Introduction to CNC Machining*, ELBS Cassell Publishers Ltd., London, 1988. With permission.)

8.9 INPUT UNITS

Data can be input in to the MCU using one of the following methods:

Manual data input (MDI). This method is normally used for setting the machine and editing the program, as well as writing complete simple programs. For NC machines with noncomputerized control units, data recording facilities are not often available. In case of CNC machines, the computer retains the data, so that it can be transferred to a recording medium such as magnetic tape or disk or transferred back to the machine when required.

Conversational MDI. This method involves the operator pressing appropriate keys on the control console in response to questions that appear in the visual display unit (VDU). This method is faster than methods that require the use of data codes.

Punched tape. Punched tapes are of standard 1 in. width that use eight 0.072 in. holes across the width of the tape and one 0.046 in. sprocket/feed hole between tracks 3 and 4. Punched tapes can be read inexpensively, are less sensitive to handling, are inexpensive to purchase, and require less equipment for manufacturing and less costly space for data storage.

The binary coded decimal (BCD), shown in Figure 8.33, is used for coding digits on the tape. Accordingly, five of the eight tracks per channel are assigned the numerical values 0, 1, 2, 4, and 8 so that any numerical value from 0 to 9 can be represented in one row of the tape. The combination of punched holes per bit in the tape establishes the values associated with that row.

The EIA RS-244-A system and the American Standard Code for Information Interchange (ASCII) RS-358 systems are available for NC and are currently used for coding the numbers on the tape as shown in Figure 8.33. The EIA RS-244-A system is the commonly used, while ASCII-coded input is optional in many of today's NC systems. It should be mentioned here that the RS-244-A coding system involves the use of odd parity, in which track 5 makes certain that an odd number of holes (not including the sprocket hole) appear on every row of the tape whereas the ASCII subset uses even parity, in which an extra hole is added to track 8 in the tape to ensure an even number of holes in each row. The BCD code format is the same in both code systems. All numerical and alphabetical codes along with some special characters and function codes are available in both systems.

Magnetic tape. Magnetic tapes, in the form of cassettes, are widely used for transmitting data. They require expensive equipment for program recording and reading. The programmer

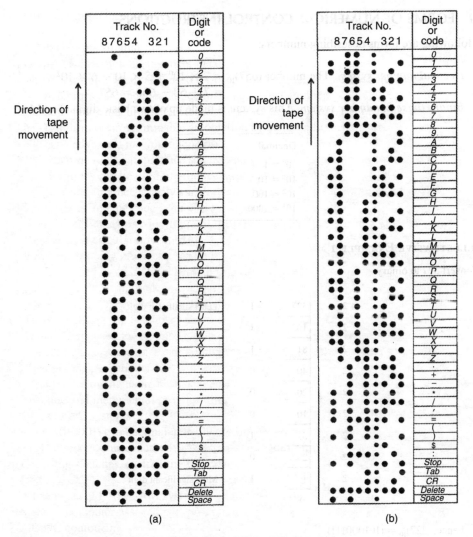

FIGURE 8.33 (a) EIA RS-244-A and (b) RS-358 (ASCII) coding systems.

cannot see the recorded data and therefore recording errors cannot be seen as punched tapes. Magnetic tape requires special storage space and must be handled carefully.

Portable electronic storage unit. In this method, the data transferred into the storage unit away from the machine shop are carried to the machine, connected to the MCU, and data are then transferred. Data transfer is high, and the capacity of such units is high, so that a number of programs can be accommodated at a time.

Magnetic disk input via computer. In this method, it is possible to transfer data stored on a floppy disk into the computer and hence into the MCU. Similarly, data on the control unit can be extracted and recorded. The rate at which the data can be transferred or retrieved using a disk is faster than when using a tape, and the storage area is also much greater.

Master computer. The prepared program stored on the memory of a master computer is transferred to the microcomputer of the MCU when required. Such an arrangement, also described earlier, is what is known as DNC.

8.10 FORMS OF NUMERICAL CONTROL INSTRUCTIONS

The following are forms describing numbers:

1. *Decimal number system.* The number $(657)_{10} = 7 \times 10^0 + 5 \times 10 + 6 \times 10^2$
$$= 7 + 50 + 600 = 657$$
2. *Binary (base 2) number system.* This system is made up of two basic digits 0 and 1:

Decimal	Binary
$10^0 = 1$	$2^0 = 1$
$10^1 = 10$	$2^1 = 2$
$10^2 = 100$	$2^2 = 4$
$10^3 = 1000$	$2^3 = 8$

ILLUSTRATIVE EXAMPLE 1

Convert 327 to binary:

2	327	1	Least significant digit
2	163	1	
2	81	1	
2	40	0	
2	20	0	
2	10	0	
2	5	1	
2	2	0	
2	1	1	Most significant digit
	0		

hence, $(327)_{10} = (101000111)_2$.

ILLUSTRATIVE EXAMPLE 2

Convert (101000111) to decimal:

$$(101000111)_2 = 1 \times 2^8 + 0 \times 2^7 + 1 \times 2^6 + 0 \times 2^5 + 0 \times 2^4 + 0 \times 2^3 + 1 \times 2^2 + 1 \times 2^1 + 1 \times 2^0$$
$$= 256 + 0 + 64 + 0 + 0 + 0 + 4 + 2 + 1$$
$$= (327)_{10}$$

Computers cannot work with the decimal system complexity because they are single electronic devices that can only sense the numbers of 0 and 1 that can represent the presence (1) or absence (0) of voltage, light, transistor or magnetic field as follows:

Voltage	On (1) or	Off (0)
Light	On (1) or	Off (0)
Transistor	On (1) or	Off (0)
Magnetic field	On (1) or	Off (0)

NC systems can understand the numbers 0 and 1, which in electrical terms correspond to on or off when sensing pressure, magnetism, light, or voltage. In NC systems, the command is given to the MCU in blocks of data where the *blocks* are made up of a collection of words, arranged in a definite sequence, to form a complete NC instruction that could be understood by the machine. A *word* is a collection of characters used to form a part of an instruction. A *character* is a collection of bits that represent a letter, number, or symbol. A *bit* is a binary digit with a value of 0 or 1 depending on the presence or absence of a hole in a certain row and column on the tape.

8.11 PROGRAM FORMAT

Tape format is the general sequence and arrangement of the coded information on a punched tape. According to the EIA standard, it appears as words made of individual codes written in horizontal lines. The most common type of tape format in current use is the word address format. However, some earlier control systems still use the fixed block or the tab sequential format.

1. *Word address format.* Each element of information is prefixed by an alphabetical character. The alphabet acts as an address that tells the NC system what it must do with the numbers that follow the prefix. If the word remains unchanged, it need not be repeated in the next block:

    N001 X2.000 Y2.500 $F_1$2.50 S573 EOB

2. *Fixed block format.* Contains only numerical data, arranged in a sequence with all codes necessary to control the machine appearing in every block. The instructions are given in the same sequence and all instructions are given in every block, including those unchanged from the preceding blocks. It has no word address letter to identify individual words such as:

    001 2.000 2.500 2.50 573

3. *Tab sequential format.* In this format, a block is given the same sequence as in case of the fixed block format but each word is separated by a tab character. If the word remains unchanged in the next block, the word need not be repeated, but a tab code is required to keep the sequence of words. Because the words are written in a set order, the address letters are not required:

    001 TAB2.000 TAB2.500 TAB2.50 TAB573 EOB

The EIA standard RS-274-A defines the various standard word addresses and describe their use as shown in Table 8.1:

1. *Sequence number function.* This is the first word of a block that is represented by letter N followed by three digits.
2. *Preparatory functions.* The word addresses or G codes relate the various capabilities or functions of particular NC machine tools. These are used as prefixes in developing the NC words used in the programs to command specific machine functions, as shown in Table 8.2.
3. *Dimensional data function.* This is represented by a symbol followed by five to eight digits, as shown in Table 8.1.
4. *Feed rate function.* This is expressed by the letter F_1 plus three digits. The digits may represent the feed rate in millimeters per minute, millimeters per revolution, or the magic-three method (explained in the following section).
5. *Tool selection.* Information regarding the tool is given by a word prefixed by the letter T followed by a numerical code for the tool in use.

TABLE 8.1
EIA RS-274-A Standard Word Addresses

Code	Function
a	Angular dimension around X-axis
b	Angular dimension around Y-axis
c	Angular dimension around Z-axis
d	Angular dimension around special axis, or third feed function[a]
e	Angular dimension around special axis, or second feed function[a]
f	Feed function
g	Preparatory function
h	Unassigned
i	Distance to arc center or thread feed parallel to X
j	Distance to arc center or thread feed parallel to Y
k	Distance to arc center or thread feed parallel to Z
l	Do not use
m	Miscellaneous function
n	Sequence number
o	Rewind application stop
p	Third rapid traverse dimension or tertiary motion dimension parallel to X[a]
q	Third rapid traverse dimension or tertiary motion dimension parallel to Y[a]
r	Third rapid traverse dimension or tertiary motion dimension parallel to Z[a]
s	Spindle speed
t	Tool function
u	Secondary motion dimension parallel to X[a]
v	Secondary motion dimension parallel to Y[a]
w	Secondary motion dimension parallel to Z[a]
x	Primary X motion dimension
y	Primary Y motion dimension
z	Primary Z motion dimension

[a] When d, e, p, q, r, u, v, and w are not used as indicated, they may be used elsewhere.

6. *Spindle speed function.* This is specified in rotational speed in revolutions per minute or the surface speed in meters per minute and is given by the letter S followed by the speed required.
7. *Miscellaneous functions.* In the word address format, miscellaneous functions are represented by the letter M followed by a numerical code for the function required. They are used to command miscellaneous or auxiliary functions of the machine, such as turning on the coolant and starting the spindle in conjunction with the first move of the machine. The standard miscellaneous functions are listed in Table 8.3.

8.12 FEED AND SPINDLE SPEED CODING

8.12.1 Feed Rate Coding

During milling operations, the feed rate is expressed in millimeters per minute or inches per minute (ipm). Additionally, the feed rate is expressed in millimeters per revolution or inches per revolution (ipr) in case of turning machines. Generally, feed rates can be expressed by one of the following methods:

1. *Four-digit field.* This coding process represents the number of digits the system can accept. Accordingly, 12.3 ipm or mm/min will be coded by $F_1 0123$, and 999.9 ipm will be coded by $F_1 9999$.

TABLE 8.2
Some Common Preparatory Codes and Functions

Code	Function
G00	PTP positioning
G01	Linear interpolation
G02	Circular interpolation arc cw
G03	Circular interpolation arc ccw
G04	Dwell
G05	Hold
G08	Acceleration
G09	Deceleration
G17	X–Y plane selection
G18	Z–X plane selection
G19	Y–Z plane selection
G33	Thread cutting, constant lead
G40	Cutter compensation cancel
G41	Cutter compensation left
G42	Cutter compensation right
G80	Fixed cycle cancel
G80–G89	Fixed cycles as selected by manufacturers

TABLE 8.3
Some Miscellaneous or Auxiliary Functions and Codes

Code	Function
M00	Program stop
M01	Optional (planned stop)
M02	End of program
M03	Spindle start cw
M04	Spindle start ccw
M05	Spindle stop
M06	Tool change
M07	Coolant no. 2 on (mist)
M08	Coolant no. 1 on (flood)
M09	Coolant off
M10	Clamp
M11	Unclamp
M13	Spindle cw and coolant on
M14	Spindle ccw and coolant on
M15	Motion +
M16	Motion –
M30	End of tape
M32–M35	Constant cutting speed

2. *Inverse time feed rate coding.* In this case, the feed rate number is expressed as the ratio of the feed rate to the distance traveled, according to the following equation:

$$\text{Feed rate number} = \frac{\text{Feed rate (ipm)}}{\text{Distance traveled (in.)}}$$

For a feed rate of 20 ipm and a distance of 2.6 in.,

$$\text{Feed rate number} = \frac{20 \text{ ipm}}{2.6 \text{ in}} \approx 7.7 \text{ min}$$

If the control system accepts inverse time feed rate coding and a four-digit feed rate field, then 7.7 ipm will be expressed by $F_1 0077$. Similarly, for a feed rate of 50 mm/min and a distance of travel 6.0 mm,

$$\text{Feed rate number} = \frac{50 \text{ mm/min}}{6 \text{ mm}} \approx 8.3 \text{ min}$$

The feed rate number becomes $F_1 0083$.

3. *Coded feed rate.* Such a code is used for low-cost PTP NC systems in which a fixed relation known from a chart supplied by the manufacturer is used. For example, 20 ipm or 50 mm/min will be coded by $F_1 10$, and so on.
4. *Magic-three method.* In this method, 3 is added to the number of digits on the left of the decimal of the numerical value of the feed or speed in metric or imperial units, the addition thus obtained providing the first digit of the feed value. Next two digits in the coded value are the first two digits of the numerical value of the feed. As an example, a feed rate of 35.5 ipm will be coded as follows: add 3 to the number of digits (2). So first digit of the feed is 5. Next, two digits of the feed code will be the first two digits of the numerical value of the feed, which is 35, so the feed rate code will be $F_1 535$. Similarly, 3.55 ipm will be coded as $F_1 435$ and 0.35 ipm will be coded as $F_1 335$. However, if the feed rate is less than 1 ipr, then the rule is modified by subtracting the number of zeros after the decimals from 3 to provide first digit of the coded value. Next two digits of the coded value will be the first two nonzero digits in the feed rate. For example, the feed rate of 0.087 ipm will be coded as $F_1 287$.

8.12.2 SPINDLE SPEED CODING

During NC spindle speeds can be coded using one of the following methods:

1. *Direct revolutions per minute.* In this case, the spindle speed code will have the same value preceded by the letter S. Hence, 1500 rpm will be coded as S1500.
2. *Coded format.* The spindle speed coding is performed through a chart that is supplied by the machine builder, for example, S10 represents a spindle speed of 146 rpm.
3. *Magic-three method.* The magic-three method is applied as described earlier for spindle speed coding. Hence, a rotating speed of 7 rpm will be coded as S470, 500 rpm as S650, and 1500 rpm will have the code of S715.

8.13 FEATURES OF NUMERICAL CONTROL SYSTEMS

NC and CNC controls can be equipped by a variety of features available as standard or optional equipment. These can add generally to the capabilities of the machine tool. The following are common features of NC and CNC control systems:

1. *Feed and spindle speed override.* This feature allows deviation from the programmed feed rate or spindle speed to increase production rate or to reduce the tool wear. Feed rates can be varied as a percentage between 0% and 125% of the programmed rate. As an example, if 250 mm/min is programmed on the tape, and the feed override ratio is 80%, the actual feed will be 200 mm/min. For spindle speeds, 80–100% of the programmed revolutions per minute is possible.
2. *Mirror imaging.* This function is also called axis inversion, and can be controlled by a simple on/off switch. It is used to machine left- and right-hand parts from the same tape, as shown in Figure 8.34.

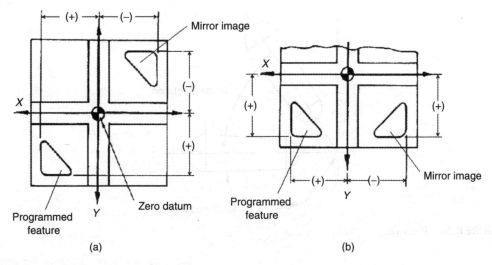

FIGURE 8.34 Mirror imaging: (a) mirror image in two axes and (b) mirror image in one axis.

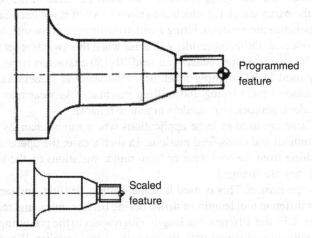

FIGURE 8.35 Scaling.

3. *Scaling.* As shown in Figure 8.35, a range of components can be machined, varying in size, from one set of programmed data.
4. *Rotation.* This technique is mainly required for milling and drilling operations. It enables the cutter path to be rotated by an angle, and repeated if required at another angle as shown in Figure 8.36. This makes programming of complex shapes relatively easy.
5. *Jog.* The jog facility enables the machine operator to manually move the machine slides through the control console. Jog is used for establishing a datum at the initial setting of the machine, stopping an automatic sequence, moving the machine slides for WP measurements, and tool changing due to breakage. It is desirable to restart the program from the point it was interrupted. Therefore, most control systems have the facility to return from jog, which returns the machine slides to their original position via a button on the control console.
6. *Position displays.* These are used for setting the machine and inspecting parts on the machine tool. Readouts can be in the imperial (inch) or metric (mm) scales.

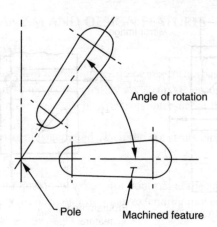

FIGURE 8.36 Rotation.

7. *Switchable input format.* Using a selector switch on the control console, the operator can choose either EIA tape format (RS-244-A) or the ASCII tape format (RS-358). A warning arises if the switch is in the wrong position. Automatic selection of the code can be made depending on the parity check for which the selector switch is not necessary.
8. *Switchable imperial or metric input.* Using a switch on the control console, imperial or metric format can be selected. Different results will arise when this switch is set incorrectly.
9. *Tape readers.* Mechanical tape readers can read 10–150 characters (rows) per second and are commonly used in NC drilling machines. Photoelectric readers can read 150–600 rows/s, in the case of NC turning and milling machines. As wear may arise in case of mechanical readers, photoelectric readers are more reliable.
10. *Tool offsets.* These are used in lathe applications where minor changes are made to the program longitudinal and cross-feed motions. In such a case, the operator avoids over- or undersize resulting from the tool wear or from minor variations on the size and shape of the tools when they are changed.
11. *Tool length compensation.* This is used in some milling, drilling, and taping machines to compensate for different tool lengths of drills, boring bars, reamers, and milling cutters. As shown in Figure 8.37, the difference in length with respect to the presetting tool is recorded and manually entered and stored with the associated tool number. Whenever these tools are called by the program instructions, their respective compensation values are activated and automatically taken into account in the tool motion.
12. *Cutter diameter compensation.* This allows the use of a cutter diameter that is different from the diameter used in developing the original part program (Figure 8.40).
13. *Operator control features.* These features include on/off, MDI, manual jog control for the machine axes, sequence number search, sequence number display, and the slide hold to inspect cut or tool conditions and then restart the cycle.
14. *Canned cycles.* These are control instructions found in many PTP control systems and in some continuous path systems. This feature allows common repetitive machining patterns, such as drilling, milling, threading, and turning, to be done automatically with a single command.

8.14 PART PROGRAMMING

The part program is a computer program containing a number of lines, instructions, or statements called NC blocks that describe the detailed plan of machining instructions proposed for the part. It is written using vocabulary understood by the MCU in terms of standard words, codes, and symbols. Part programs can also be written in higher languages such as, automatically programmed

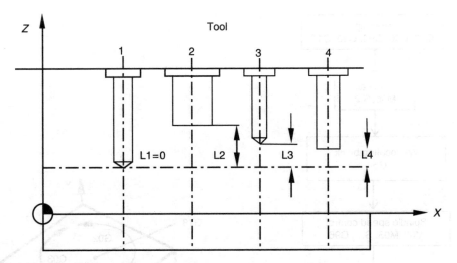

FIGURE 8.37 Tool length offset.

tools (APT), adaptation of APT(ADAPT), extended subset of APT(EXAPT), and so on. These programs can be converted into the machine tool language with the help of processors. Programming in APT is mostly processed with the help of computers and is thus known as computer-assisted part programming. Part programs can also be directly developed using CAD/CAM systems such as Unigraphics, ProEngineer, or CAM systems such as Master CAM, Surf CAM, and others.

The NC part programming procedure includes the following steps:

1. Process planning that determines the sequence of operations to be performed during machining the part.
2. Part programming that is concerned with the documentation of the planned sequence of operations in a special format known as the manuscript.
3. Tape preparation on the basis of instructions written in the program manuscript prepared by the part programmer.
4. Tape verification and checking by machining foam or a plastic material or running the tape through a computer program that indicates the contents and errors in the tape.
5. The corrected tape is then used for actual production.

The data required for part programming include the following items:

1. Machine tool specifications
2. Specifications of tools
3. WP geometry and its material specifications
4. Speed and feed rate tables

Using the part drawing, the programmer determines the sequence of operations, speeds, and feeds for various operations and determines the magnitude of various motions decided. He prepares the planning sheet and writes the instructions in a coded format for the MCU. The part programs are written in blocks using the standard codes and following the sequence shown in Figure 8.38.

Program plane identifier. As shown in Figure 8.39, G17 is used for programming in X, Y; G18 in X, Z; and G19 for programming in Y, Z planes.

Tool diameter compensation. The control unit offsets the path of the tool so that the part programmer can program the part as it appears. In this case, the same program can be

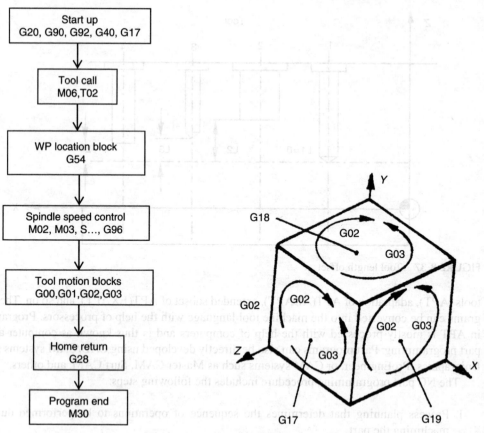

FIGURE 8.38 Part programming procedure. **FIGURE 8.39** Plane identifiers G17, G18, and G19.

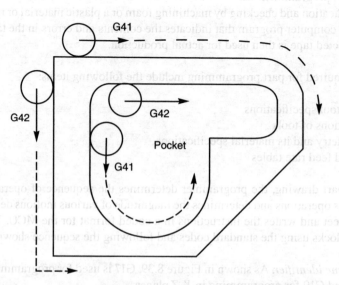

FIGURE 8.40 Tool diameter compensation: right G42, left G41, and G40 to cancel.

used with different cutters. The amount of offset equals the cutter radius (Figure 8.40). The following G codes are used:
- G41, for cutter diameter compensation to the left
- G42, for cutter diameter compensation to the right
- G40, to cancel cutter diameter compensation

WP coordinate setting G54. This setting describes the distance from the tool tip at the home position to the Z zero position of the part as shown in Figure 8.41. The use of G02 and G03 codes is shown in Figures 8.42 and 8.43. G02 and G03 are used for cutting arcs in cw and ccw directions, and G01 is used for cutting straight lines.

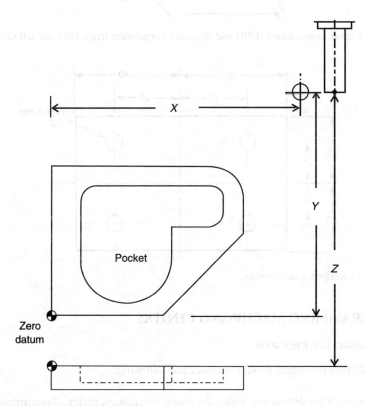

FIGURE 8.41 WP coordinate setting (G54).

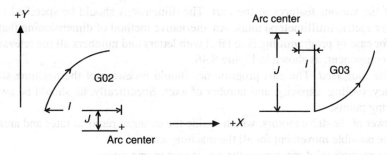

FIGURE 8.42 Cutting arcs using G02 and G03 codes.

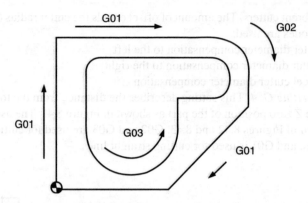

FIGURE 8.43 Linear interpolation (G01) and circular interpolation (right G02 and left G03).

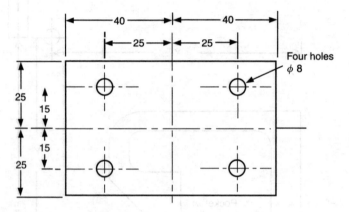

FIGURE 8.44 Centerline dimensioning.

8.15 PROGRAMMING MACHINING CENTERS

8.15.1 PLANNING THE PROGRAM

Planning for NC part programs should consider the following:

1. *Part drawing.* This drawing includes the shape, tolerances, material requirements, surface finish, and product quality. There are several methods of part dimensioning. In the centerline method (Figure 8.44), the part centerline is used as a reference point for dimensioning the component. The coordinate or base line method (Figure 8.45) enables easy transcription of the various features of the part. The dimensions should be specified in imperial (inch) or metric (millimeters) units. An alternative method of dimensioning that is convenient for ease of programming is to label with letters and numbers all the relevant features of the component, as shown in Figure 8.46.
2. *Machine tool used.* The part programmer should be aware of the machine size, power, accuracy, tooling, capacity, and number of axes. Specifically, he should be aware of the following points:
   a. Power of the drive motors, which decide the material removal rates and area of cut
   b. The possible movement for all the machine axes
   c. The possibility of machining the whole part in one setup
   d. The spindle speeds available
   e. Machine rigidity and accuracy capabilities

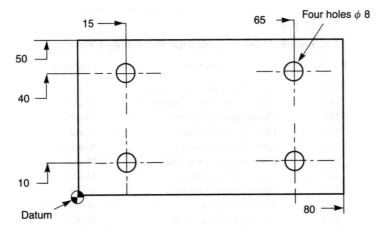

FIGURE 8.45 Coordinate or baseline dimensioning.

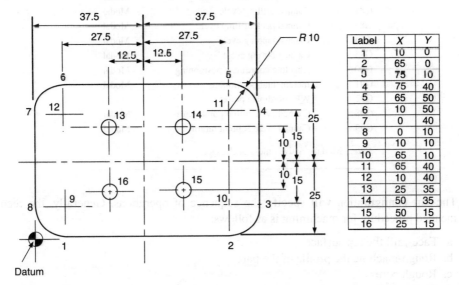

FIGURE 8.46 Labeling of positions. (From Thyer, G. E., *Computer Numerical Control of Machine Tools*, Industrial Press Inc., New York, 1988. With permission.)

    f. The number of different tools needed, and if the machine carousel will accept these tools
    g. Types of tools and the tool-holding facilities available
    h. The availability of machine when required
3. *Work holding.* Whenever possible, the use of standard work devices is recommended, with preparing fixtures to hold the part if required.
4. *Part datum location.* The datum of the drawing should be taken as the program datum. Generally, the choice of part datum depends on the type of component, required machining operations, WP holding, and the direction of cutting forces.
5. *Selecting the program tooling.* Decide the tooling ahead of time. In this regard, especially ground tools should be ready. Standard tools made of carbide or specially coated tools are recommended. Cost is a major factor that includes cycle time, tolerance, surface finish, and quality of parts needed.
6. *Program plan.* This plan is an outline of the machining steps to be done on the part. In small job shops, the programmer can do the process plan. In larger shops, the process plan comes from the engineering area and includes each step in manufacturing the part.

TABLE 8.4
Commonly Used G Codes for Machining Centers

Code	Function	Condition
G00	PTP positioning/rapid traverse	Modal
G01	Linear interpolation at a feed rate	Modal
G02	Circular interpolation cw	Modal
G03	Circular interpolation ccw	Modal
G28	Zero or home position	Nonmodal
G40	Cutter compensation cancel	Modal
G41	Cutter compensation left	Modal
G42	Cutter compensation right	Modal
G43	Tool height offset	Modal
G49	Tool height offset cancel	Modal
G54	WP coordinate preset	
G80	Fixed cycle cancel	Modal
G81	Canned drilling cycle	Modal
G83	Canned peck drilling cycle	Modal
G84	Canned tapping cycle	Modal
G85	Canned boring cycle	Modal
G90	Absolute coordinate positioning	Modal
G91	Incremental positioning	Modal
G92	WP coordinate preset	
G98	Canned cycle initial point return	Modal
G99	Canned cycle R point return	Modal

Note: Modal—active unless unchanged.

The part configuration will specify the sequence of operations. Generally, the recommended procedure for machining is as follows:

a. Face-mill the top surface
b. Rough-machine the profile of the part
c. Rough bore
d. Drill and tap
e. Finish profile surfaces
f. Finish bore
g. Finish reaming

The operator determines the most economical way to produce the WP using proper sequence, tools, cutter path, work holding devices, and machining conditions. The G and M codes for machining centers are listed in Tables 8.4 and 8.5.

8.15.2 Canned Cycles

Machining centers provide many canned cycles that include:

A. *Drilling cycle G81.* For drilling a hole, the operation sequence is as follows:
  • Rapid-positioning of the X- and Y-axis using G00 code
  • Positioning of the Z-axis to a clearance plane using G00 code
  • Feeding the tool down to the required depth using G01 code
  • Feeding back to the initial Z position

TABLE 8.5
M Codes for Machining Centers

Code	Function	Condition
M00	Program stop	Nonmodal
M01	Optional (planned stop)	Nonmodal
M02	End of program	Nonmodal
M03	Spindle start cw	Modal
M04	Spindle start ccw	Modal
M05	Spindle stop	Modal
M06	Tool change	Nonmodal
M07	Coolant no. 2 on (mist)	Modal
M08	Coolant no. 1 on (flood)	Modal
M09	Coolant off	Modal
M30	End of program and reset to the top	Nonmodal
M40	Spindle low range	Modal
M41	Spindle high range	Modal
M98	Subprogram call	Modal
M99	End subprogram and return to main program	Modal

Note: Modal—active unless unchanged.

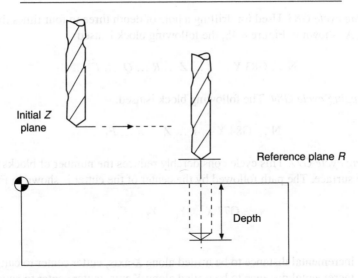

FIGURE 8.47 Drilling cycle G81.

Using canned drilling cycle G81 (Figure 8.47), a set of programmed instructions eliminates the need for many lines in programming. Canned drilling cycles are cancelled using G80 code. To drill a hole located at X, Y, and Z starting from reference point R_0, and at a feed rate F_1, use the following block:

$$N \ldots G81\ X \ldots Y \ldots Z \ldots R \ldots F_1 \ldots$$

where
R = initial level position
F_1 = NC feed rate

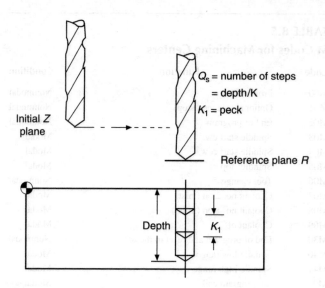

FIGURE 8.48 Peck drilling cycle G83.

B. *Peck drilling cycle G83.* Used for drilling a hole of depth three to four times the diameter, at Q steps. As shown in Figure 8.48, the following block is used:

$$N \ldots G83\ X \ldots Y \ldots Z \ldots R \ldots Q_s \ldots F_1 \ldots$$

C. *Canned tapping cycle G84.* The following block is used:

$$N \ldots G84\ X \ldots Y \ldots Z \ldots R \ldots F_1 \ldots$$

D. *Face milling cycle G77.* This cycle considerably reduces the number of blocks required to face-mill a surface. The path followed by the center of the cutter is shown in Figure 8.49.

$$N \ldots G77\ X \ldots Y_1 \ldots Y_2 \ldots F_1 \ldots$$

where
- X = incremental distance to be milled along X-axis, cutter center to cutter center
- Y_1 = incremental distance to be milled along Y-axis, cutter center to cutter center
- Y_2 = Y-axis "stepover" value. Maximum stepover is the diameter of the cutter; for efficient cutting, a more practical value is 70–80% of the cutter diameter. Last stepover is automatically adjusted by control to satisfy Y_1....
- F_1 = feed rate

E. *Slot milling cycle.* Figure 8.50 shows a slot milling cycle that is similar to the face milling cycle.

F. *Pocket milling cycle G78.* This cycle is used to mill a rectangular pocket; the path followed by the cutter centerline is shown in Figure 8.51. The cutter has to be positioned in the center of the pocket and at the desired depth before the cycle is activated. The cycle requires the following format:

$$N \ldots X_1 \ldots X_2 \ldots X_3 \ldots Y_1 \ldots Y_2 \ldots F_1 \ldots F_2 \ldots$$

Numerical Control and Computer Numerical Control Technology 325

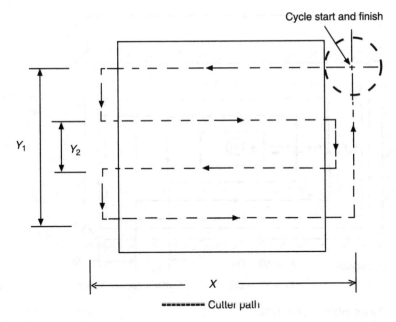

FIGURE 8.49 Face milling cycle G77.

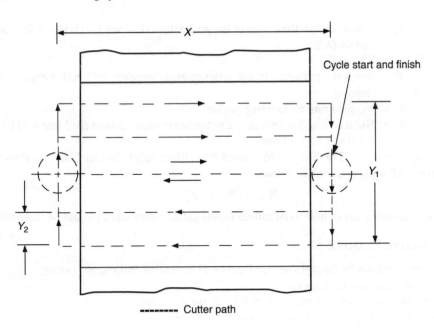

FIGURE 8.50 Slot milling cycle G77.

where

X_1 = distance from the center of the pocket to wall along X-axis (L/2), less the cutter radius (D/2).
= L/2–D/2

X_2 = step over value on X-axis. Step over is the amount the cutter moves for each cut. Maximum step over is the diameter of the cutter.

X_3 = step over for final boundary cut. If it is not programmed, default step over value for the final cut will be 0.5 mm.

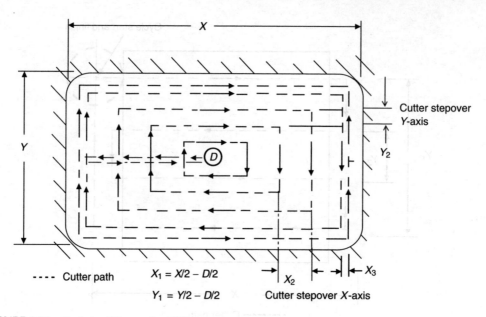

FIGURE 8.51 Pocket milling cycles G78.

Y_1 = distance from the center of the pocket to wall along Y-axis (W/2), less cutter radius D/2.
  = W/2 − D/2
Y_2 = stepover on Y-axis. If not programmed, stepover in Y will be same as X stepover.
F_1 = NC feed rate for clearing pocket.
F_2 = NC feed rate for final cut, X not programmed, default feed rate will be 1.5 times F_1 feed rate.

G. *Hole milling cycle G79*. This cycle is used for milling large circular holes, as shown in Figure 8.52, and is written as follows:

$$N \ldots G79 \; J \ldots F_1 \ldots$$

where J is the radius of hole to be milled minus cutter radius and F_1 is the NC feed rate.

ILLUSTRATIVE EXAMPLE 3

Write the part program for the part shown in Figure 8.53 for milling, drilling, and tapping.
- Use canned cycles where appropriate.
- Preset the absolute tool reference at $X = 5$, $Y = 5$, and $Z = 1$ in.
- Use letters A, B, C, D, and so on to describe the tool paths.

Solution

0010	G90 G20	Absolute or inch programming
0020	G40	Cutter compensation cancel
0030	M06 T01	Change to tool 1
0040	G54 X5.0 Y5.0 Z 1.0	WP preset position
0050	S300 M03	Rotate spindle cw at 300 rpm
0060	G00 X-1.0 Y-1.0	Rapid positioning
0070	Z0.1	
0080	Z-0.65	

0090	G41 X0.0 Y0.0	
0100	X2.5 Y5.0	
0110	G01 Y 5.0 $F_1 5$	Linear interpolation to point B
0120	X2.5 Y5	To point C
0130	Y4.25	To point D
0140	G03 X3.25 Y3.5 I .75 J0	Circular interpolation ccw to point E
0150	G01 X3.875	Linear interpolation to point F
0160	G02 X4.375 Y3.0 I0J-0.5	Circular interpolation cw to point G
0170	X 3.875 Y2.5 I-0.5 J0	Circular interpolation cw to point H
0180	G01 X3.5	Linear interpolation to point I
0190	G03 X3.25 Y 0.5 I0 J-2	Circular interpolation ccw to point J
0200	G01 X1.25 Y0.0	Linear interpolation to point K
0210	X0.0 Y0.0	Linear interpolation to point A
0220	G40 Y-1.0	Ramp off
0230	G00 Z0.1	Rapid above
0240	G00 X0.5 Y2.5	Rapid positioning
0250	G81 Z-0.65 $F_1 3$	Drill hole 1 at point L
0260	Y3.0	Drill hole 2 at point M
0270	Y3.5	Hole 3 at point N
0280	X3.75 Y3.0	Hole 4 at point O
0290	G80	Cancel
0300	G28	Home
0310	M06 T03	Change tool
0320	M03 S100	Rotate spindle
0330	G00 X3.75 Y3.0	Rapid positioning
0340	Z0.1	Rapid positioning
0350	G84 Z-0.65 $F_1 4$	Tap at point O
0360	G80	Tap cancels
0370	G28	Home
0380	M05	Spindle stop
0390	M30	Rewind the program

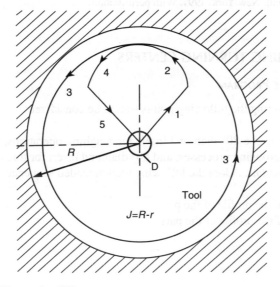

FIGURE 8.52 Hole milling cycles G79.

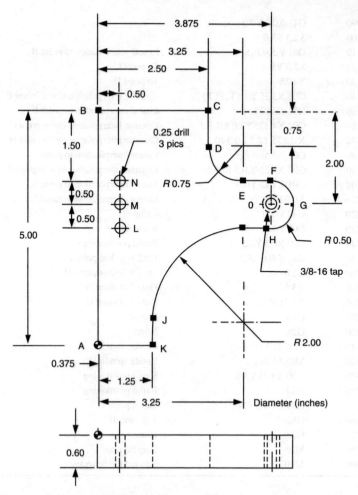

FIGURE 8.53 Part drawing. (From Senerstone, J. and Kuran, K., *Numerical Control Operation and Programming*, Prentice-Hall, New York, 1997. With permission.)

8.16 PROGRAMMING TURNING CENTERS

8.16.1 Planning the Program

For successive programming the following points should be considered:

a. *Tool consideration.* Use 80° diamond insert for roughing operations, 35° diamond insert for finishing and grooving processes, and 60° diamond insert for threading.
b. *Sequence of operations.* Adopt the following recommended sequence of operations:
   - Facing
   - Rough turning of the profile of the part
   - Finish turning of the profile of the part
   - Drilling
   - Rough boring
   - Finish boring
   - Grooving
   - Threading

TABLE 8.6
Commonly Used G Codes for Turning Centers

Code	Function	Condition
G00	Rapid positioning	Modal
G01	Linear positioning at a feed rate	Modal
G02	Circular interpolation cw	Modal
G03	Circular interpolation ccw	Nonmodal
G28	Zero or home position	Modal
G40	Tool nose radius compensation cancel	Modal
G41	Tool nose radius compensation left	Modal
G42	Tool nose radius compensation right	Modal
G50	WP coordinate setting/maximum spindle revolutions per minute setting	Modal
G20	Inch programming	Modal
G75	Grooving cycle	
G76	Threading cycle	
G90	Absolute coordinate positioning	Modal
G91	Incremental positioning	Modal
G92	WP coordinate preset	Modal
G96	Constant surface speed footage	
G77	Revolutions per minute input	Modal
G98	Feed rate per minute	Modal
G99	Feed rate per revolution	Modal

Note: Modal—active unless unchanged.

TABLE 8.7
Common M Codes for Turning Centers

Code	Function
M00	Program stop
M01	Optional (planned stop)
M03	Spindle start cw
M04	Spindle start ccw
M05	Spindle stop
M08	Coolant no. 1 on (flood)
M09	Coolant off
M30	End of program and reset to the top
M41	Spindle low range
M42	Spindle high range
M43	Subprogram call

Note: Tool code calls the tool and the offset, which accommodate for tool wear or exact sizing of the tool.

c. *Commonly used codes.* Tables 8.6 and 8.7 show the commonly used G and M codes.
d. *Return home (safe position) G28.* This allows for turret indexing (Figure 8.54) away from the WP.
e. *WP coordinate setting G92.* This tells the machine where is the part zero location with respect

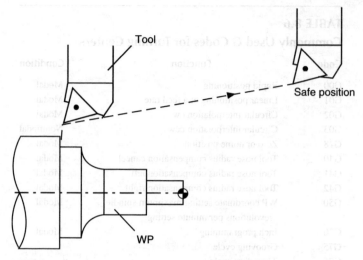

FIGURE 8.54 Return to home position G28.

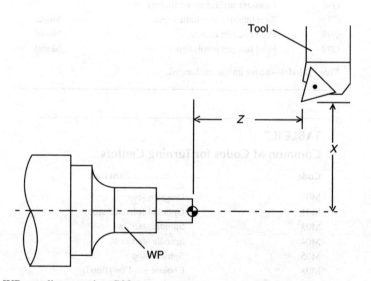

FIGURE 8.55 WP coordinate setting G92.

to the machine, as shown in Figure 8.55. For such a purpose, the following code is used:

$$N \ldots G92 \; X \ldots Z \ldots$$

  f. *Circular interpolation cw G02 and ccw G03.* As shown in Figure 8.56, G02 and G03 use the same concept used in programming machining centers, taking into consideration that the machine movements are in the X and Z directions and therefore the letters I and K will be used instead of I and J.
  g. *Tool nose radius compensation G41 and G42.* This facility permits the use of the same program for a variety of tool types. The exact radius of the tool is entered into the offset file. It may be right G42 or left G41, as shown in Figure 8.57.

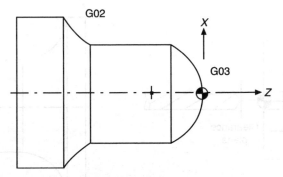

FIGURE 8.56 Turning arcs using G02 and G03 codes.

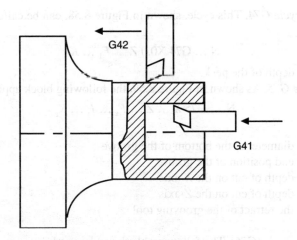

FIGURE 8.57 Tool nose radius compensation: right G42 and left G41.

8.16.2 Canned Turning Cycles

a. *Rough turning cycle G71.* For rough turning, use the following block details:

$$N \ldots G71\ P \ldots Q \ldots U \ldots W \ldots D \ldots F_1 \ldots$$

b. *Finish turning cycle G70.* For finish turning, use the following block details:

$$N \ldots G70\ P \ldots Q \ldots F_1 \ldots$$

where

P = line number for the start of profile
Q = line number for the end of profile
U = allowance left for finishing in the X-axis
W = allowance left for finishing in the Z-axis
D = depth removed per pass
F_1 = NC feed rate

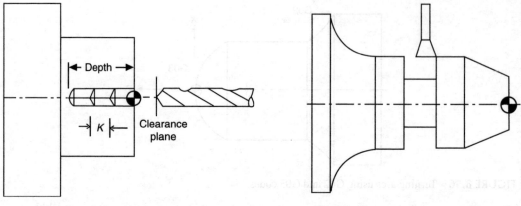

FIGURE 8.58 Peck drilling cycle G74.

FIGURE 8.59 Grooving canned cycle G75.

c. *Peck drilling cycle G74.* This cycle, shown in Figure 8.58, can be called by the following block:

$$N \ldots G74\ X0.0\ Z \ldots F_1 \ldots K$$

where K is the depth of the peck.

d. *Grooving cycle G75.* As shown in Figure 8.59, the following block applies:

$$N \ldots G75\ X \ldots Z \ldots F_x \ldots I_x \ldots K_z$$

where

X = diameter at the bottom of the groove
Z = end position of the groove
I_x = depth of cut on the X-axis
K_z = depth of cut on the Z-axis
F_x = the retract of the grooving tool

e. *Thread cutting cycle (G76).* The following block can be used:

$$N \ldots G76 \ldots X \ldots Z \ldots K \ldots D \ldots F_1 \ldots A.$$

where

X = diameter at the bottom of the thread
Z = end position of the thread
K = thread height
D = depth of cut in the first pass
F_1 = thread lead (pitch for a single start)
A = included thread angle

ILLUSTRATIVE EXAMPLE 4

Program the part shown in Figure 8.60. You are requested to

- Use canned cycles whenever possible
- Preset the absolute tool reference at $X = 6.0$ and $Z = 10.0$ in.
- Choose the proper tool from the following list:

Tool Number	Tool Description	Operation Performed
1	80° diamond	Rough turning
2	35° diamond	Finish turning
3	60° diamond	Threading
4	0.5 mm diameter	Drilling

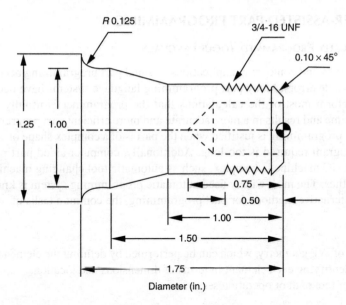

FIGURE 8.60 Part drawing. (From Senerstone, J. and Kuran, K., *Numerical Control Operation and Programming*, Prentice-Hall, New York, 1997. With permission.)

Solution

0010	G90 G20	
0020	G40	
0030	T0101	
0040	G92 X6.0 Z 10.0	
0050	G96 S400 M03	
0060	G00 G42 X1.3 Z0.1	
0070	G71 P0080 Q0120 U0.03 W 0.001 D 0.04 $F_1$0.01	Rough turning cycle
0080	G00 X .55	
0090	G01 X.75 Z-.1	
0100	G01 Z-1.0	
0110	G01 X1.0 Z-1.5	
0120	G01 Z-1.625	
0130	G02 X1.25 Z-1.75 I.125 K0	
0140	G28	
0150	T0202	
0160	G00 G42 X1.3 Z.1	
0170	G70 P080 Q 0120 $F_1$0.008	Finish turning cycle
0180	G40 G00 X2.0 Z2.0	
0190	G28	
0200	T0303	
0210	G92 X 6.0 Z10.0	
0220	G96 S300 M0S	
0230	G00 X.75 Z.2	
0240	G76 X.75 Z-.5 K0.053 D0.02 $F_1$0.065 A60	Threading cycle
0250	G28	
0260	T0404	
0270	G92 X6.0 Z 10.0	
0280	G96 S800 M03	
0290	G00 X0 Z.2	
0300	G74 X0 Z-.75 $F_1$0.01 K_1.125	Drilling cycle
0310	G28	
0320	M30	

8.17 COMPUTER-ASSISTED PART PROGRAMMING

8.17.1 AUTOMATICALLY PROGRAMMED TOOLS LANGUAGE

In complicated PTP jobs and contouring applications, manual part programming becomes tedious and is subject to possible errors. Many part programming language systems have been developed to automatically perform most of the calculations that the programmer is usually tasked with. This, in turn, saves time and results in a more accurate and more efficient part program. The use of computer-aided part programming is justified when the part is of a complex shape or when the part is simple but the program required is too long. Additionally, computer-aided part programming is justified when the NC machine is complex, such as automatic tool changing machining centers and four-axes NC lathes. The most widely used automatic programming system is known as APT. APT is one of the alternative methods for part programming, the common tasks of which include the following steps:

1. The definition of WP geometry, which can be performed by defining the elements forming the part and identifying each element in terms of dimensions and location.
2. Specification of tool path or operation sequence.

Computer-assisted part programming (CAPP) generates the cutter positions based on APT statements and the tool path is directed to various point locations and along the surfaces of the WP to carry out machining. An APT program includes language statements that fall in the following four categories:

1. *Geometry statements.* Any part is composed of basic geometric elements that can be described by points, lines, planes, circles, cylinders, and other mathematically defined surfaces. Each geometric element must be identified by the part programmer, as well as the dimensions and location of the element. In APT programming language, the following statements may be used:

Point	P1 = POINT/5.0,4.0,0.0	X, Y, Z coordinate values
	P2 = POINT/INTOF, L1, L2	Intersection of two lines
Line	L3 = LINE/P3,P4	Between two points
	L4 = LINE/P5, PARALEL, L3	From point 5 and parallel to line 3
Plane	PL1 = PLANE/P1, P4, P5	Defined by three points
	PL2 = PLANE/P2, PARALEL, PL1	From point P2 and parallel to plane PL1
Circle	CIRCLE/CENTER, P1, RADIUS, 5.0	Centers at P1 and the radius is 5

2. *Motion statements.* After defining the WP geometry, the programmer constructs the path that the cutter will follow to machine the part. The tool path specification involves a detailed step-by-step sequence of cutter moves. Such cutter moves are made along the geometric elements, which have been defined earlier. In APT language, the following statements could be used for describing the tool movements:

FROM/TARG	From target
FROM/-2.0,-0.20,0.0	From X, Y, Z
GOTO/P1	Go to point P1

Numerical Control and Computer Numerical Control Technology

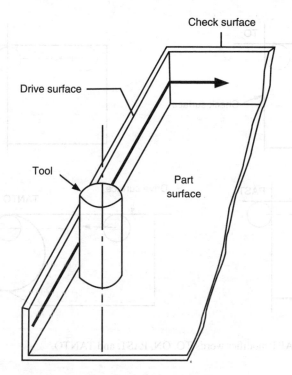

FIGURE 8.61 APT definitions of drive, part, and check surfaces.

For PTP motions the following statements are used:

GOTO/P2	Move to point P2
GOTO/2.0,7.0,00	Absolute
GODELTA/2.0,7.0,0.0	Incremental

For continuous motions as shown in Figure 8.61, the tool motion is guided by the following three surfaces:
- *Drive surface.* Guides the sides of the cutter
- *Part surface.* The bottom of the cutter rides
- *Check surface.* Stops the movements of the tool (TO, ON, PAST, TANTO), as shown in Figure 8.62.

Figure 8.63 shows the motion command that include GOLFT, GORGT, GOFWD, GOBACK, GOUP, and GODOWN

3. *Postprocessing statements.* These statements contain the machine instructions that are passed unchanged into the cutter location data (CLDATA) file to be dealt with by the postprocessor. It operates the spindle speed, feed rate and other features of the machine tool, such as adding coolant and stopping the program. Some common postprocessor statements are as follows, where the slash indicates that some descriptive data are needed:

COOLANT/	For coolant control (ON or OFF)
RAPID	To select rapid cutter motion
SPINDL/	To select spindle on/off, speed and direction of rotation
FEDRAT/	To select feed rate
TURRET/	To select cutter number

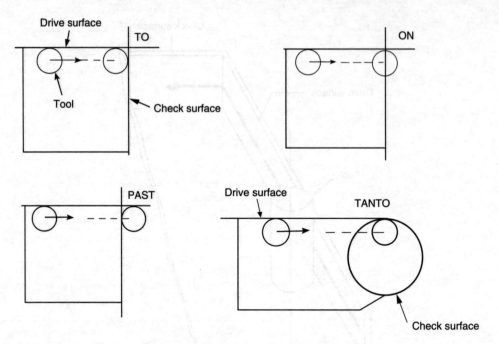

FIGURE 8.62 Use of APT modifier words TO, ON, PAST, and TANTO.

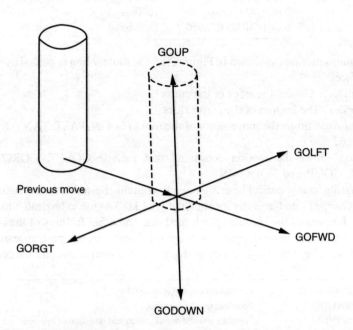

FIGURE 8.63 APT motion commands.

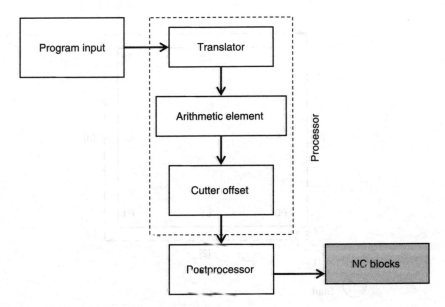

FIGURE 8.64 Steps of computer-assisted part programming.

4. *Auxiliary statements.* These statements provide additional information to the APT processor, giving part name, tolerances to be applied, and so on.

8.17.2 Programming Stages

Figure 8.64 shows the main steps of CAPP, which can be summarized as follows:

1. Identify the part geometry, general cutting motions, feeds, speeds, and cutter parameters.
2. Code the geometry, cutter motions, and general machine instructions into the part programming language (APT).
3. Process the source to produce machine-independent list of movements and ancillary machine control information, and CLDATA file.
4. Postprocess CLDATA to produce the machine control data (MCD) for a particular machine.
5. Transmit the MCD to the machine and test.

The computer's job in CAPP consists of the following steps:

1. *Input translation.* Converts the coded instructions contained in the program into computer usable form, ready for further processing.
2. *Arithmetic calculations.* Degenerates the part surface using subroutines that are called by the various part programming language statements. The arithmetic unit frees the programmers from the time-consuming calculations.
3. *Cutter offset computation.* Offsets the tool path from the desired part by the radius of the cutter. This means that the part programmer defines the exact part outline in the geometry statements.
4. *Postprocessing.* This is a separate computer program that prepares the punched tape or the program for a specific machine tool. The input to the processor is the output from the previous three steps and the output is the NC tape or program written in the correct format for the machine tool to be used.

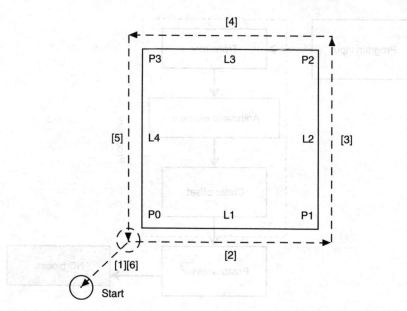

FIGURE 8.65 Part drawing.

ILLUSTRATIVE EXAMPLE 5

Program the part shown in Figure 8.65 using APT.

Solution

Initial Auxiliary and Postprocessor Statements	
PARTNO	H9253
	MACHIN/MILL,1
	INTOL/.01
	OUTOL/.01
	CUTTER/15.0
Geometry Statements	
START	= POINT/−80.0, −80.0, 40.0
P0	= POINT/0.0, 0.0, 0.0
P1	= POINT/160.0, 0.0, 0.0
P2	= POINT/160.0, 160.0, 0.0
P3	= POINT/0.0, 160.0, 0.0
L1	= LINE/P0, P1
L2	= LINE/P1, P2
L3	= LINE/P2, PARLEL, L1
L4	= LINE/P3, P0
PL1	= PLANE/P0, P1, P3
Start spindle and coolant	
	SPINDL/800
	COOLNT/ON
	FEDRAT/40.0
Motion Statements	
	FROM/START
	GO/TO, L1 TO, PL1, TO, L4 1
	GORGT/L1, PAST, L2 2
	GOLFT/L2, PAST, L3 3

Numerical Control and Computer Numerical Control Technology

	GOLFT/L3, PAST, L4	4
	GOLFT/L4, PAST, L1	5
Select rapid feed and return to start		
	RAPID	
	GOTO/START	6
Turn off coolant, and spindle, end section, and program		
	SPINDL/OSS	
	END	
	FINI	

8.18 CAD/CAM APPROACH TO PART PROGRAMMING

8.18.1 Computer-Aided Design

CAD is a technology that involves a computer in the design process. It enables the engineer to develop, change, and interact with the graphical model of a part. Computers are strong in the areas of graphics, calculations, analysis, modeling, and testing. During CAD stages, the engineer draws the part on the screen. This information can be used to create a program to machine the part. The designer must work closely with the manufacturing experts to establish some standards for the design.

CAD allows different layers to be created. This technology allows other software to take the part geometry from each layer and assign different tools to machine it. The computer allows design to be viewed and tested before manufacturing. CAD systems stress-test parts to meet the strength requirements of the application. Graphics capabilities allow three-dimensional (3D) viewing of parts from any angle. CAD systems can also export the CAD part file in Drawing Exchange File (DXF) format. Using CAD systems, the following advantages can be secured:

1. Increase the productivity of the designers
2. Create better designs
3. Reduce redundant effort
4. Allow easy and rapid modification of prints
5. Enable integration of engineering and manufacturing

8.18.2 Computer-Aided Manufacturing

CAM utilizes computers in the manufacturing stage. Such a modern technique has the following features:

- Allows the programmer to develop a model that represents the part and the machining operations.
- The programmer can interact with the model graphically to make the necessary adjustments and modifications before the CNC code is generated.
- CAM software reads the DXF that contains the part geometry and the levels or layers that the geometry exists on.
- CAM software utilizes a job plan to assign the correct tool path to each layer.
- The job plan knows the work material that will be used so that it can calculate speeds and feeds for each tool.

The integrated CAD/CAM approach prepares the part program directly from the CAD part geometry, either by using NC programming commands called in the CAD/CAM system or by passing the

CAD geometry into a dedicated CAM program. Using CAD/CAM systems, the CAD drawings can therefore be changed to CNC programs. The CAD/CAM approach has the following advantages:

1. No need to encode the part geometry and the tool motion
2. Allows the use of interactive graphics for program editing and verification
3. Displays the programmed motions of the cutter with respect to the WP, which allows visual verification of the program
4. Allows interactive editing of the tool path with the addition of the tool moves and standard cycles
5. Incorporates the most sophisticated algorithms for part programming generation

The programming steps for CAD/CAM approach are as follows:

1. The aspects of the part geometry that are important for machining purposes are identified (and perhaps isolated on a separate level or layer); geometry may be edited or additional geometry added to define boundaries for the tool motion.
2. Tool geometry is defined; for instance, by selecting tools from a library.
3. The desired sequence of machining operations is identified and tool paths are defined interactively for the main machining operations.
4. The tool motion is displayed and may be edited to refine the tool motion, or other details may be added for particular machining cycles or operations.
5. A CLDATA file is produced from the edited tool paths.
6. The CLDATA file is postprocessed to MCD, which is then transmitted to the machine.

Postprocessor

The postprocessor takes the part geometry and job plan and writes the code that the specific machine will understand.

Simulation

The program can be written and tested before it is actually run, and jigs, fixtures, and clamping can be shown during simulation to determine whether there are any potential problems. The simulation shows the machining time. Operations can then be adjusted to optimize production offline, which keeps the machine operating.

Download the CNC Programs

The program is sent to the machine manually, using a tape or disk, or electronically, using a communication cable.

8.19 REVIEW QUESTIONS

1. Mark true or false:
   [ ] Circular interpolation G02 and G03 are limited to 90°.
   [ ] Contouring NC can perform PTP and straight cuts.
   [ ] DNC allows a computer to control one machine only.
   [ ] Errors in programming using incremental positioning are more dangerous than in absolute programming.
   [ ] A fixed zero is easier for the programmer than a floating zero.

[ ] G00 requires a feed number during part programming.
[ ] G01 performs machining of straight lines at any number of axes.
[ ] G02 performs circular interpolation in ccw direction.
[ ] GOBACK is a motion statement in computer-assisted part programming.
[ ] The magic-three method is used for programming X-, Y-, and Z-coordinates.
[ ] NC machines can understand decimal and binary numbers.
[ ] NC adds many machining capabilities that are not built into the machine tool.
[ ] NC is a form of programmable automation by cams and mechanisms.
[ ] NC is an advanced machining method.
[ ] NC machine tools can produce shapes and cam profiles as easy as circular ones.
[ ] NC machines can be used to produce large numbers only.
[ ] NC user does not necessarily speak the language of the machine control system.
[ ] Parity check in tape code is helpful in winding the tape.
[ ] Polar coordinates provide the ability to generate a 360° arc in a smaller number of blocks than circular interpolation.
[ ] Rapid positioning requires a feed number during part programming.
[ ] Row 5 is used as the parity check in ISO tape code set.
[ ] Straight-cut NC can produce contours.
[ ] The character is a collection of words that forms a block.
[ ] The setup point can be chosen as the zero point on the machine.
[ ] Tape readers used in NC applications are photoelectric and electromechanical.
[ ] Turning on NC machines can be programmed at a constant surface speed or constant revolutions per minute.

2. Explain the following using neat sketches:
   - Absolute and incremental programming
   - Types of NC systems
3. Convert 101000111 to decimal and 1997 to binary.
4. Express in magic-three method: 1525 rpm–0.035 mm/min.
5. Show, using sketches, each of the following:
   - DNC
   - Advantages of CAD/CAM systems
   - Steps of computer-assisted part programming
   - 3-D NC contouring
6. Choose the right answer to finish each sentence:
   a. Machining instructions that are coded and provided to an NC machine to produce a finished WP are called:
      A. Integrated circuits
      B. Input media
      C. Output data
      D. Part programs
   b. The decimal number 92 converts to which of the following binary numbers:
      A. 1101110
      B. 1001110
      C. 1010101
      D. 1011100
   c. The magic-three method code for a desired feed rate of 42 ipm is:
      A. $F_1 542$
      B. $F_1 545$
      C. $F_1 742$
      D. $F_1 745$

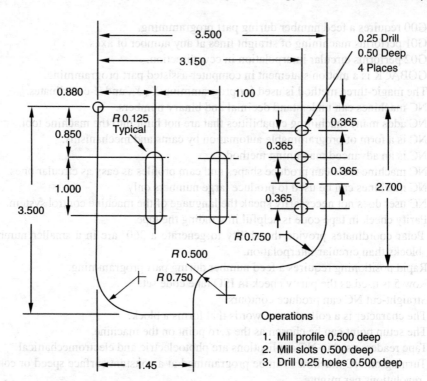

FIGURE 8.66 Part drawing. (From Senerstone, J. and Kuran, K., *Numerical Control Operation and Programming*, Prentice-Hall, New York, 1977. With permission.)

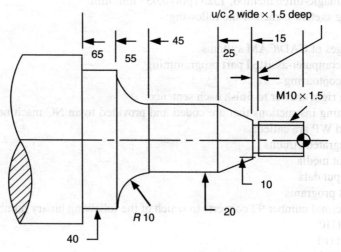

FIGURE 8.67 Part drawing. (From Thyer, G. E., *Computer Numerical Control of Machine Tools*, Industrial Press Inc., New York, 1988. With permission.)

d. Which control system would probably be used to control the path of a cutting tool in five axes?
   A. Incremental NC system
   B. Continuous-path NC system
   C. Straight-cut NC system
   D. PTP NC system
e. A prominent feature of machining centers is:
   A. Circular interpolation
   B. An automatic tool changer
   C. The BCD format
   D. Tool length compensation
7. Write the necessary part program for the part shown in Figure 8.66.
8. Program the part shown in Figure 8.67:
   - Use canned cycles where appropriate.
   - Preset the absolute tool reference at $X = 100$ and $Z = 50$ mm from the zero point.
   - Use letters a, b, c, d and so on to describe the tool paths.
   - Choose the proper tool from the following table:

Tool Number	Tool Description	Operation
1	80° diamond	Rough turn
2	35° diamond	Finish turn
3	60° diamond	Threading

REFERENCES

Gibs, D. (1988) *An Introduction to CNC Machining*, 2nd Edition, ELBS Cassell Publishers Ltd., London.
Harding Inc.
Senerstone, J. and Kuran, K. (1997) *Numerical Control Operation and Programming*, Prentice-Hall, New York.
Thyer, G. E. (1988) *Computer Numerical Control of Machine Tools*, 1st Edition, Industrial Press Inc., New York.

9 Hexapods and Machining Technology

9.1 INTRODUCTION

Most industrial machine tools, such as conventional milling machines, drilling machines, lathes, stacked axis robots, and so on, have a serial or open-loop kinematic architecture, which means that each axis supports the following one, including its actuators and joints (Figure 9.1). These machines are mainly based on the perpendicular composition of three linear axes. Two more rotary axes may be integrated to extend the ease of applying the Cartesian coordinate system to control spatial movements (AKIMA, 1997).

The kinematic analysis of this serial or stacked-axis system is easy. However, it generates cumulative errors, because inexact positioning on one axis dislocates the positioning on the next axis, which multiplies the imprecision with each subsequent station. In addition to this design drawback, the conventional stacked-axis machine, which has a bed, saddle, and so on, requires a massive concrete foundation for stability requirements. This results in cumbersome unit that takes up considerable floor space and is difficult to transport (Figure 9.1).

Machine tools based on advanced closed-loop parallel kinematics represent a promising new technology of the twenty-first century that is currently receiving a lot of focus and interest. Several prototypes, developed by famous companies all over the world, have already proved the general feasibility of the idea of the parallel kinematic system (PKS). Such systems are characterized as

> Mechanisms based on a kinematic structure which allows movements with N-Degrees of Freedom, whereby a moving base carrying the cutting tool and a fixed base is connected to each other by N independent kinematic chains. Every chain is composed of maximum two segments and the articulation between them has one degree of freedom. The motion of the structure is generated by N actuators, one for each chain (AKIMA, 1997).

Hexapods and tripods (Figure 9.2) are the most famous paradigms of the parallel kinematic mechanism (PKM). In this chapter, the hexapod positional device is introduced. Its historical background, applications in machining technology, design features, elements, and characteristics are highlighted and traced.

9.2 HISTORICAL BACKGROUND

Around 1800, the mathematician Cauchy studied the stiffness of the so-called *articulated octahedron*. The earliest known hexapod machine was a car tire tester designed by Gough of Dunlop in 1949. Stewart (1965) described the attributes of a simple hexapod, thus giving his name to the platform. Figure 9.3 shows a hexapod employed in surgical operations using a flexible movement with a laser beam for operations. This model is used nowadays to simulate the whole operation to train physicians. This platform has been successfully used as a flight simulator to train pilots. Figure 9.4 illustrates nanopods that meet stringent medical safety standards. Figure 9.4 shows a nanopod with three additional legs containing redundant position sensors. A seventh axis is added to increase the linear travel range (Figure 9.4b). In 1995, the Fraunhausen Institute in Stuttgart developed a hexapod that worked successfully as a surgical robot.

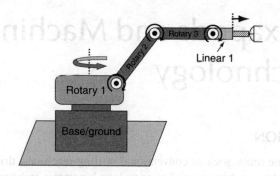

FIGURE 9.1 Conventional architecture (serial mechanism).

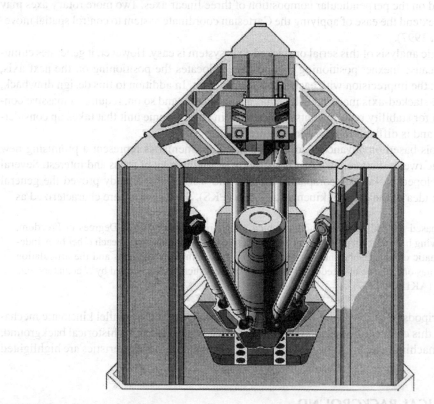

FIGURE 9.2 Tripod parallel mechanism.

Many unsuccessful attempts were made between 1970 and 1990 to make practical hexapod machines. The cost of computing strut lengths of hexapods has until recently been too high for most applications. In cases, where computing costs are justified, some interesting hexapod machines, such as flight simulators, were successfully made.

Recently the cost of computing has fallen drastically, and therefore many companies are offering hexapod-based machines for various applications that include hexapod machine tools (Geodetic, 1997). In 2006, Hitachi, Seiki, and Toyota joined the U.S. builders Giddings Lewis, Ingersoll, and Britain's Geodetic to develop hexapod technology that started in the early 1990s.

In 1990, Ingersoll announced and exhibited its own prototype of the *octahedral hexapod*. This prototype utilized a 12-node hexapod suspended from an octahedral framework, with the spindle pointing down toward the WP (Ingersoll Co., 2001).

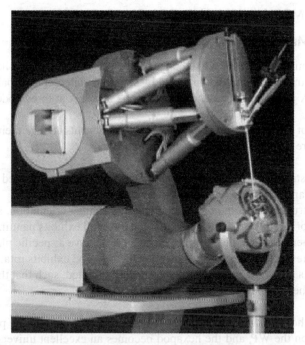

FIGURE 9.3 Hexapod for surgical operations, equipped with laser beam.

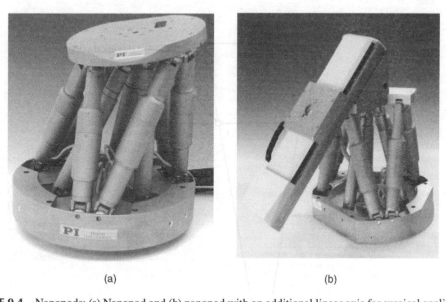

FIGURE 9.4 Nanopods: (a) Nanopod and (b) nanopod with an additional linear axis for surgical applications.

The Variax Hexacenter milling machine, introduced in 1994 for high-speed milling of aluminum, took its design inspiration from the flight simulator, with the struts crossing over and the spindle pointing downward from the platform toward the working volume enclosed by the machine.

9.3 HEXAPOD MECHANISM AND DESIGN FEATURES

9.3.1 Hexapod Mechanism

The hexapod mechanism consists of a fixed upper dome platform (base) of a hexagonal shape and a moving platform of triangular shape, connected by six struts (telescopic or ballscrew) (Figure 9.5). Starting from the six joints on the base, each two struts intersect at three nodes on the moving triangular platform.

The movement of the platform is actuated when the six struts change their lengths in a coordinated manner (Figure 9.6):

a. When the six struts simultaneously expand or contract at the same feed rate, the platform moves downward or upward horizontally in an extending movement.
b. When some struts expand and others contract and change orientation, such that the platform moves horizontally, then it is said that the hexapod exhibits panning movement.
c. When strut orientations and lengths are changed to achieve a specific platform inclination in space with respect to coordinate axes, than the hexapod exhibits rotation.
d. When all struts are of the same length and similarly rotated, such that the platform moves horizontally, then the hexapod is being twisted.

Therefore, through the rotational and axial movements of struts, the moving platform is capable of reaching any point on the WP, and the hexapod becomes an excellent universal positioning apparatus. Of course, provisions should be taken to prevent struts crossing and collision with each other or with the platforms.

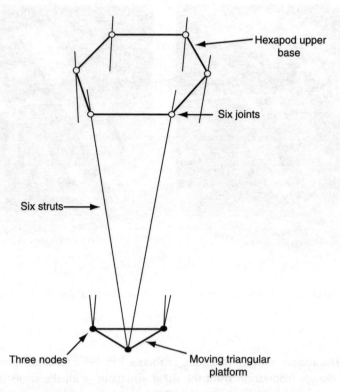

FIGURE 9.5 Hexapod mechanism.

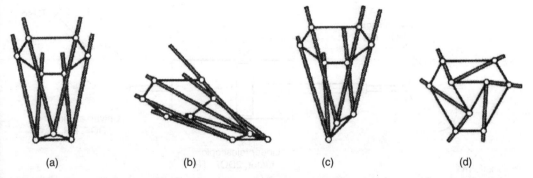

FIGURE 9.6 Models of hexapod movements: (a) extending, (b) panning, (c) rotating, and (d) twisting. (From Ingersoll, Hannover Exhibition, The Next Generation in 5-Axis Machining Technology, Octahedral Hexapod, Technical Information, Ingersoll Waldrich Siegen Werkzeugmaschinen, GmbH, 1997. With permission.)

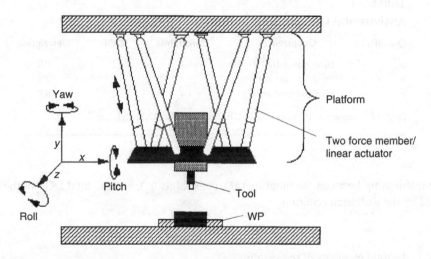

FIGURE 9.7 Hexapod of telescopic struts, Ingersoll system. (From Ingersoll, Hannover Exhibition, The Next Generation in 5-Axis Machining Technology, Octahedral Hexapod, Technical Information, Ingersoll Waldrich Siegen Werkzeugmaschinen, GmbH, 1997. With permission.)

9.3.2 Design Features

Hexapods have two main design features, as described in the following.

9.3.2.1 Hexapods of Telescopic Struts (Ingersoll System)

In this design, the hexapod consists of six hydraulic, telescopic struts. The struts may be of a circular cross section or square, and are free to expand or contract between a base and a platform. The platform represents the output element that gets the six degrees of freedom (6 DOF) of the system (Figure 9.7). Both ends of the hydraulic telescopic struts are connected to either the platform or the base by universal joints (Figure 9.8). Such a system was first introduced as a flight simulator positioning system. It is commercially available for a variety of applications that require micron and submicron accuracy.

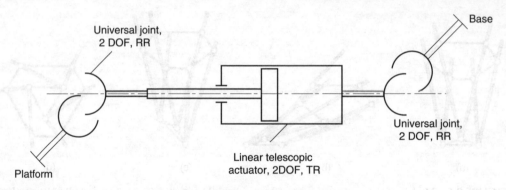

FIGURE 9.8 Telescopic struts with universal joints.

TABLE 9.1
Analysis of Telescopic Strut System

Quantity	Occurrences	Constraints	DOF	Description
6	Base: yokes 1/2 UJ	4	2	RR
6	Strut Cyl. and rod	4	2	TR
6	Strut rod/platform UJ	4	2	RR

Note: R: Rotational; T: Translational; UJ: Universal joint; Cyl.: Cylinder.

The relationship between the mobility (M), constraints (c_i), and the total DOF of the system is expressed by the Kutzbach criterion

$$M = 6(n_e - 1) - \sum_{i=1}^{j} c_i \qquad (9.1)$$

where n_e is the total elements of the system.

Calculation of DOF of the telescopic hexapod:

Total number of parts = 6 hydraulic struts, each 2 parts + 1 base + 1 platform

$$= 2 \times 6 + 1 + 1 = 14$$

Then, the total DOF = $6(n_e - 1)$

$$= 6 \times 13 = 78 \text{ DOF}$$

According to Table 9.1, the total constraints = $3 \times 6 \times 4 = 72$ constraints.

Therefore,

$$M = 78 - 72 = 6 \text{ DOF (three translations and three rotations)}$$

By combining an octahedral structural frame (Figure 9.9), with the previously described hexapod actuator, Ingersoll has created the stiffest and the most rigid machine tool possible. The stiffness and rigidity have a positive impact on the accuracy, surface quality, and cutting tool durability. The octahedron (the frame structure) consists of 12 beams of similar length that are joined at six junction points (Figure 9.9). This robust structure is self-supporting and needs minimal foundation requirements. The mechanism that guides the spindle, the hexapod with its telescopic struts, is attached to the top

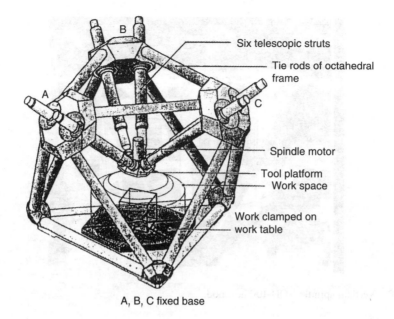

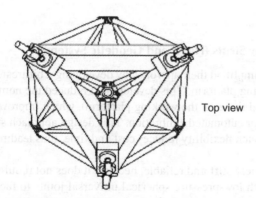

FIGURE 9.9 Hexapod with rigid frame.

of the octahedron at top corners A, B, C. Ingersoll has provided two versions of octahedral hexapods. These are horizontal spindle octahedral hexapod (HOH)-600 and vertical spindle octahedral hexapod (VOH)-1000 hexapods (Figure 9.10).

The technical specifications of the VOH-1000 model are as follows:

Axis tavels:	$X, Y, Z = 600, 600, 800$ mm
Feed rates:	Maximum feed rate (strut axis) = 30 m/min
	Maximum traverse rates (strut axis) = 30 m/min
Acceleration:	0.5 G (4.8 m/s^2) depending on X, Y, Z position
Spindle:	Speed range = 0–20,000 rpm
	Maximum torque = 49 N m
	Maximum power = 37.5 kW
Coolant pressure:	50 bar through the spindle
Tool storage magazine:	40, 80 tools, maximum tool weight = 12 kg
Volumetric accuracy:	20 µm using laser diagonal displacement facility

FIGURE 9.10 Vertical spindle VOH-100 hexapod.

9.3.2.2 Hexapods of Ball Screw Struts (Hexel and Geodetic System)

Hexel and Geodetic dramatically simplified their approach by developing a bifurcated ball between pairs of struts meeting at the working platform. This development reduced the number of nodes to nine (six nodes in the work cell and three on the working platform), which improves the stiffness, simplifies the control, and allows for automated calibration. This design approach strives for precision to be derived from software. Such flexibility reduces cost and time, thus leading to a truly soft machine (Ingersoll file).

This type of hexapod is extremely stiff and reliable, because it does not require any telescoping mechanism. It is equipped with low-pressure spherical universal joints to facilitate accurate and repeatable calibration while inherently providing excellent strut damping (Ingersoll Co., 2001).

The characteristics and working principle of this type of hexapods are summarized as follows:

1. In this case, the hexapod consists of six precise ball screw struts, each of which extends from the tool platform through an integral spherical servomotor that is mounted on the upper dome platform of the hexapod (Figure 9.11). The strut cross section must be circular to allow rotation.
2. Each of the six ball screw struts is driven by a servomotor (sphere drive) through a ball nut, which is housed in the joint. The servomotor can rotate at 3200 rpm, propelling the 5 mm pitch screw of the ball screw at speeds up to 25 m/min.
3. Through coordinate movement of ball screws, the working platform alters position according to the required contour. The struts join the lower platform at three nodes, with two struts shearing a ball and socket joint known as a bifurcated ball (Figure 9.11).
4. Each individual ball screw strut is independent from the others and possesses a personality file containing information such as
   - Error mapping (lead pitch variation)
   - Mounting offsets
   - Physical and thermal expansion characteristics

Hexapods and Machining Technology

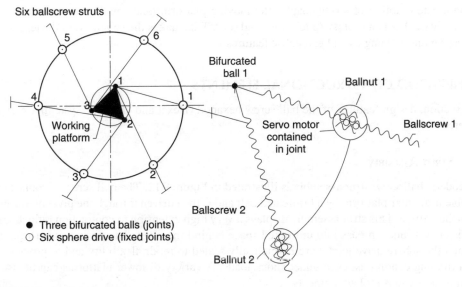

FIGURE 9.11 Ball screw hexapod.

TABLE 9.2
Analysis of a Ball Screw System

Quantity	Occurrences	Constraints	DOF	Description
6	Base/struts: ball and socket joint	3	3	RRR
6	Base struts/extensible strut, lower end	5	1	T
6	Strut upper-end/platform: ball and socket joint	3	3	RRR

Note: R: Rotational; T: Transnational.

5. The completely interchangeable drives are calibrated offline as discrete units and are automatically compensated when assembled. This concept allows hexapods to be assembled and serviced very quickly without the need for complex alignment or calibration exercises. Replacement of worn or damaged strut drives can be achieved within two hours, thereby minimizing downtime.

Referring again to Equation 9.1 expressing the Kutzbach criterion, the calculation of the DOF of ball screw hexapods is as follows:

$$n_e = \text{Total number of hexapod elements}$$
$$= 6 \text{ struts} + 6 \text{ base struts} + 1 \text{ platform} + 1 \text{ base}$$
$$= 14 \text{ parts}$$

Total DOF of the system $= 6(n_s - 1) = 6 \times 13 = 78$ DOF

According to Table 9.2,

$$\text{Total constraints} = 2 \times 6 \times 3 + 6 \times 5 = 66 \text{ constraints}$$

Therefore,

$$M = 78 - 66 = 12 \text{ DOF (6 required/basic DOF + 6 local motilities)}$$

The coordinated motion of struts enables the moving platform (and spindle) to perform 6 DOF: orthogonally (X, Y, Z) and rotary (pitch, yaw, and roll). This makes the spindle very dexterous, easily accessing unusual angles and geometric features.

9.4 HEXAPOD CONSTRUCTIONAL ELEMENTS

In this section, design features of the ball screw hexapod (Hexel and Geodetic system) are mainly considered.

9.4.1 STRUT ASSEMBLY

The Geodetic ballscrew strut assembly is illustrated in Figure 9.12. The ball screw is connected to a movable triangular platform by a bifurcated ball and is then driven through the pivot on the dome by the sphere drive. This strut assembly displays a very high extension to contraction ratio, because the ball screw struts can pass into unlimited space behind the pivot. The pivot points (bifurcated nodes and the sphere drive joints) are hermetically sealed to retain flexibility and to provide very smooth running conditions, even under shock loads. A variety of lower platforms can be used to accommodate various tool attachments.

9.4.2 SPHERE DRIVE

The sphere drive is a special mechanism that forms the heart of any ball screw strut-type hexapod. It is located in the upper base and provides the accurate positional movements of the hexapod struts. The sphere drive is a hollow ball that accommodates a high-powered, high-specification, frameless, brushless dc motor. A rotor is keyed to a spindle, which runs on two high precision bearings (Figure 9.13). A ball nut is keyed to a flange that forms a part of the internal spindle.

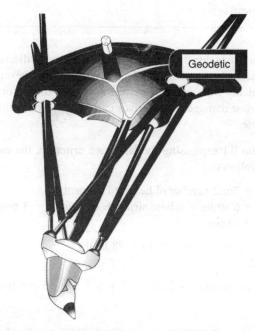

FIGURE 9.12 Geodetic ball screw strut assembly. (From Geodetic, Hannover Exhibition, Hexapod-Breakthrough, Technical Information, 1997. With permission.)

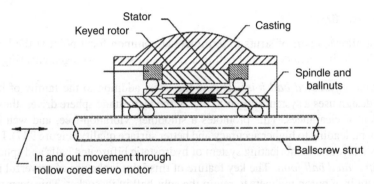

FIGURE 9.13 Sphere drive.

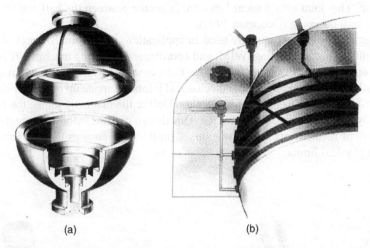

FIGURE 9.14 Casing elements of sphere drives: (a) sphere and (b) annular rings suspending the sphere drive. (From Geodetic, Hannover Exhibition, Hexapod-Breakthrough, Technical Information, 1997. With permission.)

This arrangement drives the ball screw in and out through the hollow-cored servomotor. The unit is water-cooled to maintain reliability and control of thermal inaccuracies.

A high-resolution radial incremental encoder is mounted outside the unit to provide the accurate positional information required by the controller. An integral thermocouple on the motor windings feeds back data to the controller, allowing thermal expansion to be compensated. All critical parts included in the sphere drives are effectively protected from foreign matter. The sphere drive is clamped to the top plate by six bolts; a locating dowel ensures precise and repeatable positioning.

The sphere (Figure 9.14a) containing the drive is retained in an annular ring. Figure 9.14b uses a hydrostatic system to provide the greatest rigidity combined with a low coefficient of friction, while ensuring excellent lubrication. The annular ring suspends the sphere drive within an envelope of high-pressure fluid, which is continuously recirculated. Metal-to-metal contact is minimized, thus providing greater damping and smooth operation. The sphere drive is capable of driving the ball screw at axial feed rates exceeding 40 m/min, with high forces and accelerations.

All sphere drive components such as ball screws, ball nuts, bearings, motors, and encoders are standardized. This makes the sphere drive mechanically simple, cost-effective, and reliable. Moreover, this drive can be easily tailored to meet a wide range of performance requirements for a variety of manufacturing applications.

9.4.3 BIFURCATED BALLS

This nodal joint allows a pair of struts to terminate at common focal point at the lower platform (Figure 9.15). The design includes a split ball in a variety of socket arrangements (Figure 9.16):

The hydrostatic bifurcated ball joint. This is the latest addition to the family of ball joints. The new design uses a system similar to that of the hydrostatic sphere drives, thus ensuring elimination of clearances. This promotes a smoother, vibration-free, and well lubricated environment, leading to higher accuracy and better surface quality (Figure 9.17). Figure 9.18 illustrates a layout of a lubricating system of hydrostatic bifurcated balls and sphere drives.

Magnetic bifurcated ball joint. The key feature of this joint is the use of powerful rare-earth neodymium iron boron magnets to retain the split ball in its socket. This permits the ball to be held over less than half of its surface (Figure 9.19), leaving clearance to permit a wide DOF. The magnetic socket has an inherent overload protection feature, allowing the joints to dislocate when loads exceed the holding forces without causing any damage to the mechanism. This joint uses a shear lubricant interface between the ball and the socket to provide critical damping (Geodetic, 1997).

The magnetic joint is frequently used in applications where small forces are exerted. This is common in laser cutting, WJM, and coordinate measuring machines (CMM).

Mechanical bifurcated ball joints. These have been developed for light machining applications. Similar to magnetic joints, the mechanical joints are made up of a number of stages, each contributing to its freedom of movement. Unlike the magnetic joints, the ball is kept in its cup by a retaining ring (Figure 9.20). This design permits extra forces and eliminates the problem of metal swarf contamination. A limit switch ensures that the joint does not exceed its physical limits.

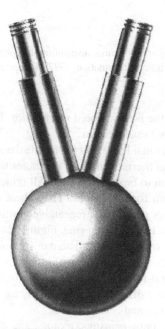

FIGURE 9.15 Bifurcated ball.

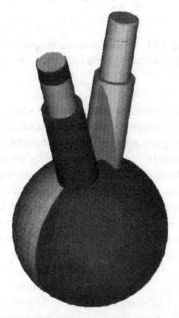

FIGURE 9.16 Split bifurcated ball arrangement. (From Geodetic, Hannover Exhibition, Hexapod-Breakthrough, Technical Information, 1997. With permission.)

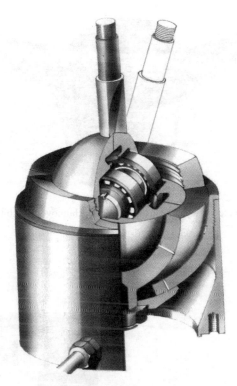

FIGURE 9.17 Hydrostatic bifurcated ball joint. (From Geodetic, Hannover Exhibition, Hexapod-Breakthrough, Technical Information, 1997. With permission.)

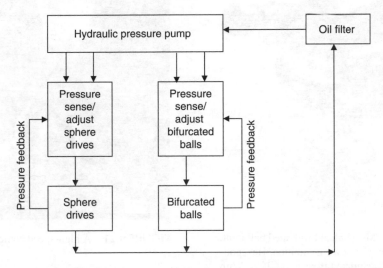

FIGURE 9.18 Layout of the lubricating system of a hydrostatic bifurcated ball joint. (From Geodetic, Hannover Exhibition, Hexapod-Breakthrough, Technical Information, 1997. With permission.)

9.4.4 Spindles

Hexapods are equipped with high-speed spindles that pack enormous power into a very compact unit. Hexapod builders make full use of the conical shape design of their spindles, which reduces interference with the WP (Figure 9.21). Geodetic machining units are equipped with motors of

FIGURE 9.19 Magnetic bifurcated ball joint. (From Geodetic, Hannover Exhibition, Hexapod-Breakthrough, Technical Information, 1997.)

FIGURE 9.20 Mechanical bifurcated ball joint. (From Geodetic, Hannover Exhibition, Hexapod-Breakthrough, Technical Information, 1997. With permission.)

FIGURE 9.21 A high-speed conical spindle.

power ratings between 3 and 20 kW at a 20,000 rpm maximum spindle speed. Geodetics also produce a range of air drive spindles for special application. A mechanical drawbar mechanism automatically clamps and releases the tool holder, making use of a high-speed tool changer. Electrically driven spindles are water-cooled. Hexapod spindles provide speed feedback to ensure precise speed control.

FIGURE 9.22 Tools accommodated by the spindle of a hexapod.

To ensure maintenance-free operation at such high speeds, the hexapod spindles should feature self-lubricating precision ceramic roller bearings. The spindles accommodate tools up to 20 mm in diameter and with a length of 200 in. High speed promotes productivity, reduces tool wear, achieves tighter machining tolerances, and provides higher surface quality. Hexapods are designed to perform milling, engraving, and drilling operations (Figure 9.22).

9.4.5 Articulated Head

An articulated head's dexterity is comparable to that of the human hand (Figure 9.23a). The two-axis head incorporates the high-speed spindle, coupled with 2 DOF, a rotary stage and a tilt stage (Figure 9.23b). This self-contained unit extends articulation to the movements attainable by the human arm (wrist and hand), allowing for machining complex surfaces as well as undercutting.

The two-axis head can tilt by over 90° and rotate over 540° (Figure 9.24). Because of the separation between the center of the platform and the additional pivot axis, the mechanism configures itself in such a way as to bend around obstructions in its working environment. This allows the task to be approached at the optimum angle. The larger the separation between the two axes, the better the reach around obstacles. A powerful positioning watercooled motor, capable of delivering high torque, delivers precise positioning in excess of 540° of rotation. Tilting is achieved by a geared sector with backlash elimination. Bearings are protected by a powerful air curtain, which prevents contamination.

9.4.6 Upper Platform

Hexapods must be mounted on a stiff base platform that does not deflect significantly under load. The Geodetic upper dome platform is shown in Figure 9.25. The dome is the most logical configuration to achieve the best angular coverage while maintaining maximum rigidity and stiffness.

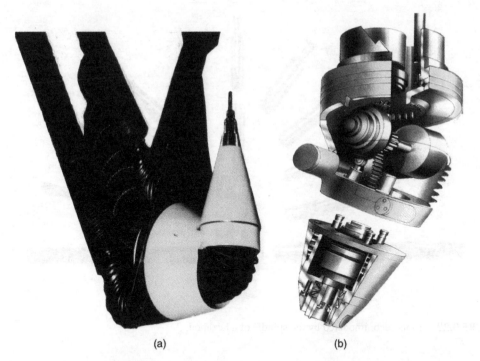

(a)   (b)

FIGURE 9.23 Articulated heads: (a) two-axis articulated head unit and (b) rotating and tilting stages of articulated head. (From Geodetic, Hannover Exhibition, Hexapod-Breakthrough, Technical Information, 1997. With permission.)

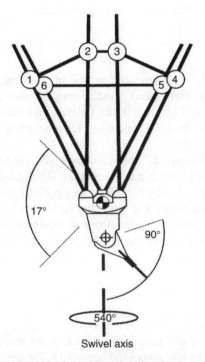

FIGURE 9.24 Tilting angles of the two-axis articulated head. (From Geodetic, Hannover Exhibition, Hexapod-Breakthrough, Technical Information, 1997. With permission.)

FIGURE 9.25 Geodetic upper dome platform. (From Geodetic, Hannover Exhibition, Hexapod-Breakthrough, Technical Information, 1997. With permission.)

The circular form ensures that the frame attachment points are as close as possible to stress points; that is, the sphere drive. Integral ribs enclose circuitry and increase stiffness while allowing for sphere drives to be plugged in without the need for complex wiring. The complete unit is embedded into resin, which further dampens vibrations and provides 100% surface contact with the frame. The cast iron dome must be designed with extensive use of finite element analysis (FEA) to achieve a lightweight yet stiff platform to support the hexapod.

9.4.7 Control System

As a machine tool, the hexapod requires sophisticated control versus conventional machine tools. In hexapods, a continuous and exact relationship of strut movements should exist to control the triangular platform movements. The contour to be followed by the cutting tool is controlled by CAD/CAM software, based on Cartesian coordinates X, Y, and Z, and orientation vectors A, B, and C of the six struts. The location coordinates X, Y, Z, A, B, and C of each strut are calculated by the controller online, which necessitates few milliseconds to be performed.

Besides the contour geometry, the calculations of the tool movements require additional data such as contouring speed and acceleration. The contour calculations are then analyzed and tested online to make sure that the dynamic limits of the machine are not exceeded to prevent the damage of the tool on the machine. The hexapod has excessive movement possibilities that possibly lead to collision between its elements. The struts may touch or even cross each other. Such possibilities must be perceived and prevented by the hexapod controller.

Contours must be corrected by the controller taking into consideration the tool diameter and length in the real-time operating system. The controller software must also be compensated for inaccuracies inherent to machine elements. The vibration, the speed profile of struts, and the positioning errors of joints and nodes, due to incorrect calculations, should also be compensated for by the controller. The developed heat due to forces and movements, which leads to complex expansion effects, should also be compensated for and corrected. Kreidler (1997) of Siemens developed a CNC control, SINUMERIK 840D, which permits the integration of different error sources to create an outstanding and efficient controller, which has been used by Geodetic and Ingersoll (Figure 9.26).

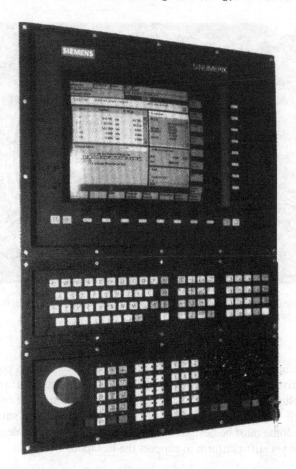

FIGURE 9.26   SINUMERIK 840D control system used by Geodetic, Inc. and Ingersoll, Inc.

Alternatively, Geodetic has used Siemens controller, a real-time, Art-to-Part that is capable of driving an unlimited number of axes simultaneously by using G and M programming codes, CLDATA, and APT.

The new high-performance PC-based controller provides a cost-effective approach to multi-axis machine control. A simple-to-use comprehensive graphical user interface guides the operator through all tedious tasks. Interaction with the machine is handled through a programmable logic controller.

Art-to-Part uses the latest forward and inverse kinematics transform algorithms. These new algorithms have been streamlined and are extremely fast. Art-to-Part is completely hardware-independent. It includes a tool management database, support for automatic tool change, automatic head change, palette changers, probes, and many other features. Written entirely in C++, Art-to-Part can be ported across to any platform with minimal effort.

9.5   HEXAPOD CHARACTERISTICS

The hexapod is a modern technology breakthrough that bridges the gap between robots and machine tools of multiaxis mechanisms that use either orthogonal or rotational movements. The current paradigm in design and manufacturing of hexapods involves integration of numerous hardwares and sophisticated software to create a unique product of extremely high rigidity and accuracy. The objective of this integrated product is to enhance quality and reliability, and to reduce the cost and overall cycle time through the dramatic departure from conventional mechanism design. As development and refinements continue, it is believed that the hexapod will eventually proliferate. A hexapod provides significant benefits to the end-user, since it offers many new attributes for the manufacturing processes.

The merits of hexapod are numerous, and include the following:

1. *Six-degree freedom.* The hexapod, with its six struts, provides the tool platform with 6 DOF. In addition to the extending motions in the orthogonal directions X, Y, Z, the platform is also able to move in other rotational compliments (pan, rotate, and twist). This advantage allows the spindle to reach unusual angles and to machine parts of difficult geometrical features such as turbine blades, plastic injection molds, stamping die, and other parts requiring high precision (Ingersoll, 1997).
2. *Flexibility and agility.* Hexapods behave according to flexible (or agile manufacturing scenarios). Flexibility is the ability to react to planned changes and agility is the ability to react to unplanned changes. Either mechanical simplicity plus its foundation independence gives the user the ability to quickly reconfigure with changes in production lines with the easy option of storing the machines disassembled when they are not needed. The agile strut-supported spindle platform positions the spindle in all 6 DOF.
3. *Productivity.* Hexapods provide higher production rates through:
   - Designing the machine to be above the worktable
   - Continuous processing capability by accommodating a pallet shuttle system that can automatically move WPs in and out of the workspace
   - Making use of a high-speed automatic tool changer
   - Reducing the mass of moving parts to achieve very fast acceleration/deceleration (up to 0.5–1 G) (Figure 9.27a). Many designs of hexapod achieve contouring feed rates up to 30 m/min (Figure 9.27b) while maintaining precision. This feed rate is much faster than that of conventional machine tools, which often have to move the WPs as well as the heavy beds
   - Using high-speed/high-power precise spindles

   These five productivity-enhancing features, together with reduced setup and processing time and consequently reduced overall cycle time, lead to the increased production rate.
4. *Stiffness and rigidity.* A well-constructed hexapod is characterized by its rigid frame, which does not deflect significantly under acting loads. An optimum design approach to a hexapod should check the tendency of struts to buckling. The critical buckling load is proportional to the fourth power of the strut diameter and is inversely proportional to the square of the strut length. Therefore, a small strut diameter is sufficient to make a stiff structure. The high stiffness and rigidity of hexapod elements result in extraordinarily high natural frequencies, which consequently allow high cutting speeds during machining. The stiffness of Ingersoll's octahedral hexapod is about three to four times that of the five-axis conventional machining center of the same rating (Figure 9.27c).
5. *Precision and accuracy.* The accuracy of a hexapod is measured volumetrically. Any loads exerted are transmitted as tension or compression. Therefore, no bending forces occur, which promotes the machining accuracy. The parallel strut arrangement lends itself to error averaging. Hexapods are lighter than conventional machines, and because sliding friction can be virtually eliminated, there is much less backlash on axis reversal, thereby promoting smoother profile movements. In this regard, it is easy to control a strut length to 2.5 μm using sophisticated software control; however, some hexapod models attain submicron accuracy.
6. *Unique installation.* By concentrating all forces of the machining process within the hexapod frame, an important advantage is offered: the lack of a need for a special foundation. Design and installing a foundation for a conventional machine tool represents a substantial

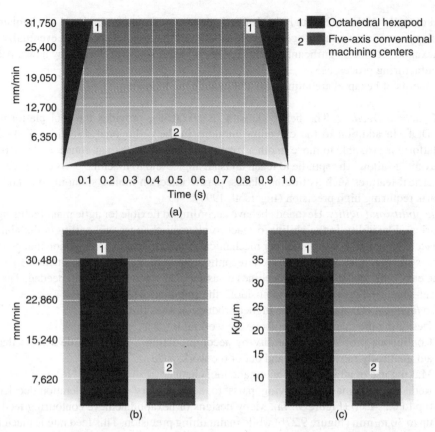

FIGURE 9.27 Hexapod characteristics. (a) Acceleration and deceleration, (b) feed rate, and (c) stiffness. (From Ingersoll Hannover Exhibition, The Next Generation in 5-Axis Machining Technology, Octahedral Hexapod, Technical Information, Ingersoll Waldrich Siegen Werkzeugmachinen, GmbH, 1997. With permission.)

cost. The hexapod foundational independence may be demonstrated by using a crane to lift one hexapod corner during machining. A hexapod could function on a ship at sea, laid on its side, or suspended from the ceiling without sacrificing the precision of its performance.

7. *Simplicity.* Another potential of hexapods is the simplicity and ease of manufacturing. The part count in a hexapod is only about 300, compared to about 1000 in conventional machine tools. The other important characteristic is that many are duplicate parts. Assembly is so easy and takes so little time that the hexapod could be sold as a kit.
8. *Portability.* Hexapod is characterized by its high potential for portability. It is a machine that could be taken to jobs like remote oil fields as well as different manufacturing plants.
9. *High load or weight ratio.* The high nominal load (power/weight) is a very important characteristic of hexapods. The cutting force acting on the moving platform is approximately equally distributed on the six parallel struts. It means that each strut suffers only from 1/6 of the total load. Furthermore, the struts are stressed longitudinally either in tension or compression; consequently, there is no need for them to be designed as massive and strongly dimensioned as in conventional machines.
10. *Scalability.* Hexapods are scaleable in size, both upward and downward, to accommodate a multitude of applications ranging from microassembly and surgery to milling, drilling, turning, welding, painting, inspection, and assembly. Versions of design varying in size from table-type models for semiconductor industry to units so large that the octahedron frame would form the building structure.

11. *Dexterity.* The hexapods have a complex working volume (a truncated hexa cone) based on the polar sweep of struts between maximum expansion and compression, and the degree of angular freedom. Dexterity extends substantially with the addition of two-axis articulated unit.
12. *Enhanced control systems.* A key step toward hexapod design is the development of a computer control system and software that are capable of processing the complex algorithmic calculations necessary to command the struts. The processor requires a calculating power equivalent to that of several fast PCs combined. Additionally, the software should be capable to compensate offset data, thermal deviations, and the like.
13. *Cost.* After conventional machine tools comes the realm of hexapods. As more hexapods are built, it is expected in the near future that their prices will be reduced to 20% or less as compared to equivalent CNC machine tools (Figure 9.28). This is so because they are simple and easy to design and assemble. Fast assembly means lower inventory, less space, and lower labor costs. Six identical struts simplify the construction, providing easy and fast assembly and reducing maintenance cost. The replacement of faulty parts subjected to wear is also easy. Control and calibration is facilitated by highly efficient software. Moreover, the power consumption of a hexapod is considerably less than that of conventional machines. They are capable of adapting to a flexible manufacturing system (FMS). Figure 9.29 illustrates comparisons of cost, number of parts, power consumption, and time elements between hexapods and conventional CNC machine tools.

However, as a new design, hexapods still have some problems that need further development and refinements. The main limitations of hexapods include:

1. *Friction.* This issue is a crucial problem for hexapods. Owing to the high coefficient of friction ($\mu = 0.8$), the accuracy and repeatability are negatively affected. In advanced designs, however, where ceramic coating and special lubricants are used, the coefficient of friction may be reduced to 0.2.
2. *Length of struts.* The hexapod accuracy is inversely proportional to the strut length due to the possibility of bending. This problem may be overcome by mapping each strut element before installing in the machine.

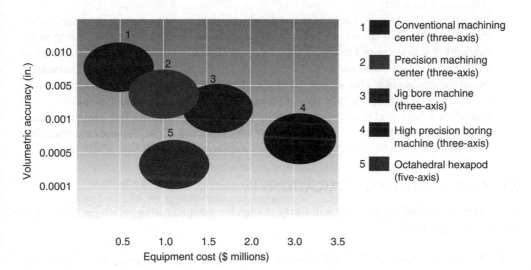

FIGURE 9.28 Cost comparison between hexapods and conventional machine tools. (From Ingersoll Hannover Exhibition, The Next Generation in 5-Axis Machining Technology, Octahedral Hexapod, Technical Information, Ingersoll Waldrich Siegen Werkzeugmaschinen, GmbH, 1997. With permission.)

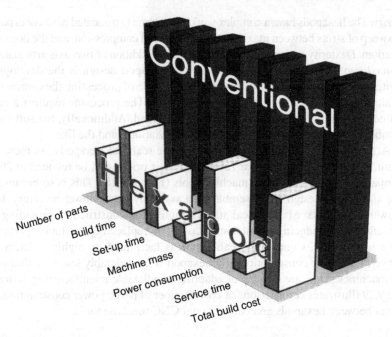

FIGURE 9.29 Total cost, number of parts, power consumption, and time elements of hexapods and conventional machine tools. (From Geodetic, Hannover Exhibition, Hexapod-Breakthrough, Technical Information, 1997. With permission.)

3. *Dynamic thermal growth.* This occurs due to the fast strut movements, as well as high spindle speeds (20,000–30,000 rpm). One way to overcome this problem is by monitoring the struts in a real-time mode, employing FEA that activates an automatic compensation routine into the software as based on the thermal growth induced in hexapods strut (Arafa, 2006). In this regard, Ingersoll has improved the thermal technique by using laser feedback that senses and eliminates deviant length changes of the hexapod struts (Ingersoll Co., 2001).
4. *Calibration.* The hexapod accuracy is not only dependent upon an accurate control of the strut length but also upon knowledge of its geometrical characteristics. According to the fabrication tolerances of hexapod struts, many factors play a role in the final accuracy of the hexapod. Many parameters must be specified to describe the geometrical characteristics of the mechanism. This is done through cumbersome calibration of the hexapod, which is still an open problem. Challenges are faced to develop a calibration system that would guarantee a high level of accuracy, enabling the hexapod to manufacture parts repeatedly within specified tolerance. A self-calibration system should allow hexapods to check their own performance and correct any detected inaccuracies (Ingersoll Co., 2001).

9.6 MANUFACTURING APPLICATIONS

The hexapods are coming, and they will likely change the manufacturing paradigm. Their applications in industry include

1. *Machining technology.* Machining is the most suitable and appropriate application making full use of hexapod's attributes. A standard hexapod offers dexterity, stiffness, and precision that is competing to conventional five-axis milling machines. Typical applications in the domain of TM include machining of press tools, mold making, turbine blade

FIGURE 9.30 A typical hexapod machining application.

cutting, and drilling at inclined angles (Figure 9.30). Stiffness and precision are expensive to achieve in conventional multiaxis grinders. The hexapod grinder offers a cost-saving upgrade and a flexible architecture suited to precise grinding. Typical application of hexapod grinders includes tool grinding and precision grinding of ceramics. Hexapods have a multitude of applications in the domain of NTM. They provide contour machining capability, so they are equipped with lasers for cutting, welding, or hardening. Similarly, they could be set up for high-pressure WJM.

In other machining domains, hexapods are used for
- Machining high-value, low-volume, and high-complexity components such as titanium for use in military aircraft
- Machining light metal and materials
- Contouring large surfaces and machining dies for precision sheet metal forming

2. *Precision assembly technology.* Hexapods are used for delicate welding in automatic assembly lines and aircraft production (Figure 9.31). Small format, low-cost hexapods, which plug into the back of a PC, are used for such applications. Coupled with the Art-to-Part control software, these smart hexapod centers are easy to operate.
3. *Measuring technology.* Hexapod is an ideal shop-floor CMM. Figure 9.32 indicates an NC-produced part for the National Aeronautics and Space Administration (NASA). A program to determine part location on the hexapod has been developed.
4. *Car-painting station.* A pair of hexapods is mounted on a simple structure (Figure 9.33). This system is expandable for different applications, such as milling, where a milling tool can be used in place of painting nozzle machine (www.//E:\Hexapodh\Feature-Hexapod-It's Working.htm, 2006).
5. *Electronic industry and fiber handling applications.* Micropositioning hexapods are used in fields that require very accurate positioning such as the electronics industry, semiconductors, and fiber-handling applications. Figure 9.34 shows a hexapod for fiber alignments. The fiber is attached to the moving platform.

FIGURE 9.31 Hexapod delicate welding in aircraft production lines.

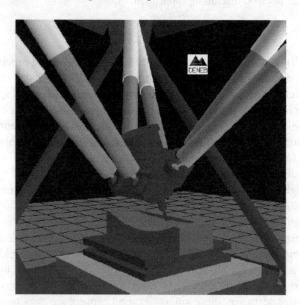

FIGURE 9.32 Verification of NASA test part by hexapod.

6. *Robotics*. With a move to offline programming, volumetric accuracy is becoming increasingly important. The ability to follow a path to within 25 μm absolute accuracy, while carrying a heavy load, is recognized as exceptional. To combine this with 1 G acceleration is unique.

9.7 REVIEW QUESTIONS

1. Differentiate using neat sketches between telescopic and ball screw/strut hexapods.
2. Draw a neat sketch to illustrate extending, panning, rotating, and twisting of a hexapod mechanism.

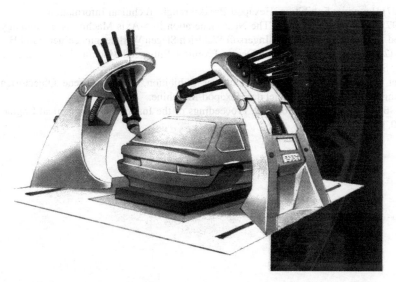

FIGURE 9.33 Hexapod car painting station.

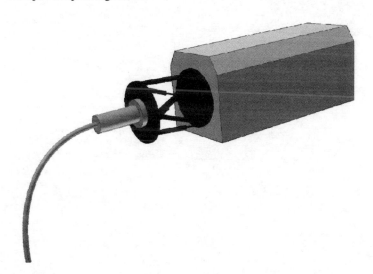

FIGURE 9.34 Fiber adjustments using micropositioning hexapod.

3. Are the bifurcated ball and ball screw elements of a telescopic strut hexapod? To what applications are the magnetic bifurcated ball hexapods best suited and why?
4. Define the following terms as applied to hexapods: flexibility–agility–scalability–dexterity–calibration.
5. What are the basic elements of a ball screw hexapod?
6. Discuss the main applications of the hexapod in manufacturing technology. List the advantages and limitations of hexapod mechanisms.
7. What is the main purpose of the Art-to-Part software?

REFERENCES

AKIMA (1997) *First European Conference on Advanced Kinematics for Manufacturing Applications*, Hannover.
Arafa H. A. (2006) Six DOF Hexapod, www.//E:\Hexapodh\Hexapodl.htm.

Geodetic (1997) Hannover Exhibition, Hexapod-Breakthrough, Technical Information.

Ingersoll (1997) Hannover Exhibition, The Next Generation in 5-Axis Machining Technology, Octahedral Hexapod, Technical Information, Ingersoll Waldrich Siegen Werkzeugmaschinen, GmbH.

Ingersoll Co. (2001) Octahedral Hexapod Design Promises Enhanced Machine Performance, Research and Data for Status Report 92-01-0034.

Kreidler V. et al. (1997) A.G. Siemens, Hannover Exhibition, Report, Offene Objectorientierte CNC-Steuerungsarchitektur am Beispiel der Hexapod-Maschine.

Stewart D. (1965) A platform with 6 DOF, Proceedings of the Institute of Mechanical Engineers, London, Vol. 180, pp. 371–386.

10 Machine Tool Dynamometers

10.1 INTRODUCTION

Machining is still one of the most important techniques for shaping metallic and nonmetallic components. During machining, the cutting tool exerts a force on the WP as it removes the machining allowance in the form of chips. Empirical values for estimating the cutting forces are no longer sufficient to reliably establish the optimum machining conditions. Depth of cut, feed rate, cutting speed, WP materials, tool material and geometry, and cutting fluid are just a few of the machining parameters governing the amplitude and direction of the cutting force.

The optimization of a machining process necessitates accurate measurement of the cutting force by a special device called a machine tool dynamometer, which is capable of measuring the components of the cutting force in a given coordinate system. It is a useful and powerful tool employed in a variety of applications in engineering research and manufacturing. A few examples of these applications are:

- Investigations into the machinability of materials
- Comparing similar materials from different sources
- Comparing and selecting cutting tools
- Determining optimum machining conditions
- Analyzing causes of tool failure
- Investigating the most suitable cutting fluids
- Determining the conditions that yield the best surface quality
- Establishing the effect of fluctuating cutting forces on tool wear and tool life

The machine tool dynamometer is not standard equipment or a device that can be used on every machine. Rather, it is equipment especially designed to fulfill some desired requirements that adapt a specific machine type operating at a specific range of machining conditions.

10.2 DESIGN FEATURES OF DYNAMOMETERS

As in most design problems, a satisfactory dynamometer design involves a compromise in which the dynamometer structure allows the highest possible sensitivity at sufficient stiffness and rigidity so that the geometry of the cutting process is maintained. At the same time, the dynamometer structure should maintain a high natural frequency to minimize chattering.

Cutting forces cannot be measured directly. Whenever a force acts on a material, it undergoes a certain deformation, which can be measured, and hence the acting forces can be accordingly derived. Therefore, the principle on which all dynamometers are designed is to measure the deflections or strains induced in the dynamometer structure caused by the resultant cutting force. Dynamometer designs differ depending on whether the deflections of the structure are directly measured with displacement transducers or whether the induced strains in the structure are measured by strain gauges and their associated equipment of high sensitivity, which allow a dynamometer structure of sufficient stiffness to be used.

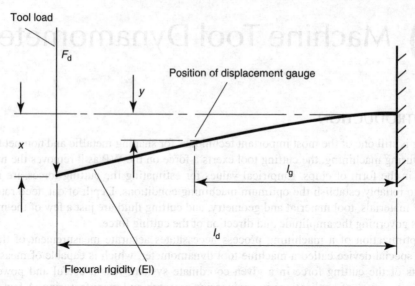

FIGURE 10.1 Displacement ratio of a cantilever dynamometer. (From Rapier, A. C., Cutting Force Dynamometers, H. M. Stationary Office, NEL Plasticity Report, 158, 1959. With permission.)

10.2.1 Rapier Parameters for Dynamometer Design

Rapier (1959) suggested two useful parameters that are used in comparing the efficiency of various dynamometer designs (Figure 10.1). These are,

1. Displacement Ratio, r_d

$$r_d = \frac{y}{x} = \frac{\text{Displacement measured by gauge}}{\text{Tool displacement by the point of application of force}}$$

2. Tool Displacement, x

In a well-designed dynamometer of high stiffness, the displacement x should be as small as possible for the reasons mentioned before. In addition, to obtain the maximum output, the displacement y measured by the gauge should be as large as possible. Thus the ratio r_d gives a guide to the dynamometer efficiency.

In most dynamometer designs, r_d does not exceed unity and thus a value of r_d approaching unity corresponds to an efficient design. For some designs, such as the slotted cantilever dynamometer, r_d may exceed unity.

In Figure 10.1, for simplicity, the dynamometer is represented by a cantilever of uniform cross section (constant flexural rigidity = EI). The deflection/unit force is given by

$$\frac{x}{F_c} = \frac{l_d^3}{3EI} \quad (10.1)$$

where

F_c = force applied (main cutting force component)
l_d = cantilever length
EI = flexural rigidity of the cantilever (assumed constant)

Machine Tool Dynamometers

The deflection per unit force at the displacement gauge arranged at a distance l_g from cantilever support is given by

$$\frac{y}{F_c} = \frac{l_d \cdot l_g}{3EI}\left(3 - \frac{l_g^2}{l_d}\right) \tag{10.2}$$

From Equations 10.1 and 10.2, r_d can be calculated as follows:

$$r_d = \frac{1}{2}\left(\frac{l_g}{l_d}\right)^2\left[3 - \frac{l_g}{l_d}\right] \tag{10.3}$$

Equation 10.3 is used to estimate the efficiency of a dynamometer design. In most cases, the ratio $l_g/l_d \approx 0.9$ and consequently $r_d \approx 0.85$.

Rigidity, sensitivity, and accuracy are the most important requirements in dynamometer design. A good dynamometer should be sensitive and accurate to within ±1%. All machine tools operate with some level of vibration. In case of milling and shaping machines, such vibrations may have large amplitudes, and therefore the dynamometer should be rigid enough to withstand such vibrations. To avoid the effect of vibrations on the measured forces by the dynamometer, its natural frequency should be at least four times as large as the frequency of the exiting vibration (f_e). The dominating stiffness criterion is the natural frequency of the dynamometer, and it is given by

$$f_n = \frac{1}{2\pi}\sqrt{\frac{K}{m}} \text{ Hz} \tag{10.4}$$

where
K = spring constant (MN/m)
m = mass of the dynamometer supported by spring (kg)

For a machine running at maximum speed of 4200 (rpm), therefore,

$$f_e = \frac{N}{60} = 70 \text{ Hz} \tag{10.5}$$

Accordingly, the natural frequency of the dynamometer (f_n) should be at least 280 Hz.

10.2.2 Main Requirements of a Good Dynamometer

The primary requirements of a quality dynamometer are as follows:

- It should possess high stiffness and rigidity.
- It should possess high sensitivity, accuracy, and reliability.
- For any cutting process, it is desirable to measure the three force components (multichannel) in a set of rectangular coordinates.
- In multichannel dynamometers, the dynamometer should be so designed so that the force in any direction should give no reading in other direction; that is, there is no cross-sensitivity (cross talk) between the channels (3% cross-sensitivity is acceptable).
- If possible, dynamometers should always be manufactured from a single block of material, as the use of clamped or bolted joints, or pivots of any kind give rise to hysteresis caused by friction. Furthermore, sliding surfaces should be avoided, because they introduce always unknown friction forces.
- It should be designed to provide the possibility of force recording using necessary bridges and multichannel recorders.
- The presence of cutting fluids makes waterproofing of the dynamometer essential.

- The dynamometer should be stable (giving consistently accurate readings with respect to time, temperature, and humidity). Once calibrated, it should only have to be checked occasionally.
- It is convenient to use a dynamometer having a linear calibration. If the calibration curve is not linear, it is then necessary to determine the zero point accurately.
- In addition, there are some special requirements that should be met, such as size, ruggedness, and adaptability to several jobs.

10.3 DYNAMOMETERS BASED ON DISPLACEMENT MEASUREMENTS

10.3.1 Two-Channel Cantilever (Chisholm) Dynamometer

This is the simplest type of two-component turning dynamometer, an advanced design proposed by Chisholm (1955), shown in Figure 10.2. The cutting tool is supported at the end of the cantilever. The deflections in the vertical (main cutting force) and horizontal (feed force) are measured by displacement gauges (dial gauge or inductive transducer). The hollow, tapered cantilever provides maximum stiffness with high natural frequency. Moreover, the dynamometer is totally machined from a solid piece of metal, as previously recommended. A Chisholm dynamometer is completely free from cross-sensitivity and its displacement ratio r_d is calculated according to Equation 10.3.

10.3.2 Two-Channel-Slotted Cantilever Dynamometer

This two-channel turning dynamometer bends the structure about the weakest points, A and B in Figure 10.3. Two displacement transducers are arranged to measure the vertical force component (main cutting force F_c) and the horizontal force component (feed force F_f), respectively. In this particular design, the displacement ratio r_d is different for each force component. For the vertical component F_c, r_d is given by

$$r_d = \frac{q}{p} \qquad (10.6)$$

Referring to Figure 10.3, it is accordingly clear that r_d can be arranged to exceed unity, which means a higher design efficiency. However, a disadvantage of this type of dynamometers is that a considerable cross-sensitivity, reaching about 15%, may result.

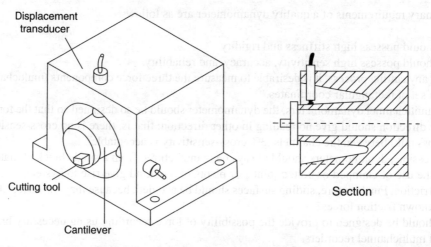

FIGURE 10.2 One-piece displacement turning dynamometer. (From Chisholm, A. W. J., Progress Report on the Wear of Cutting Tools, H. M. Stationary Office, MERL, Plasticity Report, 106, 1955. With permission.)

Machine Tool Dynamometers

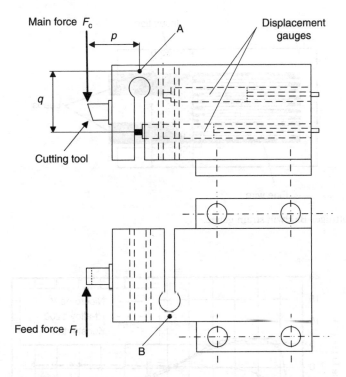

FIGURE 10.3 Two-channel cantilever-type displacement turning dynamometer. (Modified from Boothroyde, G., *Fundamentals of Metal Machining and Machine Tools*, McGraw-Hill, New York, 1981.)

10.4 DYNAMOMETERS BASED ON STRAIN MEASUREMENT

Many successful strain gauge dynamometers were developed by Shaw (1986), Hottinger-Baldwin Meβ-technik (1989), Pahlitzsch and Spur (1959), and Youssef (1971).

10.4.1 Strain Gauges and Wheatstone Bridges

Bonded-type strain gauges are sensitive sensors used to measure strains initiated from both tension and compressive static or dynamic loading. They are made of gauge wire in a flat form, of thicknesses ranging from 18 to 25 μm (Figure 10.4). The strain gauge is cemented to the sensitive member of the dynamometer, after it is properly cleaned, and suitable cement has been applied to both the surface of the member and that of the gauge. There should be no air bubble in the cement. After cementing, the gauge should be baked at about 90°C to remove moisture and then coated with wax or resin to provide some mechanical protection and prevent moisture absorption. For a proper application of the strain gauge, the resistance between the strained member and the gauge should be at least 50 MΩ. Strain gauges used for such a purpose are usually very small to limit the dynamometer size.

The two important parameters supplied by the manufacturer of a strain gauge are:

1. The electric resistance, R, which is usually 120Ω
2. The gauge factor, k_g, which varies from about 1.75 to 3.5; frequently, $k_g = 2$

The gauge factor is defined according to

$$k_g = \frac{\Delta R/R}{\Delta \ell/\ell} = \frac{\Delta R}{R} \cdot \frac{1}{\varepsilon_s} \tag{10.7}$$

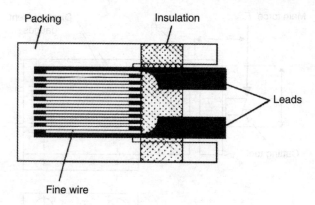

FIGURE 10.4 Hottinger-Baldwin strain gauge.

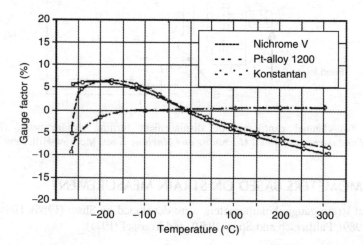

FIGURE 10.5 Effect of temperature on gauge factors. (From Hottinger-Baldwin Meβ-technik, Der Weg zum Meβgrößenaufnehmer Technical Data, Darmstadt, 1989. With permission.)

where

ε_s = elastic strain = $\frac{\Delta \ell}{\ell}$
ΔR = change of original resistance

The fine wires of the strain gauges are made of Nichrome V, Pt-alloy 1200 or Konstantan, which is an alloy of 55% Cu and 45% Ni. As an advantage, its gauge factor (k_g) is not affected by the temperature within a range of ±100°C, as shown in Figure 10.5. The change in wire resistance caused by the change of its length and cross section can be measured by the Wheatstone bridge with four active arms (Figure 10.6). No current will flow through the galvanometer (G) if the bridge is in balance when Equation 10.8 is satisfied.

$$\frac{R_1}{R_4} = \frac{R_2}{R_3} \qquad (10.8)$$

It is apparent from Equation 10.8 that although a single gauge can only be used, the sensitivity can be increased fourfold if two gauges R_1 and R_3 are used in tension, while the others R_2, R_4 are used in compression. In practice, the four gauges are selected to be of the same resistance.

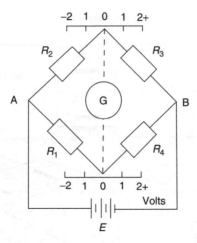

FIGURE 10.6 Balanced Wheatstone bridge.

10.4.2 Cantilever Strain Gauge Dynamometers

Cantilever strain gauge dynamometers are mainly used on lathes and drilling machines.

1. *Turning dynamometer.* Figure 10.7 shows a particular design of a two-channel strain gauge lathe dynamometer. It is convenient to locate the strain gauges on a prepared gauging section far away from the tool cutting edge. The figure also shows how the strain gauges are switched to measure the vertical main force component F_c and the horizontal feed force component F_f.
2. *Drilling dynamometer.* In any cutting process, as the force components acting on the cutting tool are equal in magnitude and opposite in direction to those acting on the WP, then it is possible to measure these components from the tool side or from the WP side. Both possibilities are widely used and each method has its advantages and disadvantages. In the first case, the dynamometer is attached to the tool and the WP may have any geometry and dimension. However, the dynamometer has a complicated construction in case of the drilling process because the drill rotates. A good design of this type has been suggested by Pahlitzsch and Spur (1959), which was produced by Hottinger, GmbH, in 1959. In the second possibility, the dynamometer is placed under the stationary WP to be drilled; the WP should be limited in size and weight.

Another drilling dynamometer has been suggested by Youssef (1971). It is composed of a star-shaped transducer (1) which is secured to the worktable (2), and simply supported on three frictionless supports (3, Figure 10.8). A U-shaped element (4) fixed to the dynamometer base (5) prevents the transducer from rotation when subjected to drilling torque. It thus enables the drilling torque M_z to be transmitted to the transducer active arms contained in the U-elements. A set of four strain gauges is applied to the sides of the active arms to measure the drilling torque M_z. Another set of gauges is applied on the top and the bottom of transducer active arms to measure the axial thrust (feed force F_y). A locator (6) enables the table (2), transducer (1), and base (5) to be aligned with each other. Figure 10.8 also illustrates the switching of a Wheatstone bridge for measuring both components M_z and T. To achieve the best results, the strain gauges should be symmetrically placed on the transducer active arms. In drilling, it is necessary to measure only the axial thrust and the torque. Therefore, a two-component dynamometer is sufficient for this purpose.

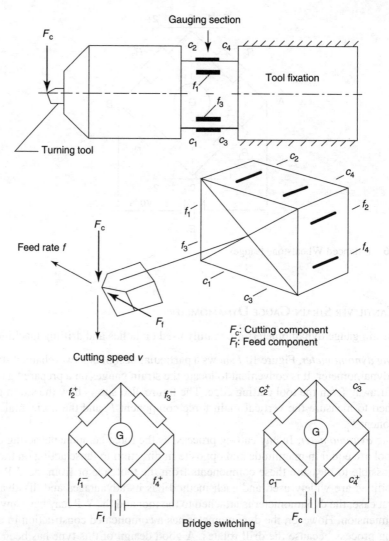

FIGURE 10.7 Two-channel strain gauge lathe dynamometer.

10.4.3 Octagonal Ring Dynamometers

10.4.3.1 Strain Rings and Octagonal Ring Transducers

A theoretical prediction of points of zero strain in circular rings under radial and tangential loads has been established and verified by many investigators. It gave best results regarding the separation of the effects of two mutually perpendicular forces. Such characteristics are very important in dynamometer design.

Consider the stress analysis of a circular ring under the action of vertical and horizontal loads F_y and F_x, respectively. The problem is solved as a statically undetermined structure, and the bending moment distributions M_y and M_x, due to the application of the vertical and horizontal force components, is tabulated as follows (Youssef, 1971). R_1 is the ring radius and θ is the location angle (Figure 10.9).

$$M_y = -\frac{F_y}{2} R_1 \sin\theta + \frac{2R_1}{\pi} \frac{F_y}{2}$$

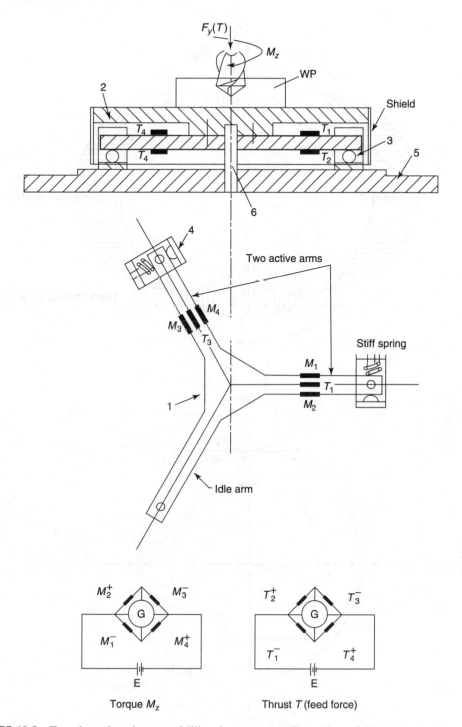

FIGURE 10.8 Two-channel strain gauge drilling dynamometer. (From Youssef, H. A., Design of machine tool dynamometers, *Bulletin of the Faculty of Engineering*, Alexandria University, Vol. X, pp. 279–303, 1971. With permission.)

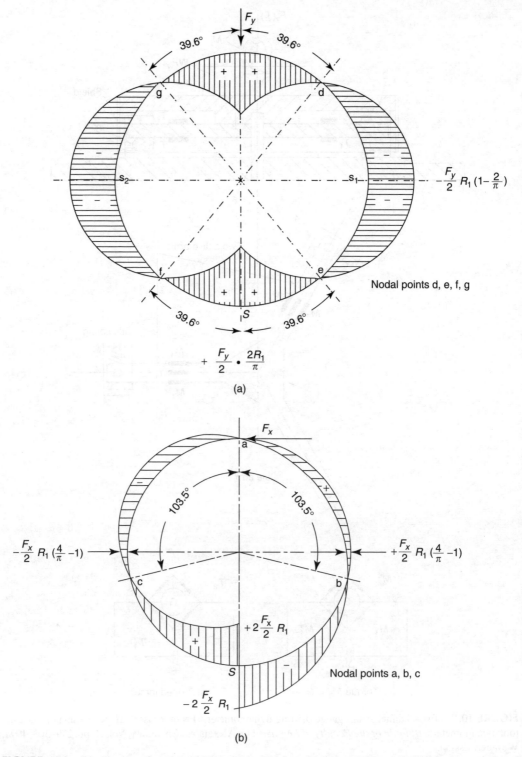

FIGURE 10.9 Distribution of moments M_y and M_x on the strain ring: (a) Stress distribution due to F_y and (b) stress distribution due to F_x. (From Youssef, H. A., Design of machine tool dynamometers, *Bulletin of the Faculty of Engineering*, Alexandria University, Vol. X, pp. 279–303, 1971. With permission.)

θ	0	$\pi/2$	39.6°	140.4°	π
M_y	$\dfrac{2R_1}{\pi} \cdot \dfrac{F_y}{2}$	$-\dfrac{F_y}{2} R_1\left(1 - \dfrac{2}{\pi}\right)$	0	0	$\dfrac{2R_1}{\pi} \cdot \dfrac{F_y}{2}$

$$M_x = -\frac{F_x}{2} R_1(1 - \cos\theta) + \frac{4}{\pi}\frac{F_x}{2} R_1\sin\theta \quad \text{for } \theta = 0 - \pi$$

$$M_x = -\frac{F_x}{2} R_1(1 - \cos\theta) - \frac{4}{\pi}\frac{F_x}{2} R_1\sin\theta \quad \text{for } \theta = \pi - 2\pi$$

θ	0	$\pi/2$	103.5°	π	π	256.5	$3\pi/2$	2π
M_x	0	$\dfrac{F_x}{2} R_1\left(\dfrac{4}{\pi} - 1\right)$	0	$-2\dfrac{F_x}{2} R_1$	$2\dfrac{F_x}{2} R_1$	0	$-\dfrac{F_x}{2} R_1\left(\dfrac{4}{\pi} - 1\right)$	0

Again referring to Figure 10.9, if the vertical force F_y is applied, maximum strain occurs along the horizontal centerline at points s_1 and s_2, whereas points of zero strain are located at positions = 39.6° from the vertical (points d, e, f, and g, respectively, in Figure 10.9a). If a horizontal force F_x is applied, the strain nodes related to this force are located at positions $\theta = 0$, 103.5°, and 256.5° (points a, b, and c, respectively) (Figure 10.9b).

Such a stress condition is very important in the design of machine tool dynamometers. The strain gauges are mounted on the ring transducer at positions corresponding to the nodes described earlier; specifically in the following manner:

1. Gauges for measuring vertical force F_y are mounted at nodal points b and c at positions $\theta = 103.5°$ and 256.5° (Figure 10.10).
2. Gauges for measuring horizontal force F_x are mounted at nodal points d, e, f, and g at position $\theta = 39.6°$ from vertical axis of the ring (Figure 10.10). Accordingly, the cross talk between the components F_x and F_y is totally eliminated. The corresponding bridge switching is shown in the same figure.

Because the nodal points are difficult to lay out accurately on circular ring transducer, an octagonal transducer with a circular hole is more practical (Figure 10.11). It is stiffer than a ring transducer of the same minimum thickness. For these reasons, most machine tool dynamometers use the octagonal transducers section. Moreover, it is easier to secure. Of course an amount of cross-sensitivity should be expected, as the angles of application of strain gauge are changed to 45° instead of 39.6° for measuring the horizontal component F_x and changed to 90° instead of 103.5° for measuring the vertical component F_y.

ILLUSTRATIVE EXAMPLE

Design a ring transducer to measure horizontal and vertical force components, each up to 250 kg.

Solution

According to Figures 10.9a and 10.9b, it is clear that the maximum stress occurs at the lowest ring cross section S.

$$M_s = \frac{2R_1}{\pi} \times \frac{F_y}{2} + 2\frac{F_x}{2} R_1 = R_1\left[\frac{F_y}{\pi} + F_x\right]$$

Assume a mean radius of transducer ring $R_1 = 40$ mm, and a width $l = 80$ mm. Therefore,

$$M_s = \frac{40}{1000}\left(\frac{250}{\pi} + 250\right) = 13.2 \text{ kg m}$$

$$= 13.2 \times 10^3 \text{ kg mm}$$

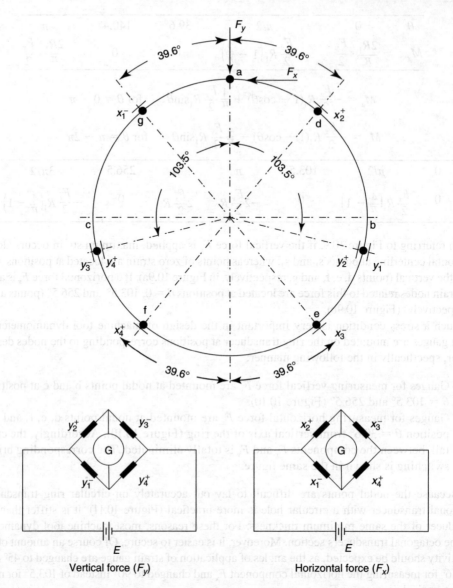

FIGURE 10.10 Application of strain gauges at nodal points of strain ring to totally eliminate the cross talk between horizontal and vertical components.

Assume that the transducer is made of steel 40 Mn 4 of allowable bending strength $f_b = 35$ kg/mm². Therefore, the minimum transducer thickness (t_r) is determined from the equation

$$f_b = \frac{M_s}{lt_r^2/6}$$

$$35 = \frac{6 \times 13.2}{80 \times t_r^2} \times 10^3$$

from which $t_r = 5.3$ mm.

10.4.3.2 Turning Dynamometer

A two-component strain gauge dynamometer with a stretched octagonal transducer is illustrated in Figure 10.12. It is used in orthogonal cutting operations. The flat sides of the stretched octagonal

Machine Tool Dynamometers

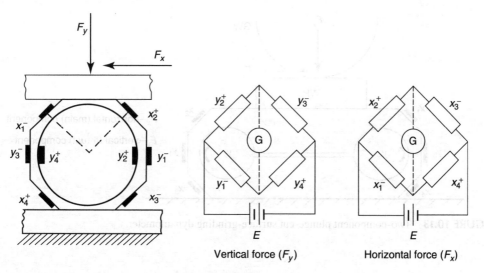

FIGURE 10.11 Octagonal transducer and related strain bridges.

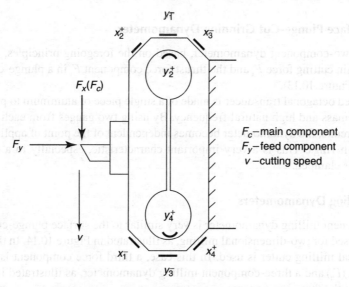

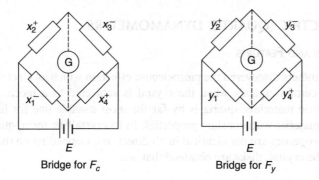

FIGURE 10.12 Two-component turning dynamometer with a stretched octagonal transducer used for orthogonal cutting.

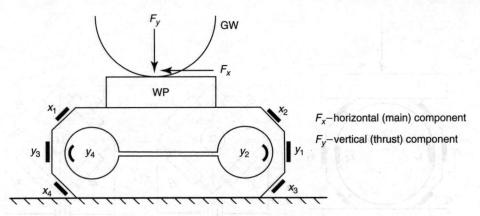

FIGURE 10.13 Two-component plunge-cut surface-grinding dynamometer.

transducer facilitate mounting of the strain gauges. In the same figure, the strain bridges for measuring the main and feed force components are visualized.

10.4.3.3 Surface Plunge-Cut Grinding Dynamometer

This is also a two-component dynamometer, based on the foregoing principles, and is applied to measure the main cutting force F_x and the thrust force component F_y in a plunge-cut surface grinding operation (Figure 10.13).

The stretched octagonal transducer is made of a single piece of aluminum to provide a sensitive element of low mass and high natural frequency. By using two gauges from each half ring in each of the bridge circuits, the dynamometer becomes independent of the point of application of the load between the half rings. This is a very important characteristic, especially in a dynamometer on which the WP is clamped.

10.4.3.4 Milling Dynamometers

The two-component milling dynamometer is very similar to the surface plunge-cut grinding dynamometer. It is used for two-dimensional milling, as illustrated in Figure 10.14. In three-dimensional milling, a helical milling cutter is used. In this case, a third force component is generated in the axial direction, (F_a) and a three-component milling dynamometer, as illustrated in Figure 10.15, is used to measure the three components.

10.5 PIEZOELECTRIC (QUARTZ) DYNAMOMETERS

10.5.1 Principles and Features

In 1880, the Cuire brothers discovered the piezoelectric effect, in which an electrical charge appears on the surfaces of certain crystals when the crystal is subjected to a mechanical load. Of the numerous piezoelectric materials, quartz is by far the most suitable one for force measurement, because it is stable material with constant properties. In its crystalline form, quartz is anisotropic, in that its material properties are not identical in all directions. Depending on the position in which they are cut out of the crystal, disks are obtained that are:

1. Sensitive only to pressure (longitudinal effect), as shown in Figure 10.16a, which measure the main force component F_z (brown).

Machine Tool Dynamometers

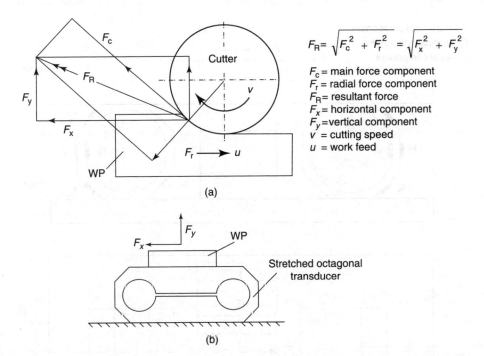

FIGURE 10.14 Two-component milling dynamometer: (a) Two-dimensional milling and (b) two-component milling dynamometer.

2. Sensitive only to shear in one particular direction (shear effect), as shown in Figure 10.16b, which measures components F_x (blue) and F_y (green), perpendicular to F_z, as well as the torque M_z (red). Figure 10.16c illustrates the generalized multicomponents with reference to a Cartesian coordinate system.

The piezoelectric force measuring principle differs fundamentally from previously discussed systems in that it is an active system. When a force acts on a quartz element, a proportional electric charge appears on the loaded surfaces, which means that it is not necessary to measure the actual deformation.

In such a system, the deflection is not more than a few micrometers at full load, whereas with conventional systems, several tenths of a millimeter may be needed. Thus piezoelectric dynamometers are very stiff systems and their resonant frequency is high, so that even rapid events can be measured satisfactorily. Moreover, the individual components of the cutting force can be measured directly, such that cross talk between measuring channels is typically less than 1%. Quartz dynamometers require no zero adjustment or balancing of the bridge circuit. It is just a matter of pressing a button, and they are ready for duty. The outstanding features of quartz dynamometers are:

- High rigidity, hence high resonant frequency
- Minimal deflections (few micrometers at full load)
- Wide measuring range
- Linear characteristic (that is, free of hysteresis)
- Lowest cross talk (typically under 1%)
- Simple in operation and without need for bridge balancing
- Compact design
- Unlimited life expectancy

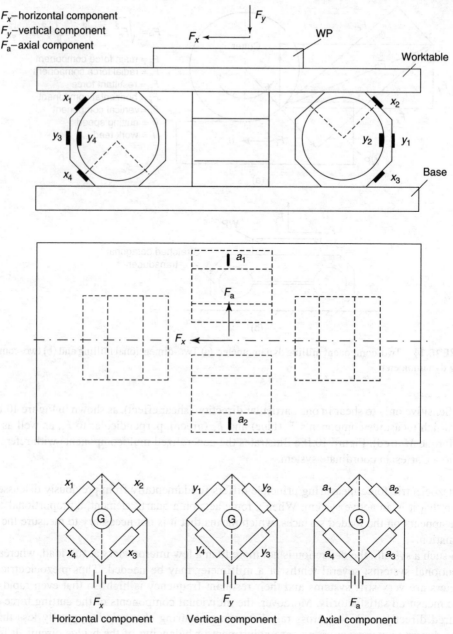

FIGURE 10.15 Three-component milling dynamometer.

10.5.2 Typical Piezoelectric Dynamometers

Piezoelectric dynamometers are efficiently used on the majority of machine tools. Three application examples are provided together with the measuring setup.

1. *Two-component piezoelectric drilling dynamometers.* Figure 10.17 shows a two-component drilling dynamometer in which shear-sensitive disks are arranged in a circle with their shear-sensitive axes oriented to respond to the torque M_z (red), whereas pressure-sensitive

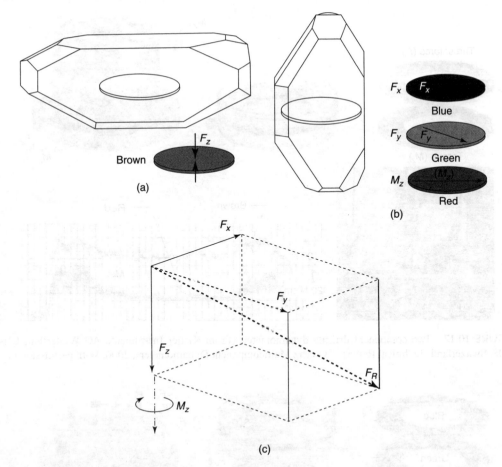

FIGURE 10.16 Disks of quartz crystals. (a) Pressure-sensitive, (b) shear-sensitive, (c) multicomponents in reference to a Cartesian coordinate system. (From Kistler Instrumente AG Winterthur, CH-8408, Switzerland, Technical Report of Quartz Multicomponent Dynamometers, 2006. With permission.)

disks are arranged and oriented to measure the thrust load F_z (brown). A high preload is necessary because the shear forces must be transmitted by friction to measure the torque.

The two-component dynamometer, shown in Figure 10.17, is suited for operations including drilling, thread cutting, countersinking, reaming, and so on. Torques and forces acting when machining holes from less than 1 mm to over 20 mm diameter can be measured satisfactorily by this dynamometer. A record of M_z and F_z is illustrated in Figure 10.17, from which it is clearly seen that F_z rises steeply at the beginning (entry of tool chisel), followed by the gradual rise of the M_z component, as the latter is more affected by the force acting on the two drill lips.

2. *Three-component piezoelectric turning dynamometer.* This model includes several shear-sensitive quartzes, with their shear-sensitive axes oriented to measure F_x (blue ring) and F_y (green ring), respectively. Their shear sensitive-axes are inclined to each other at an angle of 90°, and both are contained in a housing to form a two-component force measuring element for F_x and F_y (Figure 10.18). Pressure-sensitive quartz disks are contained in a single housing to form a single-component force-measuring element for F_z (brown ring). Another alternative is illustrated in the construction shown in Figure 10.18, where three separate elements for measuring F_x, F_y, and F_z are sandwiched under high preload

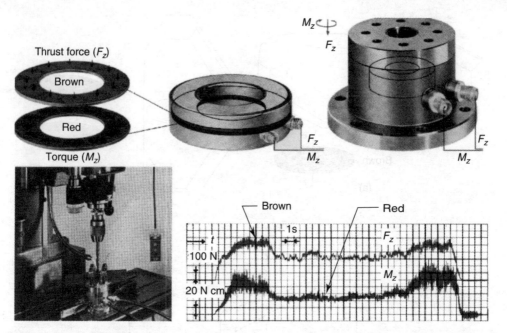

FIGURE 10.17 Two-component drilling dynamometer. (From Kistler Instrumente AG Winterthur, CH-8408, Switzerland, Technical Report of Quartz Multicomponent Dynamometers, 2006. With permission.)

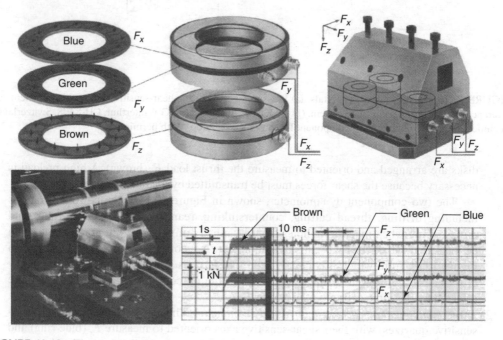

FIGURE 10.18 Three-component piezo turning dynamometer. (From Kistler Instrumente AG Winterthur, CH-8408, Switzerland, Technical Report of Quartz Multicomponent Dynamometers, 2006. With permission.)

between a base plate and a top plate. The dynamometer is mounted on the lathe slide in place of a cross-slide. A record of the three components is shown also in the same figure, from which it is clear that $F_x = F_y$, which means that the cut is performed at an approach angle $\chi = 45°$.

Machine Tool Dynamometers

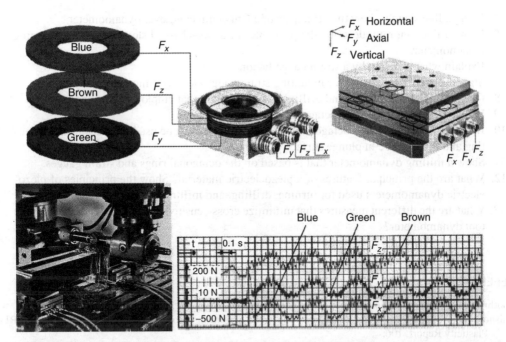

FIGURE 10.19 Three-component piezo milling dynamometer. (From Kistler Instrumente AG Winterthur, CH-8408, Switzerland, Technical Report of Quartz Multicomponent Dynamometers, 2006. With permission.)

3. *Three-component piezoelectric milling or grinding dynamometer.* Whole quartz rings may be employed. Two shear-sensitive quartz pairs, for F_x (blue) and F_y (green), and a pressure-sensitive pair for F_z (brown), can be assembled in a common housing to form a three-component force-measuring element (Figure 10.19). The pressure-sensitive quartzes are arranged in the middle so that they lie in the neutral axis under bending. During milling and grinding, the application point of the force varies a great deal. Consequently, dynamometers having four piece three-component force-measuring elements are employed. All the x, y, and z channels respectively are paralleled electrically. This makes the measurement independent of the momentary force application point. For bigger work, two dynamometers paralleled electrically and mechanically may be employed together. This system measures correctly independent of the point of force application. An output of the three-component milling dynamometer is shown in Figure 10.19. Milling is performed under the following conditions:
- Up-milling
- Cutter diameter = 63 mm, helix $\beta = 30°$, $n = 90$ rpm, $Z = 12$ teeth
- Feed $u = 53$ m/min
- Depth $t = 3.5$ mm

The severe periodic fluctuation in the measured forces is attributed to an eccentric motion of the cutter shaft. Superimposed are vibrations due to gearing of the machine. It is perfectly clear from the record that the setup shown is far from ideal. The force measure, therefore, sheds light on the machine tool behavior as well, and not just on the actual cutting operation.

10.6 REVIEW QUESTIONS

1. What are the main applications of machine tool dynamometers?
2. Explain what is meant by the Rapier parameters of dynamometer design.
3. State the main requirements of a good dynamometer.

4. Using a line sketch, show the principles of a Chisholm cantilever dynamometer.
5. Using a line diagram, describe the principles of a two-channel slotted cantilever turning dynamometer.
6. Explain what is meant by a strain gauge factor.
7. State the proper procedure for mounting strain gauges to a dynamometer body.
8. Give examples for turning and drilling dynamometers that employ strain gauges.
9. Explain the principles of an octagonal ring dynamometer.
10. Illustrate how a stretched octagonal ring dynamometer can be used for measuring the cutting and thrust force in plunge surface grinding.
11. Show a milling dynamometer that is based on the octagonal rings and strain gauges.
12. What are the principal features of a piezoelectric material? Show the principles of piezoelectric dynamometers used for turning, drilling, and milling operations.
13. What are the different measures that minimize cross-sensitivity when designing a machine tool dynamometer?

REFERENCES

Boothroyde G. (1981) *Fundamentals of Metal Machining and Machine Tools*, McGraw-Hill, New York.
Chisholm A. W. J. (1955) Progress Report on the Wear of Cutting Tools, H. M. Stationary Office, MERL, Plasticity Report, 106.
Hottinger-Baldwin Meβ-Btechnik (1989) Der Wegzum Meßgrößenaufnehmer, Technical Data, Darmstadt, Germany.
Kistler Instrumente AG Winterthur, CH-8408, Switzerland (2006) Technical Report of Quartz multi-component dynamometers.
Pahlitzsch G. and Spur G. (1959) Einrichtungen zum Messen der Schnittkräfte beim Bohren. Werkstattstechnik, 49, Heft9.
Rapier A. C. (1959) Cutting Force Dynamometers, NEL Plasticity Report, H. M Stationary Office, 158.
Shaw M. C. (1984) *Metal Cutting Principles*, Clarendon Press, Oxford, UK.
Youssef H. A. (1971) Design of machine tool dynamometers, *Bulletin of the Faculty of Engineering, Alexandria University*, Vol. X, pp. 279–303.

11 Nontraditional Machine Tools and Operations

11.1 INTRODUCTION

During recent decades, engineering materials have been greatly developed. The cutting speed and the MRR when machining such materials using traditional methods like turning, milling, grinding, and so on tend to fall. In many cases, it is impossible to machine hard materials to certain shapes using these traditional methods. Sometimes, it is necessary to machine alloy steel components of high strength in a hardened condition. It is no longer possible to find tool materials that are sufficiently hard to cut at economical speeds, such as hardened steels, austenitic steels, nimonic, carbides, ceramics, and fiber-reinforced composite materials. The traditional methods are unsuitable to machine such materials economically and there is no possibility that they can be further developed to do so, because most of these materials are harder than the materials available for use as cutting tools.

By adopting a unified program, and utilizing the results of basic and applied research, it has now become possible to process many of the engineering materials that were formerly considered to be unmachinable using traditional methods. The newly developed machining processes are often called modern machining processes or nontraditional machining processes (NTMP). These are nontraditional in the sense that traditional tools are not employed; instead, energy in its direct form is utilized.

The NTMP have specifically the following characteristics as compared to traditional processes:

- They are capable of machining a wide spectrum of metallic and nonmetallic materials irrespective of their hardness or strength.
- The hardness of cutting tools is of no relevance; especially in much of NTMP, there is no physical contact between the work and the tool.
- Complex and intricate shapes in hard and extra-hard materials can be readily produced with high accuracy and surface quality and without burrs.
- Simple kinematic movements are needed in NTM equipments.
- Micro- and miniature holes and cavities can be readily produced by NTMP.

It should be concluded that:

1. NTM methods cannot replace TM methods. They can be used only when they are economically justified or it is impossible to use TM processes.
2. A particular NTMP found suitable under given conditions may not be equally efficient under other conditions. A careful selection of the NTMP for a given machining job is therefore essential (Pandey and Shan, 1980). The following aspects must be considered in that selection:
   - Properties of the work material and the form geometry to be machined
   - Process parameters
   - Process capabilities
   - Economical and environmental considerations

TABLE 11.1
Classification of NTMP According to the Type of Fundamental Energy

Fundamental Energy	Removal Mechanism	NTMP
Mechanical	Erosion	AJM, WJM, USM
		MFM, AFM
Chemical	Ablative reaction (etching)	CHM, PCM
Electrochemical	Anodic dissolution	ECM, ECT, ECG, ECH
Thermoelectric	Fusion and vaporization	EDM, LBM, EBM, IBM, PBM

11.2 CLASSIFICATION OF NONTRADITIONAL MACHINING PROCESSES

NTMP are generally classified according to the type of energy utilized in material removal. They are classified into the following three main groups (Table 11.1):

- *Mechanical processes.* In these, the material removal depends on mechanical abrasion or shearing.
- *Chemical and EC processes.* In chemical processes, the material is removed in layers due to ablative reaction where acids or alkalis are used as etchants. The ECM is characterized by a high removal rate. The machining action is due to anodic dissolution caused by the passage of high-density dc current in the machining cell.
- *Thermoelectric processes.* In these, the metal removal rate depends upon the thermal energy acting in the form of controlled and localized power pulses leading to melting and evaporation of the work material.

An important and latest development has been realized by adopting what is called hybrid machining processes (HMP). These are new processes produced by integrating two or more NTMP to improve the performance and promoting the removal rate of the hybrid process (HP). Examples of these processes are electrochemical grinding (ECG), electrochemical honing (ECH), electrochemical ultrasonic machining (ECUSM), abrasive water jet machining (AWJM), etc.

Scientific research is still carried out in this field to check the capabilities of HMP. Some research realized remarkable success—especially the HPs that are integrated with ECM, as shown in Figure 11.1 (Rajurkar et al., 1999). A sample of some important NTMP and HMP and their relevant machines are dealt with in this chapter.

11.3 JET MACHINES AND OPERATIONS

11.3.1 Abrasive Jet Machining

11.3.1.1 Process Characteristics and Applications

In abrasive jet machining (AJM), a fine stream of abrasives is propelled through a special nozzle by a gas carrier (CO_2, Ni, or air) of pressure ranging from 1 to 9 bar. Thus, the abrasives attain a high speed ranging from 150 to 350 m/s, exerting impact force and causing mechanical abrasion of the WP (target material). The WP is positioned from the nozzle at a distance called the stand-off distance (SOD), or the nozzle-tip distance (NTD) as shown in Figure 11.2.

In AJM, Al_2O_3 or SiC abrasives of grain size ranging from 10 to 80 μm are used. The nozzles are generally made of tungsten carbides (WC) or synthetic sapphires of diameters 0.2–2 mm. To limit the jet flaring, nozzles may have rectangular orifices ranging from 0.1 × 0.5 mm to 0.18 × 3 mm.

Nontraditional Machine Tools and Operations

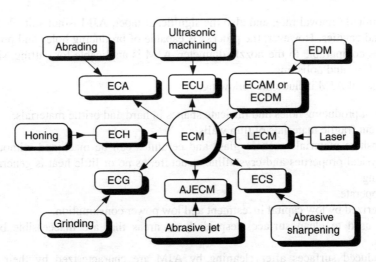

FIGURE 11.1 Hybrid NTMP integrated with ECM. (From Rajurkar, K. P., Zhu, D., McGeough, J. A., Kozak, J., and De Silva, A., *Ann. CIRP*, 48(2), 569–579, 1999. With permission.)

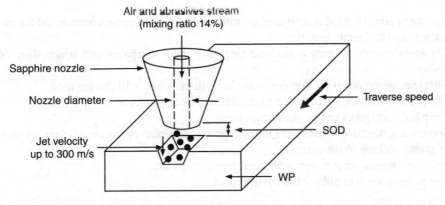

FIGURE 11.2 AJM terminology. (Modified from El-Hofy, H., *Advanced Machining Processes, Nontraditional and Hybrid Processes*, McGraw-Hill Co, New York, 2005.)

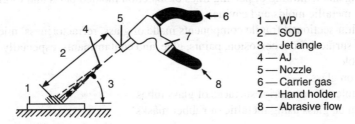

FIGURE 11.3 AJM inclination angle. (From Düniβ, W., Neumann, M., and Schwartz, H., *Trennen-Spanen and Abtragen*, VEB-Verlag Technik Berlin, 1979.)

The optimum jet angle is determined according to the ductility or brittleness of the WP material to be machined (Figure 11.3).

AJM is not considered to be a gross material removal process. Its removal rate when machining the most brittle materials, such as glass, quartz, and ceramic, is about 30 mg/min, whereas only a fraction of that value is realized when machining soft and ductile materials (Youssef, 2005).

Due to the limited removal rate, and also the significant taper, AJM is not suitable for machining deep holes and cavities. However, the process is capable of producing holes and profiles in sheets of thicknesses comparable to the nozzle diameter. AJM is applicable for cutting, slitting, surface cleaning, frosting, and polishing.

Advantages of AJM include the following:

- Capable of producing holes and intricate shapes in hard and brittle materials.
- Used to cut fragile materials of thin walls.
- Heat-sensitive materials such as glass and ceramics can be machined without affecting their physical properties and crystalline structure, as no or little heat is generated during machining.
- Safe to operate.
- Characterized by low capital investment and low power consumption.
- Can be used to clean surfaces, especially in areas that are inaccessible by ordinary methods.
- The produced surfaces after cleaning by AJM are characterized by their high wear resistance.

Limitations of AJM include the following:

- The application of AJM is restricted to brittle materials. It is not recommended for machining soft and malleable materials.
- Abrasives cannot be reused because they lose their sharpness and hence their cutting ability.
- Nozzle clogging occurs if fine grains having a diameter d_g <10 μm are used.
- The process accuracy is poor due to the flaring effect of the abrasive jet.
- Deep holes are produced by significant taper.
- Sometimes, machined parts have to undergo an additional operation of cleaning to get rid of grains sticking to the surface.
- Excessive nozzle wear causes additional machining cost.
- The process tends to pollute the environment.

Fields of Applications

AJM has been successfully applied in the following domains:

1. Deflashing and trimming of parting lines of injection molded parts and forgings
2. Cleaning metallic molds and cavities
3. Cutting thin-sectioned fragile components made of glass, refractoriness, mica, and so on
4. Cleaning surfaces from corrosion, paints, glue, and contaminants, especially those that are inaccessible
5. Marking on glass
6. Frosting interior or exterior surfaces of glass tubes
7. Engraving on glass using metallic or rubber masks

Some typical applications of AJM include the following:

- Beveling of electronic wafer disk composed of silicon disk (0.4 mm thick) welded to a tungsten disk (0.7 mm thick); see Figure 11.4a. A trimming rotating fixture is shown in Figure 11.4b. The disk rotates slowly (n = 5–10 rpm), while the nozzle is directed at an angle of 45°. 1 min is required to bevel a disk.
- Engraving registration numbers on glass windows of cars.

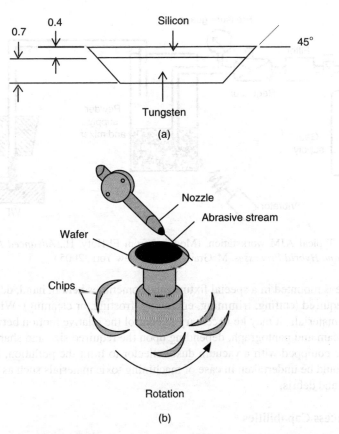

FIGURE 11.4 Edge trimming by AJM (a) Wafer disk (silicon/tungsten) and (b) trimming fixture. (From Benedict, G. F., *Non-Traditional Manufacturing Processes*, Marcel Dekker Inc., New York, 1987. With permission.)

- Deburring fine internal intersecting holes in plastic components needed for medical applications.
- Deburring of surgical needles and hydraulic valves.
- Deburring parts of nylon, teflon, and derlin.

11.3.1.2 Work Station of Abrasive Jet Machining

Figure 11.5 shows a typical workstation of AJM, which is connected to a gas supply (gas bottles or compressed air). The carrier gas must not flare excessively when discharged from nozzle to atmosphere. Furthermore, it should be nontoxic, cheap, available, and capable of being dried and filtered. Air is widely used owing to its availability. In small stations, CO_2 and N_2 gas bottles are commonly used. After filtering, the pressure of the compressed gas of 7–9 bars is regulated, to suit the working conditions. The gas is then introduced in to the mixing chamber containing the abrasives. The chamber is equipped with a vibrator providing an amplitude ξ of 1–2 mm at a frequency f_r from 5 to 50 Hz. The abrasive flow rate is controlled through the adjustment of ξ and f_r. From the mixing chamber, the gas/abrasive mixture is directed to the nozzle that directs the jet onto the target or WP. The jet velocity of 150–350 m/s depends upon the gas pressure at the nozzle, the orifice diameter of the nozzle, and the mixing ratio. The flow rate of a typical working station is about 0.6 m^3/h, which is controlled through a foot control valve.

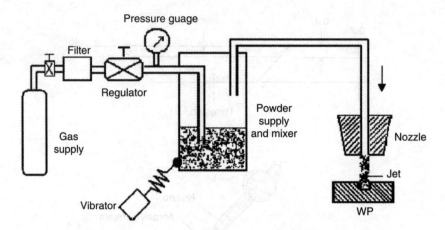

FIGURE 11.5 Typical AJM workstation. (Modified from El-Hofy, H., *Advanced Machining Processes, Nontraditional and Hybrid Processes,* McGraw-Hill Co., New York, 2005.)

The nozzle is mounted in a special fixture, and sometimes held in hand, depending on the type of operation required (cutting, trimming, engraving, frosting, or cleaning). When machining thin-walled fragile materials, it may be necessary to control the relative motion between the nozzle and the work by a cam and pantograph, depending upon the required size and shape of cut. The AJM-station must be equipped with a vacuum dust collector to limit the pollution. Strict measures and precautions should be undertaken in case of machining toxic materials such as beryllium to collect produced dust and debris.

11.3.1.3 Process Capabilities

The performance of AJM in terms of MRR and accuracy is affected by the selected machining conditions. The MRR for a certain material is mainly affected by the kinetic energy of the abrasives; that is, the speed with which the abrasive is bombarding the work material. This speed depends upon the following factors:

- Gas pressure at nozzle
- Nozzle diameter
- Abrasive grain size
- Weight mixing ratio: $\beta_m \left(\dfrac{\text{abrasive flow rate}}{\text{air flow rate}} \right)$
- SOD

MRR attains a maximum value at a mixing ratio $\beta_m = 0.15$ (Düniβ, 1979), and a SOD from 15 to 17 mm (Verma and Lal, 1984). It increases with increasing gas pressure at the nozzle. The type of material to be machined and the abrasive grain size have an influence on the MRR. The latter increases with increasing grain size (*Machining Data Handbook*, 1980). Sharp-edged abrasives that are of irregular shape, are dry, and are uniform size (noncommercial) are best suited to perform the job. The limiting size of abrasive grains that permits the grain to be suspended in the carrier gas is about 80 μm. SiC and Al_2O_3 abrasives are used for cutting and slitting operations, whereas sodium bicarbonate, dolomite, and glass beads are used for cleaning, frosting, and polishing.

When selecting the best working conditions ($\beta_m = 0.15$, abrasives Al_2O_3 of grain size 50 μm, SOD = 14 mm, nozzle pressure = 7 bar), the MRR achieved in case of machining is in the order of 30 mg/min (*Machining Data Handbook*, 1980). The accuracy improves by selecting smaller SOD (Figure 11.6), which reduces the MRR. The grain size is the decisive factor for determining the surface finish.

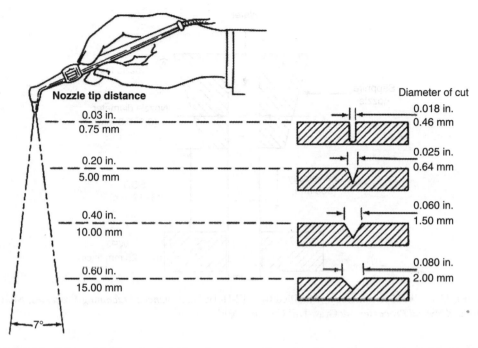

FIGURE 11.6 Effect of SOD on kerf width and accuracy. (From *Machining Data Handbook*, Machinability Data Center, Cincinnati, OH, 1980.)

11.3.2 Water Jet Machining (Hydrodynamic Machining)

11.3.2.1 Process Characteristics and Applications

For the past five decades, a number of studies using high-pressure water jets (pulsed or continuous) in mining applications have been made (Farmer and Attewell, 1965; Brook and Summers, 1969). Franz (1972) of the University of Michigan has succeeded at cutting woods using high-velocity water jets; he reported the importance of improved coherence of the water jet with the addition of polymers. Since then, the cutting capability of liquid jets has been reported for a wide spectrum of target materials, including Pb, Al, Cu, Ti, steels, and granite. It is hard to believe that a jet of water can cut steel and granite. However, in scientific terms, it is explainable, as illustrated in Figure 11.7, where a stream of water is propelled at high pressure (2000–8000 bar) through a converging nozzle to give a coherent jet of water of high speed of 600–1400 m/s. At the target, the kinetic energy (KE) of the jet is converted spontaneously to high-pressure energy, inducing high stresses exceeding the flow strength of target material, causing mechanical abrasion.

WJM has the following advantages:

- Water is cheap, nontoxic, and can be easily disposed and recirculated.
- The process requires a limited volume of water (100–200 L/h).
- The tool (nozzle) does not wear and therefore does not need sharpening.
- No thermal degrading of the work material, as the process does not generate heat. For this reason, WJM process is best suited for explosive environments.
- It is ideal for cutting asbestos, glass fiber insulation, beryllium, and fiber-reinforced plastics (FRP), because the process provides a dustless atmosphere. For this reason, the process is not hazardous and is environmentally safe.
- The process provides clean and sharp cuts that are free from burrs.

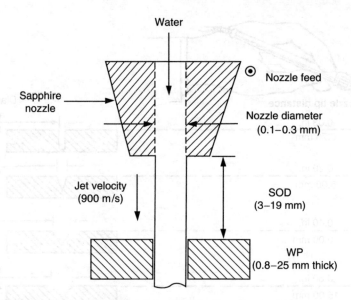

FIGURE 11.7 WJM terminology. (Modified from El-Hofy, H., *Advanced Machining Processes, Nontraditional and Hybrid Processes,* McGraw-Hill Co., New York, 2005.)

- It is applicable for laser-reflective materials such as glass, copper, and aluminum.
- Starting holes are not needed to perform the cut.
- Wetting of the WP material is minimal.
- Noise is minimized, as the power unit and intensifier can be kept away from the cutting station.
- Although AJM is commonly used to cut only brittle materials, it is able to machine both brittle and ductile materials.
- The WP is subjected to a limited mechanical stress, as the force exerted by the jet does not generally exceed 50 N.
- In WJM, cutting is performed without the need of using elaborate fixturing of the WP.
- WJM approaches the ideal single-point tool.

However, WJM has the following disadvantages:

- WJM is unsafe in operation if safety precautions are not strictly followed.
- The process is characterized by high production cost due to
  - The high capital cost of the machine
  - The need for highly qualified operators
- WJM is not adaptable to mass production because of the high maintenance requirement.

WJM is used in many industrial applications comprising the following:

1. Cutting of metals and composites applied in aerospace industries.
2. Underwater cutting and ship building industries.
3. Cutting of rocks, granite, and marble.
4. It is ideal in cutting soft materials such as wood, paper, cloth, leather, rubber, and plastics.
5. Slicing and processing of frozen foods, backed foods, and meat. In such cases, alcohol, glycerin, and cooking oils are used as alternative cutting fluids.

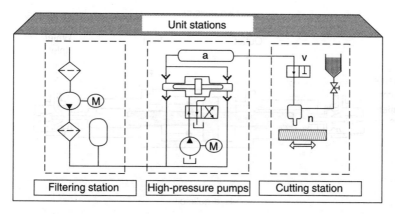

FIGURE 11.8 Simplified layout of WJM equipment. (From König, W., Fertigungsverfahren-Band 3, Abtragen, VDI Verlag, Düsseldorf, 1990.)

6. WJM is also used in
   - Cleaning, polishing, and degreasing of surfaces
   - Removal of nuclear contaminations
   - Cleaning of tubes and castings
   - Surface preparation for inspection purposes
   - Surface strengthening
   - Deburring

11.3.2.2 Equipment of WJM

Figure 11.8 visualizes a simplified layout of the WJM equipment. It consists of the following stations:

1. ***Multistage filtering station.*** The function of which is to filler the solid particles down to 0.5 µm. In this stage, it is also recommended to perform deionization and demineralization of water to allow for better performance of machine elements and extended nozzle life. After filtering, the water is mixed with polymers to obtain a coherent jet.
2. ***Oil pump and water high pressure-intensifier station.*** It consists of a hydraulic pump powered by an electric motor that provides oil at a pressure of about 120 bar. Such a pressure is needed to drive a double acting plunger pump (intensifier) that pumps water from 4 bar to about 4000 bar or more. Figure 11.9 illustrates the operation of the high-pressure intensifier that consists of two terminal small cylinders for water and a large central cylinder for oil. A limit switch, located at each end of the terminal cylinders signals the electronic controls to shift the directional control valve and reverses the piston direction. As one side of the intensifier is in the inlet position, the opposite side is generating an ultrahigh pressure output, and vice versa. The ultrahigh-pressure water is delivered to an accumulator tank (Figure 11.8) to provide the water pressure free of fluctuation and hydraulic spikes to the cutting station. During idle times, the water is stored in the accumulator under pressure to be ready at any time to perform the cutting. The intensifier offers complete flexibility for both cutting and cleaning applications. It also supports single or multiple cutting nozzles for increased productivity (El-Hofy, 2005).
3. ***Cutting station.*** The converging cutting nozzle (Figure 11.10) converts the ultrahigh pressure (about 4000 bar) into a high speed of 400–1400 m/s. The nozzle provides a coherent water jet stream for optimum cutting. The jet coherency can be enhanced by adding long

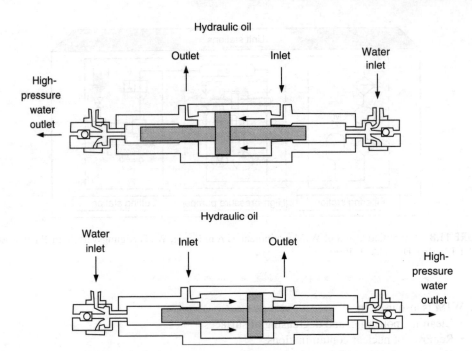

FIGURE 11.9 HP intensifier. (From Nordwood, J. A. and Johnston, C. E., *7th International Symposium on Jet Cutting Technology*, 369–388, 1984.)

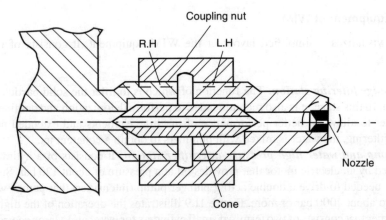

FIGURE 11.10 Nozzle assembly of WJM equipment. (From Youssef, H. A., *Non-Traditional Machining Processes—Theory and Practice*, El-Fath Press, Alexandria, 2005. With permission.)

chain polymers such as polyethylene oxide (PEO) with a molecular weight of 4 million. Such addition provides water a higher viscosity and hence increases the coherent length up to 600 d_n, where d_n is the nozzle orifice diameter that fall between 0.1 and 0.35 mm. For optimum cutting, the SOD is selected within this range. Even if the SOD is selected beyond this range, the stream is still capable of performing noncutting operations such as cleaning, polishing, degreasing, etc., due to the existence of the concentrated liquid cone in the growing spray envelop (Youssef, 2005).

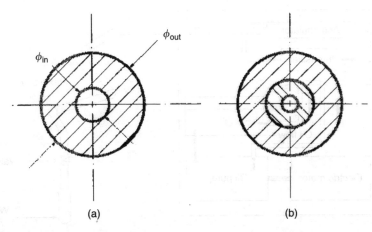

FIGURE 11.11 High-pressure tubing: (a) thick-wall tubing and (b) shrink-fit tubing.

Nozzles are generally made from very hard materials such as WC, synthetic sapphire, or diamond. Diamond provides the largest nozzle life, whereas WC gives the lowest one. About two hundred hours of operation is expected from a nozzle of synthetic sapphire, which becomes damaged by particles of dirt and the accumulation of mineral deposits if the water is not filtered and treated. High-pressure tubing (Figure 11.11) transports pressured water to the cutting nozzle. Thick tubes of diameters ranging from 6 to 14 mm and diameter ratio ranging from 1/5 to 1/10 are used (Figure 11.11a). For severe pressure, which may exceed the yielding stress of the tube material, shrink-fit tubes should be used (Figure 11.11b). To achieve the best sealing conditions, metal-to-metal, line (not surface) contact should be used in high-pressure tube fittings. The on–off valves for such machines operating at high pressures are preferred to be of the needle-type. The compact design of the nozzle head promotes integration with motions control system ranging from two-axes x–y tables to sophisticated multiaxes CNC installations.

To machine complex contours, the nozzle is mounted in a robot arm supplemented to the machine. Although the reaction forces in most cases does not exceed 50 N, the pressure waves due to the operation of the on–off valves impart vibration to the robot joints and consequently impair the machining accuracy.

The cutting station must be equipped with a catcher that acts as a reservoir for collecting the machining debris entrained in the jet. Moreover, it absorbs the rest of the energy after cutting, which is estimated to be 90% of the total jet energy. It reduces the noise levels (105 dB) associated with the reduction of the water jet from Mach 3 to subsonic levels. Figure 11.12 shows a schematic illustration of the WJM equipment.

11.3.2.3 Process Capabilities

The MRR, accuracy, and surface quality are influenced by the WP material and the machining parameters. Brittle materials fracture, while ductile ones are cut well. The material thickness ranges from 0.8–25 mm or more. Table 11.2 illustrates the cutting rates for different material thicknesses. For a given nozzle diameter, the increase in pressure allows more power to be used, which in turn increases the penetration depth or the traverse speed.

The quality of cutting improves at high pressures and low traverse speeds. Under such conditions materials of greater thicknesses can be cut.

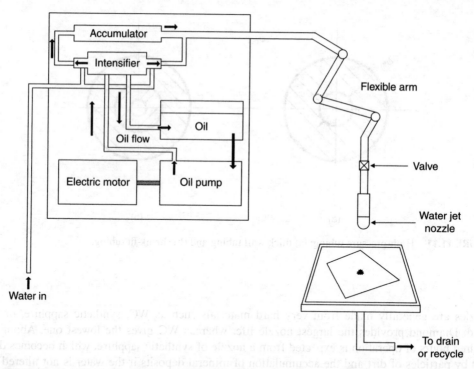

FIGURE 11.12 Schematic illustration of WJM equipment. (Modified from El-Hofy, H., *Advanced Machining Processes, Nontraditional and Hybrid Processes*, McGraw-Hill Co., New York, 2005.)

TABLE 11.2
Traverse Speeds and Thicknesses of Various Materials Cut by Water Jet

Material	Thickness (mm)	Traverse Speed (mm/min)
Leather	2.2	20
Vinyl chloride	3.0	0.5
Polyester	2.0	150
Kevlar	3.0	3.0
Graphite	2.3	5.0
Gypsum board	10.0	0.6
Corrugated board	7.0	200
Pulp sheet	2.0	120
Plywood	6.0	1.0

Source: Reproduced from Tlusty, G. in *Manufacturing Processes and Equipment*, Prentice-Hall, Upper Saddle River, NJ, 2000.

11.3.3 Abrasive Water Jet Machining

11.3.3.1 Process Characteristics and Applications

AWJM is a hybrid process (HP) since it is an integration of AJM and WJM processes. The addition of abrasives to the water jet increases the range of materials which can be cut with a water jet drastically and maximizes the MRR of this HP. The MRR is based, therefore, on using the kinetic

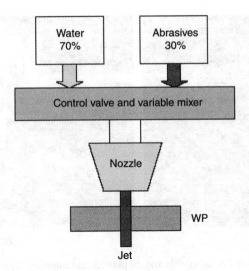

FIGURE 11.13 AWJM elements terminology. (Modified from El-Hofy, H., *Advanced Machining Processes, Nontraditional and Hybrid Processes*, McGraw-Hill Co., New York, 2005.)

energies of the abrasives and water in the jet. Intensive research works have been carried out during the last three decades to explore the capabilities of this new promising process. It has been reported that AWJM process is capable of machining both soft and hard materials at very high speeds as compared with those realized by WJM. It cuts 10–50 times faster than WJM process. Moreover, the cuts performed by AWJM have better edge and surface qualities.

AWJM uses a comparatively lower water pressure than that used by WJM (about 80%) to accelerate the AWJ. The mixing ratio of abrasives to water in the jet is about 3/7 by volume (Figure 11.13). Abrasives (garnet, sand, Al_2O_3, etc.) of grain size 10–180 μm are often used.

As previously mentioned, apart from its capability to machine soft and hard materials at very high speeds, AWJM process has the same advantages as that of WJM. However, it has the following two limitations:

- Owing to the existence of the abrasives in the jet, there is an excessive wear in the machine and its elements.
- The process is not environmentally safe as compared to WJM.

The AWJM process has many fields of application such as:

- Cutting of metallic materials: Cu, Al, Pb, Mo, Ti, W.
- Cutting carbides and ceramics.
- Cutting concrete, marble, and granite (Figure 11.14a).
- Cutting plastics and asbestos (Figure 11.14b).
- Cutting composites such as FRP, and sandwiched Ti-honeycomb without burr formation. The latter is used in aerospace industry.
- Cutting of acrylic and glass.

In the field of machining technology, the AWJM has two promising applications as illustrated in Figure 11.15. These are milling of flat surfaces and turning of cylindrical surfaces.

The process is also applicable in deburring (AWJD), sharpening of grinding wheels, and surface strengthening to increase the fatigue strength.

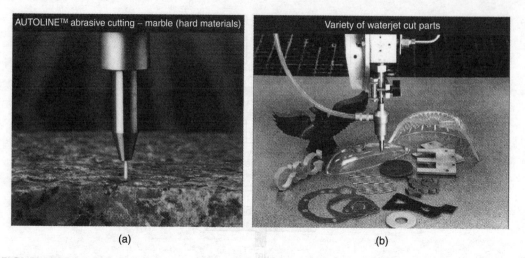

FIGURE 11.14 Cutting by AWJM: (a) marble and (b) plastics and asbestos. (From Ingersoll-Rand, *Technical Data*, Hannover Exhibition, 1996.)

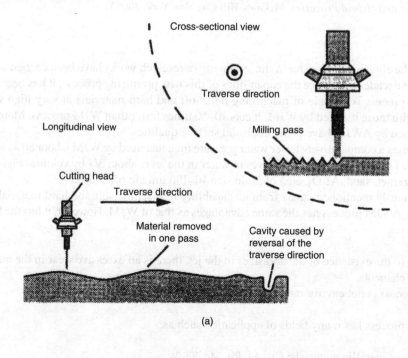

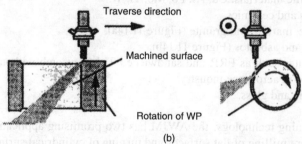

FIGURE 11.15 Two promising applications of AWJM: (a) milling and (b) turning. (From Hoogstrate, A. M., and van Luttervelt, C. A., *Ann. CIRP*, 46(2), 697–714, 1997. With permission.)

11.3.3.2 Abrasive Water Jet Machining Equipment

The AWJM equipment does not differ greatly from the basic WJM equipment. So, it is composed of

1. *Water filtering station.* It is the same as that of the WJM, but in AWJM cutting polymers are not commonly used, because the general opinion is that the increased coherence of the jet prevents the abrasive particles from being mixed with the water jet and, therefore, the accelerating process of the abrasive is less efficient, although the Swiss glass manufacturers and French crystal manufacturers have different opinions and use a polymer additive in combination with abrasives thereby increasing the cutting speed considerably and reducing the abrasive consumption drastically (Paul et al., 1998). Research in this area is therefore still needed.
2. *Pressure generation station.* A double acting intensifier is designed to deliver less pressure than that used in WJM. The usual range of pressure in AWJM is from 250 to 350 MPa at a discharge rate of 5 L/min, accordingly the pressure loss is decreased and the system piping is less stressed.
3. *Cutting station.* The cutting nozzle in the machining station of WJM equipment is replaced by what is called the jet former in AWJM equipment.

Jet Former

In the jet former, the pressure of the water is first converted into kinetic energy of the water, which in turn is partially converted into kinetic energy of the abrasive particles. Figure 11.16a illustrates a jet former of Ingersoll-Rand, while Figure 11.16b illustrates a sectional view of the same jet former. At the end of the high-pressure tubing an orifice is installed, which consists commonly of a hexagonal–rhomboedral sapphire Al_2O_3, a ruby or diamond having a hole of 0.08–0.8 mm inner diameter. Diameters under 0.25 mm are mainly used in high-pressure pure water jet cutting applications, because the total energy is too low to accelerate the abrasive particles effectively. Orifice diameters between 0.25 and 0.40 mm are used in AWJM applications. Diameters over 0.40 mm are mainly used in low-pressure cleaning applications of the AWJ (Hoogstrate and van Luttervelt, 1997).

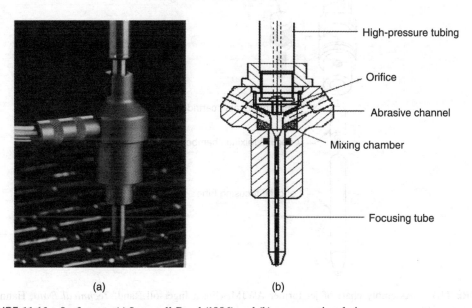

FIGURE 11.16 Jet former: (a) Ingersoll-Rand (1996) and (b) cross-sectional view.

Through the orifice (Figure 11.16b), the high-pressure water is expelled and pure water jet is formed and directed into the mixing chamber. Through the interaction of the pure water jet and the surrounding air a vacuum is created in the mixing chamber causing airflow from outside through the abrasive channels to the mixing chamber. In the mixing chamber, the jet loses its coherency; therefore a focusing tube (Figure 11.16b) is installed below the mixing chamber to restore the coherency of the AWJ. The resulting diameter of the AWJ is nearly equal to the focusing tube diameter. Figure 11.17 illustrates an assembly chart of the jet former of Ingersoll-Rand. The design of the jet former is based on the following parameters:

- Water orifice diameter
- Distance along jet axis from orifice to entrance point
- Entrance direction (angle) of abrasives
- Cross-section of the abrasive feed channel
- Mixing chamber length/diameter ratio
- Diameter of focusing tube
- Length of focusing tube

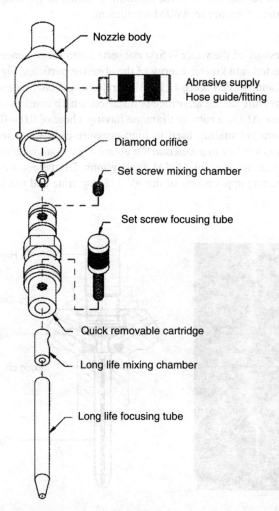

FIGURE 11.17 Assembly chart of jet former AWJM. (From Ingersoll-Rand, *Technical Data*, Hannover Exhibition, 1996.)

Focusing Tube (Also Called Abrasive Tube or Accelerator Tube)
The performance of the focusing tube depends upon the following (Hoogstrate and van Luttervelt, 1997)

- *Geometry of the inlet zone.*
- *Inner diameter of the tube.* The smaller this diameter, the more concentrated is the total energy. For reliable functioning the focusing tube diameter should be at least five times the particle diameter.
- *Length of the tube.* Longer tubes produce a more coherent jet, but cause more friction between the jet and the tube wall, resulting in lower abrasive jet velocities. Longer tubes are also more difficult to align.

Mixing Abrasives with Water
Owing to the complex turbulent nature of the mixing process not much modeling has been carried out. However, an acceptable theory for the mixing of the particles and the jet in the focusing tube was developed. It is assumed that each particle enters the water jet with a negligible velocity. It is accelerated and pushed out of the water jet, hits the inner wall of the focusing tube, rebounces, and enters the water jet again. This happens until the velocity direction of the particle is nearly parallel to the direction of the water jet (Hoogstrate and van Luttervelt, 1997) (Figure 11.18). As consequences of this acceleration process two effects are encountered.

1. The abrasive particles are fragmented due to collisions with the focusing tube and other abrasive particles (Schmelzer, 1994). This causes a significant diameter reduction of the abrasive particles after the focusing process (Figure 11.19). Recycling of abrasives seems not interesting due to this particle fragmentation. Nevertheless abrasive recycling units have been recently introduced in the market.
2. The focusing tube is exposed to extremely abrasive conditions. Therefore, it should be made of advanced wear resistant materials like ROCTEC-100, which provides a reliable, stable cutting over a longer service life as shown in Figure 11.20. An intelligent nozzle

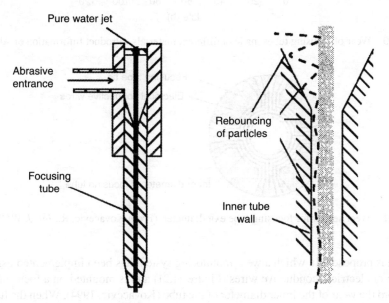

FIGURE 11.18 Abrasive acceleration in the focusing tube. (From Hoogstrate, A. M. and van Luttervelt, C. A., *Ann. CIRP*, 46(2), 679–714, 1997. With permission.)

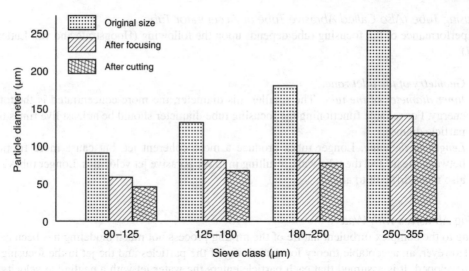

FIGURE 11.19 Wear of garnet grit in AWJM. (From Schmelzer, M., *Mechanismes für Strahlerzeugung beim Wasser-Abstrahlschneiden*, Dissertation TH-Aachen, 1994.)

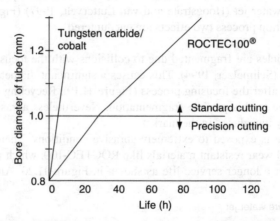

FIGURE 11.20 Wear of focusing tubes made of different materials. (Product Information of Allfi AG, 1997.)

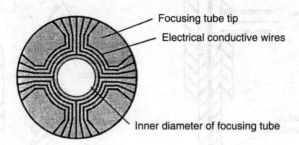

FIGURE 11.21 Wear sensor for focusing tube exit diameter. (From Kovacevic, R., *Int. J. WJ Technol.*, 2(1), 1994.)

system is proposed, in which a wear monitoring system has been implemented using a grid pattern of electrical conductive wires (Figure 11.21) and is mounted on a focusing tube-tip to record the wear of the inner diameter of the tube (Kovacevic, 1994). When the tube diameter wears out, the wires will successively be interrupted, which can be easily signaled. This diameter information can be used to trigger the end of the focusing tube life.

Important Characteristics of the AWJ

Five important AWJ characteristics that must be realized to perform effectively are as follows (Hoogstrate and van Luttervelt, 1997):

- Jet velocity determines the cutting capability.
- Jet coherence determines the kerf shape and the quality of cut.
- Abrasive/water mass ratio ensures optimum cutting efficiency.
- Rotational jet symmetry determines cutting capability in different directions.
- Establishing a time-independent jet structure gives a uniform quality along the cut in a WP.

Computerized WJ and AWJ machines are now available in the markets. They are capable of loading a CAD drawing from another system, and determin start and end points and the sequence of operations. Other CNC machines operate with a modem and CAD/CAM capabilities that permit transfer from AUTOCAD, DXF formats.

11.3.3.3 Process Capabilities

The typical machining variables of the AWJM-process are as follows:

- Water pressure
- Water nozzle diameter
- Geometry of focusing tube (length and diameter)
- SOD
- Size and type of abrasive grits
- Abrasive/water ratio
- Hardness and strength of the WP material
- Type of WP material (metallic, nonmetallic, or composite)

When machining glass by AWJ, a cutting rate of about 16–20 mm^3/min is achieved. An AWJ cuts through 360 mm thick slabs of concrete or 76 mm thick tool steel plates at a traverse speed of 38 mm/min in a single pass. When cutting steel plates (or metallic materials) the surface roughness R_t ranges from 3.8 to 6.4 μm, while tolerances of ±130 μm are obtainable. Repeatability of ±40 μm, squareness of 43 μm/m, and straightness of 50 μm per axis are expected. Sand and garnet are frequently used as abrasive materials. However, garnet is preferred because it is 30% more effective than sand. A carrier liquid consisting of water with anticorrosive additives contributes to higher acceleration of abrasives with a consequent higher abrasive speed and increased MRR (El-Hofy, 2005). The penetration depth increases with increasing water pressure and decreasing traverse velocity, provided other working conditions are being constant.

The SOD has an important effect on the MRR and the accuracy. It attains values between 0.5 and 5 mm. The smallest value (0.5 mm) realizes higher accuracy and smallest kerf width, whereas the largest value (5 mm) realizes the maximum MRR. Beyond 5 mm, the jet loses gradually its cutting capability till it reaches 50–80 mm, at which the jet is used efficiently in surface cleaning and peening.

Table 11.3 illustrates the traverse velocities when cutting different materials of different thicknesses using AWJ. Accordingly, it can be depicted that

1. Pure metals (Ti, Al) have the same machinability.
2. Glass is cut at 8–10 times faster than metals and alloys.

Surface roughness depends on the WP material, grit size, and type of abrasives. A material with a high removal rate produces large surface roughness. For this reason, fine grains are used for

TABLE 11.3
Traverse Velocity (mm/min) When Machining Different Materials by AWJM

Material Thickness	6 mm	15 mm	19 mm	25 mm	50 mm
Titanium	250	150	100	50	16
Aluminum	250	150	100	50	16
FRP	500	280	130	75	25
Stainless steel	200	90	60	40	15
Glass	2000	1000	700	500	150

Source: Reproduced from Youssef, H. A. in *Non-Traditional Machining Processes—Theory and Practice*, El-Fath Press, Alexandria, 2005.

machining soft metals to obtain the same roughness as hard ones. Additionally, the larger the abrasive/water ratio the higher will be the MRR (El-Hofy, 2005).

In the domain of composites, AWJM process is particularly good as the cutting rates are considerably high, and it does not delaminate the layered material. A comparison study is carried out by König and Schmezler (1990) to investigate the performance of WJM and AWJM when cutting a plate of FRP of 5 mm thick under optimum working conditions:

WJ and AWJ

Water pressure	= 300 MPa
Nozzle diameter	= 0.225 mm
SOD	= 5 mm

AWJ

Abrasives: garnet # 80 mesh
Abrasive flow rate = 300 g/min

The outcome of this study shows that the AWJM has realized a traverse velocity of 2000 mm/min, which is 40 times that realized by WJM. Moreover the surface roughness as obtained by AWJM ($R_a = 4.4$ μm) is about 30% less than that obtained by WJM ($R_a = 6.4$ μm).

11.4 ULTRASONIC MACHINING EQUIPMENT AND OPERATION

11.4.1 Definitions, Characteristics, and Applications

Ultrasonic machining is economically viable operation by which a hole or a cavity can be pierced in hard and brittle materials, whether electric conductive or not, using an axially oscillating tool. The tool oscillates with small amplitude of 10–50 μm at high frequencies of 18–40 kHz to avoid unnecessary noise (the audio threshold of human ear is 16 kHz). During tool oscillation, abrasive slurry (B_4C and SiC) is continuously fed into the working gap between the oscillating tool and the stationary WP. The abrasive particles are, therefore, hammered by the tool into the WP surface, and consequently they abrade the WP into a conjugate image of the tool form. Moreover, the tool imposes a static pressure ranging from 1 N to some kilograms depending on the size of the tool tip, Figure 11.22. The static pressure is necessary to sustain the tool feed during machining.

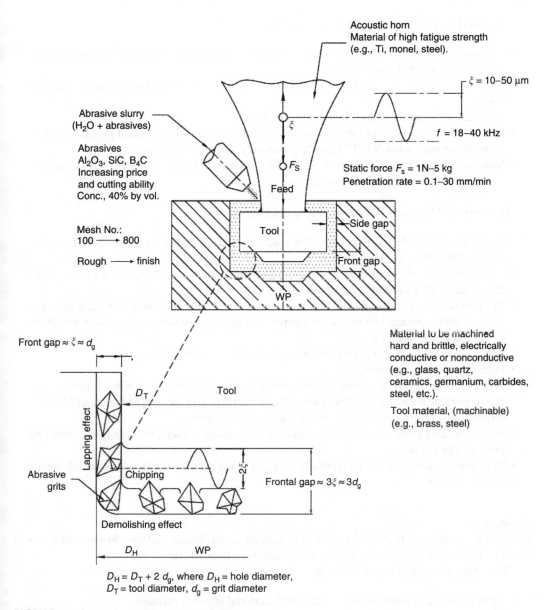

FIGURE 11.22 Characteristics of the USM process.

The process productivity is realized by the large number of impacts per unit time (frequency), whereas the accuracy is achieved by the small oscillation amplitude employed.

The tool tip, usually made of relatively soft material, is also subjected to an abrasion action caused by the abrasives; thus it suffers from wear, which may affect the accuracy of the machined holes and cavities. Owing to the fact that the tool oscillates and moves axially, USM is not limited to the production of circular holes. The tool can be made to the shape required, and hence extremely complicated shapes can be produced in hard materials. The process is characterized by the absence of any deleterious or thermal effects on the metallic structure of the WP. Outside the machining domain, US techniques are applied in nondestructive testing (NDT), welding, and surface cleaning, as well as diagnostic and medical applications.

The USM process has the following advantages:

- Intricate and complex shapes and cavities in electric or nonelectric conductive materials can be readily machined ultrasonically.
- As the tool exhibits no rotational movement, the process is not limited to produce circular holes.
- High dimensional accuracy and surface quality.
- Especially, in the sector of electrically nonconductive materials, the USM process is not in competition with other NTMP regarding accuracy and removal rates.
- Since there is no temperature rise of the WP, no changes in physical properties or microstructure whatsoever can be expected.

However, USM process has the following disadvantages:

- When machining electrically conductive materials (except carbon), a limited MRR as compared with ECM and EDM is realized.
- USM is not capable of machining holes and cavities with a lateral extension of more than 25–30 mm with a limited depth of cut.
- The tool suffers excessive frontal and side wear when machining conductive materials such as steels and carbides. The side wear destroys the accuracy of holes and cavities, and leads to a considerable conicity error.
- Every job needs a special high-cost tool, which adds to the machining cost.
- High rate of power consumption.
- When machining through holes, the WP should be supported by a pad of machinable material to prevent breaking out. Otherwise, the static force and the amplitude should be correspondingly programmed at the end of the machining stroke.
- In case of blind holes, the designer should not allow sharp corners, because these cannot be produced by USM.
- The abrasive slurry should be regularly changed to get rid of worn abrasives, which means additional cost.

Applications

It should be understood that USM is generally applied to machining shallow cavities and forms in hard and brittle materials having a surface area not more than 10 cm^2.

Some typical applications of USM are as follows:

- Manufacturing of forming dies in hardened steel and sintered carbides
- Manufacturing of wire drawing dies, cutting nozzles for jet machining applications in sapphire, and sintered carbides
- Slicing hard brittle materials such as glass, ceramics, and carbides
- Coining and engraving applications
- Boring, sinking, blanking, and trepanning
- Thread cutting in ceramics by rotating the tool or the WP

Figure 11.23 illustrates some products produced ultrasonically.

a. Engraving a medal made of agate (König, 1990)
b. Piercing and blanking of glass (König, 1990)
c. Producing a fragile graphite electrode for EDM (König, 1990)
d. Sinking a shearing die in hardened steel or WC (Lehfeld Works, 1967)

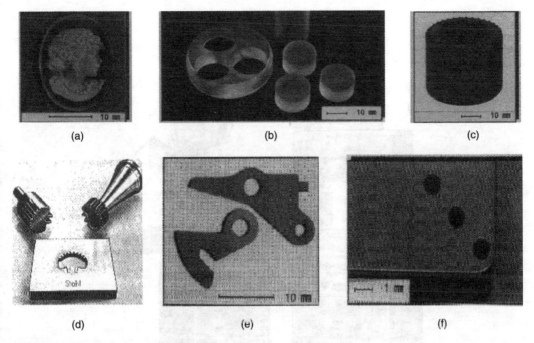

FIGURE 11.23 Typical products by USM.

e. Production of outside contour and holes of master cutters made of zirconium oxide (ZrO_2) of a textile machine (König, 1990)
f. Drilling fine holes $\phi = 0.4$ mm in glass (Kalpakjian, 1984).

11.4.2 USM Equipment

The USM equipment shown in Figure 11.24 has a table capable of orthogonal displacement in X and Y directions, and a tool spindle and carrying the oscillating system moving in direction Z perpendicular to the X–Y plane. The machine is equipped with a HF generator 2 of a rating power of 600 W, and a two-channel recording facility 3 to monitor important machining variables (tool displacement Z and oscillation amplitude ξ). A centrifugal pump is used to supplement the abrasive slurry into the working zone. Figure 11.25 shows schematically the main elements of the equipment, which consist of the oscillating system, the tool feeding mechanism, and the slurry system.

11.4.2.1 Oscillating System and Magnetostriction Effect

The oscillating system includes a transducer contained in the acoustic head, a primary acoustic horn, and a secondary acoustic horn (Figure 11.26).

1. Acoustic Transducer

This transforms electrical energy to mechanical energy in the form of oscillations. Magnetostrictive transducers are generally employed in USM, but piezoelectric ones may also be used.

The magnetostriction effect was first discovered by Joule in 1874. According to this effect, in the presence of an applied magnetic field, ferromagnetic metals and alloys change in length. The deformation can be positive or negative, depending on the ferromagnetic material. An electric signal of US-frequency f_r is fed in to a coil that is wrapped around a stack made of magnetostrive

FIGURE 11.24 USM equipment. (From Lehfeld Works, Heppenheim, Germany, 1967. With permission.)

material (iron–nickel alloy). This stack is made of laminates to minimize eddy current and hysteresis losses; moreover, it must be cooled to dissipate the generated heat (Figure 11.25a). The alternating magnetic field produced by the HF-ac generator causes the stack to expand and contract at the same frequency.

To achieve the maximum magnetostriction effect, the HF-ac current i must be superimposed on an appropriate dc premagnetizing current I_p that must be exactly adjusted to attain an optimum or working point. This point corresponds to the inflection point ($d^2\varepsilon/dI^2 = 0$) of the magnetostriction curve, (Figure 11.25b). Without the application of the premagnetizing direct current I_p, it is evident that the magnetostriction effect occurs in the same direction for a given ferromagnetic material irrespective of the field polarity, and hence the deformation will vary at twice the frequency $2f_r$ of the oscillating current providing the magnetic field (Figure 11.25b). Therefore, the premagnetizing direct current I_p has the following functions:

- When precisely adjusted, it provides the maximum magnetostriction effect (maximum oscillating amplitude).
- It prevents the frequency doubling phenomenon.

If the frequency of the ac signal, and hence that of the magnetic field, is tuned to be the same as the natural frequency of the transducer (and the whole oscillating system), so that it will be at mechanical

Nontraditional Machine Tools and Operations

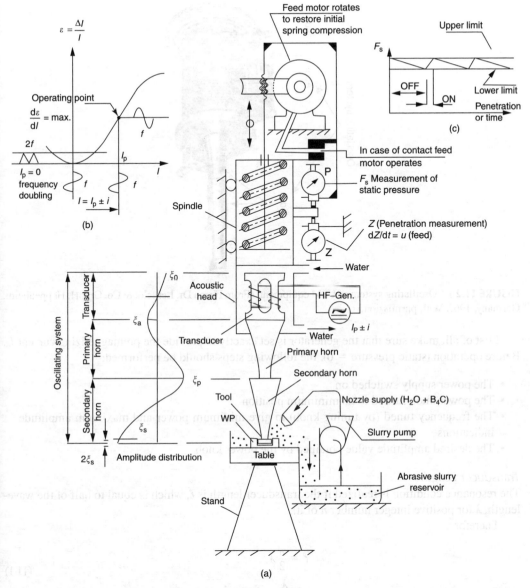

FIGURE 11.25 Schematic of complete vertical USM equipment.

resonance, then the resulting oscillation amplitude becomes quite large and the exciting power attains its minimum value.

System Tuning
For tuning purposes, the HF–ultrasonic generator (Figure 11.24) is provided with the following parts:

- An indicator for amplitude measurement
- An indicator for output power measurement of the generator
- A knob for tuning the output frequency of the generator to be the same as the natural frequency of the oscillating system
- A knob to regulate the output power of the generator

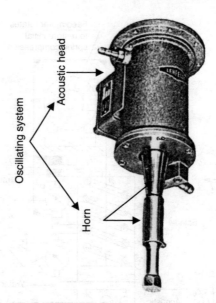

FIGURE 11.26 Oscillating system of USM equipment. (From Firma Dr. Lehfeldt & Co. GmbH, Heppenheim, Germany, 1967. With permission.)

First of all, make sure that the generator is set exactly to provide the premagnetizing current I_p. Before operation (static pressure = 0), the following steps should be performed:

- The power supply switched on
- The power knob turned to its minimum position
- The frequency tuned (by tuning knob) to give minimum power and maximum amplitude indications
- The desired amplitude value obtained by the power knob

Transducer Length

The resonance condition is realized if the transducer length is ℓ, which is equal to half of the wavelength, λ (or positive integer number n of it).

Therefore,

$$\ell = \frac{n}{2}\lambda$$
$$= \frac{\lambda}{2}, \quad \text{if } n = 1 \tag{11.1}$$

and

$$\lambda = \frac{c}{f_r} = \frac{1}{f_r}\sqrt{\frac{E}{\rho}} \tag{11.2}$$

where
c = acoustic speed in magnetostrictive material (m/s)
f_r = resonant frequency (1/s)
E, ρ = Young's modulus (MPa) and density (kg/m³) of the magnetostrictive material

Hence,

$$\ell = \frac{c}{2f_r} = \frac{1}{2f_r}\sqrt{\frac{E}{\rho}} \tag{11.3}$$

Characteristics of Some Magnetostrictive Alloys

Because the magnetostrictive materials convert magnetic energy to mechanical energy, a high coefficient of magnetomechanical coupling k_r and coefficient of magnetostrictive elongation ε_{ms} are essential.

$$k_r = \left[\frac{\text{magnituude of mechanical energy}}{\text{magnitude of magnetic energy}}\right]^{1/2} \quad (11.4)$$

and

$$\varepsilon_{ms} = \frac{\Delta \ell}{\ell} \quad (11.5)$$

Alfer (13% Al, 87% Fe) is characterized by high coefficients, k_r and ε_{ms}, as shown in Table 11.4.

1. Piezoelectric Transducers

A main disadvantage of magnetostriction transducers is the high power loss ($\eta = 55\%$). The power loss is converted into heat, which necessitates the cooling of the transducer. In contrast, piezoelectric transducers are more efficient ($\eta = 90\%$), even at higher frequencies ($f = 25$–40 kHz). Piezoelectric transducers utilize crystals like quartz that undergoes dimensional changes proportional to the voltage applied. Similar to magnetostrictors, the length of the crystal should be equal to half the wavelength of the sound in the crystal to produce resonant condition. At a frequency of 40 kHz, the resonant length l of the quartz crystal ($E = 5.2 \times 10^4$ MPa, $\rho = 2.6 \times 10^3$ kg/m^3) is equal to 57 mm. Sometimes a polycrystalline ceramic transducer like barium titanate is used.

2. Acoustic Horns (Mechanical Amplifiers or Concentrators)

The oscillation amplitude ξ_0 as obtained from the magnetostrictive transducer does not exceed 5 μm, which is too small for effective removal rates. The amplitude at the tool should therefore be increased to practical limits of 40–50 μm by fitting one or more amplifiers into the output end of the transducer (Figure 11.25a).

The acoustic horn (concentrator) should perform the following functions:

- Transmit the mechanical energy to the tool
- Amplify the amplitudes to practical limits
- Concentrate the power on a small machining area

TABLE 11.4
Coefficients ε_{ms} and k_r of a Magnetostrictive Alloys

Magnetostrictive Alloy	$\varepsilon_{ms} \times 10^6$	k_r
Alfer (13% Al, 87% Fe)	40	0.28
Hypernik (50% Ni, 50% Fe)	25	0.20
Permalloy (40% Ni, 60% Fe)	25	0.17
Permendur (49% Co, 2% V, 49% Fe)	9	0.20

Source: Reproduced from Kaczmarek, J. in *Principles of Machining by Cutting, Abrasion, and Erosion*, Peter Peregrines Ltd., Stevenage, Hertfordshire, 1976.

TABLE 11.5
Amplitudes and Magnification Factors of Individual Elements of the Cascaded Oscillating System Shown in Figure 11.25

Oscillating Element	Amplitude	Magnification Factor
Transducer	ξ_0	$R_{tr} = \xi_0/\xi_0 = 1$
Primary horn	ξ_p	$R_p = \xi_p/\xi_0$
Secondary horn	ξ_s	$R_s = \xi_s/\xi_p$

To attain resonance, the acoustic horns, like transducers, should be half-wavelength resonators whose terminals oscillate axially in an opposite direction relative to each other. The nodal points (points of zero amplitude $\xi_n = 0$) are little displaced toward the upper end in the case of tapered concentrators. Figure 11.25a illustrates the amplitude distribution of the cascaded oscillating system along its longitudinal axis. Table 11.5 shows the amplitude and magnification factors of each oscillating element.

Accordingly, the overall magnification factor R_m of the system is given by

$$R_m = R_{tr} \times R_p \times R_S$$
$$= 1 \times \frac{\xi_p}{\xi_0} \times \frac{\xi_s}{\xi_p} = \frac{\xi_s}{\xi_0} \quad (11.6)$$

The acoustic head (the transducer and the primary acoustic horn) is delivered by the manufacturer as an integral part with the machine (Figure 11.26). The tool is attached to the free end of the secondary acoustic horn by threading, brazing or press-fitting.

The oscillation amplitude of the primary horns is small enough, such that they are durable and not easily discarded.

11.4.2.2 Tool Feeding Mechanism

The tool feeding mechanism should perform the following functions:

- Bring the tool slowly to the WP
- Provide adequate static pressure and sustain it during cutting
- Decrease the pressure before the end of cut to eliminate sudden fractures
- Overrun a small distance to ensure the required hole size at the exit
- Retract the tool upward rapidly after machining

Figure 11.25c illustrates an automatic tool feed mechanism, which operates precisely through the application of roller frictionless guides. When the oscillating system is freely suspended (no contact between the tool and the WP), the static pressure on the WP equals zero. When machining starts, the tool comes into contact with the WP; the spring in the machine spindle expands giving a measure for the static pressure. The static force is indicated by the dial gauge (P). As machining proceeds, the spring is compressed and the static force decreases (Figure 11.25c) until the contact switch is actuated, allowing the feed motor to rotate, and rapidly recovers the value of static force. The dial gauge (Z) indicates the tool displacement.

11.4.3 Design of Acoustic Horns

11.4.3.1 General Differential Equation

The general differential equation for longitudinal oscillation of acoustic horns can be derived by considering the equilibrium of an infinitesimal element dx under the action of elastic and inertial forces (Figure 11.27).

$$\text{Elastic force} = F + \frac{\partial F}{\partial x} dx - F \tag{11.7}$$

$$\text{Inertial force} = A(x) \cdot dx \cdot \rho \cdot \frac{\partial^2 F}{\partial t^2} \tag{11.8}$$

where

F = elastic force = $EA(x) \cdot \dfrac{\partial y}{\partial x}$
$A(x)$ = shape function = cross-sectional area function of horn at axial position x
y = displacement, depending on x and $t = \xi \sin \omega t$
t = time
x = axial position as measured from the horn fixed end
ξ = oscillation amplitude = $f(x)$
ω = angular speed = $2\pi f$
f_r = ultrasonic frequency
E = Young's modulus of horn material
ρ = density of horn material
$\dfrac{\partial y}{\partial x}$ = strain = $f(x, t)$
$\dfrac{\partial^2 y}{\partial t^2}$ = acceleration = $f(x, t)$

Equating the elastic force (Equation 11.7) and the inertial force (Equation 11.8),

$$\frac{\partial F}{\partial x} \cdot dx = A(x) \cdot dx \cdot \rho \cdot \frac{\partial^2 y}{\partial t^2} \tag{11.9}$$

Substituting the values of F and y in Equation 11.9, the general differential equation becomes

$$\frac{d^2 \xi}{dx^2} + \frac{d \ln A(x)}{dx} \cdot \frac{d\xi}{dx} + \left[\frac{\omega}{c}\right]^2 \xi = 0 \tag{11.10}$$

where $c = \sqrt{E/\rho}$ = acoustic speed in horn material.

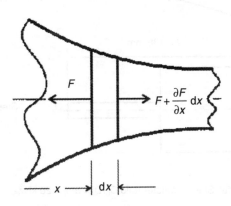

FIGURE 11.27 Equilibrium of infinitesimal element under the action of elastic and inertial forces.

The general differential Equation 11.10 can be solved after substituting the shape function $A(x)$. Four shape functions are available for acoustic horns (Figure 11.28). These are as follows:

- Cylindrical stepped horn
- Exponential horn
- Conical horn
- Hyperbolic horn

The choice of the shape function $A(x)$ controls the magnification factor R (Figure 11.29). However, exponential and stepped types are frequently used; the conical and hyperbolic horns are difficult to design.

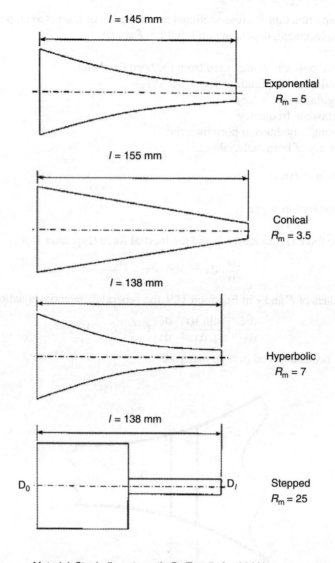

FIGURE 11.28 $\lambda/2$—Resonators of different shape functions.

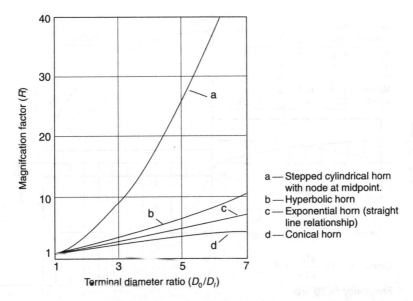

FIGURE 11.29 Effect of terminal diameter ratio and shape function on the magnification factor. (From Youssef, H. A., *Non-Traditional Machining Processes—Theory and Practice*, El-Fath Press, Alexandria, 2005. With permission.)

11.4.3.2 Design of the Cylindrical Stepped Acoustic Horns ($A(x) = C$)

Stepped horns are mainly employed in machining brittle materials such as glass, germanium (Ge), and ceramics, where there is no need to use high amplitudes. Accordingly, fatigue at nodal points, due to stress concentration, can be avoided. Moreover, stepped horns are easily designed and produced. Substituting $A(x) = C$ in the general differential Equation 11.10,

$$\frac{d^2\xi}{dx^2} + \left[\frac{\omega}{c}\right]^2 \xi = 0 \tag{11.11}$$

Figure 11.30 shows the amplitude, strain, and stress distributions of the stepped acoustic horn. Assuming $f_r = 20$ kHz, $D_0/D_l = 5$, the table in Figure 11.30 determines the resonant lengths for different horn materials. The magnification factor R_m can be calculated by

$$R_m = \frac{\xi_l}{\xi_0} = (D_0/D_l)^2 = 25 \tag{11.12}$$

11.4.3.3 Design of Exponential Acoustic Horns ($A(x) = A_0 e^{-2hx}$)

Exponential horns are mainly employed to machine hard and tough materials such as carbides and hardened steels using large oscillation amplitudes without the risk of fatigue failure. They can be easily designed and their contours can be easily produced on CNC lathes.

The area of an exponential horn varies according to the function

$$A(x) = A_0 e^{-2hx} \tag{11.13}$$

If the exponential horn has a circular cross-section, then

$$D(x) = D_0 e^{-hx} \tag{11.14}$$

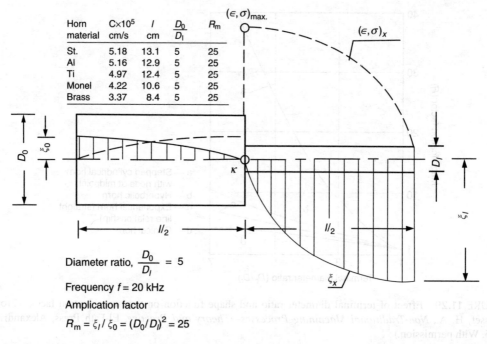

FIGURE 11.30 Amplitude, strain, and stress distribution of stepped acoustic horn. (From Youssef, H. A., *Non-Traditional Machining Processes—Theory and Practice*, El-Fath Press, Alexandria, 2005. With permission.)

where

$A(x), D(x)$ = cross-sectional area, and horn diameter at location x
A_0, D_0 = cross-sectional area, and horn diameter at $x = 0$
h = exponential ascent factor

Substituting the value of $A(x)$ according to Equation 11.13 in the general differential Equation 11.10, the differential equation of exponential horn is obtained by

$$\frac{d^2\xi}{dx^2} - 2h \cdot \frac{d\xi}{dx} + \left[\frac{\omega}{c}\right]^2 \xi = 0 \tag{11.15}$$

from which the amplitude distribution ξ is given by

$$\xi = -\xi_0 \sqrt{1 + \left(\frac{h\ell}{\pi}\right)^2} \cdot e^{hx} \cdot \sin\left(\sqrt{\left(\frac{\omega}{c}\right)^2 - h^2} \cdot x - \hat{\varphi}\right) \tag{11.16}$$

where

ξ_0 = amplitude at $x = 0$
ℓ = horn length
$\hat{\varphi}$ = arc tan $(\pi/h\ell)$

Resonance condition:

$$\sqrt{\left[\frac{\omega}{c}\right]^2 - h^2} = \frac{\pi}{\ell} \tag{11.17}$$

Horn length ℓ:

$$\ell = \frac{\pi}{\sqrt{\left[\frac{\omega}{c}\right]^2 - h^2}} \tag{11.18}$$

Nontraditional Machine Tools and Operations

Ascent factor h:

$$h = \frac{\omega}{c}\Big/\sqrt{1 + \left[\frac{\pi}{h\ell}\right]^2} \qquad (11.19)$$

The horn length ℓ can also be expressed by Equations 11.20 and 11.21:

$$\ell = \frac{c'}{2f_r} = \frac{c}{2f_r} \cdot \frac{1}{\sqrt{1 - (hc/\omega)^2}} \qquad (11.20)$$

$$\ell = \frac{c}{2f_r}\sqrt{1 + \left[\frac{\ln D_0/D_t}{\pi}\right]^2} \quad \text{for circular exponential horns} \qquad (11.21)$$

where c' is the modified acoustic speed in horn material.

Figure 11.31 shows the distribution of the oscillation amplitude along the axis of an exponential acoustic horn, where ξ_0 is the amplitude at $x = 0$, and ξ_ℓ is the amplitude at $x = \ell$ (tool amplitude). Figure 11.32a illustrates a nomogram that determines the resonant length of a circular exponential horn in terms of the acoustic speed c, frequency f_r, and terminal diameter ratio D_0/D_t.

Magnification Factor R_m:
Referring to Equation 11.16, then

$$R_m = \left|\frac{\xi_\ell}{\xi_0}\right| = \frac{e^{hl}\sin\pi - \varphi}{\sin\varphi} = e^{hl} \qquad (11.22)$$

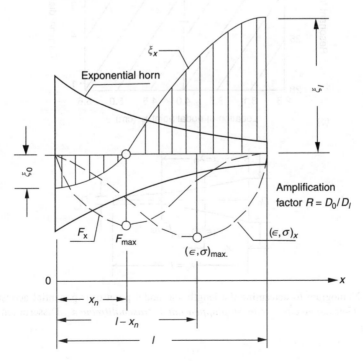

FIGURE 11.31 Amplitude, strain, and stress distribution along the axis of an exponential acoustic horn. (From Youssef, H. A., *Non-Traditional Machining Processes—Theory and Practice*, El-Fath Press, Alexandria, 2005. With permission.)

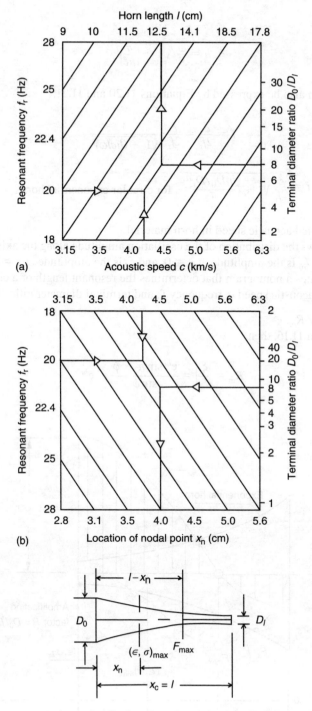

FIGURE 11.32 Nomogram to determine the length and nodal point of exponential acoustic horn. (From Blanck, D., 1961, *Getzmäßgikeiten beim Stoßläppen mit Ultraschallfrequenz*. Dissertation Braunschweig. With permission.)

From Equation 11.13

$$A_\ell = A_0 e^{-2hl}$$

$$\sqrt{\frac{A_0}{A_\ell}} = e^{hl} \qquad (11.23)$$

From Equations 11.22 and 11.23, the magnification factor R_m of noncircular acoustic horns is given by

$$R_m = e^{hl} = \sqrt{\frac{A_0}{A_\ell}} \qquad (11.24)$$

and for circular acoustic horns

$$R_m = e^{hl} = \frac{D_0}{D_\ell} \qquad (11.25)$$

From the foregoing discussion, it can be concluded that:

1. The magnification factor R_m is independent of horn material.
2. In case of an exponential circular acoustic horn, the magnification factor R_m depends upon terminal diameter ratio D_0/D_ℓ.
3. For a given diameter ratio, D_0/D_ℓ, the stepped horn possesses the highest magnification factor $R_m = (D_0/D_\ell)^2$, followed by the hyperbolic, the exponential, and finally the conical type (Figure 11.29).

Nodal Point x_n:

This is the point of zero amplitude ($\xi_{x_n} = 0$). It is important to determine this point exactly to eliminate damping of the oscillating system when the primary horn is suspended in the acoustic head (Figure 11.25a). Substituting $\xi_{x_n} = 0$ in Equation 11.16, it follows that

$$x_n = \frac{\ell}{\pi} \cdot \hat{\varphi}$$

$$= \frac{\ell}{\pi} \cdot \tan^{-1}\frac{\pi}{\ln(D_0/D_\ell)} = \frac{c}{2\pi f}\sqrt{1+\left[\frac{\ln(D_0/D_\ell)}{\pi}\right]^2} \cdot \tan^{-1}\frac{\pi}{\ln(D_0/D_\ell)} \qquad (11.26)$$

Additionally, the nodal point x_n of the exponential horns can be determined from the nomogram in Figure 11.32b in terms of the same parameters c, f_r, and D_0/D_ℓ.

Distributions of strain ε_x, stress σ_x, and force F_x along the axis of exponential horn

$$\varepsilon_x = \frac{d\xi}{dx} = \xi_0 h \sqrt{1+\left(\frac{hl}{\pi}\right)^2}\left[1+\left(\frac{\pi}{hl}\right)^2\right] \cdot e^{hx} \cdot \sin\frac{\pi}{\ell}x \qquad (11.27)$$

$$\sigma = E \cdot \varepsilon_x$$

$$= E\xi_0 h \sqrt{1+\left(\frac{hl}{\pi}\right)^2}\left[1+\left(\frac{\pi}{hl}\right)^2\right] \cdot e^{hx} \cdot \sin\frac{\pi}{\ell}x \qquad (11.28)$$

$$F_x = A_{(x)} \cdot \sigma_x$$

$$= E\xi_0 h A_0 \sqrt{1+\left(\frac{hl}{\pi}\right)^2}\left[1+\left(\frac{\pi}{hl}\right)^2\right] \cdot e^{hx} \cdot \sin\frac{\pi}{\ell}x \qquad (11.29)$$

Figure 11.31 shows the distribution of ε_x, σ_x, and F_x along the axis of the exponential acoustic horn.

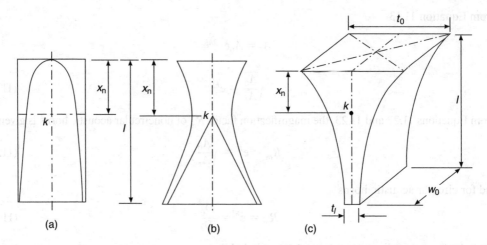

FIGURE 11.33 Different shapes of exponential acoustic horns. (From Youssef, H. A., *Non-Traditional Machining Processes—Theory and Practice*, El-Fath Press, Alexandria, 2005. With permission.)

Selection of Horn Material
Finally, the maximum induced stress as expressed by Equation 11.30 occurs at the nodal point (Figure 11.31). It should be less than the allowable fatigue strength of the horn material.

$$\sigma_{\max} = E\xi_0 h \sqrt{\left[1 + \left(\frac{hl}{\pi}\right)^2\right]\left[1 + \left(\frac{\pi}{hl}\right)^2\right]} \cdot e^{hl\left(1-\frac{\hat{\varphi}}{\pi}\right)} \sin(\pi - \hat{\varphi}) \qquad (11.30)$$

Other Shapes of Exponential Acoustic Horns
The same equations are valid for acoustic horns of the internal exponential form shown in Figure 11.33a. These are used for machining large holes. Another horn is shown in Figure 11.33b with external exponential form, and adapted with internal conical form to allow machining of the largest holes.

For machining of rectangular form or slitting, a horn of rectangular cross-section having constant width w_0, and its thickness $t(x)$ varying according to an exponential function is recommended (Figure 11.33c).

Therefore,

$$A(x) = A_0 \, e^{-2hx}$$

$$w_0 t(x) = w_0 \cdot t_0 \, e^{-2hx} \qquad (11.31)$$

$$t(x) = t_0 \, e^{-2hx}$$

ILLUSTRATIVE EXAMPLE 1

Use the chart in Figure 11.32 to design an exponential acoustic horn made of monel (acoustic speed in horn material $c = 4.22 \times 10^5$ cm/s). Its natural frequency $f_r = 20$ kHz, and its terminal diameters are $D_0 = 40$ mm and $D_l = 5$ mm.

Solution

Horn length:

Referring to Figure 11.32a,

$$D_0/D_\ell = 8, c = 4.22 \text{ km/s}, f_r = 20 \text{ kHz},$$

then, $\ell = 12.5$ cm

Location of nodal point x_n:

Referring to Figure 11.32b

$$D_0/D_\ell = 8, c = 4.22 \text{ km/s}, f_r = 20 \text{ kHz}$$

then, $x_n = 4$ cm

Magnification factor R_m:

$$R_m = D_0/D_\ell = 40/5 = 8$$

Exponential ascent factor h:

$$R_m = e^{hl}$$
$$8 = e^{12.5h}$$

from which

$$h = 0.166 \frac{1}{cm}$$

Determination of horn contour:

$$D_x = D_0 e^{-hx}$$
$$= 40 e^{-0.166x}$$

x (cm)	0	2	$x_n = 4$	7	10	12.5
$D_{(x)}$ (mm)	40	28.7	20.6	12.5	7.6	5.0

ILLUSTRATIVE EXAMPLE 2

The same acoustic horn as in Example 1 is required to be designed ($\ell = 12.5$ cm, $R_m = 8$), but it should be provided by a conical cavity as illustrated in Figure 11.34, to accommodate a tool of 20 mm diameter.

Solution

Zone I ($x = 0$–4 cm):

The horn contour is exactly the same as in Example 1.

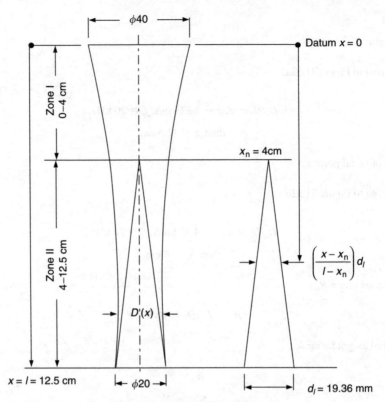

FIGURE 11.34 Exponential acoustic horn with conical cavity.

Zone II ($x = 4\text{–}12.5$ cm)

$$A_{(x)} = A_0 e^{-2hx}$$

$$D_{(x)}^2 - \left(\frac{x - x_n}{\ell - x_n} d_\ell\right)^2 = D_0^2 e^{-2hx}$$

Therefore,

$$D_{(x)} = \sqrt{D_0^2 e^{-2hx} + \left(\frac{x - x_n}{\ell - x_n} d_\ell\right)^2}$$

d_l can be calculated from the equation

$$d_l = \sqrt{(D_l)_{\text{hollow}} - (D_l)_{\text{solid}}}$$
$$= \sqrt{(20)^2 - (5)^2} = 19.36 \text{ mm}$$

If $x = 7$ cm, then

$$D_7 = \sqrt{(40)^2 e^{-2 \times 0.166 \times 7} + \left(\frac{7 - 4}{12.5 - 4} \times 19.36\right)^2} = 14.24 \text{ cm}$$

The following table illustrates the horn diameters at different lengths:

	Zone I			Zone II		
x (cm)	0	2	$x_n = 4$	7	10	12.5
$D_{(x)}$ (mm)	40	28.7	20.6	14.25	15.63	20

Nontraditional Machine Tools and Operations

ILLUSTRATIVE EXAMPLE 3

For a given piece of USM equipment, operating at a resonant frequency $f_r = 20$ kHz, it is required to design an exponential acoustic horn of ball bearing steel 100 Cr 6 ($c = 5.05 \times 10^5$ m/s), provided with a 3 mm diameter hole for the suction of the abrasive slurry from the nodal point and another one for the fixation in the primary horn. Assume $D_0 = 39$ mm, $D_l = 8$ mm, and fixation hole diameter $d_c = 16$ mm and depth 18 mm (Figure 11.35).

Solution

Magnification factor:

$$R_m = e^{hl} = \sqrt{\frac{A_0}{A_\ell}} = \sqrt{\frac{39^2 - 16^2}{8^2 - 3^2}}$$

$$= 4.8$$

Ascent factor h and horn length:

$$h = \frac{\omega}{c} \cdot \frac{1}{\sqrt{1 + (\pi/hl)^2}}$$

$$hl = \ln R = \ln 4.8 = 1.57$$

$$h = \frac{2\pi \times 20000}{5.05 \times 10^5} \cdot \frac{1}{\sqrt{1 + (\pi/1.57)^2}}$$

$$= 0.111 \frac{1}{\text{cm}}$$

$$\ell = \frac{1.57}{0.111} = 14.13 \text{ cm}$$

Nodal point $x_{\hat{n}}$:

$$x_{\hat{n}} = \frac{\ell}{\pi} \tan^{-1} \pi/\ln(D_0/D_l) = 4.98 \text{ cm}$$

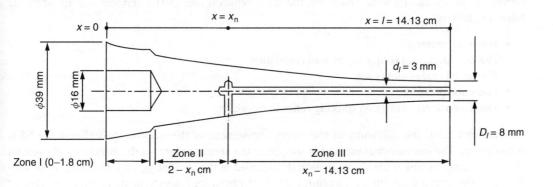

FIGURE 11.35 Exponential acoustic horn with suction and fixation holes.

Determination of horn contour:

Zone I ($x = 0$–1.8 cm)

$$D_{(x)}^2 - d_c^2 = (D_0^2 - d_c^2)\, e^{-2hx}$$

$$D_{(x)} = \sqrt{d_c^2 - (D_0^2 - d_c^2)\, e^{-2hx}}$$

$$d_c = 16 \text{ mm}, D_0 = 39 \text{ mm}, h = 0.111\, \frac{1}{\text{cm}}$$

Zone II ($x = 2 - x_n$)

$$D_{(x)}^2 = (D_0^2 - d_c^2)\, e^{-2hx}$$

$$D_{(x)} = \sqrt{(D_0^2 - d_c^2)}\, e^{-hx}$$

Zone III ($x = x_n - \ell$)

$$D_{(x)}^2 - d_\ell^2 = (D_0^2 - d_c^2)\, e^{-2hx}$$

$$D_{(x)} = \sqrt{d_\ell^2 + (D_0^2 - d_c^2)\, e^{-2hx}}$$

$$d_\ell = 3 \text{ mm}$$

The following table illustrates the horn diameters at different zones and lengths.

	Zone I			Zone II			Zone III			
x (cm)	0	1	1.8	2	4	4.98	6	8	10	12	14.13
$D_{(x)}$ (mm)	39	35.6	33.2	28.5	22.8	20.5	18.5	14.9	12.1	9.9	8.0

11.4.4 Process Capabilities

11.4.4.1 Stock Removal Rate

It seems that the dominant factor involved in USM is the direct hammering of the abrasive grains, caused by the oscillating tool. Therefore, the stock removal rate (SRR) depends mainly upon the following factors:

- The work material
- The amplitude and frequency of tool oscillation
- The abrasive size and type
- The static pressure
- The abrasive concentration (mixing ratio) in the slurry

Last and not least, the efficiency of the slurry supplement in the working gap affects the SRR considerably. The conventional method of supplying the abrasive slurry is the nozzle supply system (Figure 11.36a), in which the slurry is directly supplied at the oscillating tool. Pumping in or suction from the working gap through a central hole in the horn are found to be more effective regimes (Figure 11.36b). Figure 11.36c shows schematically a comparison between regimes A and B regarding the penetration rate u. In the nozzle supply regime A, the penetration rate is much less than that in regime B. Moreover, in regime A the penetration rate decreases continuously with the hole depth, whereas in regime B it remains unaffected by the hole depth.

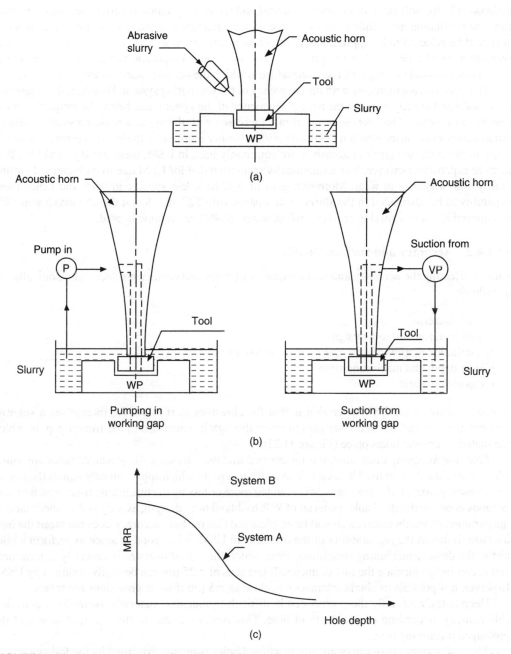

FIGURE 11.36 Slurry supplying system: (a) nozzle supply system, (b) pumping in or suction from working gap, and (c) comparison of MRR of different supply systems.

Regarding the work material, the specific removal rate is affected by the ratio of the tool hardness to the WP hardness. The higher the ratio, the lower will be the MRR. That explains why soft and tough materials are recommended for USM tools (El-Hofy, 2005). The highest machining rates are realized when machining brittle materials such as glass, quartz, ceramics, and germanium, whereas the lowest machining rates are expected when machining hard and tempered steels and carbides.

The USM process is not applicable for soft and ductile materials, such as copper, lead, ductile steels, and plastics, which absorb energy by deformation. Moreover, some of the particles become embedded in the soft faces of the work material and the cutting action is further retarded. In practice, the oscillation amplitude is mainly selected with reference to the size of abrasive grits used. It should be selected to be approximately the same as the grit size. SRR increases with increasing oscillation amplitude (or abrasive grit size). The maximum amplitude value is governed by the maximum allowable strength of the material from which the acoustic horn is designed.

The removal rate increases with the frequency of the oscillating system. However, the frequency is constant and exactly equal to the natural frequency of the system, and hence the frequency is not considered a factor. The specific removal rate increases with the applied static pressure. It attains a maximum value, after which it decreases with a further increase of the static pressure. As previously mentioned, two types of abrasives are commonly used in USM; these are B_4C and SiC. B_4C is more expensive; however, it is economically recommended for USM due to its increased cutting ability and resistance to wear. Moreover, grits of B_4C have less specific gravity, and hence more capability to be suspended in the slurry as compared with SiC. It is found that the maximum SRR is achieved if a slurry mixing ratio (abrasives/water) of 40% by volume is used.

11.4.4.2 Accuracy and Surface Quality

Factors affecting the accuracy and surface quality of holes and cavities produced ultrasonically are as follows:

- Work material
- Tool material and tool design
- Oscillation amplitude and grain size of abrasives
- Hole depth and machining time
- Cavitation effect

A main feature of the USM operation is that the abrasives start to cut for themselves a sideway between the tool and the WP (side gap) to move through it downward to the frontal gap, in which the material removal takes place (Figure 11.22).

From the foregoing discussion, it is understood that the ultrasonically produced holes are somewhat larger than the tool used by a certain oversize (overcut), which approximately equals the size of the abrasive grains used. This oversize is affected more or less by the machining time, which in turn depends upon the depth of hole, material of WP, tool land (tool design), as well as the other machining parameters. Furthermore, it should be emphasized that the hole accuracy does not mean the hole oversize. It means the repeatability of the oversize. In USM, it is a good practice to perform a pilot test at the desired machining conditions, from which the actual oversize is precisely determined, and accordingly calculate the tool diameter. Tolerances of ± 25 μm can be easily obtained by USM. However, it is possible to obtain tolerances as close as ± 5 μm if some provisions are taken.

Deep holes, especially those produced in difficult-to-machine materials, suffer from considerable conicity, depending on the depth of hole. This conicity is due to the side tool wear and the prolonged machining time.

The wall roughness of ultrasonically machined holes is mainly governed by the following:

- *The material to be machined.* The roughness is larger when machining brittle materials such as glass and germanium.
- *The oscillation amplitude, and the grain size.* The roughness increases with increasing oscillation amplitude and grain size of abrasives used.

The surface quality is deteriorated if cavitation conditions prevail. From this point of view, the use of a pumping regime in the working gap is preferred over using the suction arrangement. Moreover, the rotation of the WP in case of circular holes may improve the surface quality and the hole roundness.

11.4.5 Recent Developments

USM has been developed and new application are currently used such as:

1. *Contouring USM.* When sinking a 3-D cavity by conventional USM, a form tool is used, which is generally complex and costly (Figure 11.37a). The same cavity, however, can be produced by what is called contouring USM (Figure 11.37b), in which the machining is implemented by a simple tool in accordance to a tool path that is determined by CNC facility. Through contouring USM, the cavity volume and sinking depth may be increased above the prescribed limits of conventional USM.
2. *Rotary USM.* A modified version of USM is shown in Figure 11.38, where a tool bit is rotated at 5000 rpm against the WP. The process is therefore called rotary ultrasonic machining (RUM). This process is used for machining nonmetallic materials such as glass, ceramic, carbides, ferrite, quartz, zirconium oxide, ruby, sapphire, beryllium oxide, and some composites. RUM ensures a high removal rate, lower tool pressures for delicate and fragile parts, less breakout of through holes, and improved aspect ratio. When machining small holes, RUM allows noninterrupted drilling; conventional drilling necessitates a tool retraction, which increases machining time.
3. *Ultrasonic-assisted ECM.* Ultrasonic-assisted ECM (ECUSM) is a combination of USM process and ECM process. The machining system of this HP (Figure 11.39) is composed of normal USM equipment supplemented by a dc generator to provide the working gap with the electrolyzing current necessary for the anodic dissolution. The WP is connected as an anode,

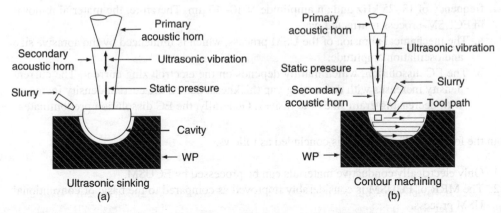

FIGURE 11.37 Sinking (conventional) and contouring USM: (a) ultrasonic sinking and (b) contour machining. (Modified from El-Hofy, H., *Advanced Machining Processes, Nontraditional and Hybrid Processes,* McGraw-Hill Co., New York, 2005.)

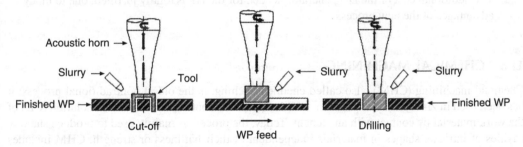

FIGURE 11.38 Rotary ultrasonic machining (RUM). (Modified from El-Hofy, H., *Advanced Machining Processes, Nontraditional and Hybrid Processes,* McGraw-Hill Co., New York, 2005.)

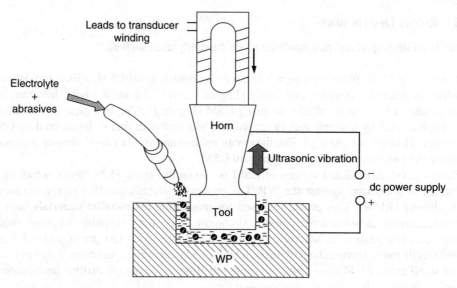

FIGURE 11.39 The ECUSM hybrid process.

whereas the tool is connected as a cathode. The supply voltage is 3–15 V, which ensures current densities between 5 and 30 A/cm². An electrolyte of $NaCl$, $NaNO_3$, or KNO_3 is used instead of water as an abrasive carrier liquid. The tool oscillates ultrasonically at a frequency of 18–25 kHz and an amplitude of 10–40 μm. Therefore, the material removal in ECUSM-process is affected by

a. The mechanical abrasion of the USM process, which is influenced by the abrasive size and oscillation amplitude.
b. The EC dissolution, which mainly depends on the electrolyzing current. The current density increases with decreasing gap thickness; that is, the current density increases with decreasing grain size of abrasives. Generally, the EC dissolution predominates.

From the foregoing discussion, it is concluded as follows:

1. Only electrically conductive materials can be processed by ECUSM.
2. The MRR of ECUSM is considerably improved as compared to the basic or conventional USM process.
3. The tool wear decreases considerably, which has a positive impact on the accuracy of the HP.
4. The surface quality of ECUSM improves due to the EC effect.
5. The additional cost of the DC generator needed for the HP is totally justified, due to many advantages of the new process.

11.5 CHEMICAL MACHINING

Chemical machining (CHM), also called chemical etching, is the oldest nontraditional process; it has been used in zincograph preparation. CHM depends on controlled chemical dissolution (CD) of the work material by contact with an etchant. Today, the process is mainly used to produce shallow cavities of intricate shapes in materials independent of their hardness or strength. CHM includes two main applications. These are chemical milling (CH-milling, shown in Figure 11.40a), and photochemical machining (PCM), also called spray etching (Figure 11.40b).

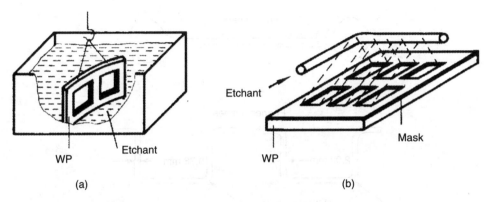

FIGURE 11.40 CHM processes: (a) CH-milling and (b) PCM (spray etching).

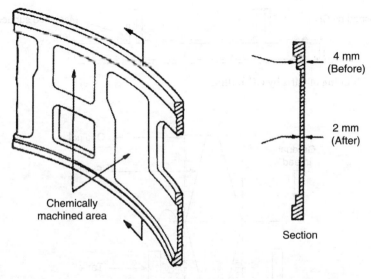

FIGURE 11.41 CH-milling striving to improve the stiffness-to-weight ratio of Al-alloy plates for space vehicles. (Adapted from *Advanced Materials and Processes*, ASM International, Materials Park, OH, 1990.)

11.5.1 Chemical Milling

This process has a special importance in airplane and aerospace industries, where it is used to reduce the thicknesses of plates enveloping walls of rockets and airplanes, striving at improving stiffness to weight ratio (Figure 11.41). CH-milling is used also in metal industries to thin out walls, webs, and ribs of parts that have been produced by forging, sheet metal forming, or casting (Figure 11.42). Furthermore, the process has many applications related to improving surface characteristics, such as the following:

- Elimination of Ti oxide (α-case) from Ti forgings and superplastic formed parts
- Elimination of the decarburized layer from low-alloy steel forgings
- Removal of the recast layer from parts machined by EDM
- Removal of burrs from conventionally machined parts of complex shapes

Figure 11.43 shows the production of a tapered disk by gradual immersion of the disk in the etchant while it is rotating. The process is also capable of producing burr-free printed circuit boards (PCBs).

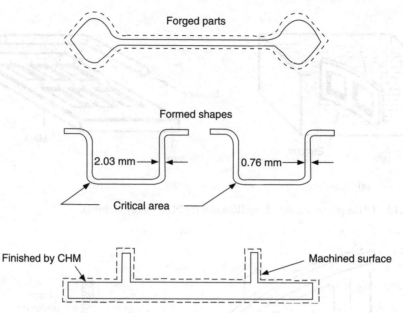

FIGURE 11.42 Thinning of parts by CH-milling.

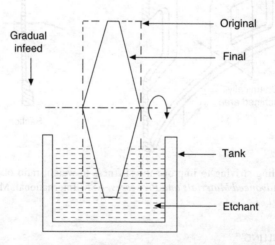

FIGURE 11.43 CH-milling of tapered disk by gradual immersion in etchant while rotating.

In CH-milling, a special coating called maskant protects areas from which the metal is not to be removed. The process is used to produce pockets and contours. CH-milling consists of the following steps:

1. Preparing the WP surface by cleaning, mechanically or chemically, to provide good adhesion of the masking material.
2. Masking using a strippable mask that adheres to the surface and withstands chemical abrasion during etching.
3. Scribing of the mask using special templates to expose areas to be etched. The type of selected mask depends on the work size, number of parts, and the desired resolution of details. Silk screens are recommended for shallow cuts of close dimensional tolerances.

4. After etching, the work is rinsed, and the mask is stripped manually, mechanically, or chemically.
5. The work is washed by deionized water and then dried by nitrogen.

During CH-milling (Figure 11.44), the etching depth is controlled by the time of immersion. The etchants used are very corrosive, and therefore must be handled with adequate safety precautions. Vapors and gases produced from the chemical reaction must be controlled for environmental protection. A stirrer is used for agitation of fluid. Typical reagent temperatures range from 37°C to 85°C, which should be controlled within ±5°C to attain a uniform machining. Faster etching rates occur at higher etchant temperatures and concentrations.

When the mask is used, the machining action proceeds both inwardly from the mask opening and laterally beneath the mask, thus creating the etch factor (EF), which is the ratio of the undercut d_u to the depth of etch T_e (EF = d_u/T_e), as seen in Figure 11.45. This ratio must be considered when scribing the mask using templates. A typical EF of 1:1 occurs at a cut depth of 1.27 mm. Deeper cuts can reduce this ratio to 1:3.

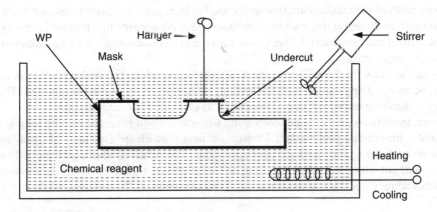

FIGURE 11.44 CHM setup.

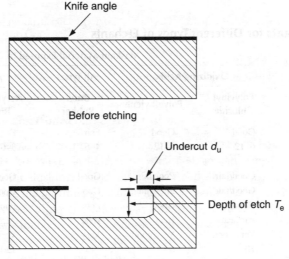

FIGURE 11.45 EF after CHM.

Tooling for Chemical Milling
Tooling for CH-milling is relatively inexpensive and simple to modify. Four types of tools are required: maskants, etchants, scribing templates, and accessories.

1. Maskants
Synthetic or rubber base materials are frequently used as maskants. They should possess the following properties:

- Tough enough to withstand handling
- Inert to the chemical reagent used
- Able to withstand heat generated by etching
- Adhere well to the work surface
- Scribe easily
- Able to be removed easily after etching

Table 11.6 shows the recommended maskants used for different types of etchants as well as the characteristics of these maskants.

Multiple coats of the maskant are frequently used to increase the etchant resistance and to avoid the formation of pinholes on the machined surfaces. Also deeper cuts that require longer exposure time to the etchant can be achieved. Dip, brush, roller, and electrocoating, as well as adhesive tapes, can be used to apply masks.

Spraying the mask on the WPs through a silk screen on which the desired design is imposed combines the maskant application with the scribing operation, as no peeling is required. The product quality is therefore improved, due to the ability to generate finer details.

However, the thin coating applied when using silk screens will not resist etching for a long time as compared to the cut-and-peel method. Photoresist masks, which are used in PCM (spray etching), also combine both the coating and scribing operations. The relatively thin coats applied as dip or spray coats will not withstand rough handling or long exposure times. However, photoresist masks ensure high accuracy and ease of modification. Typical tolerances for different masks are given in Table 11.7.

TABLE 11.6
Recommended Maskants for Different Types of Etchants

	Maskant Material Exposed to				
	Oxidizing Acids		All Types	Acids and Alkalines	
Property	Polyvinyl Chloride	Polyethylene	Butyl Rubber	Acrylonitrile Rubber	Neoprene Rubber
Ease to manufacture	Good	Good	Fair	Fair	Good
Shelf life (months)	6–12	6–12	4–6	3–6	6–8
Ease of application					
Dipping	Good/fair	Poor/fair	Good	Good	Good
Flow coating	Good/fair	Poor/fair	Good	Good	Good
Air spraying	Good	Good	Poor	Poor	Fair
Type of cure	Air/heat	Heat	Heat	Air/heat	Air/heat
Resist to etchant	Very good	Very good	Very good	Very good	Ver good
Up to temperature limit (°C)	70	60	145	120	80

Source: Reproduced from *Machining Data Handbook*, Machinability Data Center, Cincinnati, OH, 1980.

TABLE 11.7
Tolerances of Different Masks

Maskant	Tolerance (μm)
Cut-and-peel masks	±180
Silk screen	±75
Photoresist	±13

TABLE 11.8
Machined Materials and Recommend Etchants in Chemical

Etchant	Concentration	Temperature (K°)	Etch rate (μm/min)	Etch factor (EF = d_u/T_e)	Metal to be Machined
$FeCl_3$	12–18° Be[a]	320	20	1.5:1	Al-alloys
$HCl:HNO_3.H_2O$	10:1:9	320	20–40	2:1	
$FeCl_3$	42° Be[a]	320	20	2:1	Cold rolled steel
HNO_3	10–15% (volume)	320	40	1.5:1	
$FeCl_3$	42° Be[a]	320	40	2.5:1	Cu and Cu alloys
$CuCl_2$	35° Be[a]	325	10	3:1	
HNO_3	12–15% (volume)	300–320	20–40	—	Magnesium
$FeCl_3$	42° Be[a]	320	10–20	(1–3):1	Nickel
$FeCl_3$	42° Be[a]	325	20	2:1	Stainless steel, Tin
HNO_3	10–15% (volume)	320–325	20	—	Zinc

[a] Baume specific gravity scale.

Moreover, the tolerance depends also on the etching depth and the material of WP machined. Cu and Cu alloys provide the closest tolerances, whereas Al alloys are machined with large tolerances (*Machining Data Handbook*, 1980).

2. Etchants

Etchants are highly concentrated acidic or alkaline solutions maintained within a controlled range of chemical composition and temperature. They are capable of reacting with the WP material to produce a metallic salt that dissolves in the solution. Table 11.8 shows the machined material, the recommended etchant, its concentration and temperature, and the EF etch rate. When machining glass or germanium, the acidic solutions HF or HF + HNO_3 are used as etchants. When machining tungsten (W), it is recommended to use either of the following (Kalpakjian, 1984):

Alkaline solution. $K_3Fe(CN)_6$:NaOH = 20:3 (by volume)
Acidic solution. $HF:HNO_3$ = 30:70 (by volume)

A suitable etchant should provide the following requirements:

- Good surface finish of the work
- Uniformity of metal removal
- Control of selective and intergranular attack (IGA)
- Low cost and availability
- Ability to regenerate, or readily neutralize and dispose off its waste products

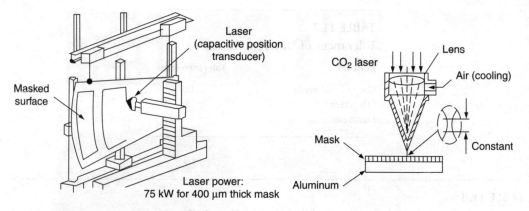

FIGURE 11.46 Laser cutting of masks for CH-milling of large surfaces.

- Nontoxic
- Control of hydrogen absorption in case of Ti alloys

3. *Scribing Templates*

These are used to define the areas for exposure to CD. The most common scribing method is to cut the mask with a sharp knife, followed by careful peeling. The EF allowance must be included. Figure 11.46 shows NC laser scribing of masks for CH-milling of a large surface area.

4. *Accessories*

These include tanks, hooks, brackets, racks, and fixtures.

Advantages and Disadvantages of Chemical Milling

Advantages
- Weight reduction is possible on complex contours that are difficult to machine conventionally.
- Several parts can be machined simultaneously.
- Simultaneous material removal from all surfaces, improves productivity and reduces wrapping.
- No burr formation.
- No induced stresses, thus minimizing distortion and enabling machining of delicate parts.
- Low capital cost of equipment, and minor tooling cost.
- Quick implementation of design changes.
- Less skilled operator is needed.
- Low scrap rate.

Disadvantages
- Only shallow cuts are practical. Deep narrow cuts are difficult to produce.
- Handling and disposal of etchants can be troublesome.
- Masking, scribing, and stripping are repetitive, time-consuming, and tedious.
- Surface imperfections, if any, are reproduced.
- For best results, metallurgical homogeneous surfaces are required.
- Porous castings yield uneven etched surfaces.
- Welded zones are frequently etched at rates that differ from base metal.

Process Capabilities

Using fresh solutions, and depending on the working conditions (etchant concentration and temperature), the etch rate ranges from 20 to 40 μm/min (Table 11.7). Etch rates are high for hard materials and low for soft ones (*Metals Handbook*, 1989). Generally, the high etch rate is accompanied by a low surface roughness and hence closer machining tolerances. Typically, surface roughness of 0.1–0.3 μm (R_a value), depending on the initial roughness, can be obtained. However, under special conditions, surface roughness of 0.025–0.05 μm becomes possible (*Machining Data Handbook*, 1980).

11.5.2 Photochemical Machining (Spray Etching)

PCM (spray etching) is a variation of CH-milling where the resistant mask is applied to the WP by photographic techniques. The two processes are quite similar, because both use etchant to remove material by CD. CH-milling is usually used on the 3-D parts originally formed by other manufacturing processes such as forging and casting of irregular shapes. However, PCM is a promising method for machining foils and sheets of thicknesses ranging from 0.013 to 1.5 mm to produce accurate and micro shapes. So, the PCM process becomes a realistic alternative to shearing and punching operations performed by mechanical presses.

Additionally, a main difference between CH-milling and PCM is that in CH-milling, the depth of etch is controlled by the time the component is immersed in the etchant, whereas in PCM, the etch depth is controlled by the time the component is sprayed by fresh etchant through upper and lower nozzles, thus improving the performance of the PCM process by activating the etch rate and enhancing the quality. Visser et al. (1994) claimed that the etch rate of PCM is 5–10 times that achieved by CH-milling. Of course, in PCM, highly developed expensive equipment is needed to provide high pressure/high temperature (Figure 11.47). This machine is equipped with the following units:

1. System of upper and lower nozzles
2. Multispeed conveyor for serving the WP

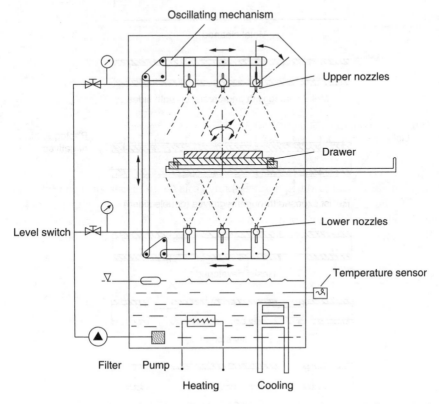

FIGURE 11.47 Schematic of PCM equipment. (From Visser, A., Junker, M. and Weißinger, D., Sprühätzen metallischer Werkstoffe, 1st Auflage, Eugen G. Leuze Verlag, Saulgau, Württ, Germany, 1994. With permission.)

3. A unit for cleaning the worksheet by water and drying it by hot air
4. A unit for measuring and controlling the density and concentration of etchant
5. A unit for product inspection

In PCM, the following steps are carried out (Figure 11.48):

1. The required part shape that is considered as a primary image for the phototool is created by CAD. The drawing is then laser-printed for accurate work, such as PCB.
2. Two photographic negatives, called artwork, are produced at the actual size of the work.
3. The sheet metal is chemically cleaned and then coated with a highly sensitive photoresist (called sensitive emulsion). The coating is performed by spraying, dipping, or rolling. The work is allowed to dry. The photoresist adheres to the surface, protecting it during etching.
4. After coating, the work is sandwiched between the two negatives (artwork), then exposed in vacuum, to an ultraviolet (UV) light. The coating is solidified in exposed areas and is removed from the exposed area by dissolving into developer.
5. The worksheet is exposed once to a powerful water jet to remove the soft photoresist. The worksheet is then rinsed by deionized water, then dried by nitrogen gas.
6. The worksheet is then spray-etched from the top and bottom. This permits the material to be etched from both sides, thus minimizing the undercutting, reducing the machining time, and producing straighter sidewalls (Figure 11.48).
7. After etching, the hard photoresist is removed and the worksheet is rinsed to avoid any reactions with suspended etchant.

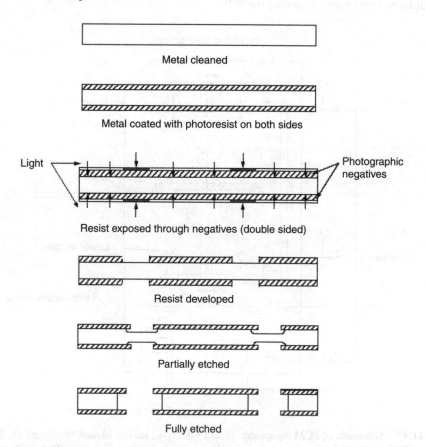

FIGURE 11.48 PCM steps.

Applications of Photochemical Machining Process

Aluminum, copper, zinc, steels, stainless steels, lead, nickel, titanium, molybdenum, glass, germanium, carbides, ceramics, and some plastics are photochemically machined. The process also works well on springy materials that are difficult to punch. The materials must be flat so that they can later be bent to shape and assembled into other components. Products made by PCM are generally found in the electronic, automotive, aerospace, telecommunication, computer, and other industries. Typical products such as PCB, fine screens, flat springs, and so on machined from foils are illustrated in Figure 11.49.

Figure 11.50 illustrates a spray-etched stainless steel of 1 mm thickness sprayed from the top and bottom, whereas Figure 11.51 shows the development of a corner radius, which increases with the etching time.

Advantages and Disadvantages of Photochemical Machining

Advantages
In addition to the previously mentioned advantages of CH-milling, PCM is characterized by the following:

- The accuracy and etch rate are considerably greater than those realized by EC milling.
- Because tooling is made by photographic techniques, they can be easily stored, and patterns can be reproduced easily.

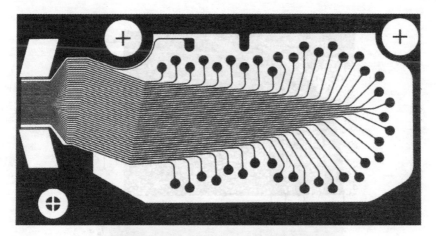

FIGURE 11.49 Typical products produced by PCM. (From Visser, A., Junker, M., and Weiβinger, D., Sprühätzen metallischer Werkstoffe, 1st Auflage, Eugen G. Leuze Verlag, Saulgau, Württ, Germany, 1994. With permission.)

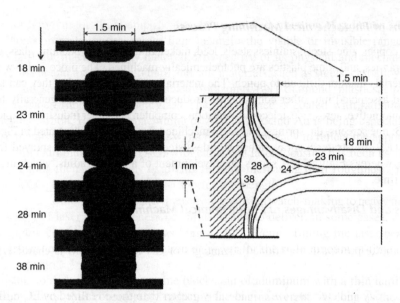

FIGURE 11.50 Cavity development with time by PCM. (From Visser, A., Junker, M., and Weißinger, D., Sprühätzen metallischer Werkstoffe, 1st Auflage, Eugen G. Leuze Verlag, Saulgau, Württ, Germany, 1994. With permission.)

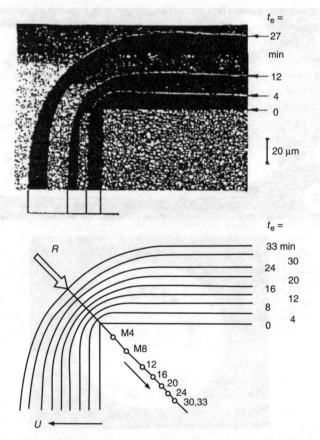

FIGURE 11.51 Corner development with time by PCM. (From Visser, A., Junker, M., and Weißinger, D., Sprühätzen metallischer Werkstoffe, 1st Auflage, Eugen G. Leuze Verlag, Saulgau, Württ, Germany, 1994. With permission.)

- Lead times are small compared to those required by processes that require hard tooling.
- Delicate and fragile parts of small thicknesses are produced by PCM without any deformation and warpping.

Disadvantages

Apart from the disadvantages of CH-milling, the PCM has also the following limitations:

- Requires highly skilled operators.
- Requires more expensive equipment.
- The machine should be protected from the corrosive action of etchants.

11.6 ELECTROCHEMICAL MACHINES AND OPERATIONS

11.6.1 Process Characteristics and Applications

ECM is one of the most effective NTMP; in this process, the metal removal is based on the anodic dissolution governed by Faraday's principle of electrolysis (1833). According to this principle, the anodic dissociation rate in the machining (depleting) cell is directly proportional to the dc electrolyzing current and the chemical equivalent ε of the anode material (ε = atomic weight/valence).

In the machining cell, the WP is connected to the anode, while the tool is connected to the cathode of a dc source of 5–30 V. Both the tool and WP electrodes must be electrically conductive. They are separated by a gap of 0.1–1 mm thickness, into which an electrolyte (NaCl, KCl, or $NaNO_3$) is pumped rapidly to sweep away the reaction products (sludge) from the narrow machining gap. Depending on the gap thickness, a machining current of high density (20–800 A/cm^2) is passed, causing a high anodic dissolution rate. The shape of the cavity formed in the WP is the female mating image of the tool shape. The tool advances axially toward the WP by means of a servomechanism at a constant feed rate v_f ranging from 0.5 to 10 mm/min (Figure 11.52). The anodic dissolution rate of the WP v_A adjusts itself to match with the selected tool feed ($v_f = v_A$). This matching characteristic or the self-adjusting feature of ECM process controls and stabilizes its performance. Consequently, during machining, the gap thickness attains a constant value known as the equilibrium gap.

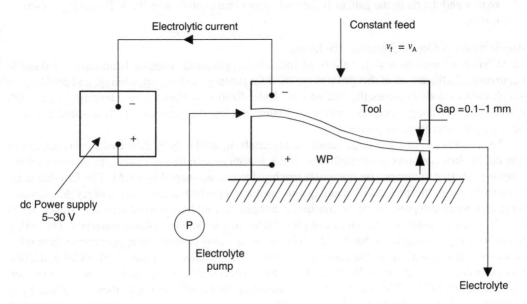

FIGURE 11.52 ECM process.

According to the vision described earlier, Gusseff introduced his patent on EC sinking in 1930 (BP-335003, 1930). However, the first industrial application of this patent was realized in the 1950s and 1960s. ECM is characterized by the following advantages and disadvantages.

Advantages
- Three-dimensional surfaces with complicated profiles can be easily machined in a single operation, irrespective of the hardness and strength of the WP material.
- ECM offers a higher rate of metal removal as compared to traditional and nontraditional methods, especially when high machining currents are employed.
- There is no wear of the tool, which permits repeatable production.
- No thermal damage or HAZ.
- High surface quality and accuracy can be achieved at the highest MRR (R_a = 0.1–1.2 µm).
- Labor requirements are low.
- The surfaces produced by ECM are burr-free and free from stresses.

Disadvantages
- Nonconductive materials cannot be machined.
- Inability to machine sharp interior corners or exterior edges of less than 0.2 mm radius.
- The machine and its accessories are subjected to corrosion and rust, especially when the NaCl electrolyte is used. Less corrosive but more expensive electrolytes like $NaNO_3$ can also be used.
- The endurance limit of parts produced by ECM is lowered by about 10–25%. In such a case, shot peening after ECM is recommended to restore the fatigue strength.
- Metal removal rates are slow compared to traditional methods.
- Specific power consumption of ECM is considerably higher than that required for TM.
- Cavitation channels may form, which deteriorates the surface quality.
- Pumping electrolytes at high pressures into the narrow gap gives rise to large hydrostatic forces acting on the tool and WP, which necessitates a rigid machine frame.
- The machined parts need to be cleaned and oiled immediately after machining.
- There is a danger of explosion if the hydrogen generated during machining is not safely disposed off.
- The tool and the WP may be damaged if arcing is initiated due to the contamination of oxides and debris in the gap, or if the tool comes into contact with the WP, causing a short circuit.

Applications of Electrochemical Machining

ECM has been used in a wide variety of industrial applications ranging from cavity sinking to deburring. Modifications of this process are used for turning, slotting, trepanning, and profiling, in which the electrode becomes the cutting tool. Hybridization performed by integrating ECM with conventional finishing processes leads to the highly developed electrochemically assisted grinding, honing, and superfinishing processes.

The process can handle a large variety of materials, limited only by their electrical conductivity and not by their hardness or strength. The MRR is high especially for difficult-to-machine alloys. Fragile parts that are otherwise not easily machinable can be shaped by ECM. The fact that there is no tool wear in this process is advantageous, as it has a positive impact on accuracy. Moreover, a large number of components can be machined without the tool having to be replaced. Hence, ECM is well-suited for mass production of complex shapes in hard and extra-hard materials. The ability to machine high-strength and hardened steels has led to many cost-saving applications where other processes are impractical. In the sector of electrically conductive materials, both ECM and EDM processes compete each other. However, ECM has a special attraction, due to the absence of thermal stresses and HAZ. This characteristic is useful in the manufacturing of dies in hardened and tempered steel blocks (Figures 11.53 and 11.54).

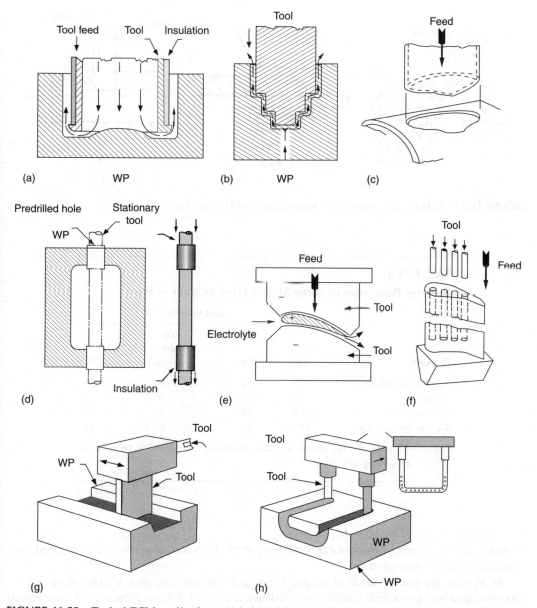

FIGURE 11.53 Typical ECM applications: (a) hole sinking with insulated tool, (b) EC sinking of stepped through hole, (c) EC trepanning, (d) ECM of internal cavity by stationary electrode, (e) ECM of turbine blade, (f) EC deep hole drilling, (g) EC surfacing, and (h) EC hogging.

11.6.2 Elements of Electrochemical Machining

The basic elements of ECM process are the tool, the WP, and the electrolyte.

11.6.2.1 Tool

During ECM, the tool performance depends upon the characteristics described in the following sections.

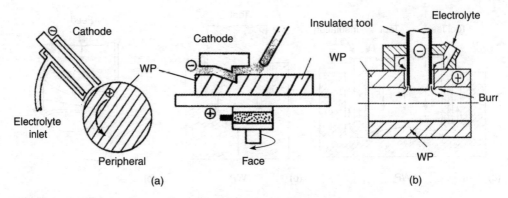

FIGURE 11.54 Other applications: (a) EC turning and (b) EC deburring.

TABLE 11.9
Relative Properties of Some Metals Used as Tools in ECM

	Tool Material			
Property	Cu	Brass	316 Stainless Steel	Ti
Electrical resistivity	1.0	4.0	53.0	48.0
Stiffness	1.1	1.0	1.9	1.1
Machinability	6.0	8.0	2.5	1.0
Thermal conductivity	25.0	7.5	1.0	2.6

Source: Reproduced from *Machining Data Handbook*, Machinability Data Center, Cincinnati, OH, 1980.

1. *Suitable Choice of Its Material*

The tool material should be machinable, stiff, and possess high corrosion resistance and good electrical and thermal conductivity.

ECM tools are usually made of copper, brass, and 316 stainless steel. Carbon steels are not recommended because of their low corrosive resistance. Table 11.9 shows properties of some metals used for such ECM tools.

2. *Tool Design*

The tool is shaped to be not exactly a mirror image of the machined cavity; its dimensions must be slightly different from the nominal dimensions of the cavity to allow for an overcut.

The tool design must permit electrolyte flow at a rate sufficient to dissipate heat generated to eliminate boiling of the electrolyte in the interelectrode gap. To produce smooth surfaces on the WP, tool design must enable a uniform flow over the entire machining area. Ideally, flow should be laminar and free from eddies.

On machining complex cavities, a good designer performs pilot tests under the same machining conditions. Accordingly, corrections to the tool form are carried out to realize the required accuracy and surface quality.

11.6.2.2 Workpiece

In ECM, there is no restriction on the WP whatsoever, except being electrically conductive. The machinability depends only on the chemical composition of the WP. Carbon is passive in EC reactions; consequently, CI, which contains free carbon, cannot be machined satisfactorily by ECM.

11.6.2.3 Electrolyte

Electrolytes are highly conductive solutions of inorganic salts—usually NaCl, KCl, and $NaNO_3$, or their mixtures to meet multiple requirements. The main functions of the electrolytes are as follows:

a. Complete the electric circuit between the tool and the WP
b. Allow desirable reactions to occur, and create conditions for anodic dissolution
c. Carry away heat generated during chemical reactions
d. Remove products of reaction (sludge) from the machining gap

Effective and efficient electrolytes should therefore have the following properties:

a. High electrical conductivity to ensure high current density
b. Low viscosity to ensure good flow conditions in an extremely narrow interelectrode gap
c. High specific heat and thermal conductivity to be capable of removing the heat generated from the gap
d. Resistance to the formation of a passive film on the WP surface
e. High chemical stability
f. High current efficiency and low throwing power
g. Nontoxic and noncorrosive to the machine parts
h. Inexpensive and available

Electrolytes also play an important role in the dimensional control. As shown in Figure 11.55a, a $NaNO_3$ solution is preferable, because the local metal removal rate is high at the small gap locations where both the current density and current efficiency are high. However, the local removal rate is low in the side gap, where both current density and current efficiency are low. This results in a smaller side gap (Figure 11.55b) (see McGeough, 1988). Several methods of supplying electrolytes to the machining gap are shown in Figure 11.56. The choice of the electrolyte supply method depends on the part geometry, machining method, required accuracy, and surface finish. Typical electrolyte conditions in the gap include a temperature of 90–110°C, a pressure between 10 and 20 ATM, and a maximum electrolyte velocity of 25–50 m/s.

11.6.3 ECM Equipment

ECM equipment includes a dc power generator and an EC machine. The power generator should be powerful enough to supply the necessary machining current to the working gap. EC machines equipped with power generators of current capacities ranging from 50 to 40,000 A are on the market. The power sources supply constant voltages ranging from 5 to 30 V. They are generally characterized by a high power factor, high efficiency, and should be equipped with short-circuit protection to prevent catastrophic short circuits across the electrodes within a small fraction of a second.

Figure 11.57 illustrates schematically a typical EC sinking machine. The machine must be rigid enough to withstand the hydrodynamic pressure of the electrolyte in the machining gap, which tends to separate the tool from WP. A servomechanism is necessary to control the tool

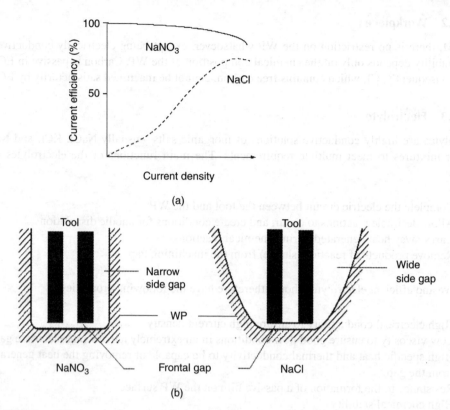

FIGURE 11.55 (a) Current efficiency and (b) side gap of NaCl and $NaNO_3$.

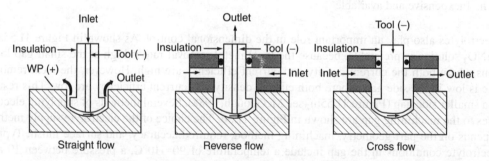

FIGURE 11.56 Methods of electrolyte feeding in ECM. (Modified from El-Hofy, H., *Advanced Machining Processes, Nontraditional and Hybrid Processes*, McGraw-Hill Co., New York, 2005.)

movement in such a way that the material dissolution is balanced by the feed rate of the tool. The rate of current change is monitored and tool feeding is stopped when an abnormal rise in current is detected.

In contrast to conventional machine tools, the EC machines are designed to stand up to corrosion attack by using nonmetallic materials. For high strength and rigidity, metals with nonmetallic coatings are recommended. To eliminate the danger of corrosion on other machinery, EC machines should be perfectly isolated in separate rooms in the workshop. The electrolyte feeding unit supplies electrolyte at a given rate, pressure, and temperature. Facilities for electrolyte filtration, temperature control, and sludge removal are also included.

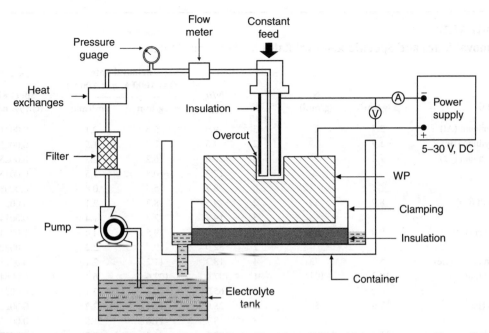

FIGURE 11.57 ECM setup. (Modified from El-Hofy, H., *Advanced Machining Processes, Nontraditional and Hybrid Processes*, McGraw-Hill, New York, 2005.)

11.6.4 PROCESS CAPABILITIES

In ECM, the cutting rate is solely a function of the ion exchange rate, irrespective of the hardness or toughness of the work material. The process provides metal removal rates in the order of 1.5 cm³/min/1000 A. Penetration rates up to 2.5 mm/min are routinely obtained when machining carbides, and steels, either hardened or not. Table 11.10 shows the EC removal rate of most common metals, assuming an electrolyzing current of 1000 A and a current efficiency of 100%. The table includes also a formula to determine the chemical equivalent of alloys, necessary to calculate their removal rates (Youssef, 2005).

A well-known and unique characteristic of ECM, among all traditional and nontraditional processes, is that the accuracy and surface quality improve when applying higher removal rates (i.e., higher current densities).

A major problem of ECM is the overcut (side gap), which affects the accuracy. Roughly speaking, the side gap is governed by a complex set of parameters, of which the type of electrolyte and the electrolyte flow are most crucial. It is important to have a small side gap, because the dimensional tolerance is proportional to the gap width. A typical dimensional tolerance of ECM is ±0.13 mm; however, through proper control of the machining parameters, tight tolerance of ±0.025 mm can be achieved. It is difficult to machine internal radii smaller than 0.8 mm. A typical overcut of 0.5 mm and taper of 1 µm/mm are possible (*Metals Handbook*, 1989). Typical surface roughness (R_a value) of ECM ranges from 0.2 to 1 µm (0.4–0.8 µm is common) that decreases with increasing machining rate.

The principal tooling cost is due to the preparation of the tool electrode, which can be time-consuming and costly, requiring several cut-and-try efforts, except for simple shapes. There is no tool wear, and the process produces stress-free surfaces. The capability to cut the entire cavity in one stroke makes the process very productive, but the complicated tool form increases the tool cost.

TABLE 11.10
Removal Rates and Specific Removal Rates for Different Metals

Metal	ρ (g/cm³)	N (g/mol)	n	N/n (g)	Removal Rate (I = 1000 A, η = 100%) g/min	cm³/min	Spec. RR (cm³/A min)
Aluminum (Al)	2.7	27	3	9.0	5.6	2.1	0.0021
Beryllium (Be)	1.9	9	2	4.5	2.8	1.5	0.0015
Chromium (Cr)	7.2	52	2	26.0	16.2	2.3	0.0023
			3	17.3	10.8	1.5	0.0015
			6	8.7	5.4	0.8	0.0008
Cobalt (Co)	8.9	59	2	29.5	18.3	2.1	0.0021
			3	19.7	12.3	1.4	0.0014
Copper (Cu)	9.0	64	1	64.0	39.5	4.4	0.0044
			2	32.0	19.7	2.2	0.0022
Germanium (Ge)	5.3	73	4	18.3	11.2	2.1	0.0021
Gold (Au)	19.3	197	1	197.0	122.6	6.4	0.0064
			3	65.7	40.8	2.1	0.0021
Iron (Fe)	7.9	56	2	28.0	17.4	2.2	0.0022
			3	18.7	11.6	1.5	0.0015
Lead (Pb)	11.4	207	2	103.5	64.4	5.7	0.0057
			4	51.7	32.2	2.8	0.0028
Magnesium (Mg)	1.8	24	2	12.0	7.6	4.4	0.0044
Manganese (Mn)	7.5	55	2	27.5	17.1	2.3	0.0023
			3	18.3	11.3	1.5	0.0015
			4	13.8	8.5	1.2	0.0012
			6	9.2	5.7	0.8	0.0008
			7	7.8	4.9	0.7	0.0007
Molybdenum (Mo)	10.2	96	3	32.0	20.0	2.0	0.0020
			4	42.0	14.9	1.5	0.0015
			6	16.0	10.0	1.0	0.0010
Nickel (Ni)	8.9	59	2	29.5	16.2	2.1	0.0021
			3	19.7	12.2	1.4	0.0014
Platinum (Pt)	21.5	195	2	97.5	60.6	2.8	0.0028
			4	48.7	30.3	1.4	0.0014
Silver (Ag)	10.5	108	1	108.0	67.1	6.4	0.0064
Tantalum (Ta)	16.6	181	5	36.2	22.5	1.3	0.0013
Tin (Sn)	7.3	119	2	59.5	37.0	5.0	0.0050
			4	29.7	18.5	2.5	0.0025
Titanium (Ti)	4.5	48	3	16.0	10.0	2.2	0.0022
			4	12.0	7.5	1.6	0.0016
Tungsten (W)	19.3	184	6	30.7	19.0	1.0	0.0010
			8	23.0	14.3	0.7	0.0007
Vanadium (V)	6.1	51	3	17.0	10.6	1.7	0.0017
			5	10.2	6.4	1.0	0.0010
Zinc (Zn)	7.2	65	2	32.5	20.0	2.9	0.0029

$$\left(\frac{N}{n}\right)_{\text{alloy}} = \frac{1}{100}\left[x_P\frac{N_P}{n_P} + x_Q\frac{N_Q}{n_Q} + \ldots\right]^a$$

[a] This equation is used to calculate the chemical equivalent of an alloy composed of elements $x_P, x_Q, \ldots$ of atomic weights $N_P, N_Q,$ and valences $n_P, n_Q,$ respectively.

11.7 ELECTROCHEMICAL GRINDING MACHINES AND OPERATIONS

ECG is one of the most important HPs, in which metal is removed by a combination of EC dissolution and mechanical abrasion. The equipment used in ECG is similar to a traditional grinding machine, except that the GW is metal-bonded with diamond or borazon abrasives; the wheel is the negative electrode and is connected to the dc supply through the spindle insulated from the machine frame, whereas the work is connected to the positive terminal of the dc supply (Figure 11.58).

A flow of electrolyte, usually $NaNO_3$ is provided in the direction of wheel rotation for the ECM phase of the operation. The wheel rotates at a surface speed of 25–35 m/s.

The abrasives in the wheel are always nonconductive, and thus they act as an insulating spacer, maintaining a separation (electrolytic gap) of 12–80 μm between electrodes. The abrasives also mechanically remove the reaction residue from the working gap and cut chips from the WP. Therefore, removal of WP material occurs due to the flow of current (100–300 A/cm^2) through the gap and the rate of decomposition is sped up by the grinding action of the abrasive grains. With proper operation, typically 95% of material removal is due to electrolytic dissolution, and only about 5% is due to the abrasion effect of the grinding wheel. Consequently, the wear of the wheel is very low, thus eliminating or considerably reducing the need for redressing, which reduces the sharpening costs approximately by 60%.

The lack of heat damage, distortions, burrs, and residual stresses in ECG is very advantageous, particularly when coupled with high MRR, in addition to far less wear of the grinding wheel. That is why the process has been applied most successfully in sharpening cutting tools, die inserts, and punches made of hardened high-strength steel alloys. ECG is suitable for applications similar to those for milling, grinding, cutting off, and sawing. It is not adaptable to cavity-sinking operations such as die making. The process offers the following specific advantages over traditional diamond wheel grinding:

- Increased MRR due to the added EC effect
- Reduced tool wear and sharpening costs
- Less risk of thermal damage and distortion
- Absence of burrs
- Reduced wheel pressure, which improves accuracy

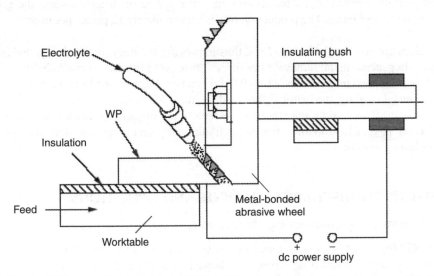

FIGURE 11.58 ECG setup. (Modified from El-Hofy, H., *Advanced Machining Processes, Nontraditional and Hybrid Processes*, McGraw-Hill, New York, 2005.)

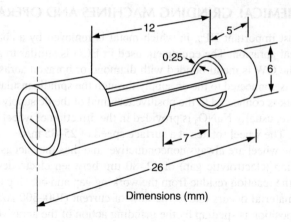

FIGURE 11.59 ECG of thin-walled fragile tube made of 316 stainless steel. (Adapted from *Advanced Materials and Processes*, ASM International, Materials Park, OH, 1990.)

However, ECG has the following disadvantages:

- Higher capital cost of the equipment.
- Limited to electrically conductive materials.
- Hazard due to corrosive nature of electrolyte; for that reason $NaNO_3$ of limited corrosion nature is used.
- Necessity of electrolyte filtering and disposal.

Some typical applications of ECG operation are as follows:

1. Sharpening of carbide cutting tools. ECG provides saving of about 75% in wheel costs and about 50% in labor cost when sharpening carbide tools (Pandey and Shan, 1980).
2. Grinding fragile parts such as honeycomb, thin-walled tubes, and skin hypodermic needles. Figure 11.59 shows a thin-walled fragile tube made of 316 stainless steel that is electrochemically ground. After machining, the tube is free of distortions and the ground edges are free of burrs. The production rate is approximately 12 pieces per hour.

ECG machines are now available with NCs, thus improving accuracy, repeatability, and increased productivity. In general, metal removal rates of ECG process of the order 1 cm³/min/100A are realized. Surface roughness in the range of 0.2–0.6 µm can be easily obtained generally, the higher the hardness of metal or alloy, the better is the finish. Typical tolerances achieved by ECG are of the order of ±10 µm. If more tight tolerances are essential, the majority of stock can be removed by ECG and a final pass of 10–100 µm can be taken traditionally, on the same equipment, by switching off the machining current.

11.8 ELECTRICAL DISCHARGE MACHINES AND OPERATIONS

11.8.1 Process Characteristics and Applications

Of all the NTMP, none has gained greater industry-wide acceptance than EDM. It is well known that when two current-conducting wires are allowed to touch each other, an arc is produced. Although this phenomenon has been detected by Priestly since 1790, it was not until the 1940s that a machining process based on this principle was developed by Lazerenkos in Russia. The principle

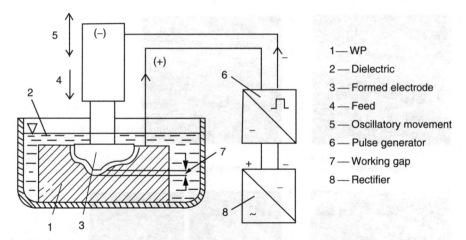

FIGURE 11.60 Concept of EDM. (From Düniβ, W., Neumann, M., and Schwartz, H., *Trennen-Spanen and Abtragen*, VEB-Verlag Technik, Berlin, 1979.)

of EDM (also called spark erosion machining) is based on erosion of metals by spark discharges between a shaped tool electrode (usually negative) and the WP (usually positive).

The tool and the WP are separated by a small gap of 10–500 μm. Both are submerged or flooded with electrically nonconducting dielectric fluid. When a potential difference between the tool and the WP is sufficiently high, the dielectric in the gap is partially ionized, so that a transient spark discharge ignites through the fluid, at the closest points between the electrodes. Each spark of thermal power concentration, typically 10^8 W/mm^2, is capable of melting or vaporizing very small amounts of metal from the WP and the tool (Figure 11.60). A part of the total energy is absorbed by the tool electrode, yielding some tool wear, which can be reduced to 1% or less if adequate machining conditions are carefully selected.

The instantaneous vaporization of the dielectric produces a high-pressure bubble that expands radially. The discharge ceases with the interruption of the current, and the metal is ejected, leaving tiny pits or craters in the WP and metal globules suspended in the dielectric (Figure 11.61). A sludge of black carbon particles, formed from hydrocarbons of the dielectric, produced in the gap and expelled by the explosive energy of the discharge, remains in suspension until removed by filtering. Immediately following the discharge, the dielectric surrounding the channel deionizes and, once again, becomes effective as an insulator.

The capacitor discharge is repeated at rates between 0.5 and 500 kHz, at a voltage between 50 and 380 V and currents from 0.1 to 500 A. EDM has the following advantages:

- It is applicable to machine metals, alloys, and carbides, irrespective of their hardness and toughness.
- Because there is no mechanical contact between the tool and the WP, very delicate work can be machined.
- The process is widely used to produce cavities and profiles of complex and intricate shapes accurately in extra-hard materials.
- The process leaves no burr on the edges.
- EDM electrodes can be accurately produced from machinable materials, which have a positive impact on the accuracy of machined holes and cavities.

However, the process is hampered by the following limitations:

- It cannot be used if the WP material is a bad electric conductor.
- On machining materials, the process produces HAZ (Figure 11.62), which is characterized by hairline cracks and thin, hard recast layer. The surface integrity (SI) is improved if

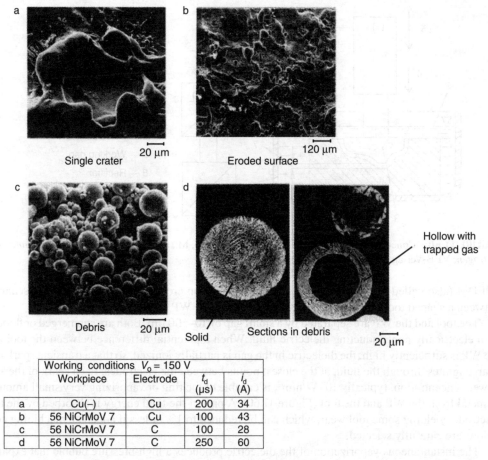

FIGURE 11.61 SEM surfaces and debris produced by EDM. (From König, W. Fertigungsverfahren-Band 3 Abtragen, VDI Verlag, Düsseldorf, 1990.)

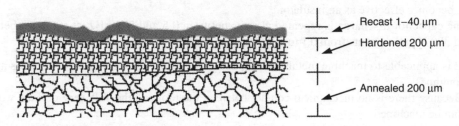

FIGURE 11.62 HAZ of surface produced by EDM.

other machining processes such as CHM, shot peening, and so on, are used subsequently to EDM. This measure is done if the product will be operating in a fatigue environment.
- EDM cannot produce sharp concerns and edges.
- The process is characterized by high specific energy.

As compared to TM, a major problem of EDM is the low MRR.

Nontraditional Machine Tools and Operations

Applications

EDM has become an indispensable process in the modern manufacturing. It is used in numerous applications such as producing die cavities for automotive body components, connecting rods, and various intricate shapes, to a high degree of accuracy. About 80–90% of EDM work is the manufacture of tool and die sets for the production of castings, forgings, stampings, and extrusions. Micromachining of holes, slots, texturing, and milling are also typical applications of the process. Figure 11.63 illustrates some typical products made by EDM (Nassovia-Krupp, 1967):

a. Sinking of two connecting rods with electrode
b. Sinking of extrusion insert plate of motor anchor with single electrode
c. Machining of embossing die
d. Machining of forging die
e. Sinking of injection mold insert for light alloys

Figure 11.64 illustrates the machining of spherical internal cavity using a specially designed mechanically rotating electrode that is equipped with a hinged tip (Lusiesa, France).

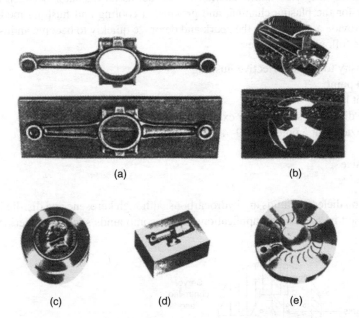

FIGURE 11.63 Typical products produced by EDM: (a) forging die of two connecting rods, machined with one electrode, (b) extrusion insert plate of motor anchor and electrode, (c) embossing die, (d) forging die, and (e) injection mold insert for light alloys.

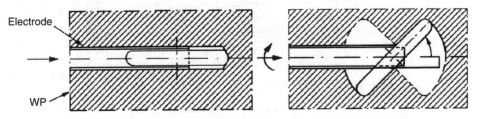

FIGURE 11.64 Internal cavity produced by mechanically rotating electrode that is equipped with a hinged tip to produce the spherical internal cavity.

11.8.2 ED Sinking Machine

A typical setup of an ED sinking machine is illustrated in Figure 11.65. Most machines are of the ram type, in which the sinking head is actuated by a hydraulic cylinder. In EDM, the gap between the electrode and the WP is critical, thus the down feed of the tool should be controlled by a servomechanism that automatically maintains a constant gap. The servo gets its input signal from the difference between the selected reference voltage and the actual voltage across the gap. This signal is amplified and the tool is advanced by hydraulic control. A short circuit across the gap causes the servo to reverse the motion of the tool until proper control is restored.

The WP is clamped within the tank containing the dielectric fluid and its movements are numerically controlled. The machine is equipped with a working table mounted on orthogonal slide ways, often provided with accurate optical scales, similar to that of a jig boring machine. The machine is also equipped with a pump and a filtering system for the dielectric fluid. The dielectric is flushed through the spark gap, supplied either through a hollow tool or from external jets, or both. It may also be supplied through holes in the WP.

Dielectric Fluids

The dielectric fluids have four main functions. They act as an insulator between tool and work, spark conductor for the plasma channel, and provides a cooling and flushing medium. The fluid must ionize to provide a channel for the spark and deionize quickly to become an insulator. A good dielectric has the following properties:

1. Low viscosity to ensure effective flushing
2. High flash point
3. High latent heat
4. A suitable dielectric strength; for example, 180 V/25 μm
5. Rapid ionization at a potential 40–400 V followed by rapid deionization
6. Nontoxic
7. Noncorrosive
8. Nonexpensive

The most common dielectric fluids are hydrocarbons, although kerosene and distilled and deionized water may be used in specialized applications. Polar compounds such as glycerine water (90:10)

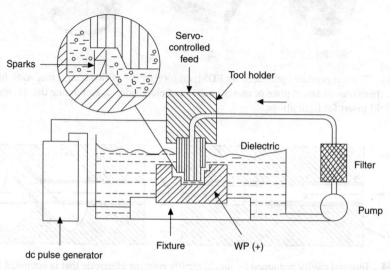

FIGURE 11.65 ED sinking machine.

with triethylene oil as an additive have proved to improve the MRR and decrease the tool wear as compared to kerosene.

Flushing Techniques

Dielectric flushing techniques are of vital importance in EDM. At the beginning, the fresh supply of dielectric fluid is clean and has a higher insulation strength than a supply that contains particles (contaminated). When spark discharges commence, debris is created and insulation strength is diminished by particles acting as stepping stones in the interelectrode gap. If too many particles are allowed to remain, a bridge is formed, resulting in arcing across the gap and causing damage to the tool and WP. Therefore, the contamination in the gap must be controlled to provide optimum conditions (Figure 11.66). In the same figure, the effect of optimum flushing on the MRR with the tool advance is schematically illustrated.

Figure 11.67 differentiates between the techniques of flushing:

1. Injection flushing in which a slight taper is produced on the sides of the cavity due to lateral discharges as debris pass up the side of the tool (Figure 11.67a)
2. Suction flushing through which the side taper is avoided (Figure 11.67b)
3. Side flushing in which a slight taper is produced on the side of the cavity at the outlet (downstream) of the dielectric (Figure 11.67c)

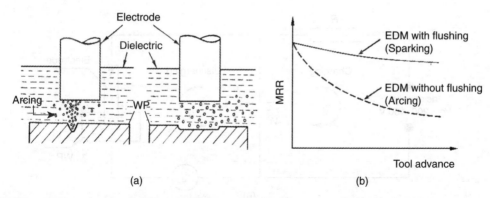

FIGURE 11.66 (a) Contamination in EDM. (From Lissaman, A. J. and Martin, S. J., *Principles of Engineering Production*, Hodder and Stoughton Educational, London, UK, 1982. With permission.) and (b) effect of flushing on MRR. (From Youssef, H. A., *Non-Traditional Machining Processes-Theory and Practice*, El-Fath Press, Alexandria, 2005. With permission.)

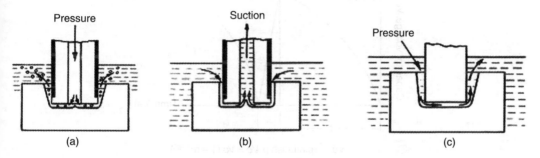

FIGURE 11.67 Dielectric flushing modes: (a) injection flushing, (b) suction flushing, and (c) side flushing. (From Lissaman, A. J. and Martin, S. J., *Principles of Engineering Production*, Hodder and Stoughton Educational, London, UK, 1982. With permission.)

11.8.3 EDM-Spark Circuits (Power Supply Circuits)

The ED machine is equipped with a spark-generating circuit that can be controlled to provide optimum conditions for a particular application. This generator should supply voltage adequate to initiate and maintain the discharge process, and provides necessary control over the process parameters such as current intensity, frequency, and cycle times of discharge. The cycle time ranges from 2 to 1600 μs (Lissaman and Martin, 1982). Two main types of generators are applicable for this purpose. These are the resistance-capacitance generator (RC circuit) and the transistorized pulse generator.

11.8.3.1 Resistance-Capacitance Circuit

It is also called the Lazerenko circuit, which is basically a relaxation oscillator. It is simple, reliable, rigid, low-cost power source that is ordinarily used with copper or brass electrodes. It provides a fine surface texture of 0.25 μm R_a, but the machining rate is slow, because the time required to charge the capacitors prevents the use of high frequencies. The relaxation circuit operates at selectively high input voltages and is difficult to operate. The reversed polarity encountered in a relaxation circuit leads to an additional tool wear.

The basic form of the RC circuit is shown in Figure 11.68a. On commencing operation, the capacitor is in the uncharged condition. Then it is charged with a dc voltage source Vo, usually

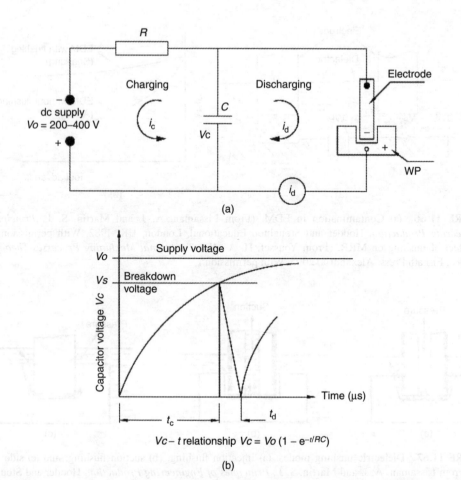

FIGURE 11.68 (a) RC circuit and (b) capacitor voltage-charging time exponential relationship.

200–400 V, via the resistor R, which determines the charging rate. The capacitor voltage Vc increases exponentially as charging proceeds (Figure 11.68b).

$$Vc = Vo(1 - e^{-t/RC}) \tag{11.32}$$

where,
t = time (s)
RC = time constants = resistance (Ω) × capacitance (Farad)

When Vc attains the level of breakdown voltage Vs existing in the working gap, the capacitor charges across the gap eroding both WP (causing material removal) and the tool electrode (causing wear). The spark is not sustained, because the capacitance is discharged more quickly than it can recharge via the resistor,

$$t_d = 0.1 t_c \tag{11.33}$$

where,
t_c = charging time
t_d = discharging time

The cycle charging and discharging is repeated until the cut is performed.
For maximum production rate (Barash, 1962):

$$Vs = 0.73\, Vo \tag{11.34}$$

The energy of each individual spark discharge in joule is given by

$$E_d = \tfrac{1}{2} CVs^2 \tag{11.35}$$

Therefore, the increase in Vo, Vs, and C leads to an increase in machining rate; however, it leads to poor surface texture.

A reduction of Vs enables a smaller gap to be used, improving finish and accuracy, but reducing machining rate. High rates of machining are obtained by reducing the time constant RC in Equation 11.32 to give rapid charging. However, as R is reduced, the frequency increases and may reach a point at which deionization is prevented from taking place and arcing occurs. Arcing causes effective machining to cease and creates thermal damage to the machined surface. It follows that in an RC circuit, the machine setting for optimum performance in a given set of machining conditions involves a compromise in selecting the process parameters.

ILLUSTRATIVE EXAMPLE 4

In an EDM operation using Lazerenko's generator, $Vo = 250$ V, $R = 10\ \Omega$, and $C = 3\ \mu F$.

If the cut is required to be performed at maximum removal rate condition, calculate:

1. Discharge voltage
2. Charging time, t_c
3. Cycle frequency, f_r
4. Energy/individual discharge of the capacitor, E_d
5. If the dielectric used has a strength of 180 V/25 μm, estimate the expected gap thickness to realize this cut.

Solution

The cut is performed at maximum removal rate condition, then:

1. $Vs = 0.73\ Vo$
   $= 0.73 \times 250 = 182.5$ V
2. Charging time, t_c

$$Vs = Vo\,(1 - e_c^{-t/RC})$$

$$\frac{Vs}{Vo} = 0.73 = 1 - e_c^{-t/30} \quad (C \text{ in } \mu F, \text{ and } t_c \text{ in } \mu s)$$

from which,

$$t_c = 39.2\ \mu s$$

3. Cycle frequency, f_r

$$f_r = \frac{t}{t_c + t_d}$$

$$t_d = 0.1\ t_c = 3.9\ \mu s$$

$$t_c + t_d = 39.2 + 3.9$$
$$= 43.1\ \mu s = 43.1 \times 10^{-6}\ s$$

$$f_r = \frac{10^6}{43.1} = 23200\ Hz = 23.2\ kHz$$

4. Energy/individual discharge, E_d

$$E_d = \frac{1}{2}\,CVs^2$$

$$= \frac{1}{2} \times 3 \times 10^{-6} \times (182.5)^2 = 0.05\ J$$

5. Expected gap thickness, h_g

$$h_g = \frac{Vs}{180} \times 25$$

$$= \frac{182.5}{180} \times 25 = 25.5\ \mu m$$

11.8.3.2 Transistorized Pulse Generator Circuits

Among the disadvantages of the RC relaxation circuits are interdependence (lack of control of parameters), the restricted choice of electrode material, and their high wear rate. The adoption of the transistorized pulse generators in the 1960s allowed the process parameters (frequency and energy of discharges) to vary with a greater degree of control, in which charging takes only a small portion of the cycle. Furthermore, the voltage of these machines is reduced to 60–80 V range, permitting low discharge current pulses of a square profile. This results in shallower and wider craters, which means better surface texture. Alternatively, when required, they provide high MRRs at the expense of surface quality by permitting high discharge currents. Moreover, this type of generators provide considerably lower electrode wear as compared to simpler RC circuits.

In the simple form of the transistorized pulse generators, the parameters are selected and pre-adjusted according to the machining duty. The selected parameters remain constant; that is, not influenced by the variation of working conditions in the gap during machining.

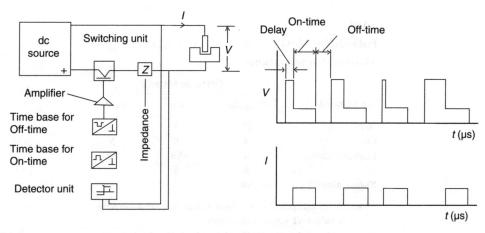

FIGURE 11.69 Pulse generator of Charmilles Technologies, Geneva, Switzerland.

An improved circuit incorporating feedback (Isopulse generator, Charmilles) is illustrated in Figure 11.69. In such a circuit, the conditions into the spark gap are monitored by a detector unit, which determines the exact moment of current flow after the ignition lag. The time base for the on-time then becomes effective, providing a constant discharge period. The time base for the off-time ensures a constant interval for deionization and flushing away the debris by the dielectric. The following are the specifications of a typical isopulse generator, 25 A, produced by Charmilles:

> Power. 2 kW
> Open gap voltage. 80 V
> Discharge energy. 0.18–1 J
> Maximum discharge current. 25 A
> Discharge duration. Off-time 2–1600 μs; on-time 2–1600 μs
> Achieved roughness. $R_a = 0.4$ μm

11.8.4 EDM-Tool Electrodes

In ED sinking, electrodes are often the most expensive part of an EDM operation. Material fabrication, wear, and redressing costs must be carefully weighed to determine the best electrode material and EDM machine setup. The ideal electrode material should have high electrical conductivity and high melting point, and be easy to fabricate and strong enough to stand up to EDM without deformation.

Most electrodes for EDM are usually made of graphite, although brass, Cu, or Cu/W-alloys may be used. These electrodes are shaped by forming, casting, and powder metallurgy, or, frequently, by machining. EDM tool wear is an important factor, as it affects the dimensional and form accuracy. It is related to melting point of the tool material involved—the higher the melting point, the lower the wear rate. Consequently, graphite electrodes have the highest wear resistance, as graphite has the highest melting point of any known material (3600°C); moreover, it is low in cost and readily fabricated. Tungsten (3400°C), and W alloys are next in melting temperature, followed by molybdenum (2600°C); however, these metals are expensive and difficult to fabricate. The tool wear can be minimized by reversing the polarity, which depends on the tool/WP combination. Table 11.11 illustrates the recommended polarity for various electrode/WP material combinations.

The wear may reach a zero value during the so-called no-wear EDM process. Work material machinable by no-wear EDM can be steels, satellites, Ni-base alloys, and aluminum. However, no-wear EDM is not recommended for machining carbides.

TABLE 11.11
Polarity for Most Common Electrode/WP Material Combinations

	Electrode Material		
WP Material	Graphite	Cu	Cu–W
Steel	SR	S	S
Cu	R	R	R
Cemented carbide	R	SR	SR
Al	S	S	S
Ni-base alloys	SR	S	S

Note: S—straight polarity (WP positive electrode) and R—reverse polarity (WP negative electrode).
Source: Modified and reclassified from Metals Handbook (1989).

No-wear EDM requires pulse generators and equipment capable of attaining the following conditions:

- Reverse polarity of the tool electrode.
- Low-pulse frequency ranging from 0.4 to 20 kHz. Generally, about 2 kHz is recommended.
- Graphite, Cu, Cu/W, or Ag/W electrodes.
- High duty cycle of more than 90%.
- High-intensity discharge current.
- Smooth control of servomechanism.
- Supply voltage of not more than 80 V.
- Temperature of dielectric of not above 40°C, and dielectric recycled at low pressure.
- Dielectric flow must not trap particles.
- No capacitance across the spark gap, and no inductance in series with it.

11.8.5 Process Capabilities

EDM is a slow process compared to conventional methods. It produces matte and pitted surfaces composed of small craters, which are characterized by a nondirectional, randomly distributed nature due to succession of individual sparks of the process. The metal removal rates usually range from 0.1 to 600 mm³/min. In EDM, there is a proportional relationship between the MRR and the surface roughness. High removal rates produce a very rough finish, having a molten and recast structure with poor SI and low fatigue strength. The finish cuts are made at low removal rates, and the recast layer during rough cuts is removed later by finishing EDM operations. Table 11.12 illustrates a proportional interrelation between the removal rate and surface roughness R_a when cutting steel using metal and graphite electrodes at different processing levels.

The MRR depends not only on the WP material but also on the machining variables, such as pulse conditions (voltage, current, and duration), electrode material and polarity, and the dielectric. The results in Figure 11.70 show the machining rate and surface roughness (R_t-value) for EDM of different materials (El-Hofy, 2005). The same figure depicts how the material of a low melting point (Al) has the highest MRR and hence the highest surface roughness, and vice versa (graphite).

In EDM, the surface finish varies widely as a function of the spark frequency, voltage, current, and other parameters, which also control the MRR. New techniques use an oscillating electrode to provide very fine surface quality. Alternatively, is characterized by EDM using graphite electrodes bad surfaces and surface defects.

TABLE 11.12
Proportional Interrelation between Volumetric Removal Rate and Surface Roughness R_a

Processing	Volumetric Removal Rate (mm³/min)	R_a (μm)
Metal electrodes		
Gentle	0.75	<1.6
Finishing	0.75–1.5	1.6–3.2
Normal	1.5–110	3.2–6.3
Graphite electrodes (usually of +ve polarity)		
Roughing	110–400	6.3–12.5
Abusive	>400	>12.5

Source: Reproduced from Youssef, H A. in *Non-Traditional Machining Processes—Theory and Practice*, El-Fath Press, Alexandria, 2005.

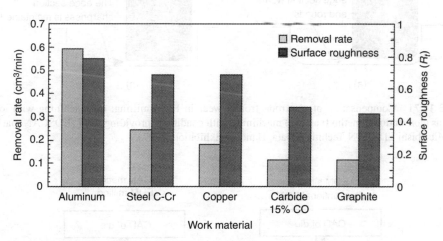

FIGURE 11.70 Machining rate and surface roughness (R_t) for EDM of different materials.

The size of the cavity cut by the tool electrode is larger than the tool by a side overcut between the surfaces of the electrode and WP. This overcut is due to the sparking in the side gap. It is equal to the length of the spark, which is essentially constant all over electrode areas, regardless of size or shape. Typically, overcut values vary from 10 μm to 300 μm, depending on the breakdown voltage, and the size of debris flowing in the side gap. In this respect, suction flushing is preferred, because the debris is not drawn past the side gap, and thus lateral sparking is minimized, leading to a smaller overcut and side taper. EDM equipment manufacturers publish overcut charts to their machines; however, these values should be used only as a guide for the tool designer.

Typical taper varies from 1 to 5 μm/mm per side, depending on the machining conditions, and especially on the flushing technique used. The minimum corner radius more or less equals the size of the overcut. Tolerance of ±50 μm can be easily achieved; however, close control tolerance of ±5 to ±10 μm can be obtained.

11.8.6 Electrical Discharge Milling

The conventional ED sinking, discussed in the previous sections, requires a preliminary phase for producing specially shaped electrodes. These electrodes are very expensive, as they are difficult to design and manufacture, and therefore they add more than 50% to the total machining cost of the product. Recently, a revolutionary breakthrough in the EDM realm has been achieved—a new ED milling technology that makes use of simple and cheap standard rotating pipe electrodes.

In this process, 3-D cavities are machined by successive sweeps of the electrode down to the desired depth, while the NC automatically compensates by means of powerful algorithms the electrode's front wear to ensure product accuracy along the three axes (Figure 11.71). Therefore, there is no need to manufacture the specially shaped electrodes as in the case of conventional EDM, which means saving of time and money (Figure 11.72).

The theory of ED milling (also termed ED scanning) is shown in Figure 11.73. The thickness of the layer removed per each path ranges from 0.1 mm to several mm on rough paths and from 1 µm to 100 µm on finished paths.

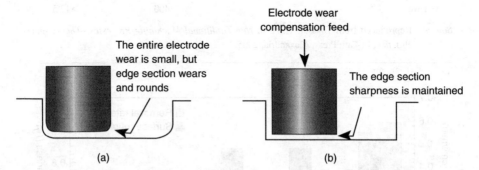

FIGURE 11.71 Compensation of electrode frontal wear in EDM-milling: (a) machining with low wear conditions (conventional method) and (b) machining with conditions providing wear (EDSCAN machining). (From Mitsubishi EDSCAN Technical Data, Hanover Exhibition, 1997.)

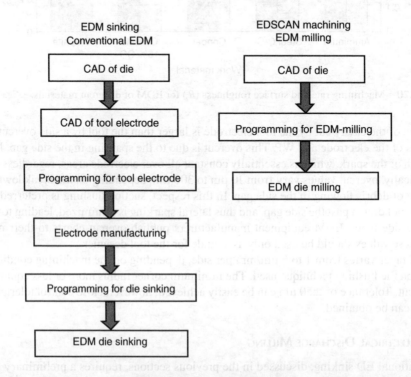

FIGURE 11.72 ED sinking and ED milling steps.

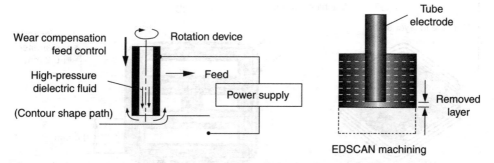

FIGURE 11.73 Theory of ED milling. (From Mitsubishi EDSCAN Technical Data, Hanover Exhibition, 1997.)

Advantages and Limitations of Electrical Discharge Milling
The advantages of EDM include the following:

- Design and manufacture of electrodes is totally omitted.
- Fine shapes can be readily produced.
- Electrode wear need not be considered.
- The surface roughness and waviness is favorable even for large areas.
- Estimating machining time is easy.
- NC data can be directly generated from the EDM die data.
- Sharp edges and corners can be readily produced.

The disadvantages and limitations of EDM are as follows:

- The removal rate may be less than that achieved by conventional EDM.
- If there is large side taper (10° or more), it is difficult to maintain side accuracy.

Fields of Applications of Electrical Discharge Milling
ED milling technology is particularly applicable for machining cavities with or without taper, including 3-D shapes. It is used notably for making molds of parts in electrical and electronic industries, household appliances, and automotive and aeronautical industries. Another technological breakthrough of ED milling is that the process has entered the domain of micromachining. It becomes possible to produce fine and intricate shapes with sharp corners using this process.

When machining with conventional ED sinking, the process is adjusted to ensure low wear conditions for the electrode. Simultaneously, the edges wear and become round. However, when using conditions that provide front wear to a certain degree on the pipe electrode, sharp edges of the electrode base can be maintained, while the electrode front wear is automatically compensated for. This compensation occurs at a micro level when the depth is periodically measured to ensure final dimensions, keeping the waviness within prescribed limits rendering flat surfaces (Figure 11.71).

The path of the 2-D shape is created by the NC system built into the machine beforehand to allow the target shape to be stored. Layered machining is then carried out by executing the NC path program several times, until the required depth is achieved, as shown in Figure 11.74. In the Z-direction, electrode wear caused by machining is automatically compensated for. A Cu/W pipe electrode ($\phi = 0.12/0.2$ mm) is used. Figure 11.75 illustrates a fine part produced by ED scanning using a fine graphite electrode of 10 μm diameter or less, which is dressed by a dressing unit mounted on the EDSCAN8E of Mitsubishi machine. ED milling technology was announced by the Japanese (Mitsubishi), and Charmilles at the micromachine exhibit in October 1996. Since then, this technology has gained attention in applications such as fabrication of microdies and others.

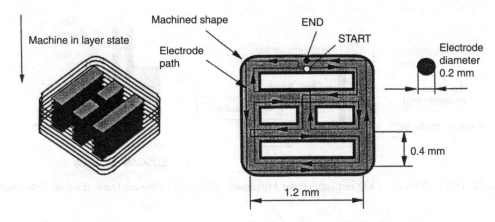

FIGURE 11.74 Micro EDM. (From Mitsubishi EDSCAN Technical Data, Hannover Exhibition, 1997.)

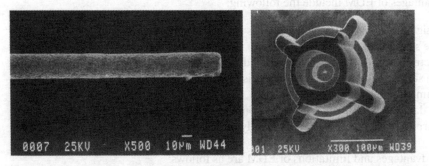

FIGURE 11.75 A fine part produced by ED milling using a graphite electrode of 10 μm diameter. (From Mitsubishi EDSCAN Technical Data, Hannover Exhibition, 1997.)

11.8.7 Electrodischarge Wire Cutting

Electrodischarge wire cutting (EDWC) is a variation of EDM that is similar to contour cutting with a band saw. A continuously moving wire travels along a prescribed path, cutting the WP, with discharge sparks acting like cutting teeth. The tensioned wire is used only once, traveling from a take-off spool to a take-up spool while being guided to provide an accurate narrow kerf. The horizontal movement of the worktable is numerically controlled to determine the path of the cut (Figure 11.76).

EDWC is used to cut plates as thick as 300 mm, and for making punches, tools, stripper plates, and extrusion dies in hard materials. It is also used to cut intricate shapes in the electronics industry. Figure 11.77 shows typical products that are cut by EDWC. EDWC machines are now available with a CNC facility in which a taper of the WP up to ±30° is fully integrated.

Generators for EDWC

EDWC machines are equipped only with pulse generators, where peak current and on-time are the major variables controlling spark energy. Modern machines of EDWC are equipped with sophisticated isopulse generators (Figure 11.69), where the previously mentioned variables, along with off-time and spark frequency, can be set independently while mentoring the gap. The wire has a limited current capacity, so that the current rating rarely exceeds 30 A. The potential difference between the wire electrode and the WP is usually set between 50 and 60 V. Because wire electrode wear is of

Nontraditional Machine Tools and Operations

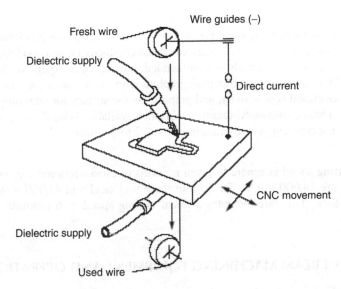

FIGURE 11.76 The EDWC process.

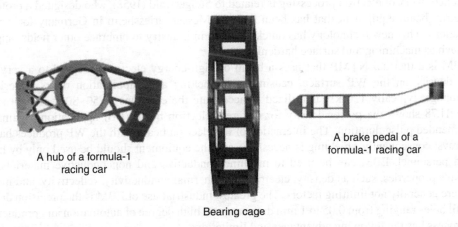

FIGURE 11.77 Typical products cut by EDWC. (From AGIE Charmilles Group, Charmilles, The Solutions, When to EDM, Geneva, 2004.)

little importance, negative wire (straight WP) polarity is always used. A larger wire diameter can handle higher energy of sparks and therefore cuts at higher machining rates.

Dielectric Flushing in Electrodischarge Wire Cutting

Effective flushing in EDWC is very important. Flushing nozzles should be as close as possible to the working gap (Figure 11.76). WP of large variations in thickness are especially troublesome as they prevent effective dielectric flushing. The usual result of poor flushing is wire breakage, which may be avoided by decreasing the on-time, a solution that is accompanied by a slower cutting rate.

Deionized water is used almost exclusively as a dielectric. The low viscosity is ideal for the difficult flushing conditions found in EDWC. Additives are sometimes used as antirust compounds or ethylene glycol-based compounds to make the dielectric slippery. Light oils are sometimes used as dielectrics for EDWC. Good filtration is also important, because contaminates affect the gap distance and consequently the machining accuracy. The contaminates cause arcing and wire breakage. Dielectric cooling and temperature control are essential for stable operation.

Wire Electrodes

Brass is the most commonly used wire, because it satisfies most of the requirements needed for EDWC. It possesses high tensile strength, high electrical conductivity, and good wire drawability to close tolerances. Layered wires are also recommended, but are more expensive; however, they cut faster than brass. One example is steel/copper/graphite wire, with a steel core for tensile strength, a copper layer for electrical conductivity, and graphite on the surface for attaining high machining speeds. Zinc-coated brass, with molybdnum-core, is also available. Wire diameters range from 5 to 300 µm. It travels at a constant velocity ranging from 0.2 to 9 m/min.

Cutting Speed

In EDWC, the cutting speed is generally given in terms of cross-sectional area cut per unit time. Typical examples are 18,000 mm^2/h for 50 mm thick tool steel and 45,000 mm^2/h for 150 mm thick aluminum block. This rate indicates a linear cutting speed of 6 mm/min and 5 mm/min, respectively.

11.9 ELECTRON BEAM MACHINING EQUIPMENT AND OPERATIONS

11.9.1 Process Characteristics and Applications

The pioneer work of electron processing is related to Steigerwald (1958), who designed a prototype of electron beam equipment that has been built by Messer-Griessheim in Germany for welding applications. This new technology has quickly spread in industry to embrace other fields' applications such as machining and surface hardening.

EBM is a thermal NTMP that uses a beam of high-energy electrons focused to a very high power density on the WP surface, causing rapid melting and vaporization of its material. A high voltage, typically 120 kV, is utilized to accelerate the electrons to 50–80% of light speed. Figure 11.78 shows the power density for different electron beam (EB) applications against the recommended pulse duration. The interaction of the electron beam with the WP produces hazardous x-rays; consequently, shielding is necessary, and the equipment should be used only by highly trained personnel. EBM can be used to machine conductive and nonconductive materials. The material's properties, such as density, electrical and thermal conductivity, reflectivity, and melting point, are generally not limiting factors. The greatest industrial use of EBM is the precision drilling of small holes ranging from 0.05 to 1 mm diameter, to a high degree of automation and productivity. The process has the following advantages and limitations.

Advantages and Limitations of Electron Beam Machining

The advantages of EBM include the following:

- Drilling of fine holes is possible at high rates (up to 4000 holes/s)
- Machining any material irrespective of its properties
- Micromachining economically at higher speeds than that of EDM and ECM
- Maintaining high accuracy and repeatability of ±0.1 mm for position and ±5% of the diameter of the drilled hole
- Drilling parameters can easily be changed during machining even from row to row of holes
- Producing best finish compared to other processes
- Providing a high degree of automation and productivity
- No difficulty encountered with acute angles

EBM has also the following limitations:

- High capital cost of equipment
- Time loss for evacuating the machining chamber

Nontraditional Machine Tools and Operations

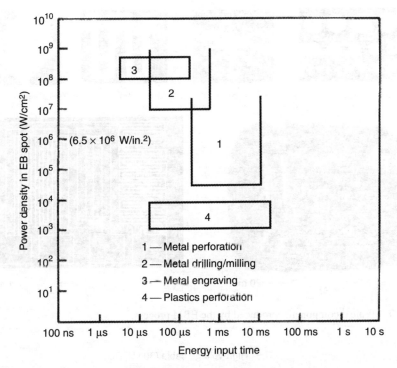

FIGURE 11.78 Power density for different EB applications against recommended pulse duration. (Adapted from *Advanced Materials and Processes*, ASM International, Materials Park, OH, 1990.)

- Presence of a thin recast layer and HAZ
- Necessity for auxiliary backing material
- Need for qualified personnel to deal with CNC programming and x-ray hazard

Applications of Electron Beam Machining

EBM is almost exclusively used in drilling and slitting operations. Drilling is preferred when many small holes are to be made, or when holes are difficult to be drilled because of the hole geometry or material hardness. Textile and chemical industries use EB drilling as a perforating process to produce a multitude of holes for filters and screens.

Figure 11.79 shows typical components machined by the EBM process:

a. Drilling of holes starting from 20 μm diameter in metallic and nonmetallic sheets
b. Machining of spinnerets of synthetic fibers, 60 μm width, 0.5 mm depth, and 1.2 mm arm length
c. Machining of a vaporization mask with 60 μm trace, in a tungsten strip of 50 μm thick
d. Multihole drilling (1600 holes) in a tungsten strip 3.5 mm × 3.5 mm and 0.1 mm thick
e. Multihole drilling of a cylindrical filter shell made of V2A stainless steel, 0.1 mm thick at a drilling rate of 3000 holes/s
f. Hybrid circuit engraved with 40 μm traces at a traverse speed of 5 m/s

11.9.2 Electron Beam Machining Equipment

A typical EBM equipment (also called an electron beam gun) is shown schematically in Figure 11.80. The electrons are released from a heated tungsten filament. A high potential voltage, typically 120 kV, is necessary to accelerate the electrons from the cathode (filament) toward the hollow anode, and the electrons continue their motion in vacuum toward the WP. A bias cup (Wehnelt electrode)

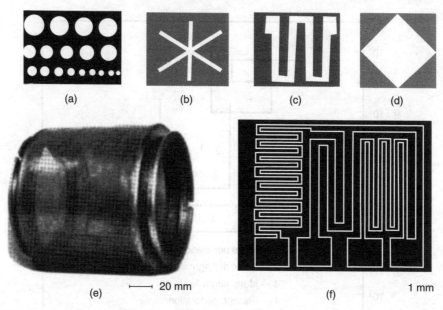

FIGURE 11.79 Typical components produced by the EBM process.

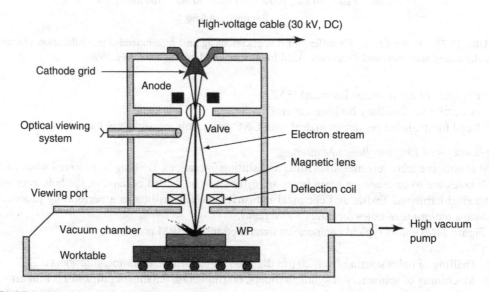

FIGURE 11.80 Electron beam gun.

located between the cathode and the anode acts as a grid that controls the beam current (1–80 mA) by controlling the number of electrons. The bias cup also acts as a switch of the beam current.

A magnetic lens is intended to focus the beam on a spot of diameter ranging from 12 to 25 μm, whereas deflection coils are used to deflect the beam within an angle of not more than 5° to extend the machining range. The beam loses its speed and circularity. Through beam deflection, standard shapes and configurations similar to those shown in Figure 11.81 can be produced without WP manipulation. For drilling purpose, the beam is pulsed once per hole.

Another alternative in WP manipulation occurs when the beam is deflected, so that it moves in sequence with the part, thus allowing drilling while the part is moving. This is called an on-the-fly drilling operation. The manipulator shown in Figure 11.82 includes a rotary axis with translation

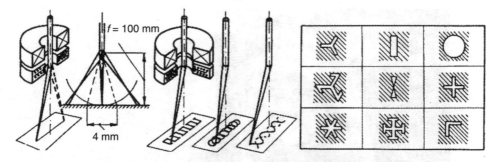

FIGURE 11.81 Standard configurations produced by EB without WP manipulation. (From Visser, A., Werkstoftabtrag mittels Electronenstrahl, Dissertation, T. H. Braunschweig, 1968.)

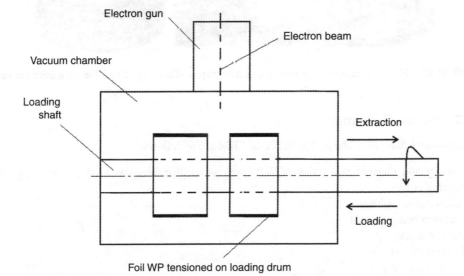

FIGURE 11.82 Rotary axis WP manipulator for the on-the-fly drilling EB operation (Adapted from *Metals Handbook*, Machining, Vol. 16, ASM International, Materials Park, OH, 1989.)

motion capability. The work sheet formed as a cylinder is clamped over the tensioning drum together with the backing material. A flying drilling operation is performed. As each beam pulse starts hole drilling, the beam is deflected, while the drum rotates at a constant speed. When the hole is completed, the pulse is turned off, and the beam instantaneously goes back to its original position to drill the next hole. This on-the-fly drilling process is continuously repeated. The translational axis is moved in synchronization with the beam to advance the WP under the beam. In this design, the chamber with its powered sliding door is usually a part of the manipulator. A more sophisticated multiaxes manipulator (Figure 11.83) can be used for drilling holes in complicated shapes. The WP is held in a chuck while the motion of the axes is computer numerically controlled. It is worth noting that the bearings must be carefully sealed to protect them against damage by metal vapor and drilling debris.

The electron beam is confined with the WP in the evacuation chamber to prevent:

- Oxidation of filament and other elements
- Collision of electrons with the massive molecules of O_2 and N_2 to eliminate loss of their kinetic energy
- Contamination of metal vapor and debris

FIGURE 11.83 Multiaxis manipulator for complicated shapes. (Courtesy of MG Industries/Steigerwald.)

11.9.3 Process Capabilities

The parameters that affect the performance of EBM are as follows:

- Density and thermal properties of WP (such as specific heat, thermal conductivity, and melting point)
- Accelerating voltage
- Electron beam current
- Pulse energy
- Pulse duration
- Pulse frequency
- Spot diameter
- Traverse speed of the WP

Equation 11.36 expresses the traverse speed v_f in terms of the machining parameters of the process (Kaczmarek, 1976).

$$v_f = \frac{C_d}{d_f}\left[0.1\frac{P_e}{\theta_m \cdot t_1 \cdot k_t}\right]^2 \text{ m/s} \tag{11.36}$$

where
  t_1 = plate thickness (m)
  P_e = power of electron beam (N m/s) = $i_b \times V_b$ = beam current × acceleration voltage
  d_f = beam focusing diameter (m)
  k_t = thermal conductivity (N/s °C)
  C_d = coefficient of thermal diffusivity = $k_t/\rho \cdot c1$ (m²/s)
  ρ = density of WP material (kg/m³)
  $c1$ = specific heat of WP material (Nm/kg °C)
  θ_m = melting point of WP material (°C)

Table 11.13 illustrates typical parameters of the EBM process.

EBM is characterized by its minor volumetric removal rate that reaches a maximum value of 0.1 cm³/min. The volumetric removal rate increases by increasing pulse energy (power intensity),

TABLE 11.13
Typical Machining Parameters of EBM

Parameter	Value
Acceleration voltage (V_b)	50–150 kV
Beam current (i_b)	0.1–40 mA
Power (P_e)	1–150 kW
Pulse duration (t_i)	4–60,000 μs
Pulse frequency (f_r)	0.1–16,000 Hz
Vacuum	10^{-3} to 10^{-5} torr (1 torr = 1 mmHg)
Minimum spot (d_f)	12–25 μm
Beam deflection	6 mm
Beam intensity	10^6–10^9 W/cm²
Tolerance	Depends on plate thickness or hole diameter, otherwise, ±10% of lateral dimension.
Roughness (R_a)	1 μm

Source. Reproduced from Machining Data Handbook, 1980.

provided that the same number of pulses is used. The machinability depends on the melting point of the material to be machined. Therefore, tin and cadmium have the highest machinability, whereas W and Mo are of low machinability.

In EB hole drilling, the achieved tolerance depends on the hole diameter and WP thickness, whereas the surface roughness depends mainly on the pulse energy. Depending on the working conditions, the tolerances of the drilled holes (and slits) may attain values between ±5 and ±125 μm. The surface roughness R_a ranges from 0.2 to 6.3 μm.

ILLUSTRATIVE EXAMPLE 5

Calculate the traverse speed v_f in mm/s for cutting a tungsten sheet of 2.5 mm thick, if EB equipment of 10 kW is used. The equipment is capable of focusing the beam to a diameter of 0.5 mm. The following WP data are given: $\theta_m = 3400°C$, $C_d = 8.1 \times 10^{-5}$ m²/s, $k_t = 214$ N/s °C.

Solution

$$v_f = \frac{C_d}{d_f} \left[0.1 \frac{P}{\theta_m \cdot t_1 \cdot k_t} \right]^2$$

$$= \frac{8.1 \times 10^{-5}}{0.0005} \left[0.1 \frac{10 \times 1000}{3400 \times 0.0025 \times 214} \right]^2$$

$$= 0.049 \text{ m/s} = 49 \text{ mm/s}$$

11.10 LASER BEAM MACHINING EQUIPMENT AND OPERATIONS

11.10.1 PROCESS CHARACTERISTICS

After the age of electricity, the age of photons began, in the early 1960s. The photons of zero charge have replaced the negatively charged electrons as a favorite tool in modern industry. The photon beam (laser) is currently used in every thing from eye surgery to communication and metal processing. The earliest industrial applications in welding occurred in the 1970s, and the 1980s saw the beginning of LBM and micromachining.

Laser is an acronym for "light amplification by stimulated emission of radiation." It is a highly collimated monochromatic and coherent light beam in the visible or invisible range. LBM is a promising NTMP for machining any material, irrespective of its physical and mechanical properties. It is used to cut and machine both hard and soft materials, such as steels, cast alloys, refractory materials, ceramics, tungsten, titanium, nickel, Borazon (CBN), diamond, plastics, cloth, alumina, leather, wood, paper, rubber, and even glass when its surface is coated with radiation-absorbing material such as carbon. However, machining of Al, Cu, Ag, and Au is being especially problematic, as these metals are of high conductivity and have the tendency to reflect the applied light. But recently, yttrium aluminum garnet (YAG) with enhanced laser focusing has been used to cut such metals after treating their surfaces by oxidizing them or increasing their roughness. YAG is more superior than a CO_2 laser, because it emits a shorter wavelength.

Laser is a versatile tool, useful in many areas from precision watch-making to heavy metalworking industry. The key of laser's effectiveness lies in its ability to deliver, in some cases, a tremendous quantity of highly concentrated power—as high as 10^{12} W/cm^2. Tuning the laser beam makes it possible to deliver just the right concentration of power for the right amount of time to perform a specific piece of work.

It is possible to make automotive engine blocks out of aluminum with a thin hard layer inside the cylinder by lasers, thus considerably reducing the engine weight. As long as the beam is not obstructed, it can be used to machine inaccessible areas.

One of the main advantages of laser beam is that it does not take up time for the evacuation of the machining area, as does EB; a laser can operate in transparent environments like air, gas, vacuum, and in some cases even liquids. However, LBM is quite inefficient and cannot be considered as a mass metal removal process. A significant limitation of laser drilling is that the process does not produce round and straight holes. This can, however, be overcome by rotating the WP as the hole is being drilled. A taper of about 1/20 is encountered. HAZ is produced in LBM, and heat-treated surfaces are also affected.

High capital and operating cost and low machining efficiency ($\eta = 1\%$) prevent LBM from being competitive with other NTM techniques. Protective measures are absolutely necessary when working around laser equipment. Extreme caution should be exercised with lasers; even a low power of 1 W can cause damage to the retina of the eye. In all cases, safety goggles should be used, and unauthorized personnel should not be allowed to approach the laser working zone.

Process Advantages and Limitations

Advantages
- A wide variety of metallic and nonmetallic materials can be machined.
- No mechanical contact, and therefore no deformations of the WP and no tool wear are encountered.
- Work fixation is easily performed.
- Laser beam can travel without diffraction and can be branched to different workstations working at the same time.
- It can reach inaccessible areas on the WP.
- It produces microholes in difficult-to-machine and refractory materials.
- It is easy to control the beam characteristics to adapt to a specific machining duty.
- There is no need for time-consuming evacuation as in EBM.
- Holes can be drilled at an acute angle to the surface (10°).
- The process can be automated easily.
- The operating cost is low.
- As compared to other thermal NTMP, LBM is characterized by HAZ of smaller thickness.

Limitations
- High equipment cost.
- Unsafe operation.

- Blind holes of precise depth are difficult to achieve.
- Laser produces tapered holes.
- Holes produced are of limited dimensional and form accuracy, and of bad surface quality.
- HAZ cannot be avoided.
- The process is of low efficiency and is not considered to be a mass removal process.
- Material thickness is restricted to 50 mm in the case of drilling.
- Adherent materials at hole exists need to be removed.

11.10.2 Types of Lasers

11.10.2.1 Pyrolithic and Photolithic Lasers

A laser beam is of high power density, especially when focused to a small spot on the surface of the WP. Depending on its wavelength, the beam interacts with the WP material either pyrolithically (thermally) or photolithically.

In the pyrolithic laser, the material removal occurs by melting and vaporization of the material spontaneously. This laser is used mainly in applications such as cutting, drilling, welding, and surface hardening. The removal or processing rate depends on the material being machined, its thickness, and its physical and optical properties, such as specific heat, latent heat of melting and vaporization, and surface reflectivity. The machinability of materials increases by decreasing the previously mentioned properties. It depends also on the beam characteristics, especially its power density.

In the domain of machining, the pyrolithic laser is used to machine a wide spectrum of metallic and nonmetallic engineering materials, taking into account that the thermal diffusivity does not allow heat to be transmitted beyond the machined surface.

In the photolithic laser, the material is not removed thermally, but it is affected by the dissociation and breaking of the chemical bond between the material molecules, when its bond energy is below the photon energy of the beam. The photon energy of the beam is inversely proportional to its wavelength.

The fluorine excimer laser is a beam of ultra-short wavelength ($\lambda = 157$ nm); consequently, it possesses a high photon energy of 7.43 eV (1 eV = 1.6×10^{-19} J), whereas the CO_2 laser is an infrared (IR) laser beam of long wavelength ($\lambda = 10,600$ nm) and with low photon power of 0.12 eV. It follows that an excimer laser is capable of machining plastic and Teflon photolithically, as its photon energy is greater than the chemical bond energy, which ranges from 1.8 to 7 eV for most of the plastics. A CO_2 laser is not capable of machining plastics photolithically, but pyrolithically.

In photolithic material removal, three phases are necessary:

1. The ultra-short wave photons are absorbed into the surface to a depth of about 200 nm.
2. The chemical bond between molecules is broken.
3. The reaction products escape as gas and small particle ashes.

11.10.2.2 Industrial Lasers

Industrial lasers comprise in most cases the solid-state lasers such as neodymium: yttrium aluminum garnet (Nd:YAG), neodymium:glass, Nd:glass, and the ruby and gas lasers (CO_2, excimer, and He/Ne). Basically, four types prevail in metal working processes; namely, the CO_2, Nd:YAG, Nd:glass, and excimer lasers. Out of these, the CO_2 and YAG are considered the most dependable workhorses.

In most cases, the same laser can be applied in cutting, welding, machining, and surface treatments. It is necessary to vary only the laser beam parameters, such as power density, focus diameter, and pulse duration to suit the specified machining process. The four types of industrial lasers are described in brief:

> *CO_2 lasers.* In these lasers, the active lasing material is the CO_2 gas. However, a mixture of gases is used ($CO_2:N_2:He = 0.8:1:7$). Helium acts a coolant of the gas cavity. CO_2 lasers are

characterized by their long wavelength of 10,600 nm; thus, the material removal depends on only the thermal interaction with the WP. CO_2 lasers yield the highest depth-to-diameter ratio in most metals using gas-jet assistance. However, these lasers are bulky but economical. There are two types of CO_2 lasers:

1. *Axial-flow CO_2 laser.* This laser can operate in pulsed (P) and continuous wave (CW) modes. In this type, the typical average output for (CW) ranges from 250 to 5000 W, while in the (P) operation, the average power output is reduced to 100–2000 W, and the pulse frequency ranges from 1 to 10,000 p/s.
2. *Transverse-flow CO_2 laser.* This laser operates only in the (CW) mode. This type is used when high power outputs between 2500 and 15,000 W are required.

Nd:YAG lasers. This laser is a single crystal of YAG doped by 1% neodymium as an active lasing material. This laser is compact and economical, its wavelength is 1060 nm, and it can operate in either P or CW mode. It is characterized by relatively high efficiency and high pulsating frequency, and operates with simple cooling system. Its pulsating frequency ranges from 1 to 10,000 p/s and pulse energy 5–8 J/p. It has an average power output close to 1 kW.

Nd:glass laser. This laser is glass rod doped by 2–6% neodymium as the acting lasing material. This laser is often uneconomical. It has the same wavelength as the Nd:YAG. It operates only in the P mode. Owing to the low thermal conductivity of glass, the pulse rate should be limited. Consequently, it is only used in drilling and welding necessitating higher energy output and low pulse frequency (1–2 p/s).

Excimer lasers. Excimer lasers are a family of pulsed lasers operating in the UV region of the spectrum. Excimer is an abbreviation of "excited dimmer." The beam is generated due to fast electric discharges in a mixture of high-pressure dual gas, composed of one from the halogen gas group (F, H, Cl) and another from the rare gas group (Kr, Ar, Xe). The wavelength of the excimer laser attains a value from 157 to 351 nm, depending on the dual gas combination. Excimer lasers have low power output, so they removes the material photolithically, and have a remarkably efficient application in the machining of plastics and micromachining as previously mentioned. The main characteristics of important industrial lasers are given in Table 11.14.

11.10.2.3 Laser Beam Machining Operations

As previously mentioned, industrial lasers operate either in CW mode or in P mode. Generally, CW lasers are used for processes like welding, soldering, and surface hardening that require an uninterrupted supply of energy for melting and phase transformation. Controlled pulse energy is desirable for cutting, drilling, marking, and so on, striving for less heat distortion and the minimum possible HAZ. The applications of beam in machining are described in the following sections.

1. Drilling

Laser beam drilling is done either by percussion or trepanning.

a. Percussion Drilling

Percussion drilling is used for relatively small holes (<1.3 mm diameter), through metal sections up to 25 mm thick, and is most often performed using pulsed Nd:YAG lasers because of their higher pulse energy.

Operating parameters:

- Pulse power. 100–250 W (average).
- Pulse energy. Up to 40 J/p depending on hole diameter and material thickness; higher energy provides faster drilling rate but lower hole quality.

TABLE 11.14
Main Characteristics of Important Industrial Lasers

Laser Type, Mode		λ (nm)	W_{ave} (W)	f_r (p/s)	d_f (μm)	Comments
Nd:YAG	(P)	1,060	100–500	1–10,000	13	Compact, economical
	(CW)	1,060	10–800	—	13	
Nd:glass	(P)	1,060	1–2	0.2	25	Often uneconomical
CO_2 Axial	(P)	10,600	100–2,000	400	75	High efficiency, bulky but
	(CW)	10,600	250–5,000	—	75	economical
Transverse (CW)		10,600	2,500–15,000	—	75	
Excimer	(P)	157–531	~100	10–500	N/A	Micromachining, plastic, ceramic

Note: λ = wavelength, W_{ave} = average power, f_r – pulses/s, d_f = focus diameter, N/A = not available.

- Pulse duration. 0.5–2 ms.
- Pulse frequency. 5–20 Hz with Nd:YAG; 100 Hz with pulsed CO_2.
- Focal length. 100–250 mm, larger focal length for large hole diameters and depths.
- Focal position. Optimized above, below, or on surface, depending on the desired results. Most often, focus at a depth 5–15% of metal thickness, determined empirically for best quality as judged through roundness, taper, microcracking, and recast layer.

b. *Trepanning*

Trepanning is used for large holes (>1 mm diameter). It is commonly performed with both CO_2 and Nd:YAG lasers. Trepanning is essentially a machining process that can be performed by operating the laser in the (CW) or (P) mode. Trepanning requires a percussion-drilled pilot hole using operating parameters for percussion drilling. As the metal thickness increases, the trepanning speed decreases.

Operating parameters are as follows:

- Pulse power. Established at levels that maximize the trepanning rates; that is, maximizing the part or beam translation without sacrificing quality.
- Pulse duration. Less than 2 ms with both pulsed CO_2 and Nd:YAG lasers.
- Pulse frequency. Lower pulse frequencies are used as metal thickness increases.
- Focal length. Similar to those used in percussion drilling. If a CO_2 laser is used, the recommended focal length is 125 mm or less.

2. *Cutting*

Similar to trepanning, cutting can be accomplished by laser operating in either CW or P mode.

CO_2 laser: CW mode is used for thicker metal sections, while P mode is used for thinner metal sections.

Nd:YAG (P): lasers are used to cut thick sections of superalloys.

More often, cutting is done by a CO_2 laser, because of its faster machining rate.

Operating parameters are as follows:

- *Power*
  250–5,000 W for (CW) CO_2
  100–2,000 W for (P) CO_2

<800 W for (CW) Nd:YAG
<500 W for (P) Nd:YAG
Lower power is sufficient for pulsed lasers because of their higher peak instantaneous powers.
- *Pulse duration*
  <0.75 ms is used for intricate cutting of thin metals. Longer duration, up to 2 ms, provides greater pulse energy for thicker metals.
  Similar durations for both CO_2 and Nd:YAG.
- *Pulse frequency:*
  200–500 Hz for (P) CO_2.
  30–100 Hz for (P) Nd:YAG.
  Higher frequency is used to cut thinner metals, and lower frequency to cut thicker metal sections.
- *Pulse energy:*
  2 J/p at longer pulse durations and lower pulse frequency for (P) CO_2.
  Up to 80 J/p (pulse frequency limited by maximum power rating for (P) Nd:YAG for thicker material.
- *Focal length:*
  CO_2 laser
   l_f = 65 mm for materials up to 6 mm thick
   = 125 mm for materials from 6 to 16 mm thick
   = 190 mm for materials thicker than 13 mm
  Nd:YAG lasers
   l_f = 100 mm for materials thinner than 3 mm
   = 150–250 mm for materials up to 25 mm thick
   >250 mm for materials thicker than 25 mm
  A wider kerf is obtained when using longer focal lengths. Al requires large kerf for good results.

3. *Marking*

The laser system is used to imprint characters (letters and symbols) to a depth of 5–250 μm. Laser marking arrangement is made up of a low-power pulsating laser system, lasting only for nanoseconds, and a CNC beam scanning system. The necessary information to generate different characters is stored in the computer. Accordingly, 30 characters can be imprinted per second. Table 11.15 provides a general selection guide of industrial lasers for different applications.

4. *Gas-Assisted Laser Cutting*

In an important development, a coaxial nozzle can now be supplemented with continuous jet of air, O_2, or one of the inert gases (N_2, Ar, or He). The selection of the gas depends upon the type of work material, its thickness, and type of cut (see Table 11.16).

In a gas-assisted laser, the gases have the following functions:

1. Providing oxidizing atmosphere (in case of O_2 and air), thus reducing reflectivity and improving absorption of beam energy.
2. Promoting exothermic reactions (in case of O_2), thus improving process efficiency by providing 80% of the energy needed for cutting.
3. In all cases, the pressurized gas expels vapors and molten metal from the machining zone or the bottom of the hole in drilling.

O_2 is the most commonly used assisting gas for steels and most metals. When an oxide-free surface of high quality cut is desired, an inert gas is used. Additionally, the oxide-free edges can improve the weldability. Inert gas is used also to prevent plastics and other organic materials from charring.

TABLE 11.15
Laser Beam Selection Guide for Different Applications, Modified

Application	Type of Laser Beam
Cutting, trepanning	
Metals	(CW) CO_2, (P) CO_2, (P) Nd:YAG
Plastics	(CW) CO_2
Ceramics	(P) CO_2
Drilling, percussion drilling	
Metals	(P) Nd:YAG, (P) CO_2
Plastics	Excimer
Marking, micromachining	
Metals	(P) Nd:YAG, (P) CO_2
Plastics	Excimer
Ceramics	Excimer
Welding, soldering	(CW) CO_2, (P) CO_2, (P) Nd:YAG
Surface hardening	(CW) CO_2

Source: Reproduced from Kalpakjian, S. and Schmidt, S. R. in *Manufacturing Processes for Engineering Materials*, Prentice Hall, New York, 2003.

TABLE 11.16
Gas Selection of Gas-Assisted Laser Cutting and Marking

Material	Best	Optional
Gas-assisted laser cuttings		
Steels and stainless steel	O_2	N_2
Ti	He	He/Ar, CO_2
Non ferrous alloys	O_2	Air
Nickel alloys	O_2	Air
Nonmetals	O_2	Air
Composites	O_2	N_2, Ar
Plastics	O_2	Air, Ar
Wood	N_2	Air
Gas-assisted laser marking (scribing)		
Aluminum	O_2	(CW) CO_2, 1000 W, no burrs
Ceramic alumina	N_2	(P) CO_2, 50 W, 0.5 ms, no HAZ

Source: Reproduced from *Machining Data Handbook*, Machinability Data Center, Cincinnati, OH, 1980.

Operating parameters include the following:

- *Laser used.* (CW) Nd:YAG, (CW) CO_2 for cutting, (P) Nd:YAG for drilling
- *Power.* 0.25–16 kW, depending on the type of material and its thickness
- *Feed rate.* 0.25–7.5 m/min, depending on the power and type of material and its thickness
- *Focus diameter.* 50 μm (minimum), focal point on the surface or slightly lower
- *Gas pressure.* 1–3 ATM for O_2; 2–6 ATM for inert gases
- *SOD.* 1–15 mm for CO_2 laser, and 5 mm for Nd:YAG laser

11.10.3 LBM Equipment

Figure 11.84 shows schematically three important elements of LBM equipment; namely, a lasing material (solid state or gas), a pumping energy source required to excite the atoms of the lasing

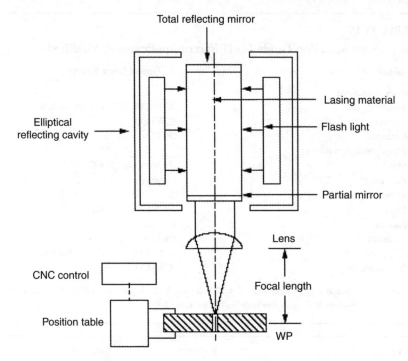

FIGURE 11.84 Schematic of LBM.

material to a higher energy level, and a mirror system. One of these mirrors is fully reflective, while the other one is partially transparent to provide the laser output (output mirror). It allows the radiant beam to either pass through, or bounce back and forth repeatedly through the lasing material. To make the laser beam useful for machining, its power density should be increased by focusing, thus attaining power density values between 10^5 and 10^7 W/mm². The laser beam is usually delivered to the WP in the transverse excitation mode (TEM). The common mode of optical configuration is TEM_{00} mode, which is a Gaussian output beam with lowest beam divergence, and consequently highest power density. This mode provides the most uniform beam profile. The other widely used modes are TEM_{10}, in which a broader and less intense spot is required in cutting and welding, and TEM_{01}, which is useful for machining large holes and surface hardening.

To improve the process performance, most of the LBM equipment is provided with a Q-switching facility (Q means quality factor) to amplify the power. It provides the beam, despite energy loss due to magnification with enormous power (hundreds or thousands of its normal pulsing power), acting on an extra-short pulse duration on the order of a nanosecond). Therefore, Q-switching enhances the beam capabilities regarding the removal rate and the quality of cut. The Q-switched beam is capable of evaporating the material in no time. In case of Nd:YAG and CO_2 lasers operating in (CW) mode, Q-switching also converts the continuous wave into a train of pulsating power. Most of the new lasers are computer-controlled to take the advantage of their high-speed processing. During machining, motion can be provided to the WP, the beam, or both.

In recent LBM developments, significant progress has been made in integrating robot technology and CNC facility with lasers in a setup called a flexible machine station (Figure 11.85). A single laser beam travels to the processing locations without diffraction or loss of power, where it is divided to perform many functions simultaneously.

Sometimes, LBM equipment is provided with a laser beam torch (LBT), which may be used in cutting, welding, and surface hardening operations (Figure 11.86).

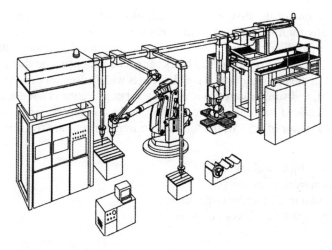

FIGURE 11.85 Flexible laser beam machine station. (From König, W., Fertigungsverfahren-Band 3 Abtragen, VDI Verlag, Düsseldorf, 1990.)

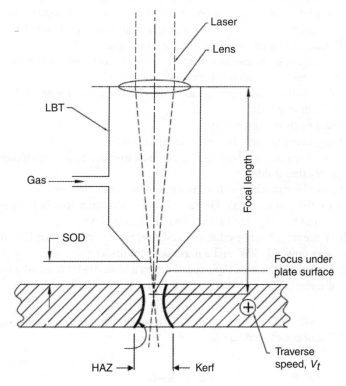

FIGURE 11.86 Laser beam torch.

11.10.4 Applications and Capabilities

Laser and gas-assisted cutting techniques are best suited for applications demanding high accuracy and for machining jobs in which the HAZ is to be as narrow as possible to avoid distortions and obtaining cuts of high-quality edge finish.

Laser drilling was one of the first practical applications of laser technology. The demand for laser drilling, especially for microholes of 75 μm diameter, is increasingly emphasized.

With increasing hole diameter and depth, ejected liquid metal gets deposited on the walls and the bottom such that perfectly cylindrical holes cannot be obtained, and that is why laser is not used for producing deep holes. In industry, laser drilling is widely used for drilling of watch jewels, diamond drawing dies, and similar jobs where a high level of precision is not demanded.

Traverse speeds attained using lasers are impressive. In this regard, when using a CO_2 laser capable of providing a power of 0.5 kW, the following traverse speeds are realized:

4.5 m/min for 1 mm thick steel sheet
0.5 m/min for 6 mm thick steel sheet
2.0 m/min for 1.5 mm thick stainless steel sheet
0.9 m/min for 3 mm thick stainless steel sheet

Using 1 kW power, the machining speeds are doubled.
Some typical applications of LBM are listed here:

1. Machining of microholes in filter screens, carburetors, and fuel injection nozzles.
2. Machining of miniature holes of diameter 0.1–0.5 mm at rates of 1–10 holes/s.
3. Microdrilling of diamond wire drawing die (50 μm), using a Q-switched microsecond pulse Nd:YAG laser, and a nanosecond-pulse excimer laser.
4. Laser drilling of rubber cups. Because there is no forces involved, there is no deformation, and the lack of stiffness is of no significance (Figure 11.87).
5. Scribing to widths 5–10 μm at speeds up to 12 m/min. Ultra-short wave excimer lasers now produce cuts 0.5 μm in width.
6. Trimming of flashes from plastic parts.
7. Marking and engraving in metallic and nonmetallic WPs.
8. In the aircraft turbine industry, laser drilling is used to make holes for air bleeds, air cooling, or passage of other fluids.
9. Lasers may be used for machining hard materials (white CI, Inconel) in combination with a TM process (milling or turning). The laser is directed onto a spot in front of the turning tool in laser-assisted turning (LAT), as shown in Figure 11.88.
10. CO_2 lasers have recently been used to cut and peel masks, needed for CH-milling of airplane wings (laser power 75 kW, and a mask 0.4 mm thick).
11. Restoring of dynamic balance of high-speed rotors and shafts by removing infinitesimally small pieces of material during rotation on the dynamic balancing machine.

Other applications include making holes in hypodermic needles, ceramic substrates for electronic circuits, holes in WC tool plates, and so on.

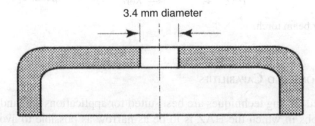

FIGURE 11.87 Laser drilling of rubber cup.

FIGURE 11.88 Laser-assisted turning

11.11 PLASMA ARC CUTTING SYSTEMS AND OPERATIONS

11.11.1 Process Characteristics

When a gas is heated to a high temperature on the order of 2000°C, its molecules separate out as atoms. If the gas temperature is raised above 3000°C, the electrons of some of the atoms dissociate and the gas becomes ionized; that is, consisting of free electrons, positive charged ions, and neutral atoms. This state of ionized gas is known as plasma gas, and is characterized by high electrical conductivity.

The plasma arc is initiated in a confined gas-filled chamber by a HF spark. The dc from a high voltage source sustains the arc and the plasma stream that exits from the nozzle at sonic speed.

The source of heat generation in the plasma is due to the recombination of electrons and ions into atoms and recombination of atoms into molecules. The liberated bonding energy is responsible for increased kinetic energy of atoms and molecules formed by the recombination. The temperature associated with the recombination can be of the order 20,000–30,000°C. Such a temperature melts out and even vaporizes any work material subjected to machining or cutting. Plasma arc cutting (PAC) is a thermal NTMP that was adopted in the early 1950s as an alternative method for oxy-fuel cutting of stainless steel, aluminum, and other nonferrous metals. Recently, cutting of conductive and nonconductive materials by PAC has become much more attractive. The main attraction is that PAC is the only method that cuts faster in stainless steel than it does in mild steel.

Advantages of PAC
- The process provides smooth cuts, free of contaminants.
- It can cut exotic metals at high rates.
- The process has the least-specific cutting energy among all NTMPs.

Disadvantages of PAC
- Reduced accuracy and surface quality are expected.
- The process requires high power.
- It produces toxic fumes.
- Owing to high thermal effects, the WP is highly distorted and HAZ of large depth reduces the fatigue resistance.
- The plasma arc produces IR and UV radiations that cause eye injuries (cataracts) and loss of sleep. UV radiation leads to skin cancer. Therefore, gloves, goggles, and earplugs should be used.

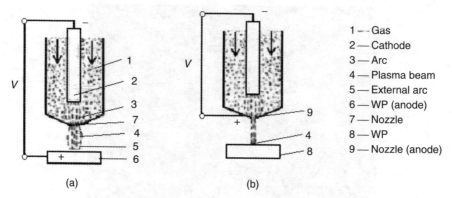

FIGURE 11.89 (a) Transferred and (b) nontransferred plasma torches.

11.11.2 Plasma Arc Cutting Systems

PAC systems operate either in transferred arc mode or nontransferred jet mode. In transferred arc mode (Figure 11.89a), the arc is struck from the rear negative electrode of the plasma torch to the conductive WP (+ve electrode), causing temperatures as high as 33,000°C. Owing to the greater efficiency of the transferred systems, they are often used in the cutting of any electrically conductive material, including those of high electrical and thermal conductivity that are resistant to oxy-fuel cutting, as aluminum.

In nontransferred jet mode (Figure 11.89b), the arc is struck with the torch itself. The plasma is emitted as a jet through the nozzle orifice, causing a temperature rise of about 16,000°C. Because the torch itself is switched as the anode, a large part of the anode heat is extracted by cooling water, and therefore is not effectively used in the material removal processes. Nonconductive materials that are difficult to cut by other methods are often successfully cut by plasma nontransferred systems.

A constructional assembly of a typical transferred plasma torch is illustrated in Figure 11.90. The nozzle diameter depends on the arc current and the flow rate of the working gas. It ranges from 1.2 to 6 mm. Fine nozzles of 50 µm diameter are especially used to cut metals with a kerf width of 0.1 mm and operate at low power of 1 kW. Multiple torch cuts are possible on tracer-controlled cutting tables for plates up to 150 mm thick stainless steel.

The commonly used working gases are He, Ar, H_2, N_2, or a mixture of them. The gas flow rate ranges from 0.5 to 6 m^3/h, depending on the arc power and the plate thickness.

The nonconsumable electrodes are made of 2% thoriated tungsten to resist wear. Shielded plasma torches may be gas- or water-shielded:

> *Gas-shielded plasma.* During cutting of aluminum, stainless steel, and mild steels, shielding gases are used to obtain cuts of acceptable quality. An outer gas shield (N_2 or Ar/H_2) is added around the main stream of plasma. A CO_2 shield is favorable for ferrous metals. For mild steel, air or O_2 may also be used as shielding gases.
>
> *Water-shielded plasma.* In this case, N_2 is used as the main working gas, while the shield is a water curtain (Figure 11.91). It is reported that the cooling effect of water reduces the kerf width and improves the quality of cut; however, the cutting rate is not improved.

11.11.3 Applications and Capabilities of Plasma Arc Cutting

The PAC has many applications. Some examples are listed below:

1. PAC has a special attraction in the case of profile cutting of metals such as stainless steel and aluminum that are difficult to tackle by the oxy-fuel technique.

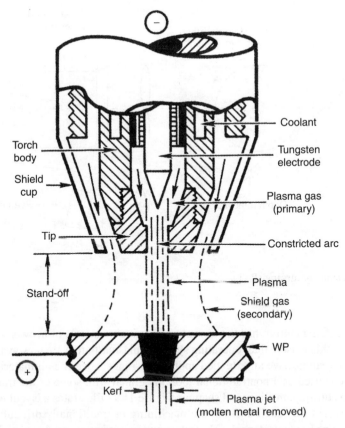

FIGURE 11.90 Constructional assembly of transferred plasma torch. (From *Machining Data Handbook*, Vol. 2, 3rd Edition, Machinability Data Center, Cincinnati, Ohio, USA, 1980.)

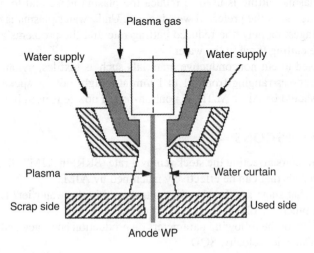

FIGURE 11.91 Water-shielded plasma.

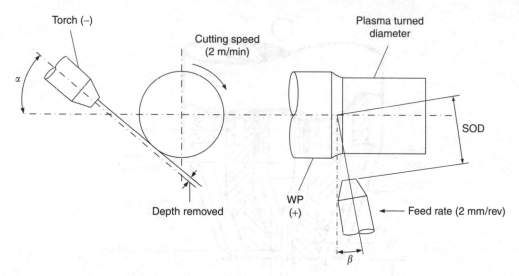

FIGURE 11.92 Plasma arc turning (PAT).

2. Oxyacetylene flame cutting has the advantage that it cuts metals of heavier sections than PAC does. For this reason, dual operating systems (plasma/flame) are now available on the market. Dual systems have an extended application range, covering all material.
3. A plasma arc is used as a nontraditional tool, integrated with some of the traditional processes such as turning, milling, and shaping. Figure 11.92 illustrates a layout of the plasma arc turning (PAT) for rough and smooth machining of traditionally difficult-to-machine materials such as Inconel, Rene 41, Hastalloy, and precipitation-hardened stainless steels. According to the machined material and the degree of finish required, cutting speeds are selected from a range of 10–100 m/min, and a nozzle feed between 1 and 5 mm/rev, depending on the required finish and the depth of cut. The nozzle is fixed at a suitable SOD.
4. Underwater plasma cutting is used to reduce the plasma noise and to get rid of plasma fumes and glare. N_2 is the preferred working gas. Underwater plasma is characterized by two disadvantages; namely, the reduced cutting rate and the problems associated due to immersing the cutting torch in the water.
5. PAC is also used to cut nonconductive materials such as textiles, nylon, and polypropylene with thicknesses ranging from 0.1 to 1 mm at a high traverse speed of 1000 m/min. Working gas should be Ar or Ar/H_2. A nontransferred nozzle is used in this case.

11.12 REVIEW QUESTIONS

1. What are parameters affecting the stock removal rate (SRR) in AJM? Give three examples of typical materials that can be effectively machined by AJM.
2. In AWJM, at what stage are the abrasives introduced in the water jet? Draw a schematic outline of equipment involved.
3. Discuss the effect of the following parameters on production accuracy and the removal rate in AJM: grain size, jet velocity, SOD.
4. Describe at least three typical applications of AJM.
5. Using a block diagram or a line sketch, show the main components of a WJM plant.
6. Explain how material is removed in USM.

Nontraditional Machine Tools and Operations

7. Show diagrammatically the main elements of an USM machine.
8. Explain the advantages and disadvantages of USM.
9. What are the main applications of USM?
10. Explain the effect of USM parameters on the removal rate.
11. Define the magnetostriction effect as applied in USM. What are the aims of using a premagnetizing dc current in magnetostrictive transducers?
12. Give three examples of materials that can be machined economically by USM.
13. Mention three types of abrasive materials that are frequently used in USM and arrange them according to their cutting ability.
14. Design and construct the exponential horns for an ultrasonic drilling machine operating at a resonant frequency of 20 kHz for the following cases:
    a. Horn material: steel 100 Cr6, $c = 5.05 \times 10^5$ cm/s. Resonant length $l = 14$ cm. Initial diameter $D_0 = 40$ mm.
    b. Same as (a), but the horn has a rectangular cross-section of constant width $b = 50$ mm and $A_0 = 40 \times 50$ mm^2.
    c. Horn material: brass, $c = 3.3 \times 10^5$ cm/s. Resonant length $l = 10$ cm, initial diameter $D_0 = 40$ mm.
    d. Same as (c), but the horn is hollow and provided with a conical hole starting from the nodal point to secure an outside horn diameter. $D_\ell = 50$ mm to accommodate a large trepanning tool.
15. An exponential horn is made of Monel of acoustic speed $c = 4.2 \times 10^5$ cm/s; the horn has the following features: maximum diameter $D_0 = 36$ mm, resonant length $\ell = 12$ cm, operating frequency = 20 kHz. The horn should be provided with a central hole for slurry suction of 2.5 mm diameter from nodal point to $x = 12$ cm. Calculate magnification factor, location of the node, horn diameter at nodal point, horn outside diameter at $x = 2$, 7, 12 cm.
16. What are the advantages and disadvantages of ECM? Show diagrammatically the main elements of an ECM machine.
17. What are the advantages and disadvantages of PCM?
18. What are the advantages and limitations of CHM?
19. Explain what is meant by EC deburring.
20. What are the factors on which the selection of a resist (maskant) for use in CHM depends? Distinguish between cut and peel resists and photographic resists.
21. What measures should be considered to achieve maximum dimensional control in ECM.
22. Describe the PCM process, and list its fields of application.
23. What are the functions of an electrolyte? What factors need to be considered while selecting it? Discuss the advantages and limitations of some common electrolytes.
24. Write down the Faraday's law of electrolysis.
25. What is the "self-adjusting feature" in ECM?
26. Use a neat sketch to explain briefly the principles of ECG process. What are the main advantages of ECG over conventional grinding?
27. Explain briefly the principle of ECUSM (ECU).
28. Explain using a neat sketch the principle of material removal in EDM. Draw a typical relaxation circuit used for EDM power supply. Explain the main disadvantages of the relaxation circuits used in EDM.
29. Show diagrammatically the main elements of an EDM machine.
30. State the main functions of a dielectric used for EDM. Show the different modes of dielectric feeding to the EDM gap.
31. Explain the advantages and disadvantages of EDM.
32. Compare wire EDM and milling by EDM.

33. In an EDM operation employing a relaxation circuit, discuss the effects of charging resistance, gap setting, and capacitance on the rates of metal removal. How does this type of machine compare with EDM-machine equipped with pulse generator?
34. What is the difference between sparking and arcing? Which condition leads to arcing in EDM?
35. What are the main applications of graphite electrodes in EDM?
36. Explain the term "No-wear EDM."
37. Draw a neat sketch to show the difference between injection and suction flushing as applied in EDM. What type do you recommend in the following cases? Give reasoning.
    - Production of true cylindrical holes
    - Production of forming tools and dies
38. For an EDM operating on RC circuit under optimum working conditions, determine the average power output, given that the resistance $R = 3.2\ \Omega$; capacitance $C = 150\ \mu F$; supply voltage $V_0 = 200$ V, and the breakdown voltage V_s.
39. Define EBM.
40. State the important parameters that influence the MRR in EDM, LBM, and PBM.
41. What are the advantages and limitations of PAC?
42. How does a laser operate, and what does it consist of?
43. Explain why the CO_2 laser is particularly effective for machining nonmetals.
44. List the important advantages and limitations of LBM.
45. Make a comparison between LBM and EBM on the basis of their applications and limitations.
46. Precision engineering is a term that is used to describe manufacturing of high-quality parts with close tolerances and good surface finishes. Based on their process capabilities, make a comprehensive list of NTMPs with decreasing order of quality of parts produced. Include brief commentary on each method.
47. Make a list of nontraditional machining processes that may be suitable for the following materials: ceramics, CI, thermoplastics, thermosets, diamond, and annealed copper.
48. Explain this statement: NTM should not be considered as a replacement for TM.
49. Which of the NTMP causes thermal damage? What is the consequence of such damage to the WP?
50. Arrange the following NTMP in descending order of maximum metal rate, indicating their values in mm³/min; also arrange them in descending order of penetration rate in mm/min: USM, ECM, CHM, EDM, LBM, PAC.
52. Name factors governing: tool feed rate in ECM, and accuracy in USM.
53. Mention the parameters affecting the SRR of the processs: USM, ECM, WJM.
54. Mention briefly the purpose of :
    - Adding 1% of a long-chain polymer to the water in WJM
    - Applying a premagnetizing current in addition to HF current in USM
    - Applying suction of abrasive slurry in USM
55. Give typical values for the following parameters:

Statement	Value	Units
Depth of HAZ in ECM		mm
Optimum breakdown voltage V_s using a source $V_0 = 150$ V (assume RC circuit)		V
Surface roughness of ECMed surface, R_t		µm
Penetration rate in case of USM of glass		mm/min
Gap voltage in ECM		V

Gap thickness in EDM	mm
Tool oscillation amplitude in USM	mm
Hole over size in USM	mm
Nozzle diameter in AJM	mm
SOD in WJM	mm
Oscillation frequency in USM	Hz
Gap thickness in ECM	mm
Frontal gap thickness in USM	mm
Side gap thickness in USM	mm
Tool wear in ECM	mm
Depth of HAZ in EDM	mm
Accelerating voltage of an EBM	kV
Power density in EBM	kW/mm^2

56. Mark true (T) or false (F) for each of the following statements:

Statement

[ ] To produce accurate cylindrical holes by EDM, injection flushing and not suction flushing should be used.
[ ] A main advantage of ECM is that tool wear is equal to zero.
[ ] While a relaxation oscillator is highly desirable in that it is simple and rugged, it is severely limited in metal removal capability.
[ ] EBM and EDM are thermal NT processes, which are used to machine only electrically conductive materials.
[ ] It is possible to set working conditions in EDM to obtain zero or minimal tool wear.
[ ] Controlled pulse circuits have the disadvantages of low metal removal rate and high tool wear.
[ ] In USM, the removal rate depends upon the mechanical properties of the WP.
[ ] In ECM, the removal rate is proportional to the valence, and inversely proportional to the atomic weight of the anode material.
[ ] In USM, the only function of the premagnetizing current is to avoid frequency doubling.
[ ] WJM is suitable for metallic WPs only.
[ ] ECM produces a white layer just like EDM.
[ ] The surface quality of electrochemically machined holes and cavities improves with increasing feed.
[ ] USM is economical in machining only electrically conductive hard materials to complex shapes with high accuracy and reasonable surface finish.
[ ] In AJM, it is always recommended to reuse the abrasive powder.
[ ] In WJM, material is removed by the mechanical action of a high-velocity stream impinging on a small area, whereby its pressure exceeds the yield strength of material.
[ ] In ECM, the passage of current through the gap results in material transfer from anode to cathode in the form of anions.
[ ] NaNO$_3$ has desirable characteristics as an electrolyte and is less corrosive than NaCl. However, it has a tendency to passivate chemical reactions.
[ ] A CO$_2$ laser is a kind of solid-state laser.
[ ] A laser is usable for metal removal only.
[ ] Mechanical shear is the main mechanism of material removal in EBM.
[ ] MRR in AJM is greater than that in AWJM.

[ ] In USM, for the same static load, the larger the tool diameter, the greater will be the penetration rate.
[ ] Complex shapes are produced in glass using EDM.
[ ] The current used in EDM is an AC.
[ ] PAM produces more accurate parts than EDM.

REFERENCES

Advanced Materials and Processes (1990) ASM International, Materials Park, OH, p. 43.
AGIE Charmilles Group (2004) Charmilles, the Solutions, When to EDM, Geneva.
Barash, M. M. (1962) Electric spark machining, *International Journal of Machine Tool Design and Research*, 2, 281.
Benedict, G. F. (1987) *Non-Traditional Manufacturing Processes*, Marcel Dekker Inc., New York.
Blanck, D. (1961) *Getzmäßgikeiten beim Stoßläppen mit Ultraschallfrequenz*. Dissertation, Braunschweig.
Brook, N. and Summers, D. A. (1969) The penetration of rock by high speed water jets, *International Journal of Rock Mechanics and Mineral Science*, 6(3), 249.
Düniβ, W., Neumann, M. and Schwartz, H. (1979) *Trennen-Spanen and Abtragen*, VEB-Verlag Technik, Berlin.
El-Hofy, H. (2005) *Advanced Machining Processes, Nontraditional and Hybrid Processes*, McGraw-Hill Co., New York.
Farmer, I. W. and Attewell, P. B. (1965) Rock penetration by high speed water jets, *International Journal of Rock Mechanics and Mineral Science*, 2(2), 165–169.
Firma Dr. Lehfeldt & Co. (1967) GmbH, Heppenheim, Germany.
Franz, N. C. (1972) Fluid additives for improving high velocity jet cutting, *First International Symposium of Jet Cutting Technology*, England.
Gusseff, W. (1930) *Method and Apparatus of Electrolytic Treatment of Metals*, BP, Nr. 335003.
Hoogstrate, A. M. and van Luttervelt, C. A. (1997) Opportunities in AWJM, *Annals of CIRP*, 46(2), 679–714.
Ingersoll-Rand (1996) *Technical Data*, Hannover Exhibition.
Kaczmarek, J. (1976) *Principles of Machining by Cutting, Abrasion, and Erosion*, Peter Peregrines Ltd, Stevenage, Hertfordshire.
Kalpakjian, S. (1984) *Manufacturing Processes for Engineering Materials*, Addison Wesley Publishing Co, Reading, MA.
Kalpakjian, S. and Schmidt, S. R. (2003) *Manufacturing Processes for Engineering Materials*, 4th Edition, Prentice Hall Publishing Co, New York.
König, W. (1990) *Fertigungsverfahren-Band 3 Abtragen*, VDI Verlag, Düsseldorf.
König, W. and Schmelzer, M. (1990) Schneiden mit feststoffbeladenen Wasserstrahlen als leistungsfähiges Bearbeitungsverfahren für faserverstärkte Kunststoffe, Industrie-Anzeiger, 109 H91, 70–71.
Kovacevic, R. (1994) Sensing the AWJ nozzle wear, *International Journal of WJ Technology*, 2(1).
Lissaman, A. J. and Martin, S. J. (1982) *Principles of Engineering Production*, Hodder and Stoughton Educational, London, UK.
Machining Data Handbook (1980) Vol. 2, 3rd Edition, Machinability Data Center, Cincinnati, OH.
McGeough, J. A. (1988) *Advanced Methods of Machining*, Chapman and Hall, London.
Metals Handbook (1989) Machining, Vol. 16, ASM International, Materials Park, OH.
Mitsubishi EDSCAN Technical Data (1997) Hannover Exhibition.
Nassovia-Krupp, Werkzeugmaschinenfabrik, Langen/Frankfurt (1967).
Nordwood, J. A. and Johnston, C. E. (1984) New adaptations and applications for water knife cutting, *7th International Symposium on Jet Cutting Technology*, pp. 369–388.
Pandey, P. C. and Shan, H. S. (1980) *Modern Machining Processes*, Tata McGraw Hill Co, New Delhi.
Paul, S., Hoogstrate, A. M., van Luttervelt, C. A. and Kals, H. J. J. (1998) Analytical and experimental modeling of AWJC of ductile materials, *Journal of Materials Processing Technology*, 73, 189–199.
Product Information of Allfi AG (1997).
Rajurkar, K. P., Zhu, D., McGeough, J. A., Kozak, J. and De Silva, A. (1999) New developments in ECM, *Annals of CIRP*, 48(2), 569–579.

Schmelzer, M. (1994) *Mechanismes für Strahlerzeugung beim Wasser-Abstrahlschneiden*, Dissertation TH-Aachen.

Steigerwald, K. H. (1958) Materialbearbeitung mit Elektronenstahlen. *Fourth International Kongress für Elktronenmikroskopie*, Springer, Berlin.

Tlusty, G. (1999) *Manufacturing Processes and Equipment*, Prentice Hall, Upper Saddle River, NJ.

Verma, A. P. and Lal, K. G. (1984) An experimental study of AJM, *International Journal of Machine Tool Design and Research*, 24(1), 19–29.

Visser, A. (1968) *Werkstoftabtrag mittels Electronenstrahl*, Dissertation, T. H. Brauanschweig, Germany.

Visser, A., Junker, M. and Weiβinger, D. (1994) *Sprühätzen metallischer Werkstoffe*, 1st Auflage, Eugen G. Leuze Verlag, Saulgau, Württ, Germany.

Youssef, H. A. (2005) *Non-Traditional Machining Processes-Theory and Practice*, 1st Edition, El-Fath Press, Alexandria.

12 Environment-Friendly Machine Tools and Operations

12.1 INTRODUCTION

In the early 1970s, public discussion of the consequences and measures necessary to conserve the environment has been stimulated by citizens, action groups, and parliamentary movements. In the early 1980s, all parties had integrated environmental protection into their political programs. Ever since, environmental protection has been ranked high in public opinion and its importance is growing constantly.

There is an increasing awareness of the people of the industry toward the ecological aspects of the machining processes. The increasing sensitivity to environment and health issues is reflected in increasingly stringent legislation and national and international standards. The restrictions resulting from such legislation pose a challenge to scientists and engineers to develop new and alternative manufacturing technologies.

Despite various advantages achieved by the machining processes, they may generate solid, liquid, or gaseous by-products that present hazards for workers, machine, and the environment. The large-scale and long-term environmental threats created by the machining processes lead to direct environmental consequences. It is now appropriate to take direct and immediate actions toward understanding the environmental hazards created by the machining processes and analyze their impacts on the environment, with the following suggestions as a guide:

1. Each company's environmental protection policy should keep hazards within acceptable limits through the following steps:
   - Effective management and reduction of emissions in air, water, and land
   - Compliance with relevant legislations covering manufacturing operations
   - Pollution prevention
   - Efficient use of energy
   - Minimization of consumption of natural resources
   - Minimization of waste streams
2. Ensure compliance with the environmental management system standard EN ISO 14001.
3. Set environmental objectives and targets.
4. Monitor and review the environmental performance against the set objectives and targets.
5. Raise staff awareness of the environmental implications of their work.
6. Provide the necessary instructions and resources that help in implementing the environmental policy.

Environmental impacts are measured by the degree of hazard, which is a function of the chemical/physical properties of the substance(s), the quantities involved; the potential effects are classified under three categories:

- Class A—Major immediate environmental effect
- Class B—Intermediate environmental effect (may be serious but not immediate)
- Class C—Minor environmental effect

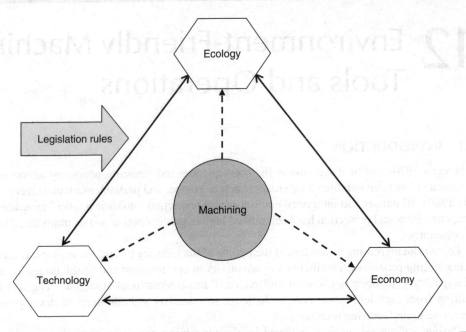

FIGURE 12.1 Interaction between technology, economy, and ecology for machining processes. (From Byrne, G. and Scholta, E., *Ann. CIRP*, 42(1), 471–474, 1993. With permission.)

In considering a clean machining process, the interaction between economy, ecology, and technology has to be considered, as shown in Figure 12.1. Conflicts may arise among these three factors and a good compromise has to be made. To improve the quality of the machining processes, it is essential to adopt innovative methods that achieve the minimum environmental contamination in addition to their stability, reliability, and acceptable economic conditions. In this regard, the application of near net shape technology to manufacture parts with complex shapes by substituting cutting operations with forming activities provides advantages such as reduced chip volumes lower cutting forces, reduced volumes of cutting fluids and cutting fluid losses, and simpler machine tools with lower power requirements.

As shown in Figure 12.2, one of the possibilities for minimizing environmental contamination is to modify and develop existing processes and replace conventional processes with alternative ones (rapid prototyping, laser machining, new cutting tool materials, and so on). It is essential that the new technologies do not lead to environmental hazards of a different nature.

Most machining processes use chemicals and liquids in the form of coolants, lubricants, etchants, electrolytes, abrasive slurry, dielectric liquids, gasses, and anticorrosive additives as shown in Table 12.1. These chemicals can be transported by a variety of agents and in a variety of forms, which are defined by Hughes and Ferrett (2005) as shown in Table 12.2.

Hazardous substances cause ill health for people at work. These are classified according to their severity and type of hazard that they may present to the workers. The most common types are summarized in Table 12.3. The effect of these hazards may be acute or chronic:

Acute effects. These effects are of short duration and appear fairly rapidly, usually during or after a single or short-term exposure to a hazardous substance.

Chronic effects. These effects develop over a period of time that may extend for many years. Chronic health effects are caused by prolonged or repeated exposures to hazardous substances resulting from the machining processes. Such effects may result in a gradual, latent, and often irreversible illness that may remain undiagnosed for many years. During that period, the individual may experience symptoms. Cancers and mental diseases fall into chronic category.

Environment-Friendly Machine Tools and Operations

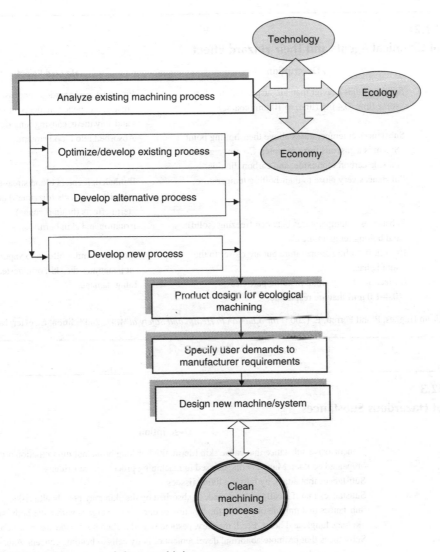

FIGURE 12.2 Achievement of clean machining process.

TABLE 12.1
Machining Liquids

Liquid/Medium	Application
Lubricants	Machine tools
Coolants	Cutting operations
Liquid nitrogen	Ecological machining
Etchants	CHM, PCM
Electrolytes	ECM
Dielectric liquids	EDM
Vegetable oil	EDM/lubricants/coolants
Deionized water	EDM wire cutting
Abrasive slurry	USM
Air + abrasives	AJM
Water + abrasives	AWJM

TABLE 12.2
Forms of Chemical Agents and Their Hazard Effect

Form	Description	Hazard Effect
Dust	Solid particles heavier than air suspended in it for some time (0.4 μm (fine) to 10 μm (coarse)).	Fine dust are hazardous because they are respirable, penetrate deep into lungs, and stay there, causing lung disease.
Gas	Substances at temperatures above their boiling point. Steam is a gaseous form of water. Common gasses include carbon monoxide, and carbon dioxide.	Absorbed into blood stream.
Vapor	Substances very close to their boiling temperature.	If inhaled, it enters blood stream causing short-term effects (dizziness) or long-term effects (brain damage).
Liquid	Substances at temperatures between freezing (solid) and boiling temperatures.	Irritation and skin burn.
Mist	Exist near boiling temperature but are closer to the liquid phase.	Produces similar effects to vapors where it penetrates the skin or ingested.
Fume	Collection of very small metallic particles (0.4–1.0 μm) that are respirable.	Lung damage.

Source: From Hughes, P. and Ferrett, E. (2005) *Introduction to Health and Safety at Work*, 2nd Edition, Elsevier, New York.

TABLE 12.3
Effects of Hazardous Substances

Effect	Description
Irritant	Noncorrosive substance that cause skin (dermatitis) or lung bronchial inflammation after repeated contact. Many chemicals used in machining processes are irritants.
Corrosive	Substances that attack by burning living tissues.
Harmful	Substances that if swallowed, inhaled, or absorbed by the skin may pose health risks.
Toxic	Substances that impede or prevent the function of one or more organs within the body, such as kidney, lungs, and heart. Lead, mercury, pesticides, and carbon monoxide are toxic substances.
Carcinogenic	Substances that promote abnormal development of body cells to become cancers. Asbestos, hard wood dust, creosote, and some mineral oils are carcinogenic.

12.2 TRADITIONAL MACHINING

Machining by cutting is the process of removing the machining allowance from a WP in the form of chips. It requires a tool that is harder than that the WP, a relative motion between the WP and the cutting tool, and it also requires penetration of the tool in the WP. One of the main machining drawbacks is its impact on environment in terms of noise, leakage, and flying chip. Heat and machining waste cause great hazards and carries a high risk of injuries and disorders. Therefore, safety precautions must be considered to reduce the negative effect of the machining process.

The main hazards created by metal cutting and abrasion processes are shown in Figures 12.3 and 12.4. These include the following:

a. *Noise*. During machining, vibration components of different frequencies that are numerous and not harmonically related to one another are generated and produce noise. Noise levels at 85 dB can damage hearing permanently. That is why the National Health and Medical Research Council has proposed 85 dB as the maximum noise level regarded as

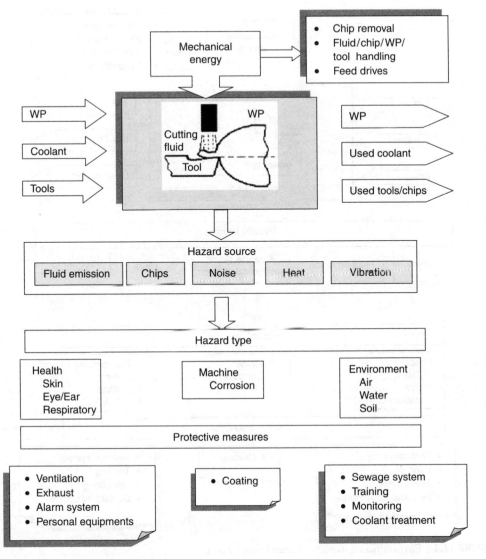

FIGURE 12.3 Environmental impacts of metal cutting processes.

safe and tolerable for an eight-hour exposure. When noise levels exceed 90 dB in any work area, ear plugs must be worn. Many machines cause noise due to vibration. Figure 12.5 shows the expected level of noise from some machine tools such as the center lathe, milling, and grinding machines, and Figure 12.6 shows the typical allowable noise levels and their permissible safe exposure times.

b. *Flying chips.* All metal cutting processes produce chips. Flying chips are considered to be a major environmental impact of machining that causes great hazards. These chips have two main effects on the environment. The first is the danger and the risk they pose to the operator when flying from the machine during the cutting process. The second one is the dust and other flying particles, such as metal shavings, that may result in eye or skin injuries or irritation. Grinding, cutting, and drilling of metal and wood generate airborne particles. Under such circumstances, it is always recommended to wear safety glasses, goggles, or shields.

c. *Cutting tools.* From an environmental perspective, the most significant waste stream is mainly generated from the remaining portion of the tool that is disposed off after its useful life.

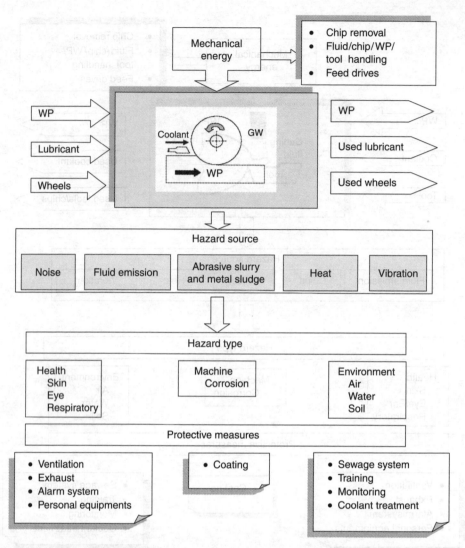

FIGURE 12.4 Environmental impacts of abrasive processes.

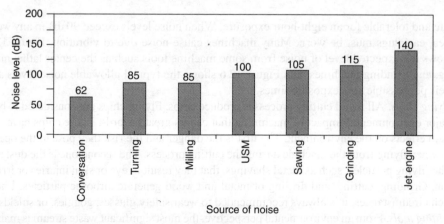

FIGURE 12.5 Noise levels for different machine tools.

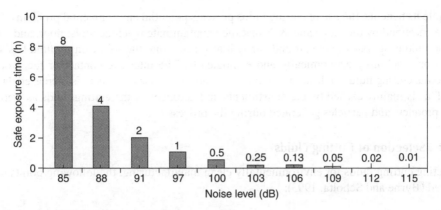

FIGURE 12.6 Safe exposure times for different noise levels.

Environmental impacts of machining by cutting and abrasion processes can be minimized at the input stage by:

1. Using the minimum material through the near net shape technology
2. Using process parameters for optimum tool life
3. Avoiding dangerous substances and substances hazardous to water
4. Avoiding residuals and emissions (noise, heat, and vibrations)

12.2.1 Cutting Fluids

The main functions of cutting fluids are to:

1. Cool the cutting tool and the WP (cooling effect)
2. Reduce friction between the tool and WP (lubricating effect)
3. Remove the chips (flushing effect)
4. Provide a safe working environment (nonmisting, nontoxic, nonflammable, and nonsmoking)

12.2.1.1 Classification of Cutting Fluids

Practically, all cutting fluids currently in use fall into one of the following types:

1. *Air.* Compressed air can be used for cooling the machining zone using a pure air jet or air mixed with a fluid.
2. *Water-based cutting fluids.* Water, soluble oil, emulsions, and chemical solutions or synthetic fluids form the cutting fluid.
3. *Neat oils.* Mineral oils, fatty oils, composed oils, extreme pressure (EP) oils, and multiple-use oils.
4. *Liquid nitrogen.* Liquid nitrogen, having a temperature of −196°C, is used as a cutting fluid for difficult-to-cut materials. It is also used to cool WPs of tubular shapes or to cool the tool through internal channels that supply the nitrogen under pressure or by flooding the cutting area.

Cutting fluids contain many additives such as emulsifiers, antioxidants, bactericides, tensides, EP additives, corrosion inhibitors, or agents for preventing foaming. Cutting fluids account for 15% of the shop production cost. The costs of purchase, care, and disposal of the cutting fluids are more than twice as high as tool costs. The tooling cost may be increased by the use of dry cutting or due to the increased tool wear; however, the manufacturing cost may be reduced when compared to the conventional process in which cutting fluids are used.

Despite its benefits, the use of cutting fluids presents potential environmental problems. Cutting fluids are entrained by the chips and WPs and they contaminate machine tools, floor, and workers. Partly the fluid evaporates to the air and partly it also flows into the soil. Cutting fluid has a direct influence on machining economically and ecologically. The intensive contact of the production worker to the cutting fluid can lead to skin and respiratory diseases and there is increased danger of cancer. This is mainly caused by the constituents and additives of the cutting fluids as well as the reaction products and particles generated during the process.

12.2.1.2 Selection of Cutting Fluids

For selecting cutting fluids for environmentally clean manufacturing, the following points must be considered (Byrne and Scholta, 1993):

- The constituents of the cutting fluid must not have negative effects on the health of the machinist or on the environment.
- During use, cutting fluids should not produce contaminants.
- Multifunction oils that can be used for hydraulic systems, for slide way lubrication, and as a coolant and lubrication in machining should have minimum vaporization characteristics.
- Cooling and lubrication should be applied in a manner that minimizes the volume of the fluid used.
- Continuous monitoring of the cutting fluids and machine tool environment with online sensors is desirable.
- Through adequate care and maintenance of cutting fluids, the amount of water required for emulsions can be reduced, thus leading to cost savings.

From the point of view of environmental conservation and protection, complete elimination of the cutting fluids is desirable. The problem of cutting fluid disposal is an important issue in relation to environmental protection. Cleaning components is commonly time-consuming and cleaning agents have to be replaced by environment-friendly agents to reduce possible hazards to the environment and the employees. Through the use of dry machining, there is a complete elimination of hazards due to cutting fluids; the extent of cleaning can be significantly reduced with resulting economical and ecological benefits.

12.2.1.3 Evaluation of Cutting Fluids

The goal of machining is to make an acceptable part as quickly and as cost effectively as possible. Many factors affect the machining cost such as the machine tool, the tooling selected, and the cutting fluid utilized. The issues discussed in the following sections are to be considered in evaluating cutting fluids in machining processes.

1. Lubricant Stream

The cutting fluid is assumed to diverge into four paths during the machining process, as shown in Figure 12.7. Unfortunately, spoiled or contaminated cutting fluids are the most common wastes from the machining process; these are considered hazardous wastes due to their oil content as well as other chemical additives they may contain. Therefore, it is essential to prevent or reduce this waste to avoid the cost of having to frequently replace and dispose off them. Contaminants of these cutting fluids may include the chips, machined parts, dust and other particulates, and moisture in nonaqueous solutions. However, the most troublesome contaminant is the tramp oils that interfere with the cooling fluid, promote bacterial growth, and contribute to unwanted residue on cutting tools. All of these factors contribute to premature fluid degradation and the need for replacement.

During machining, high cutting speeds are used (>3500 m/min). Such a speed produces HT in the machining zone, which in turn results in high thermal stresses for the cutting tool. To achieve

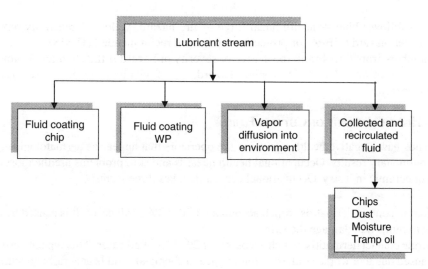

FIGURE 12.7 Lubricant streams.

an acceptable tool life, cutting fluids must be used. Vaporization of the fluids and particles that are exposed to elevated temperatures takes place. The emissions entering the atmosphere thus represent a complex mixture of:

- Vapors due to elements of the WP and cutting tool material
- Vapors caused by deposits on the surface of the component being machined
- Vapors resulting from cutting fluids

Regulations under the U.S. Occupational Safety and Health Administration (OSHA), Environmental Protection Agency (EPA), and the Department of Transportation (DOT) are developed to provide the users of metal removal fluids with a user-friendly guide for assessing and minimizing the environmental health and safety (EHS) impacts in the selection, use, and disposal of metal removal fluids. Information on material safety data sheets (MSDS) and labeling of metal removal fluids, a matrix for rating metal removal fluid, and other issues that affect the use of metal removal fluids are also available.

2. *Disposal*

Disposal is rapidly becoming the second most important factor in selecting new metal removal fluids. This importance is primarily due to the increasing cost and liability associated with these fluids. Whether onsite waste treatment and disposal or a pay-to-dispose method is used, the waste treater should provide details on product components that may adversely affect the waste treatment process, resulting in increased cost to the company.

12.2.2 Hazard Ranking of Cutting Fluids

The following are the main types of hazards that may occur by the cutting fluids:

1. Physical hazards (flammable, explosive, reactive, radioactive)
2. Health hazards (irritant, toxic, carcinogenic)

Two commonly used labeling formats with colors and numbers are supplied by the National Fire Protection Association (NFPA) system and the Hazardous Material Identification System (HMIS). Both the NFPA and HMIS systems use the same colors and numbers to identify the degree of hazard if the chemical is not handled properly. Each color on the label stands for a different type of hazard as follows: blue = health hazard, red = fire hazard, yellow = reactivity hazard, and white = special hazard (NFPA) or protective equipment recommended (HMIS).

The numbers from 0 to 4 rank the degree or severity of hazard if the chemical is not handled correctly: 0 = minimum hazard, 1 = slight hazard, 2 = moderate hazard, 3 = serious hazard, 4 = severe hazard.

12.2.3 Health Hazards of Cutting Fluids

Cutting fluids have negative health effects on the operators that appear as dermatological, respiratory, and pulmonary results. Occupational health hazards and skin problems are the most common for metal or ceramic industry. Occupational dermatitis takes three forms:

1. Irritation contact dermatitis, which accounts for 50–80% of all cases. It is caused by exposure to fluids that damage the skin.
2. Allergic contact dermatitis, which accounts for 20–50% of all cases. This type of dermatitis is caused due to a worker's allergic intolerance to chemicals and is generally noncurable.
3. Exposure to mists caused by the cutting fluids raises a worker's susceptibility to respiratory problems. This depends on the level of chemicals and particles contained in generated mists.

Factors such as operator's hygiene, plant cleanliness, and air quality contribute to the likelihood of dermatitis among workers. Several factors leading to skin irritation include the following:

- The high pH of the cutting fluids (8.5) compared to 5.5–6 for the human skin
- Certain metals such as nickel, chromium, and cobalt
- Microbial contamination of cutting fluids
- The use of surfactants in cutting fluids

Deterioration of the cutting fluid through use causes adverse effects on process performance:

1. Contamination with small metal particles enhances tool wear and WP surface deterioration.
2. Bacterial growth, additive absorption, demulsification, and oxidation cause fluid failure.
3. Contamination with tramp oil and hydraulic oil provides breeding grounds for microorganisms.
4. Corrosion of the machine components, paint stripping, and foaming caused by improperly formulated cutting fluids.

Several cleaning methods are used to clean the used fluids, including centrifugal and hydrocyclene separators. In this regard, the vibratory-enhanced shear processing (VESP) membrane filter (Figure 12.8) can be used to filter and clean the cutting fluids at a temperature of 80°C, significantly higher than competitive membrane technology. The vibration amplitude and corresponding shear rate can also be varied, which directly affects filtration rates. This system has many applications, such as coolant recycling and metal hydroxide recovery filtration.

12.2.4 Cryogenic Cooling

High-speed machining is characterized by the generation of high cutting temperatures. Such temperature levels adversely affect the tool life, dimensional and geometrical accuracy, and the surface integrity of the machined parts. The HT generated in the cutting zone has been controlled, conventionally by employing flood cooling. In high-speed machining, cutting fluids fail to penetrate the tool–chip interface and thus cannot remove the heat effectively. However, a high pressure jet of soluble oil reduces the temperature at the tool-chip interface and improves the tool life to some extent.

Environment-Friendly Machine Tools and Operations

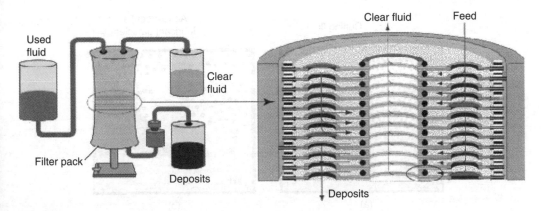

FIGURE 12.8 VESP membrane filter pack. (From New Logic Research, Inc., Emeryville, CA, www.vsep.com/solutions/technology.html.)

The application of conventional cutting fluids causes several environmental problems:

a. Environmental pollution due to chemical dissociation of the cutting fluid at the high cutting temperature and thus formation of harmful gases and fumes
b. Biological problems to operators due to physical contact with cutting fluids
c. Water pollution and soil contamination during disposal
d. Requirement of extra floor space and additional systems for pumping, storage, filtration, recycling, chilling, and so on
e. The high cost of disposal of used coolants under tougher environmental laws

Coolant injection at high pressure has reduced the consumption of the cutting fluid by 50%, as well as the cutting temperature and the cutting forces. Cryogenic cooling by a liquid nitrogen jet has been used in the machining and grinding of steels. Under such circumstances, better surface integrity, lower cutting forces, and longer tool life have been achieved. The experimental results of Dhar et al. (2002) indicated the possibility of a substantial reduction in cutting forces by cryogenic cooling, which enabled a reduction in cutting forces and increased tool life due to reduced cutting temperature. Cryogenic cooling is therefore a potential environment-friendly clean technology for control of the cutting temperature.

12.2.5 Ecological Machining

Pending OSHA and EPA regulations in metal cutting fluids have made dry/ecological machining an important issue. Promising alternatives for the commonly used flood cooling (Figure 12.9) include:

1. Minimum quantity lubrication (MQL) in which a very low amount of fluid (<50 mL/h) is pulverized in the flow of compressed air during high-speed machining
2. Dry/ecological machining
3. Liquid gases like N_2
4. Super-hard tools such as CBN or diamond that do not require any coolant due to their outstanding wear resistance
5. Ionized air

Minimum quantity lubrication (MQL) is an alternative for normal flood cooling application. It uses a mist or a minimum quantity of neat oils or emulsions in the flow of compressed air, thus generating

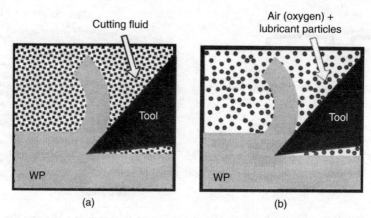

FIGURE 12.9 Conventional and MQL supply of cutting fluids: (a) conventional and (b) MQL supply. (From Weinert, K., Inasaki, I., Sutherland, J. W., and Wakabayashi, T., *Ann. CIRP*, 53(2), 17, 2004. With permission.)

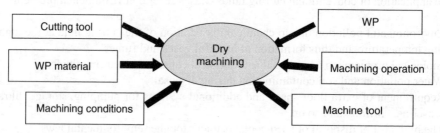

FIGURE 12.10 Factors that affect dry machining.

a spray that is directed to the cutting zone to serve as a lubricant and coolant. This technique is suitable for drilling and some grinding operations.

Factors Affecting the Use of MQL

Despite the advantages of using MQL, its wide acceptance in the machining technology is affected by the following drawbacks:

1. *Environmental pollution.* Although this technique replaces the flood fluid method, it causes pollution because the pulverization of the oil in the air flow causes the suspension of a lot of oil particles in the air. This problem requires special encapsulation, protection guards, and good exhaust systems with particle control for the machine.
2. *Noise.* A line of compressed air must be used with MQL, which works intermittently during the machining process. These air lines produce a lot of noise, usually higher than the human ear can handle (>80 dB). It also makes communication between persons more difficult, which is also bad for the working environment.

MQL is considered to be an intermediary solution between the conventional use of cutting fluid and dry cutting. Figure 12.10 shows the different factors that affect dry machining. Dry machining benefits are shown in Figure 12.11. To enhance the capabilities of dry cutting, multiple-layer coated-carbides, ceramics, and CBN are recommended. Such tool materials withstand the intensive heat and achieve satisfactory results without the use of cutting fluids. Properly applied coatings offer longer tool life at high cutting speeds and feeds. Recent developments in multilayer coatings on carbide cutting tools are important for dry cutting applications. Figure 12.12 shows the main requirements of cutting tool materials for dry cutting.

Environment-Friendly Machine Tools and Operations

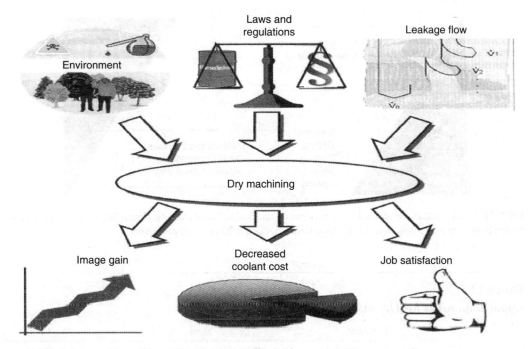

FIGURE 12.11 Benefits of dry machining. (From Weinert, K., Inasaki, I., Sutherland, J. W., and Wakabyashi, T., *Ann. CIRP*, 53(2), 17, 2004. With permission.)

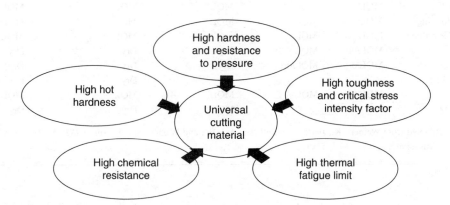

FIGURE 12.12 Optimal cutting tool materials for dry machining. (From Weinert, K., Inasaki, I., Sutherland, J. W., and Wakabyashi, T., *Ann. CIRP*, 53(2), 17, 2004. With permission.)

Applications of Ecological Machining

Figure 12.13 shows the influence of machining process on the cooling liquid supply; Table 12.4 shows the different applications of dry and MQL machining. Dry machining applications include the following areas:

Turning. Dry cutting of hardened steel with super-hard cutting tools such as CBN or polycrystalline diamond (PCD) showed good results as an ecological alternative that reduces the operation time and hence improves productivity. The rationale is related to the tools'

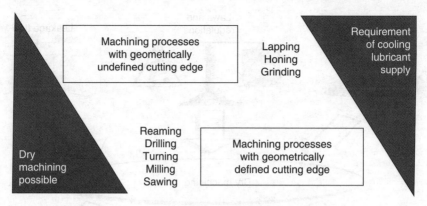

FIGURE 12.13 Influence of machining process on the cooling liquid supply. (From Weinert, K., Inasaki, I., Sutherland, J. W., and Wakabyashi, T., *Ann. CIRP*, 53(2), 17, 2004. With permission.)

TABLE 12.4
Application Areas for Dry Machining

	Aluminum		Steel		CI
Material Process	Cast Alloy	Wrought Alloys	High-Alloyed Bearing Steel	Free Cutting Quench and Tempered Steel	GG20 to GGG70
Drilling	MQL	MQL	MQL	MQL/dry	MQL/dry
Reaming	MQL	MQL	MQL	MQL	MQL
Tapping	MQL	MQL	MQL	MQL	MQL
Thread forming	MQL	MQL	MQL	MQL	MQL
Deep-hole drilling	MQL	MQL		MQL	MQL
Milling	MQL/dry	MQL	Dry	Dry	Dry
Turning	MQL/dry	MQL/dry	Dry	Dry	Dry
Gear milling			Dry	Dry	Dry
Sawing	MQL	MQL	MQL	MQL	MQL
Broaching			MQL	Dry	Dry

Source: Adapted from Weinert, K., Inasaki, I., Sutherland, J. W., and Wakabyashi, T., *Ann. CIRP*, 53(2), 17, 2004. With permission.

lower coefficient of friction, higher abrasion resistance, higher thermal conductivity, and better chemical and thermal stability, which make these materials suitable for high-speed machining applications. Moreover, in case of hardened steels, dry cutting with CBN tools provides a suitable cheaper substitution for grinding.

Drilling. Drilling without coolants or lubricants sticks the chip to the cutting tools and breaks them in a short time. MQL provides good lubrication for drilling using naturally dissolving oils such as vegetable oil or ester oil. It is impossible to perform drilling using dry cutting or using dry compressed air. MQL is normally used with an oil flow of 10 mL/h. However, the roughness, roundness, diameter accuracy, and cylindericity of drilled holes are identical to the flood technique of soluble oil.

Milling. For minimizing the negative effect on the environmental drawbacks when using ample amount of flood coolant, MQL using 8.5 mL/h is used instead of the 42,000 mL/h used in the normal flood technique. It is clear that MQL achieves a drastic reduction (1/5000 times) in lubricant consumption. There is a considerable reduction in cutting force components for MQL when compared to both dry cutting and flood cooling techniques.

Hobbing. Conventional hobbing requires coolant, which complicates the process, and the disposal of which affects the environment. Dry hobbing is also possible using high-speed steel hobs coated by TiAlN, which makes the hob operate at a higher speed than is normally used in the wet cutting of gears.

Gear shaping. Cutting conditions for gear shaping allows dry machining with TiN, which doubles the tool life.

Grinding. The main functions of grinding fluids are cooling, lubrication, and flushing of chips and abrasives. Currently, soluble coolants containing chemical additives are used. These fluids are environmentally hazardous and their disposal is expensive. Other methods of cooling in grinding include:

1. Cryogenic grinding using N_2 is uneconomical and is therefore not applicable.
2. Compressed cold air provides significant roughness due to lack of cooling at large depths.
3. Dry grinding induces high thermal wear and grinding forces (and is thus not recommended).
4. Chilled air can be used in precision grinding.
5. Cold air and oil mist reduces the grinding force and surface roughness, and therefore can be used for precision grinding.
6. Solid lubricants such as graphite, calcium fluoride, and barium fluoride are mixed in oil to form a paste that requires a flushing for this process.

Vegetable oils are considered potential candidates to substitute for conventional mineral oil–based lubricating oils and synthetic esters for the following reasons:

- Vegetable oils are nontoxic, have low volatility, pose no workplace hazards, and are biodegradable.
- The polar ester groups that are present in the vegetable-based lubricants are able to adhere to metal surfaces and therefore possess good boundary lubrication properties.
- Vegetable oils have high solubility power for polar contaminants and additive molecules.
- Vegetable oils have poor oxidation stability, so they are highly susceptible to radical attack.
- Vegetable oils maintain their environmental friendliness using additive packages or recently by chemical modification. The use of additives allows an increase the overall performance of oil and improvement of its physical properties, but they also may raise the cost of lubricants and, in some cases, may even be harmful, as in the case of antioxidant lubricants commonly composed by phenolic and amine molecules.

The use of renewable materials benefits the environment by reducing greenhouse gases, and the use of natural resources improves the economic competitiveness of industry through the development of new markets and produces social benefits by stimulating rural communities.

Vegetable oils have been considered as potential substitutes for lubricant production when used in applications like hydraulics or metalworking fluids, reporting benefits for environment. The evaluation has been performed through a life cycle assessment approach of the comparative environmental impact of a traditional mineral dielectric fluid and a renewable resources–based one. From this analysis, a clear benefit has been obtained with regard to the safety of human health by using environment-friendly fluids.

12.3 NONTRADITIONAL MACHINING PROCESSES

NTM provides alternatives for many machining by cutting operations. It is used for the production of complex profiles in hard-to-cut materials. Most lubricants and machining liquids (Table 12.1) come from petroleum, which is toxic to the environment and difficult to dispose of. That is the reason why the chemical industry people are trying to increase the ecological friendliness of their products. NTM processes have many advantages; however, they generate solid, liquid, or gaseous by-products that present hazards for the workers, environment, and the equipment. The increasing awareness of environmental aspects has led the industry to become more sustainable and to adopt the use of environment-friendly products. In this section, the type and quantity of hazardous substances for the NTM methods are discussed.

12.3.1 Chemical Machining

CHM and PCM are important machining processes that have environmental impacts that must be minimized within the manufacturing company. CHM depends on chemical acids that have severe effects on the surrounding environment, difficulties in handling and storage, and damaging effects on different materials. An acid is defined as any substance that when dissolved in water dissociates to yield corrosive hydrogen ions. The acidity of an etchant dissolved in water is commonly measured in terms of pH number (defined as the negative logarithm of the concentration of hydrogen ions). Solutions with pH values of less than 7 are described as being acidic.

Acid deposition influences the environment by attacking structures that are made from steel, and fading paint on machine tools. Chemical acids also cause air pollution, which causes significant corrosion of metals. There are many acids involved in CHM, such as:

$FeCl_3$	Ferric chloride
NaOH	Sodium hydroxide
HNO_3	Nitric acid
Hf	Hafnium
$CuCl_2$	Cupric chloride
HCl	Hydrochloric acid

The majority of PCM is carried out with aqueous solutions of $FeCl_3$ used at temperatures over 50°C. $FeCl_3$ is acidic, relatively cheap, and readily available; it is also versatile, as it attacks the majority of commonly used engineering metals and alloys. Environmentally, it is attractive as it is of low toxicity and relatively easy to filter, replenish, and recycle. However, the spent etchant and its rinse water contain heavy metal ions such as nickel and chromium, which are hazardous to the environment. Methods that can be employed to reduce the overall consumption of ferric chloride include prolonging the life of the etchant before disposal, with the drawback of reduced etch rate, and regenerating the spent etchant by *in situ* oxidation, thus maintaining a constant etch rate.

During CHM, exposure to Hf can occur through inhalation, ingestion, and eye or skin contact. Overexposure to Hf and its compounds may cause mild irritation of the eyes, skin, and mucous membranes. Figure 12.14 shows the typical impacts of CHM, which include the following:

1. *Labor.* CHM causes health effects on labor, which include the following points:
    a. Irritation causing inflammation and chemical burns.
    b. Corrosive injuries and burns due to heat as acids and alkaline come in contact with water.
    c. Rapid, severe, and often irreversible damage of the eyes.
    d. Acid fumes may also corrode the teeth.

Environment-Friendly Machine Tools and Operations 511

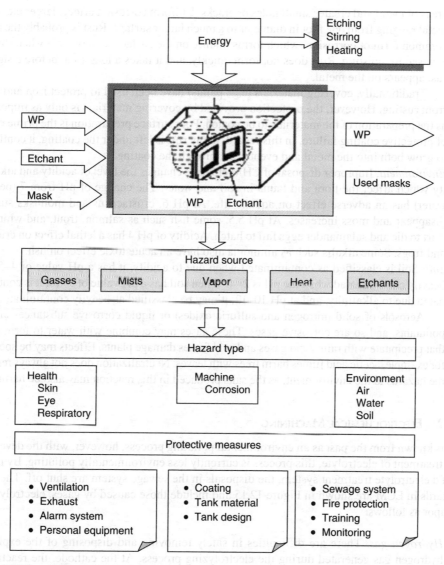

FIGURE 12.14 CHM environmental impact.

  e. Direct contact of many organic anhydrides with skin, mucous membranes, eyes, or the respiratory system causes irritation and sensitization.
  f. The exposure may also increase the risk of cancer.
  g. Inhaling mists of inorganic acids containing sulfuric acid involves an elevated risk of larynx and lung cancer of occupational origin.

  The effect of CHM hazards depends on the properties of the substance, the concentration, and time of contact with acids and alkalis. Although a dilute solution may cause irritation, the effect of strong acids and alkalis is experienced within moments of exposure. Depending on the substance and the concentration, the hazard effect may also be delayed.

2. *Machine*. Corrosion is the aging of unprotected materials that are formed from the chemical effects of a gas or liquid. Many metals are often affected by corrosion of many types, which include localized attack and uniform corrosion. A localized attack affects small

areas of metal and creates small holes or cracks. Uniform corrosion affects larger pieces of metal ranging from one foot in diameter to a much larger surface. Rust is probably the most common form of corrosion, which forms mainly on the surface area of steel when steel is exposed to moisture. Rust does not form quickly, and it takes a long time before a sign of rust appears on the metal.

Traditionally, covering materials (e.g., paints) have been used to protect iron and steel from rusting. However, the protection provided by covering materials is only as important as the preparation of the materials to be coated. Poor surface preparation is the prime cause of protective coating failure. In this regard, if any rust is left under the coating, it continues to grow both into the metal and eventually through the coating.

3. *Environment.* Improper disposal of CHM etchants changes the level of acidity and alkalinity which affect the flora and fauna in soil and water. The change of pH from 7 (neutral water) has an adverse effect on aquatic life. At pH 6, crustaceans and mollusks start to disappear and moss increases. At pH 5.5, some fish such as salmon, trout, and whitefish start to die and salamander eggs fail to hatch. Acidity of pH 4 has a lethal effect on crickets and frogs. Some alkalis such as ammonia also have an acute toxic effect on fish.

4. *Soil.* Soil is classified as contaminated when due to acidity, it has a pH value of 4–5 and heavily contaminated when the pH is 2–4. When soil has a pH value of 9–10, it is contaminated due to alkalinity, and at pH 10–12, it may be classified as heavily contaminated.

Aerosols of solid (nitrogen and sulfuric oxides) or liquid corrosive substances are air pollutants, and so are corrosive gases. These gases may combine with water to form acids that precipitate with rain. Acid gases and acid fumes damage plants. Effects may be specific; for example, acetic acid fumes harm trees with leaves. Neutralization does not always remove the hazards to the environment, as the salts produced in this reaction may also be harmful.

12.3.2 Electrochemical Machining

ECM is known from the past as an environmental polluting process; however, with the development in the treatment of electrolyte, this process is currently less environmentally polluting. By realizing a closed electrolyte treatment system, the disposals in the sewage system are shut off. The sources of hazards in ECM are shown in Figure 12.15 and include those caused by gases, electrolyte, mist, and vapor as follows:

A. *Hydrogen gas.* There are difficulties in safely removing and disposing of the explosive hydrogen gas generated during the electrolyzing process. At the cathode, the reaction is likely to be the generation of hydrogen gas and the production of hydroxyl ions:

$$H_2O + 2e^- \rightarrow H_2 + 2OH^-$$

The net reaction is thus

$$Fe + 2H_2O \rightarrow Fe(OH)_2 \text{ (s)} + H_2$$

Local exhaust must be provided to prevent the hydrogen gas from reaching its lower flammability limit and remove the mists from the workers breathing zone.

B. *Electrolyte splash.* Electrolytes are pumped through the gap between the WP and the tool. As the dc flows, metal ions are removed from the WP and swept away by the electrolytes. Electrolytes are aqueous solutions of sodium chloride, sodium nitrate, and other salts that become insoluble hydroxides that deposit out of the solution as a sludge. Skin contact

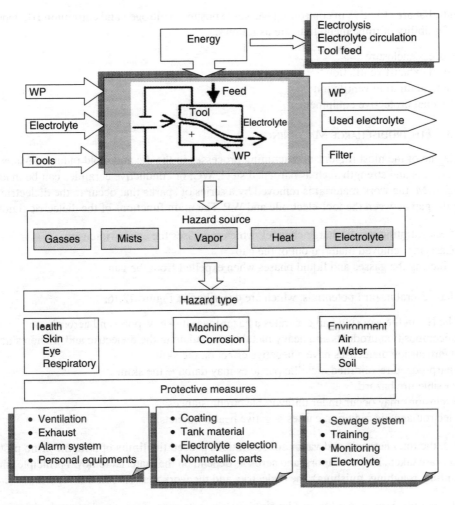

FIGURE 12.15 ECM impacts.

with the electrolyte contents must be controlled by good work practices. The door/cover shall be interlocked with the electrolyte supply system where there is electrolyte splashing. Machining process should be stopped or not started when the interlock of the safeguarding device is open.

C. *Chromate.* Exposure to chromium III compounds occurs through inhalation, ingestion, and eye or skin contact. Chromium III compounds can affect the skin, liver, and kidneys in humans. Dermal contact with trivalent chromium compounds has resulted in contact dermatitis.

D. *Nitrate.* Nitrates and nitrites are known to cause several health effects, including the following:
- Reactions with hemoglobin in blood, decreasing the oxygen-carrying capacity of the blood (nitrite)
- Decreased functioning of the thyroid gland (nitrate)
- Shortages of vitamin A (nitrate)
- Fashioning of nitrosamines, which are known as one of the most common causes of cancer (nitrates and nitrites)

Methods that are effective in controlling worker exposure to nitrogen and chromium III, depending on the feasibility of implementation, are as follows:

- Process enclosure
- Local exhaust ventilation
- General dilution ventilation
- Personal protective equipment

12.3.3 Electrodischarge Machining

EDM is one of the most important machining processes in the die and mold industry. Materials of high hardness and strength such as hardened steel, WC, or conductive ceramics can be machined. During EDM, the work material is removed by a series of sparks that occur in the dielectric liquid filling the gap between the tool-electrode and WP. The main functions of the dielectric fluids are

1. Concentrating the discharge channel, which increases the energy intensity
2. Carrying removed material out of the gap
3. Filtering the gasses and liquid phases when expelled from the gap

EDM has several hazard potentials, which are described in Figure 12.16:

- The HT in the working gap generates a hazardous smoke, vapors, and aerosols.
- Decomposition products and heavy metals accumulate in the dielectric and erosion slurry.
- Hydrocarbon dielectrics have a negative effect on the skin.
- Sharp-edged resolidified metallic particles may damage the skin.
- Possible fire hazard.
- Explosions may occur under unfavorable circumstances.
- Electromagnetic radiation causes negative health impacts.

In EDM, the total aerosols and vapor concentrations exceed the limits of 5 mg/m^3 if no protective measures are taken. Fumes, vapors, and aerosols depend on the material removal principle (sinking, wire cutting, roughing, finishing), the dielectric, and the work material. In this regard,

1. Rough machining conditions (die sinking) generate more fumes and aerosols than finish cut by wire EDM.
2. The material composition is of interest when they contain toxic or health attacking substances such as nickel.
3. The type of the dielectric, its composition, and viscosity have high influence on the fumes and vapor developed under the HTs of the plasma channel created in the working gap.
4. Lower viscosity produces less fumes and vapors.
5. The level of the dielectric has to be higher than 40 mm over the erosion spot to condense and absorb a considerable part of the vapor and fumes in the dielectric itself (80 mm is recommended).

When using mineral oils or organic dielectric fluids for EDM die sinking, the following hazardous fumes are generated:

- Polycyclic aromatic hydrocarbons (PAH)
- Benzene
- Vapor of mineral oil
- Mineral aerosols
- Products generated by dissociation of oil and its additives

For hydrocarbon dielectrics, all of these vapors and aerosols appear except for PAH and benzene.

Environment-Friendly Machine Tools and Operations

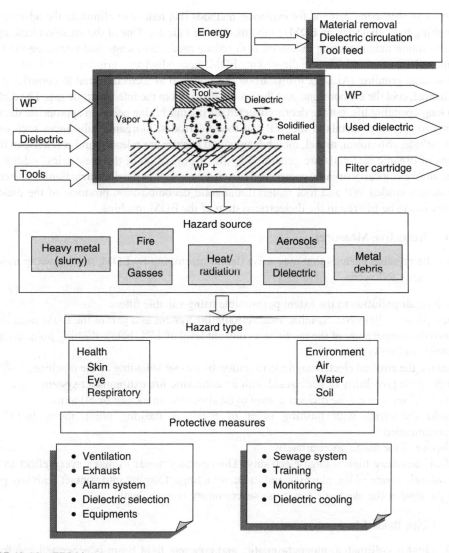

FIGURE 12.16 EDM impacts.

For water-based solvents, normally used in wire EDM, the following hazardous materials are formed:

- Carbon monoxide
- Nitrous oxide
- Ozone
- Harmful aerosols

Improvements in refining processes have led to more acceptable mineral dielectric oils. Research in the field of dielectric fluids includes the use of water-based dielectrics, gas as dielectric, and sometimes solid dielectrics. Although there are synthetic fluids in the market, the highly refined mineral dielectrics are still widely used. That contributes to the fact that 40,000 tons of mineral lubricants are lost every year all around the world, causing important environmental problems and leading to the need for alternative products with minimal environmental impacts.

There is an increasing demand for exploring methods that reduce or eliminate the adverse effect of the working fluid (dielectric in EDM) in the machining industry. One of the greatest challenges for current and future manufacturing industries is to reduce production waste and minimize the related environmental impact to near zero. In die-sinking EDM, electrodischarge grinding (EDG) and abrasive electrodischarge grinding (AEDG), hydrocarbon oils are used as dielectric fluid to constrict the discharge channel, cool the erosion zone, and flush away debris from the interelectrode gap. The dielectric oil has a long operating life, but the debris collected during machining needs an appropriate disposal.

Owing to sparking, metals of the WP and tool electrodes, inorganic substances such as WC, titanium carbide, chromium, nickel, molybdenum, and barium are released and condense in the air. Emissions of organic materials are generated by the vaporization of the dielectrics. Additionally, the rising smoke can carry organic components from substances in the dielectric liquid. The erosion slurry contains eroded WP and tool material and solid decomposition products of the dielectric. This slurry has to be filtered in the dielectric system of the EDM machine.

12.3.3.1 Protective Measures

To reduce the possible hazards that may arise due to machining by EDM, the following measures should be strictly followed:

1. Reduce air pollution to the extent permissible using suitable filters.
2. Incorporate a dielectric cleaning and recirculating system as a part of the EDM machine.
3. Keep the temperature of the media at a constant level of 15°C below flashing point using a proper cooling system.
4. Reduce the emitted electromagnetic radiation by proper shielding of the machine.
5. Reduce the possibility of fire hazard with an automatic fire extinguisher system.
6. Use level sensors for the dielectric level to be above the spark gap by 80 mm.
7. Avoid dielectrics with flashing point of 65°C. A flashing point above 100°C is recommended.
8. Dispose of the waste appropriately.
9. EDM uses very high-voltage electricity. The operator needs to make every effort to be constantly aware of the machine and its surroundings. Coming into contact with any part of the fluid or the electrode can cause severe injury or even death.

12.3.4 Laser Beam Machining

In LBM, a highly collimated, monochromatic, and coherent light beam is generated and focused to a small spot. High power densities (10^6 W/mm^2) are then obtained. The unreflected light is absorbed, thus heating the surface of the specimen. Additionally, heat diffusion into the bulk material causes phase change, melting, and vaporization. Depending on the power density and the time of beam interaction, the mechanism progresses from one of heat absorption and conduction to one of melting and then vaporization. Machining by laser occurs when the power density of the beam is greater than what is lost by the conduction, convection, and the radiation. Moreover, the radiation must penetrate and be absorbed into the WP material. LBM has many impacts related to:

a. *Environment*. Global (greenhouse effect, ozone layer), regional (acidification, entrophication, toxicity), and local impacts (acute toxicity, odor, and noise) occur.
b. *Occupational health*. Chemophysical risks for the workers (process, gas, fumes, laser machine emissions), light emission risks (laser light, secondary light emission), and noise arise due to LBM.

Figure 12.17 shows the possible process impacts on the environment. During machining by lasers, the material is heated, partly vaporized, and chemically transformed.

Environment-Friendly Machine Tools and Operations

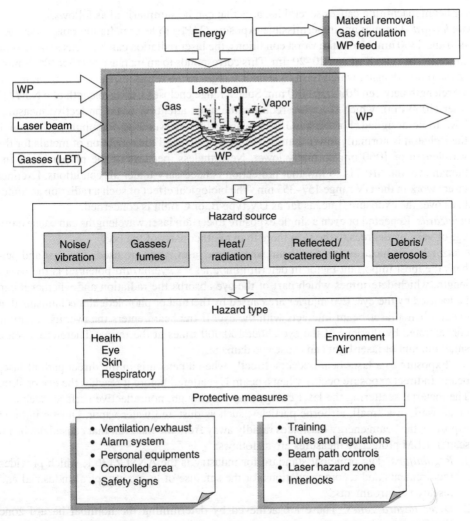

FIGURE 12.17 LBM hazards.

The hazardous materials have the consistency of gasses or aerosols. Aerosols are solid or liquid substances in the gas with particle sizes in the range of 10–16 nm. They can be characterized by the chemical content of the particles, the size distribution, and the material flow rate. During laser machining of thermoplastics, Tönshoff, et al. (1996) reported that 99% of the particles generated have a diameter less than 10 μm and more than 90% are smaller than 1 μm. Most particles are in the range 0.03–0.5 μm. High alloyed steels emit four to five times more aerosols than carbon steel does. Almost all particles generated by LBM have a diameter in the range 0.042–0.35 μm.

The permissible concentrations are quickly reached, especially when the laser machining process generates toxic or dangerous emissions. Consequently working enclosures and exhaust systems have to be applied. A suitable method to extract the hazardous by-products in addition to filtration must be also considered.

Safety aspects are concerned with impacts on the human body; that is, the skin, and especially the eyes. The hazard potential is influenced by power density/intensity, wavelength, exposure time, and whether the beam is visible or invisible.

Laser beam machining causes several hazards that can be summarized as follows:

Beam hazard. The maximum permissible exposure (MPE) to laser radiation ranges between 400 and 1400 nm. Under the worst conditions, the laser radiation can be focused by lens on the retina in a small spot of 10–20 µm. This corresponds to an increase of 0.5×10^6 W/mm² in the power density, which forms a serious danger to the retina, especially in the range of wavelength between 400 and 1400 nm. Such a range encloses the wavelength of a Nd:YAG laser (1060 nm), which therefore requires the most attention when protective measures have to be designed. For Nd:YAG lasers, protective measures are compulsory, although their photon is normally lower than that of CO_2 lasers, and the reflection of metals for the wavelength of 1060 nm normally lower. Nevertheless, because of the properties of the human eye, the MPE is so low that protection is necessary under all conditions. Excimer lasers work in the UV range 157–353 nm. The biological effect of such a radiation accumulates over the exposure time, as far as UV type B, or C light is concerned.

Skin hazards. Repeated or even a single exposure to certain laser wavelengths can cause damage of varying degrees to the skin more than to other parts of the body.

Eye hazards. Eye hazards are significant when some laser types are used. The first and perhaps the most important factor in determining a laser's eye hazard potential is the wavelength, which determines which part of the eye absorbs the radiation and whether it can be focused by the eye. Eye injuries are caused by thermal or photoletical mechanisms that occur when a laser beam interacts with the eye. If the beam enters the eye, its energy is concentrated by the lens of the eye about 100,000 times at the retina. Therefore, even a small amount of laser light can cause eye damage.

Exposure to a laser beam occurs directly when a person is in the direct path of laser beam. Indirect exposure occurs when a beam is scattered before it reaches the eye or skin. The material scattering the laser energy may be a rough, nonreflective surface, such as a break wall, or a small, airborne particles, such as dust and water vapor. During indirect exposure, the beam energy dissipates rapidly away from the material that caused the beam scatter. LBM protective measures are as follows:

1. *Regulations.* The principal standard for industry is the ANSI Z136.1, which provides requirements and recommendations for the safe use of lasers in typical industrial and research environments.
2. *Laser hazard zones.* These are achieved by determining the nominal hazard zone (NHZ), defined as the space within which the level of direct, reflected, or scattered radiation exceeds the level of the applicable MPE.
3. *Beam path controls.* Most industrial lasers fall into the higher classifications of a source of potential eye, skin, and even fire hazards, unless they are totally enclosed, interlocked, and there is no beam access during normal system operation.
4. *Controlled area.* When the beam path is not sufficiently enclosed, the exposure to radiation above the MPE limit occurs. Hence, a laser controlled area is required. During periods of service, a temporary controlled area may be established.
5. *Engineering controls.* Engineering controls are features designed into the laser machine to minimize the risk of exposure to hazardous beams. The most common engineering controls are:
    a. Protective housings and enclosures that cover the equipment or the beam path.
    b. Interlocks are often placed on the protective housings so that if they are removed, the beam is shut off.
    c. Beam stops that provide safe termination of the beam path.
    d. Labels and signs that give notice of lasers operating in a given area.
6. *Protective equipment.* Protective equipment such as barriers or curtains, clothing, or eyewear should be relied upon only if the other control measures do not provide adequate protection.

7. *Administrative and procedural controls.* Administrative and procedural controls consist of a series of rules, regulations that are designed to minimize the risk of laser beam exposure. One of the most effective administrative controls is training.

12.3.5 Ultrasonic Machining

USM is the removal of hard and brittle materials using an axially oscillating tool at ultrasonic frequency (18–20 kHz). During that oscillation, the abrasive slurry of B_4C, Al_2O_3, or SiC (100–800 grit) is continuously fed into the machining zone, between a soft tool (brass or steel) and the WP. The abrasive particles are therefore hammered into the WP surface and cause chipping of fine particles from it. The oscillating tool, at amplitude ranging from 10 to 40 µm, imposes a static pressure on the abrasive grains and feeds down as the material is removed to form the required tool shape (Figure 12.18).

The abrasive slurry is circulated between the oscillating tool and WP through a nozzle close to the tool/WP interface at an approximate rate of 25 L/min. The process finds many industrial

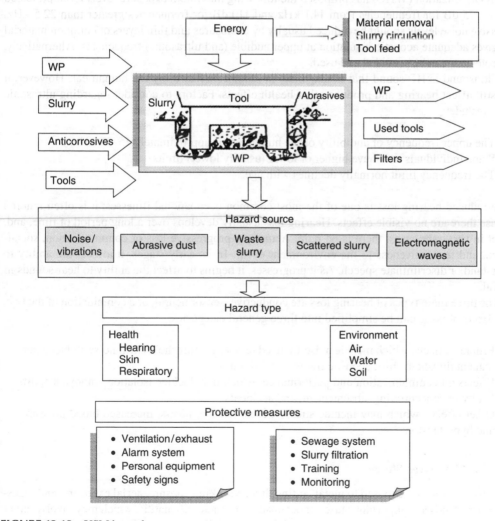

FIGURE 12.18 USM hazards.

applications when machining hard and brittle materials such as ceramics, glass, and carbides. However, it has many environmental and health hazards that include the electromagnetic field (EMF), ultrasonic waves, and abrasives slurry.

12.3.5.1 Electromagnetic Field

In spite of the absence of valid evidence based on solid scientific research, the effects of the EMF on the health of individuals and environments is still of concern to some people. Until research data suggests a need for more extreme action, individuals are advised to stay away from the EMF sources. The strength of a magnetic field drops quickly just few feet from the source. Additionally, it is not recommended to sleep or sit for long periods of time near electrical devices, especially ultrasonic generators and motors.

12.3.5.2 Ultrasonic Waves

Ultrasonic frequencies have been found to produce sound in the audible range from 96 to 105 dB, although it may not appear noisy to older persons and those with imperfect hearing. The World Health Organization (WHO) has proposed the following maximum exposure levels for unprotected persons: 75 dB for frequencies from 14.1 kHz and 110 dB for frequencies greater than 22.5 kHz. Excessive noise levels should be reduced usually by enclosures and thin layers of common material that gives adequate acoustic insulation at upper audible (and ultrasonic) frequencies. Alternatively, ear protectors can be provided and used.

Ultrasound is HF sound that is inaudible, or cannot be heard, by the human ear. However, it may still affect hearing and produce other health effects. Factors to consider regarding ultrasonic waves include:

- The upper frequency of audibility of the human ear is approximately 15–20 kHz.
- Some individuals may have higher or lower (usually lower) limits.
- The frequency limit normally declines with age.

Noise-induced hearing loss is one of the most common occupational illnesses; it is often ignored because there are no visible effects. Hearing loss usually develops over a long period of time, and, except in very rare cases, there is no pain. Actually a progressive loss of communication, socialization, and responsiveness to the environment occur. In its early stages, it affects the ability to understand or discriminate speech. As it progresses, it begins to affect the ability to hear sounds in general.

The three main types of hearing loss are conductive, sensor neural, or a combination of the two. The effects of noise can be simplified into three general categories:

- Primary effects, which include noise-induced temporary threshold shift, noise-induced permanent threshold shift, acoustic trauma, and tinnitus
- Effects on communication and performance, which may include isolation, annoyance, difficulty in concentrating, absenteeism, and accidents
- Other effects, which may include stress, muscle tension, ulcers, increased blood pressure, and hypertension

12.3.5.3 Abrasives Slurry

Basically, workers are exposed to metal cutting fluids via skin exposure, aerial exposure, and ingestion. Slurry fluids are important causes of occupational contact dermatitis, which may involve either irritant or allergic mechanisms. Water-mixed fluids generally determine irritant contact dermatitis and allergic contact dermatitis when they are in touch with workers' skin. Nonwater-miscible fluids usually cause skin disorders such as folliculitis, oil acne, keratoses, and carcinomas.

Mists are aerosols comprised of liquid particles less than 20 μm, which may accommodate abrasive dust. The nonaqueous components of the abrasive slurry, such as the biocide additives, then become a fine aerosol that can enter the workroom air. Mist also may be generated by the spray from the slurry application. Small droplets may be suspended in the air for several hours or several days, possibly in the workers breathing zones. Inhaled particles (with aerodynamic diameters less than 10 μm) deposit in the various regions of the respiratory system by the complex action of the different deposition mechanisms. Particulates below 2.5 μm aerodynamic diameter deposit primarily in the alveolar regions, which is the most sensitive region of a lung. The size of particulates range from 2.5 to 10 μm deposits, primarily in the airways. The mist droplets can cause throat, pancreas, rectum, and prostate cancers, as well as breathing problems and respiratory illnesses. One acute effect observed is mild and reversible narrowing of airways during exposure to the slurry mist.

Several other epidemiological studies have also suggested that exposure to fluid mist may be associated with increased risk of airway irritation, chronic bronchitis, asthma, and even laryngeal cancer. The OSHA standard for airborne particulate (largely due to fluid mist) is 5 mg/m^3, and the International Union of United Automobile, Aerospace, and Agricultural Implement Workers of America (UAW) has proposed a reduction in the standard to 0.5 mg/m^3.

Antimisting compounds, such as a polymethacrylate polymer, polyisobutylene, and poly-N-butane in concentrations of 0.2% as well as poly (1,2-butene oxide) have been suggested to be added to cutting fluids. But consideration must be given to the effects of these chemicals upon humans. The most effective way to control mist exposure is to use a mist collector that prevents mist from entering the plant air. Many collectors use several stages of filters in series and other collectors use centrifugal cells or electrostatic precipitators as intermediate stages.

12.3.5.4 Contact Hazards

If fingers or hands are put into a ultrasonic machine, a tickling sensation is instantly experienced on the skin surface followed 2–3 s later by pain in the joints.

12.3.5.5 Other Hazards

Some scientists have asserted that effects such as nausea, dizziness, tiredness, and tinnitus can be caused by exposure to sound in the ultrasonic frequencies and others that the upper audible frequencies are also implicated.

12.3.6 ABRASIVE JET MACHINING

Atmospheric contaminants from abrasives are a major hazard in AJM. The prime hazard in AJM is dust; in particular, silica dust. Prolonged inhalation of crystalline silica dust can cause silicosis. Particles of other materials may also be present in the dust to which the employee is exposed. The nature of these materials depends on the material being machined. Sometimes heavy concentrations of iron oxide are produced during the cleaning of iron castings. Machining toxic metals such as lead, mercury, arsenic, zinc, and cadmium, and toxic dusts may constitute a significant hazard. Employers must ensure that their employees are provided with appropriate respiratory protective equipment to ensure that they are protected against atmospheres containing substances that may be toxic and harmful if breathed.

There are four risk factors with dust:

- Type of particulate involved and its biological effect
- Concentration of airborne particulate in the breathing zone of the work
- Size of the particles present in the breathing zone
- Duration of exposure

Dust generated by AJM falls into three categories:

a. *Inspirable dust* is the dust which a worker can inhale from the air in the work area. It can contain a wide range of particles of different sizes, including particles that are too heavy to be inhaled and captured by the respiratory system to very small particles of respirable size.
b. *Respirable dust* is that portion of inspirable dust consisting of very small particles of dust that can easily penetrate the lung down to the lower bronchioles and alveolar regions. In this regard, AJM produces high levels of respirable dust.
c. *Silica dust* is a major hazard in AJM, generated when
   - The abrasive medium contains silica
   - Machining materials that contain silica

The major risk from silica dust is silicosis, which is a chronic disease that causes stiffening and scarring of the lung. Symptoms usually take a number of years to appear. They include shortness of breath, coughing, and chest pain. This leads to degeneration in the individual's health. The risk of disease is directly related to the amount of dust inhaled. There is some evidence that people with silicosis have an increased risk of developing lung cancer.

Measures to control silica dust generated by AJM include using a jet machining medium that does not contain silica, and ensuring the process is isolated or appropriate administrative strategies are implemented. Operators must wear personal protective equipment and wet AJM techniques must be used.

Isolating the process where silica dust is produced may be achieved by:

- Machining in a closed chamber
- Enclosing an area with plastic or other forms of impervious protective sheeting to limit the movement of dust from the source where jet machining in a blast chamber is not practicable
- Setting up designated exclusion zones by:
  1. Installing physical barriers and warning signs to prevent unprotected persons from entering the area
  2. Shrouding the area where jet machining is to occur
  3. Restricting entry of unprotected persons into the AJM area while jet machining is running

If isolation is by means of an exclusion zone, signs must be posted at least 25 m from the perimeter of the exclusion zone of the AJM site to warn workers and others that:

- AJM is in progress
- Access to the work area is restricted to authorized persons
- Respiratory protection must be worn in the work area

Particular care must be taken in the cleanup process. Dust must be cleaned and collected in an appropriate manner to ensure that the level of silica dust does not exceed the exposure standard. Methods of cleaning include the wetting down of dust prior to cleanup. Hazardous impacts associated with AJM include:

1. Worker: The hazardous environment associated with AJM can affect the worker's health and performance through the following:
   - Health hazard
   - Low worker productivity
   - High physical/mental fatigue
   - Low job satisfaction
   - High error rates

TABLE 12.5
Hazardous Materials by Different Machining Processes

Process	Mode	Solid		Chemical						Noise	Radiation	Vibrations	Magnetic Field
		Chips	Dusts	Gasses	Vapors	Liquids	Mists	Fume	Slurry				
Turning	MQL	X		X	X		X	X		X		X	
	Dry	X										X	
Drilling	MQL	X		X			X	X		X		X	
Milling	MQL	X		X	X					X		X	
	Dry	X										X	
Reaming	MQL	X		X	X		X			X		X	
Grinding	MQL		X			X	X	X	X	X		X	
CHM				X		X							
ECM				X	X	X				X			
EDM				X	X	X				X			
LBM				X	X					X	X		X
EBM				X									
PBM						X							
USM			X							X	X		X
AJM			X	X								X	
WJM			X			X				X		X	

Worker condition improvement can be achieved by wearing suitable protective clothing and using safety tools and by replacing the worker regularly from time to time to decrease health problems.

2. Product quality is impured as the abrasives may get impeded in the work surface that may be also covered with dust.
3. Workplace is affected also by the mixture of abrasives and WP particles that may lead to the following:
   - Producing dusty air
   - Abrasive particles make problems in other machines exist in the same station
   - Other work parts on the other m/cs may be polluted
   - Health problems to other workers exist in the same station

   Workplace condition improvement can be achieved through good ventilation and implementing air filtration system. For better machine conditions, isolation of the machined part and the use of a suitable dust collection system are recommended. Table 12.5 summarizes the possible hazards of machining processes.

12.4 REVIEW QUESTIONS

1. What are the main steps that should be followed by a company's environmental policy?
2. Show by a line sketch how a clean machining technology could be achieved.
3. Illustrate the different machining liquids used for cooling, for lubricating, and as a machining medium.
4. What are the major chemical agents that may arise during the various machining processes?
5. Discuss the effect of the various hazardous substances associated with the machining technology.
6. Explain using diagrams the hazardous sources and type of hazards for machining by the cutting, abrasion, and erosion processes.

7. Show by a sketch the possible streams of cutting liquids.
8. State the different machining processes that adopt the dry machining method.
9. What is meant by the hazards ranking system of cutting fluids?
10. State the possible hazards of cutting fluids.
11. Explain what is meant by cryogenic cooling and ecological machining.
12. What are the major health effects of CHM and ECM?
13. State the hazards potentials and the protective measures for machining by cutting and abrasion.
14. State the hazards potentials and the protective measures for machining by ECM, CHM, LBM, EDM, and USM.

REFERENCES

Byrne, G. and Scholta, E. (1993) Environmentally clean machining processes—a strategic approach, *CIRP*, 42(1), 471–474.

Dhar, N. R., Nada Kishore, S. V., Paul, S., and Chattopadhay, A. B. (2002) The effect of cryogenic cooling on chips and cutting forces in turning AISI 1040 and AISI 4320 steels, Proceeding of Institution of Mechanical Engineers, *Journal of Engineering Manufacture,* 216(Part B), 713–724.

Hughes, P. and Ferrett, E. (2005) *Introduction to Health and Safety at Work*, 2nd Edition, Elsevier, New York.

New Logic Research, Inc., Emeryville, CA, www.vsep.com/solutions/technology.html.

Tönshoff, H. K., Egger, R., and Klocke, F. (1996) Environmental and safety aspects of electrophysical and electrochemical processes, *Ann. CIRP*, 45(2), 553–567.

Weinert, K., Inasaki, I., Sutherland, J. W., and Wakabyashi, T. (2004) Dry machining and minimum quantity lubrication, *Ann. CIRP*, 53(2), 17.

13 Design for Machining

13.1 INTRODUCTION

Manufacturing cost is the key factor to the economic success of a product. Economic success depends on the profit margin earned on each sale of the product and how many units the firm can sell. The number of units sold and the sales price depends on the product quality. Successful design therefore is ensured by maintaining high product quality while minimizing the manufacturing cost. Design for manufacturing (DFM) is one method of achieving this goal. Effective DFM practices lead to low manufacturing costs without sacrificing product quality. The following principles aid designers in specifying components and products that can be produced at minimum cost:

1. *Simplicity of the product.* This means the minimum number of parts, the least intricate shape, the fewest precision adjustments, and the shortest production sequence.
2. *Standard material and components.* This enables benefits of mass production and simplifies inventory management, avoids tooling and equipment investment, and speeds up the manufacturing cycle.
3. *Standard design of the product.* When several similar products are to be produced, specify the same materials, part, and subassemblies for each as much as possible.
4. *Specify liberal tolerances.* The higher costs of tight tolerance arise due to:
    i. Extra machining operations, such as grinding, honing, or lapping after primary machining operations
    ii. Higher tooling cost
    iii. Longer operating cycles
    iv. Higher scrap and rework costs
    v. The need for more skilled and highly trained workers
    vi. Higher materials cost
    vii. High investment for precision equipments

Table 13.1 shows the approximate relative cost for achieving certain tolerances and surface finishes. Accordingly, it is recommended to consider the following:

1. Use the most machinable materials available.
2. Avoid secondary operations such as deburring, inspection, plating, painting, and heat treatment.
3. Design should be suitable for the production method that is economical for the quantity required.
4. Use special process capabilities to eliminate many operations and the need for separate costly components.
5. Avoid process restrictiveness and allowing manufacturing engineers to choose a process that produces the required dimensions, surface finish, and other characteristics.

13.1.1 GENERAL DESIGN RULES

The general design rules for economic production are as follows:

1. Simplify the design by reducing the number of parts required.
2. Design for low labor cost operations wherever possible.

TABLE 13.1
Approximate Relative Cost for Machining Tolerances and Surface Finishes

	Tolerance		Roughness (R_a)	
Machining Process	± (mm)	Relative Cost	µm	Relative Cost
Rough machining	0.770	100	6.25	100
Standard machining	0.130	190	3.12	200
Fine machining (rough grinding)	0.030	320	1.56	440
Very fine machining (ordinary grinding)	0.010	600	0.80	720
Fine grinding, shaving, honing	0.005	1100	0.40	1400
Very fine grinding, shaving, honing, lapping	0.003	1900	0.20	2400
Lapping, burnishing, superhoning, polishing	0.001	3500	0.18	4500

3. Avoid generalized statements on drawings that may be difficult for the production personnel to interpret.
4. Dimensions should be made from specific points or surfaces on the part itself.
5. Once the functional requirements are met, designers should strive at minimum weight.
6. Dimensions should be made from one datum line rather than from a variety of points to simplify tooling, and gauging, and to avoid overlap of tolerances.
7. Design to use general purpose tooling rather than special ones.
8. Avoid sharp corners for ease of production and avoid stress concentration on the part.
9. Design a part so that many operations can be performed.
10. Space holes in machined parts so that they can be made in one operation without tooling weakness.
11. Whenever possible, cast, molded, or powder-metal parts should be designed without stepped parting line and with uniform wall thickness.

13.2 GENERAL DESIGN RECOMMENDATIONS

Because of the highly competitive nature of machining processes, the question of finding ways to reduce cost is ever-present. A good starting point for cost reduction is in the design of the product. The design engineer should always keep in mind the possible alternatives available in making his design. Unfortunately, designers often consider that their job is to be to design the product for performance, appearance, and reliability and think that it is the manufacturing engineer's job to produce whatever has been designed. Of course, there is often a natural reluctance to change a proven design for the sake of a reduction in the machining cost. As a subject, design for machining hardly exists as compared with design for strength.

For obvious reasons, machining is considered as a wasteful process and many engineers will feel that the main concern should be to design components that do not require machining. As 80–90% of manufacturing machines are designed to machine metal, the view that machining should be avoided must be considered rather impracticable for the immediate future. However, the trend toward the use of processes that conserve material is clearly increasing, and when large-volume product is involved, this approach should be foremost in the designer's mind. In this chapter, certain design principles that can help to simplify the machining of components and reduce costs are introduced. Machined parts are used in applications for which precision is required. Machining is also involved if surface finish, flatness, roundness, circularity, parallelism, or close fit is involved. Additionally, if the part is in motion, or fits precisely with another part, machining operations will be employed. Machined parts can be as small as miniature screws, shafts, or gears. They can be as large as huge

Design for Machining

turbines, turbine housings, and valves found in hydroelectric power stations. Machined components are made from ferrous and nonferrous materials. However, plastics, rubber, carbon, graphite, and ceramics are also employed.

The following are some important regulations, that should be followed by the part designers.

1. Avoid machining operations if the surface or the feature required can be produced by casting or forming.
2. Specify the most liberal surface finish and dimensional tolerances consistent with the function of the surface to avoid costly grinding, lapping, and other finishing operations (Figure 13.1).
3. Design the part for ease of fixation and secured clamping during the machining operation.
4. Avoid sharp corners and sharp points in cutting tools to avoid their breakage.
5. Use stock dimensions whenever possible (Figure 13.2).
6. Avoid interrupted cuts during single-point machining operations.
7. Design parts that are rigid enough to withstand clamping and cutting forces.
8. Avoid tapers and contours that simplify tooling and setups.
9. Reduce the number and the size of shoulders, as they require extra materials and operations.

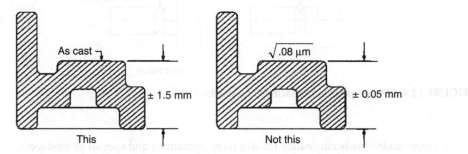

FIGURE 13.1 Avoid tolerances that necessitate machining if as-cast, as-forged, or as-formed dimensions and surface finishes are satisfactory for the parts function.

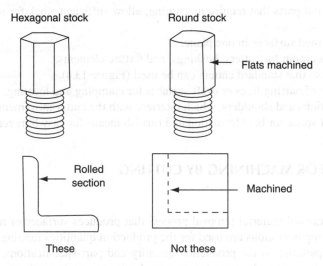

FIGURE 13.2 Use stock dimensions and minimize the machining allowance.

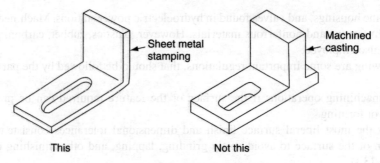

FIGURE 13.3 Metal-formed parts are better than machined castings.

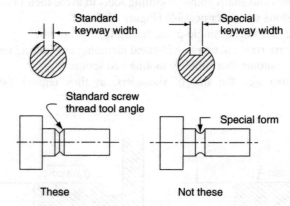

FIGURE 13.4 Design parts to be machined by standard tools.

10. Avoid undercuts because they involve more operations and special ground tools.
11. Substitute a stamping operation for the machined component (Figure 13.3).
12. Avoid the use of hardened or difficult-to-machine materials unless their functional properties are required.
13. For thin and flat parts that require machining, allow sufficient stock for rough and finish operations.
14. Put the machined surfaces in one plane.
15. Provide access room for cutters, bushings, and fixture elements.
16. Design parts so that standard cutters can be used (Figure 13.4).
17. Avoid the use of parting lines or draft surfaces for clamping and locating.
18. Avoid projections and shoulders, which interfere with the cutter movement.
19. Provide relief space for burr formation and furnish means for easy burr removal.

13.3 DESIGN FOR MACHINING BY CUTTING

13.3.1 Turning

Turning is a conventional material removal process that produces surfaces of rotation on the WP (Figure 13.5). Turning operations are used for the production quantities ranging from one piece to many millions. Depending on the production quantity and part specifications, turning operation ranges from manual, numerical, computer, or completely automatic mechanical control. Tooling costs for parts machined on engine lathes are very low. Turret lathes are used for 10–25 or more parts.

Design for Machining

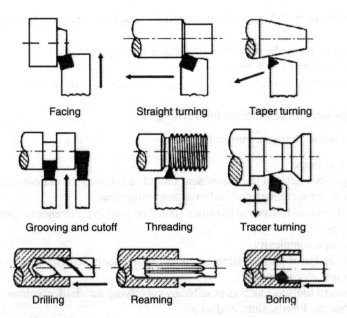

FIGURE 13.5 Basic turning operations.

In general, turning machine operations are more frequently used for lower ranges of production quantities. The most economic range for each machine tool is as follows:

- Engine lathes: very low to low quantity
- Turret lathes: low to medium quantity
- Tracer lathes: low to medium quantity
- NC and CNC lathes: low to medium quantity
- Single spindle (chucking type): medium to high quantity
- Multispindle chucker: high to very high quantities

Automatic screw machined parts are generally cylindrical in shape and may have several outside diameters as well as a hexagonal or square-surfaced portion. They may include threads on one or both ends, and may have an internal axial hole with more than one diameter. The hole may be chamfered or tapered. Threads may be different in size and pitch and may be both external and internal. The diameter ranges from the smallest watch parts to 200 mm. Thread length can be as short as 1 mm and as long as 1 m. Such parts can be produced using one of the following machines:

- Swiss-type (shafts, pinions, contacts for electrical devices, pins, and valves)
- Turret-type (rivets, nuts, bolts, shafts, spacers, washers, pulleys, valve stems, spools gear blanks, rollers, and push rods)
- Multispindle automatic machine

13.3.1.1 Economic Production Quantities

The economic production quantity depends on the machine tool used. In this legand, the following considerations should be made:

1. The output rate ranges from few seconds to about 5 min per piece for single-spindle machines.
2. Higher rates are possible using multispindle automatic machines.

3. One operator can tend 4 or more single-spindle machines and may be assigned to 10 or more.
4. Two or more multispindle machines are assigned per operator.
5. Setup times are from 1 to 8 hours.
6. Multispindle machines require more tooling than single-spindle machines.

13.3.1.2 Design Recommendations for Turning

Designers should follow these recommendations:

A. Stock size and shape
   1. The largest diameter of the component should be taken as the diameter of the bar stock in order to conserve material and save machining time.
   2. Standard sizes and shapes of bar stock should be used in preference to special diameters and shapes.
B. Basic part shape complexity
   1. Keep the design of parts as simple as possible to reduce the number of tool stations and gauging processes required.
   2. Use standard tools as much as possible by specifying standard, common sizes of holes, screw threads, knurls, slots, and so on.
C. Avoiding secondary operations
   1. The part should be complete when cut off from the bar material.
   2. Secondary operations such as slots and flats should be small and performed when the part is held in the pickoff attachment.
   3. Internal surfaces and screw threads should be located at one end so that they can be performed before cutoff and without the need for rechucking (Figure 13.6).
D. External forms
   1. The length of the formed area should not exceed two and half times the minimum WP diameter (Figure 13.7).
   2. Sidewalls of grooves and other surfaces that are perpendicular to the axis of the WP should have a slight draft (Figure 13.8) of 1/2° or more to prevent tool marks when the tool is withdrawn.
   3. When turning from square or hexagonal stock, the turned diameter is the distance between two opposite flats of the stock. It is advisable to design turned parts to be about 0.25 mm or smaller than the bar stock size.
   4. Avoid deep narrow grooves and sharp corners.

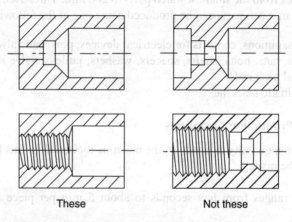

FIGURE 13.6 Operations should be finished without the need for rechucking.

Design for Machining

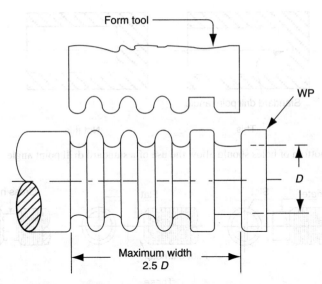

FIGURE 13.7 Form tool width limitations.

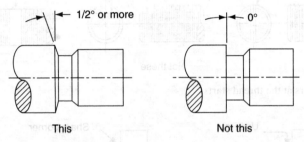

FIGURE 13.8 Provide a slight angle to sidewalls and faces to prevent tool marks when the tool is withdrawn.

E. Undercuts
 1. Avoid angular undercuts and use undercuts obtainable with traverse or axial tool movements.
 2. External grooves are machined more economically than internal recesses.
F. Holes
 1. The bottom shape of blind holes should be that made by a standard drill point (Figure 13.9).
G. Screw threads
 1. Avoid the formations of burrs in threaded parts (Figure 13.10).
H. Knurls
 1. Knurled width should be narrow (≤WP diameter).
 2. Specify the approximate number of teeth per inch, type of knurl, general size, and use of knurl.
I. Sharp corners
 1. Avoid sharp corners (external and internal) as they cause weakness or more costly fabrication of form tools.
 2. Provide a commercial corner break of 0.4 mm by 45°.
 3. An internal sharp corner can be made by providing an undercut at the corner (Figure 13.11).

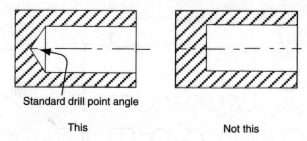

FIGURE 13.9 The bottom of holes should allow the use of a standard drill point angle.

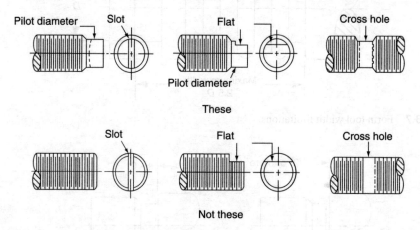

FIGURE 13.10 Avoid burrs at the thread starts.

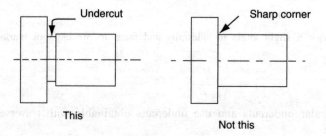

FIGURE 13.11 Undercuts avoid the problems of sharp corners.

  J. Spherical ends
    1. Design the radius of the spherical end to be larger than the radius of the adjoining cylindrical surface (Figure 13.12).
  K. Slots and flats
    1. Slots are produced with a concave surface at the bottom or end (milling cutter radius) (Figure 13.13).
  L. Marking
    1. Position impression marking so that roller marking tools can be used (Figure 13.14).

Designers should also follow these additional recommendations:

1. Incorporate standard tool geometry at diameter transitions, exterior shoulders, grooves, and chamfer areas.

Design for Machining

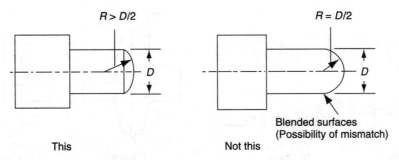

FIGURE 13.12 Avoid blended surfaces formed by a separate cutter.

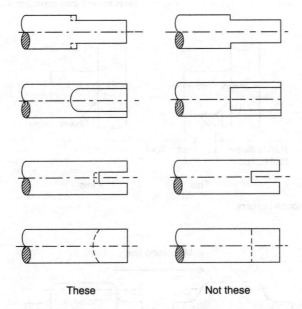

FIGURE 13.13 Permit curved bottoms of slots and flats if possible.

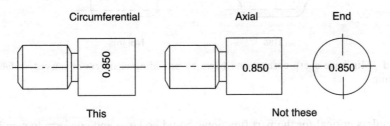

FIGURE 13.14 Marking should allow the use of roller tools.

2. Minimize unsupported, delicate, small-diameter work whenever possible to reduce work deflection. In this regard, short, stubby parts are easier to machine (Figure 13.15).
3. Avoid interrupted cutting actions that may be caused by hole intersections, slant surface drilling, and hole or slotting operations before turning.
4. Castings and forgings with large shoulders or other areas to be faced should have 2°–3° from the plane normal to the axis of the part to provide edge relief for the cutting tools (Figure 13.16).

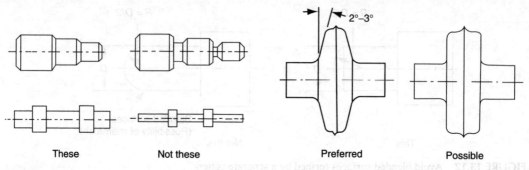

FIGURE 13.15 Avoid slender and long parts to avoid deflection.

FIGURE 13.16 Allowing a relief on cast or forged parts to be faced provides tool clearance.

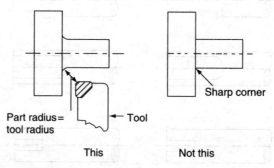

FIGURE 13.17 Avoid sharp corners.

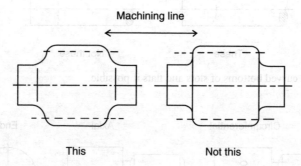

FIGURE 13.18 Minimize sharp corners and burrs by providing chamfers or curved surface to the part before machining.

5. Radii, unless critical for the part function, should be large and conform to standard tool-nose radius specifications (Figure 13.17).
6. Specify a break of sharp corners where sharpness or burs may be hazardous or disadvantageous to the function of the part (Figure 13.18).
7. Avoid clamping or locating the part using parting lines, draft angles, and forging flash (Figure 13.19).
8. For tracer-controlled parts, easy tracing is necessary with a minimum number of changes of the stylus and cutting tool. Grooves with parallel or steep sidewalls are not possible and undercuts should be avoided.

Design for Machining

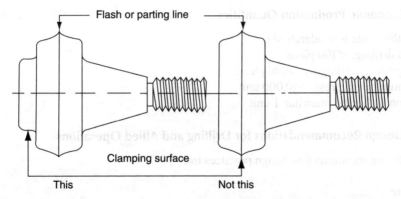

FIGURE 13.19 Avoid designs that require clamping on parting lines or flash areas.

13.3.1.3 Dimensional Control

The dimensional tolerances are inversely proportional to the part size. In this regard, the produced tolerances depend on:

- Machine construction
- Control of operational disturbances such as vibrations, deflection, thermal distortion, and wear of operating parts
- Part deflection, tool wear, measuring tool accuracy, and operator skill

Surface finish is also related to the aforementioned factors. It is directly related to the tool feed rate, tool sharpness, tool and WP geometry, and tool materials. Typical machining tolerances are specified by Bralla (1999) for well-maintained turning machines.

13.3.2 DRILLING AND ALLIED OPERATIONS

Hole-making operations include drilling, boring, trepanning, and gun drilling. Counterboring and countersinking are secondary operations for existing holes. Drilling and reaming can produce holes in the following range:

- Usual diameter range: 1.5–38 mm
- Minimum diameter range (spade-type microdrills): 0.025 mm
- Minimum diameter (reaming): 0.3 mm
- Maximum diameter: 80–90 mm
- Usual maximum depth: eight times diameter
- Hardness of drilled material: usually less than HRC 30
- Maximum hardness: HRC 50 rarely to HRC 60

Boring is used when a particular accuracy of diameter, location, straightness, or direction is required. The normal dimensional limits for bored holes are:

- Minimum diameter: 2.5 mm, with fishtail-type solid cutting tool
- Maximum diameter: 1.2 m
- Maximum hardness of material: 60 HRC
- Maximum length of conventional boring bars: five times diameter
- Maximum length of solid boring bars: eight times diameter

13.3.2.1 Economic Production Quantities

- NC drilling: one to moderate size
- Normal drilling: <100 piece
- Drilling using drilling jig: >100 piece
- Multispindle drill press: >10,000 unit
- Precision jig boring machine: 1 unit

13.3.2.2 Design Recommendations for Drilling and Allied Operations

The following are recommended design practices for,

A. *Drilling*
   1. The drill entry surface should be perpendicular to the drill bit to avoid starting problems and to ensure proper location (Figure 13.20).
   2. The exit surface of the drill should be perpendicular to the axis of the drill to avoid drill breakage when leaving the hole (Figure 13.20).
   3. For straightness requirements, avoid interrupted cuts to avoid drill deflection and breakage (Figure 13.21).
   4. Use standard drill sizes whenever possible.
   5. Through holes are preferable than blind holes, as they provide easier clearance for tools and chips.
   6. Blind holes should not have flat bottoms because they require a secondary machining operation and cause problems during reaming.
   7. Avoid deep holes (over three times diameter) because of chip clearance problems and the possibility of straightness errors (Figure 13.22).
   8. Avoid designing parts with very small holes if they are not truly necessary (3 mm is the desirable minimum diameter).
   9. If large holes are required, it is desirable to have cored holes (casting) in the WP before drilling.

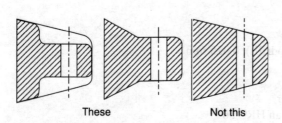

These   Not this

FIGURE 13.20 Entrance and exit surfaces should be perpendicular to the drill axis.

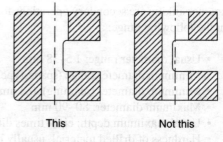

This   Not this

FIGURE 13.21 Keep the center of the drill point in the work throughout the drilling operation.

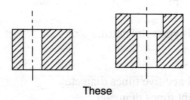   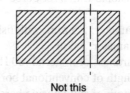

These   Not this

FIGURE 13.22 Avoid deep, narrow holes (depth <3 diameter or consider stepped diameter).

Design for Machining

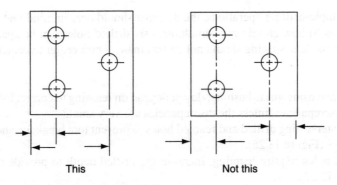

FIGURE 13.23 Locate holes from one datum surface.

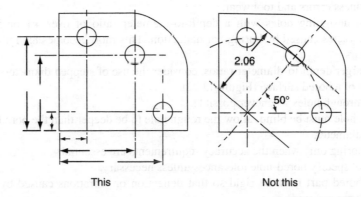

FIGURE 13.24 Rectangular coordinates are preferable to angular coordinates for describing the point location of holes.

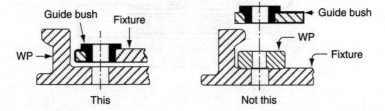

FIGURE 13.25 Allow room for drill bushes close to the drilled surface.

10. If the part requires several drilled holes, dimension them from the same surface to simplify fixturing (Figure 13.23).
11. Rectangular rather than angular coordinates should be used to designate hole locations (Figure 13.24).
12. Design parts so that all can be drilled from one side or from the fewest number of sides.
13. Design parts so that there is a room for the drill bushing near the surface where the drilled hole to be started (Figure 13.25).
14. Standardize the size of holes, fasteners, and screw threads as much as possible.

15. For multiple-drilling operations, the designer should bear in mind that there are limitations as to how closely two simultaneously drilled holes can be spaced (for 6 mm diameter or less, spacing should not be less than 19 mm center to center).

B. *Reaming*
1. Even when using guide bushing, do not depend on reaming to correct location or alignment discrepancies unless the discrepancies are very small.
2. Avoid intersecting drilled and reamed holes to prevent tool breakage and burr removal problems (Figure 13.26).
3. If blind holes require reaming, increase the drilled depth to provide room for chips (Figure 13.27).

C. *Boring*
1. During boring, avoid designing holes with interrupted surfaces, as they cause out-of-roundness errors and tool wear.
2. Avoid designing holes with a depth-to-diameter ratio of over 4:1 or 5:1 to avoid inaccuracies caused by boring-bar deflection. This ratio becomes 8:1 for carbide boring bars.
3. For larger depth-to-diameter ratios, consider the use of stepped diameters to limit the depth of a bored surface (Figure 13.22).
4. Use through holes whenever possible.
5. If the hole must be blind, allow the rough hole to be deeper than the bored hole by 1/4 hole diameter.
6. Use boring only when the accuracy requirements are essential.
7. Do not specify bored-hole tolerances unless necessary.
8. The bored part must be rigid so that deflection or vibrations caused by the cutting forces are reduced.

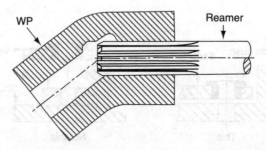

FIGURE 13.26 Avoid intersecting drilled and reamed holes.

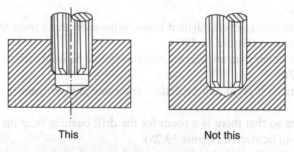

FIGURE 13.27 Provide extra hole depth when reaming blind holes.

Design for Machining

13.3.2.3 Dimensional Control

The following factors affect the dimensional tolerances of drilling, reaming and boring operations:

- The drill sharpness affects the accuracy of the diameter and straightness of drilled holes.
- The play and lack of rigidity in the drill spindle.
- Drill bushing reduces the bell mouthing of holes as well as the possible hole oversize.
- Thermal expansions of the material to be drilled.
- The location and direction of reamed holes are affected by the previous drilling operation even when using guide pushing.
- Temperature changes of the WP caused by heat generated by cutting or from other causes affect the accuracy of bored holes.
- WP distortion from clamping.
- Machine condition and rigidity and boring-bar rigidity are essentially important in jig boring operations.

Bralla (1999) provided recommended dimensional tolerances for drilled, reamed, and bored holes.

13.3.3 MILLING

13.3.3.1 Design Recommendations

The following general product design rules apply to other machining and milling operations:

1. Sharp inside and outside corers should be avoided.
2. The part should be easily clamped.
3. Machined surfaces should be accessible.
4. Easily machined material should be specified.
5. Design should be as simple as possible.

Additional recommendations particularly applicable to milling are as follows:

1. The product design should permit the use of standard cutter shapes and sizes rather than special ones (Figure 13.28).
2. The product design should permit manufacturing preference as much as possible to determine the radius where two milled surfaces intersect or where profile milling is involved (Figure 13.29).

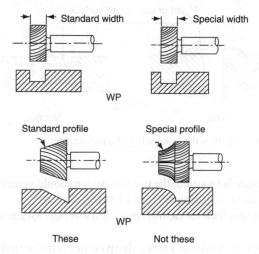

FIGURE 13.28 Allow the use of standard cutter shapes and sizes rather than special ones.

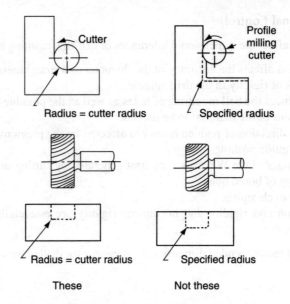

FIGURE 13.29 Allow the use of radii generated by the milling cutter.

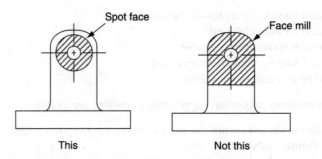

FIGURE 13.30 Spot facing of small surfaces is preferred over face milling.

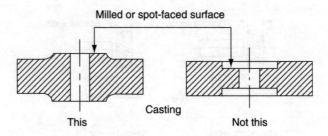

FIGURE 13.31 A low boss simplifies the machining of a flat surface.

3. When small flat surface is required, the product design should permit the use of spot facing, which is quicker than face milling (Figure 13.30).
4. When spot faces are specified for casting, provide a low boss for the surface to be machined (Figure 13.31).
5. When the outside surfaces intersect and a sharp corner is not desirable, the product design should allow a bevel or chamfer rather than rounding (Figure 13.32).

Design for Machining

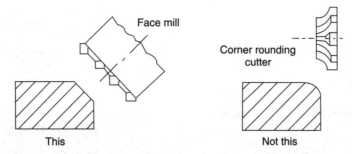

FIGURE 13.32 Allow beveled rather than rounded corners for economical milling.

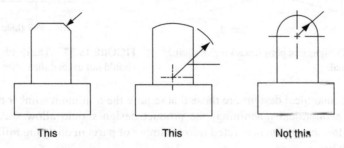

FIGURE 13.33 Do not specify a blended radius on machined parts.

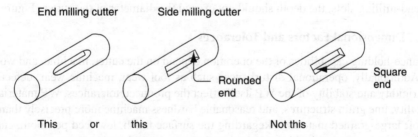

FIGURE 13.34 Design keyways so that a standard milling cutter finishes its sides and ends in one operation.

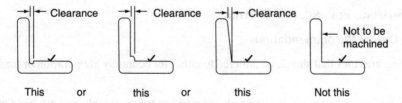

FIGURE 13.35 Provide clearances for milling cutters.

6. When form-milling or machining rails, do not blend the formed surface to an existing milled surface (Figure 13.33).
7. Keyway design should permit the keyway cutter to travel parallel to the center axis of the shaft and form its own radius at the end (Figure 13.34).
8. A design that requires the milling of surfaces adjacent to a shoulder should provide clearance to the cutter path (Figure 13.35).
9. A product design that avoids the necessity of milling at parting lines, flash areas, and weldments will generally extend the cutter life.

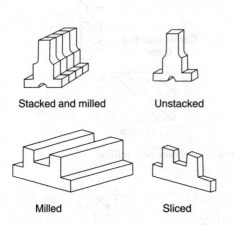

FIGURE 13.36 Designs that permit stacking or slicing are more economical.

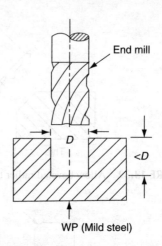

FIGURE 13.37 The depth of end-milled slots should not exceed the cutter diameter.

10. The most economical designs are those that require the minimum number of operations.
11. For more economical machining, the product design should allow staking so that a milled surface can be incorporated into a number of parts in one gang milling operation (Figure 13.36).
12. The product design should provide clearance to allow the use of larger-size cutters rather than small-size ones to permit high removal rates.
13. In end-milling slots, the depth should not exceed the diameter of the cutter (Figure 13.37).

13.3.3.2 Dimensional Factors and Tolerances

The tolerance-holding capabilities of the operation depend on the cutter, machine, and work-holding devices. Additionally, operational disturbances such as tool wear, machine wear, defection, vibration, and rigidity and stability of the WP itself affect the produced tolerances. WP materials of good machinability, fine grain structures, and reasonable hardness machine more precisely than very hard or very soft large-grained materials. Regarding the surface finish, low feed per tooth, cutting fluid, high cutting speed, and machinable materials produce the desired surface smoothness. Surface finish ranges from 1.5 to 3.8 μm (R_a-value) for milling free-machining steels and nonferrous materials.

13.3.4 Shaping, Planing, and Slotting

13.3.4.1 Design Recommendations

The following are rules that should be adhered to either for economy of operation or for dimensional control:

1. Design parts so that they can be easily clamped to the worktable and are rigid enough to withstand deflection during machining (Figure 13.38).
2. It is preferable to put machined surfaces in the same plane to reduce the number of operations required.
3. Avoid multiple surfaces that are not parallel to the direction of tool reciprocation, which would need additional setups.
4. Avoid contoured surfaces unless a tracer attachment is available and then specify gentle contours and generous radii as much as possible.
5. With shapers and slotters, it is possible to cut to within 6 mm of an obstruction or the end of a blind hole (Figure 13.39). If possible, allow a relived portion at the end of the machined surface.

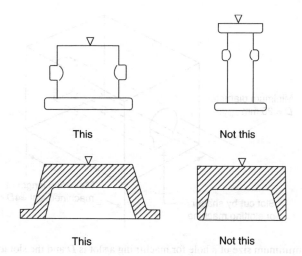

FIGURE 13.38 Shaped and planed parts should withstand cutting forces and provide solid clamping.

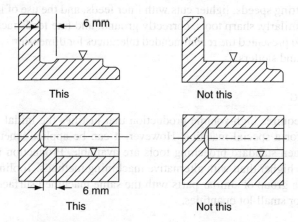

FIGURE 13.39 Avoid machining close to an obstruction at the end of a stroke.

6. For thin, flat WPs that require surface machining, allow sufficient stock for a stress-relieving operation between rough and finish machining or, if possible, rough machine equal amounts from both sides to allow 0.4 mm for finish machining on both sides.
7. The minimum size of hole in which a keyway or a slot can be machined with a slotter or a shaper is about 25.54 mm (Figure 13.40).
8. Because of the lack of rigidity of long cutting tool extensions, it is not feasible to machine a slot longer than four times the hole diameter (Figure 13.40).

13.3.4.2 Dimensional Control

Dimensional variations occur from human factors, the design and condition of the part itself, and the clamping method. The squareness and flatness of the clamping surface is affected by spring back after machining. Additionally, the dimensional variations are affected by

- Deflection of the part by the cutting forces
- WP warping as a result of the release of the internal stresses in the material during machining
- Tool rigidity, especially in slotting operation or shaping of internal surfaces

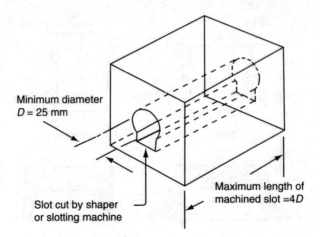

FIGURE 13.40 The minimum size of a hole for machining a slot is D and the slot length should be $<4D$.

Generally, slower cutting speeds, lighter cuts with finer feeds, and the use of lubricants improve the product accuracy. Similarly, sharp tools, correctly ground, and fine feeds facilitate smooth surface finishes. Bralla (1999) presented the recommended tolerances for dimensions and surfaces produced by shapers, planers, and slotters.

13.3.5 Broaching

Broaching usually requires high-volume production that justifies the initial cost of the broaching tools and the need for a special machine. However, it can be applied when there is no machining alternative or when standard broaching tools are available. Production rate range from 15 to more than 100 times higher than with alternative machining methods. Tooling a machine for more than one part or for a group of similar parts with the same machined surface makes broaching an economical choice for small-lot quantities.

13.3.5.1 Design Recommendations

Designers should follow specific recommendations regarding the following issues (Bralla, 1999):

A. Entrance and exit surfaces
  1. The product design should allow easy location and holding of the part.
  2. Surface configuration to the machined area should be square and relatively flat.
  3. Avoid location of parting lines and gates to prevent poor support during machining.
  4. The designer should visualize how the part is supported and retained and avoid the possibility of uneven or inconsistent surfaces in these areas (uneven or inclined surfaces cause side forces that affect the accuracy of finished holes in internal broaching).
B. Stock allowance
  1. Forgings should be held to as close dimension as possible, allowing the minimum stock for finishing to avoid overloading of broach tools during machining (Figure 13.41).
  2. Castings and cold-punched (pierced) holes require greater stock allowance to make sure that clean surfaces are produced.
C. Wall sections
  1. It is advisable to avoid thin wall sections and to maintain a uniform thickness for any wall subjected to the machining forces.

Design for Machining

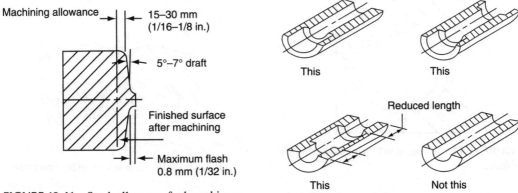

FIGURE 13.41 Stock allowance for broaching a forged part.

FIGURE 13.42 Long holes should be recessed.

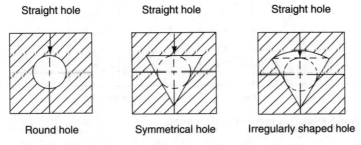

FIGURE 13.43 Irregularly shaped broached holes are started from round holes.

D. Families of parts
   1. The designer should attempt to design parts so that a group of parts use the same broaching tool and, if possible, the same holding fixture.
E. Round holes
   1. Starting holes may be cored, punched, bored, drilled, flame cut, or hot-pierced.
   2. For drilled or bored starting holes, consider 0.8 mm stock for 38 mm in diameter and 1.6 mm on larger holes.
   3. When cored holes are broached, draft angles, surface texture, and size variation must be taken into consideration in determining the size to assure part cleanup.
   4. Long holes should be recessed to improve accuracy and reduce cost (Figure 13.42).
F. Internal forms
   1. Symmetrically shaped internal forms are usually broached from round holes.
   2. Irregularly shaped internal forms may be started from round holes (Figure 13.43), cored, punched, pierced or machined holes, or machined irregular holes.
   3. For casting, stamping, or forging, it is always advisable to leave a minimum amount of stock in addition to the draft, mismatch, and out of roundness errors for complete part cleanup.
G. Internal keyways
   1. Whenever possible, design keyways to the ASA specifications so that standard keyway broaches can be used.
   2. Pilot holes for internal keys should be on the same centerline as the finished hole.
   3. Balanced designs having more than one key equally spaced are preferred to prevent broach drifting (Figure 13.44).

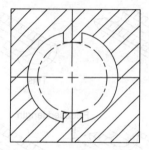

FIGURE 13.44 Balanced WP shapes avoid broach drifting.

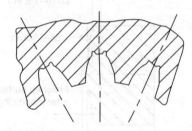

FIGURE 13.45 Allow a room for upset burr.

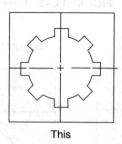   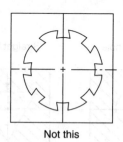

This   Not this

FIGURE 13.46 Avoid dovetail or inverted angle splines.

- H. Straight-splined holes
  1. Parallel or straight-sided holes should be designed to the SAE standard.
  2. Involute splines should be designed to SAE, Deutsches Institut für Normung (DIN), or the American Gear Manufacturers Association (AGMA) standards.
  3. Fine diametral pitches and stub-tooth forms are advisable to reduce the length of broach required.
  4. Long holes should be recessed or relieved.
  5. The designer is allowed to modify the spline profile to allow room for the upset burr (Figure 13.45).
  6. Dovetail or inverted-angle splines should be avoided whenever possible (Figure 13.46).
- I. Spiral splines
  1. In addition to the preceding guidelines, spiral splines with helix angles greater than 40° cannot be broached by conventional methods; use the lowest helix angle possible.
  2. Splines with helix angles >10° require the broach rotation while traveling; the designer should consider preventing the WP rotation.
- J. Tapered splines
  1. Tapered splines cannot be produced by broaching (Figure 13.47).
- K. Square and hexagonal holes
  1. It is advantageous to use slightly oversized starting holes, particularly for square holes (Figure 13.48).
  2. Avoid sharp corners at the major diameter to reduce broach costs (Figure 13.49).
- L. Saw cut or split splinded holes
  1. When the part will have an interesting cut into the splined hole, the splinded hole should be designed with an omitted space, as shown in Figure 13.50. This allows room for burr produced by the saw cut.

Design for Machining

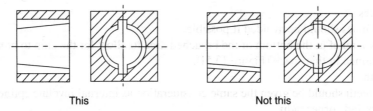

FIGURE 13.47 Avoid tapered splines.

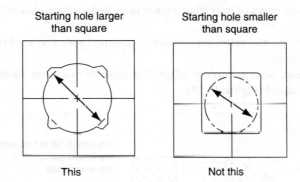

FIGURE 13.48 Use a slightly oversized starting hole.

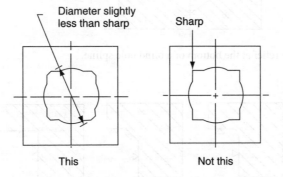

FIGURE 13.49 Avoid sharp corners on major diameters.

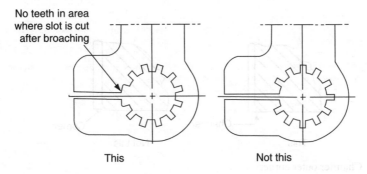

FIGURE 13.50 Allow room for the burr produced by the saw cut.

M. Blind holes
   1. Blind holes should be avoided if possible.
   2. Allow a relief at the bottom of the broached area to allow for the chip to break off and be retained in that space (Figure 13.51).
N. Gear teeth
   1. Gear teeth should be given the same consideration as internal involute splines.
O. Chamfers and corner radii
   1. In all situations that require breaking the corner, chamfers are preferred over radii.
   2. Sharp internal corners should be avoided to eliminate stress concentration and minimize broach tooth edge wear (Figure 13.52).
   3. Sharp corners or edges of intersecting outer broached surfaces should be avoided whenever possible.
   4. Castings, forgings, and extrusions should be designed with a corner break that does not require machining.
   5. After broaching, outer corners or edges that must be machined should be chamfered rather than rounded (Figure 13.53).

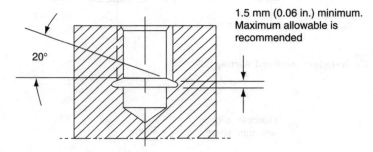

FIGURE 13.51 Allow a relief at the bottom of a blind hole spline.

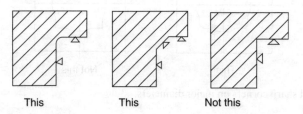

FIGURE 13.52 Internal corner design.

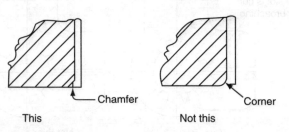

FIGURE 13.53 Chamfer outer corners.

Design for Machining

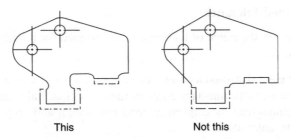

FIGURE 13.54 Reliefs of undercuts in the corners simplify the broaching operation of external surfaces.

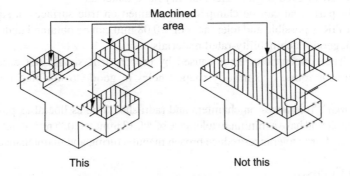

FIGURE 13.55 Break large surfaces into a series of bosses.

- P. External surfaces
    1. External surfaces should be relieved to reduce the area that must be broached.
    2. Reliefs of undercuts in the corners simplify the broaching operation (Figure 13.54).
    3. Large surfaces should be broken into a series of bosses whenever possible (Figure 13.55).
- Q. Undercuts
    1. Machined undercuts should be as shallow as possible.
    2. Avoid sharp or narrow undercut configurations.
- R. Burrs
    1. Chamfers or reliefs at the exit edge of the surface to be broached are recommended to contain the burr produced and eliminate the need for deburring operation.
- S. Unbalanced cuts
    1. Unbalanced stock conditions caused by cross holes or other interruptions that cause tool deflection should be avoided.

13.3.5.2 Dimensional Factors

1. Internal broaching, especially single-pass operations, produces tighter tolerances than external applications.
2. Broaching applications that require tool guiding or multiple passes are subject to residual stresses.
3. Uniformity of material, consistency of datum faces, and strength of the part affect tolerance control.
4. Tool maintenance and resharpening, machinability of material, and proper design of the tool are essential for controlling size and finish required.

13.3.5.3 Recommended Tolerances

According to Bralla (1999), the range of tolerance recommended for broaching operation can be summarized as follows:

Surface finish. Surface finish produced by broaching does not match the grinding finish. However, it is superior to the finish produced by most other machining methods. Burnished finishes can be guaranteed by employing good tool design and proper cooling oils for highly machinable materials.

Flatness. Parts of uniform sections having sufficient strength to withstand cutting forces can be machined within ±0.025 mm total indicator reading (TIR).

Parallelism. Parallelism of machined surfaces machined in the same stroke should be within ±0.025 mm TIR in good-to-fair machinability rated materials.

Squareness. For parts that can be clamped and retained on true surfaces, a squareness of ±0.025 mm TIR is possible and tolerances of ±0.08 mm can be obtained under controlled conditions in good-machinability-rated materials.

Concentricity. The concentricity error caused by broach drift should not exceed ±0.025 to ±0.05 mm for round or similarly shaped holes in good- to fair-machinability-rated materials.

Chamfers and radii. Tolerances on chamfers and radii should be as liberal as possible. Radii under 0.8 mm should have a minimum tolerance of ±0.13 mm; ±0.025 mm should be allowed for larger sizes. Large tolerances reduce broach manufacturing and maintenance cost.

13.3.6 THREAD CUTTING

Cutting screw threads in free cutting materials raises the rate of production and thus reduces the machining and tool cost, while non-free-machining metals are difficult to thread. In this regard, brasses and bronzes cut better and at higher speeds than steels. Cast aluminum is quite abrasive and causes excessive tool wear. It is difficult to cut threads in steels having <160 hardness Brinell (HB). Materials of HRC > 34 are not suitable for die chasers and taps, which are made from HSS. Carbide tools are used for single-point cutting for materials above HRC 34. The most suitable materials for thread grinding are the hardened steels and any material having HRC > 33. Aluminum and comparable soft materials tend to load the GW and cause burning and therefore are difficult for thread grinding.

13.3.6.1 Design Recommendations

Designers should follow these recommendations:

1. Provide a space (1.5–19 mm) for the thread cutting tool (Figure 13.56).
2. Allow chip clearance space when cutting internal threads (through holes are best) (Figure 13.57).
3. Consider the use of a reduced height thread form, which machines more easily (Figure 13.58).

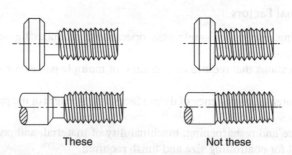

FIGURE 13.56 Allow thread relief at the end of a threaded length.

Design for Machining

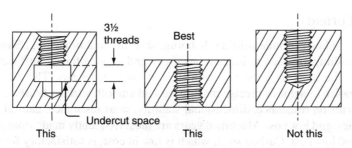

FIGURE 13.57 Allow chip clearance for internal threads.

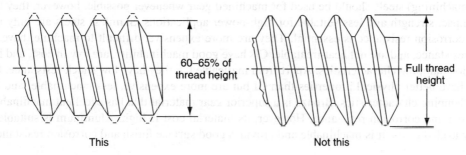

FIGURE 13.58 Reduce thread height for easier machining.

4. Keep the thread as short as possible, which machines quicker and provides longer tool life.
5. Include a chamfer at the top and the end of external threads and a countersink at the top and the end of internal threads.
6. The surface of the starting thread must be flat and perpendicular to the thread's center axis.
7. Avoid slots, cross holes, and flats that intersect with the cut threads.
8. When cross holes are unavoidable, consider countersinking of such cross holes.
9. Do not specify closer tolerances than required (class 2 is commonly satisfactory).
10. Ground threads should be provided with corners of 0.25 mm at the root.
11. The length of centerless ground threads should be larger than the thread diameter.
12. Coarse threads are more economical to produce and assemble faster than fine threads.
13. Tubular parts must have a wall thickness that withstands the cutting forces.

13.3.6.2 Dimensional Factors and Tolerances

The following factors affect the accuracy of machined thread:

- The accuracy and condition of the tooling and equipment
- The skill of the operator
- The suitability of the WP material
- The feed rate of the threading tool

Threads of classes 4 and 5 can be produced by machining while dies are capable of producing threads of classes 1 and 2 only. Surface finishes smoother than 1.6 μm are not obtained by thread cutting. Using thread milling, dimensional tolerances of ±0.025 mm are obtainable. It is capable of producing accurate classes of thread in materials of poor machinability. Thread grinding achieves threads of classes 4 and 5.

13.3.7 Gear Cutting

Gear machining methods include milling, hobbing, shaping, and broaching, Shaving, grinding, lapping, and burnishing are used to improve accuracy and surface finish. Sufficient strength and machinability are the most important prerequisites of machined gears. The high machinability ratings make it easier to achieve precise machining and smooth surface finishes. Other requirements include corrosion resistance, dimensional stability, wear resistance, natural lubricity, noise damping properties, and low cost. Machined gears are most frequently made from steel, which has a high strength and low cost. Carbon steel, which is low in cost, is satisfactory for machining and case hardening, and is very commonly used for commercial gears. Alloy steels have the advantages of strength, heat treatment, and corrosion and wear resistance. However, they also have poor machinability and are higher in cost compared to plain carbon steels. Leaded and resulfurized (free-machining) steels should be used for machined gear whenever possible; however, they have low impact strength and less suitable for high-power applications. Stainless steels are only used when corrosion resistance is essential. They are more expensive, difficult to machine, have low wear resistance, and are not heat-treatable. CIs have good machinability, are low in cost, and have vibration damping characteristics. Apart from malleable CI, they have low shock resistance. Cast steels have better physical properties than CI but are more expensive, less machinable, and lack good damping characteristics. Bronze is a superior gear material that has excellent machinability and wear and corrosion resistance. However, its material cost is high. Aluminum is suitable for lightly loaded gears. It is machinable and provides good surface finish and corrosion resistance.

13.3.7.1 Design Recommendations

When designing gears for machining, the designer must consider choices that have a significant effect on the cost and performance of gears. Some points of consideration described by Bralla (1999) include the following:

1. The coarsest pitch that performs the required function is the most economical to cut (Figure 13.59).
2. Helical, spiral, and hypoid gears are difficult to machine than spur gears.
3. Dimensional tolerances and surface finishes should be as liberal as the function of gears permit.

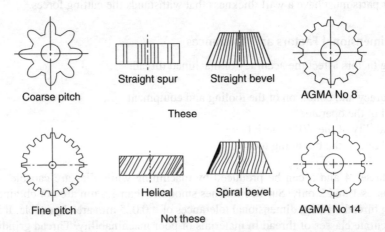

FIGURE 13.59 Use economical gear designs of coarse pitch, straight teeth, and small AGMA number.

Design for Machining

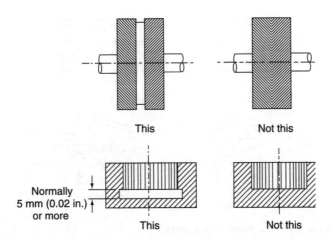

FIGURE 13.60 Allow a groove between halves of herringbone gears and clearance or undercut for internal gears for the cutting tool.

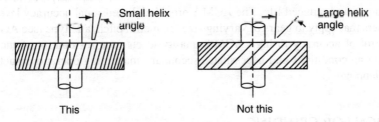

FIGURE 13.61 Avoid large helix angles whenever possible.

4. Shoulders, flanges, or other portions of the WP larger than the root diameter of the gear should be located away from the gear teeth to allow sufficient clearance for the gear-cutting tool.
5. Herringbone gears should have a groove between halves; internal gears in blind holes require an undercut groove or other recessed space for cutter over-travel (Figure 13.60).
6. Use nonheat-treated gears of larger size instead of heat-treated ones.
7. Heat-treated gears should be of uniform cross section to minimize heat treatment distortion using nitriding.
8. It is advisable to machine gears as separate parts and assemble them afterward onto shafts or other machine components.
9. Use as small helix angles as possible (Figure 13.61).
10. Avoid wide-faced gears, which are more difficult to machine to the given tolerance (Figure 13.62).
11. Design gear blanks that withstand the clamping and cutting forces without distortion.
12. When gears are press-fitted to a shaft or other components, the fitting surface should not be too close to the teeth.
13. Use standard pitches to minimize tooling costs.
14. When gears are subjected to a finishing process, specify the proper stock allowance that minimizes the production cost.
15. The involute form of the tooth should be the standard specified tooth form for all normal gearing.

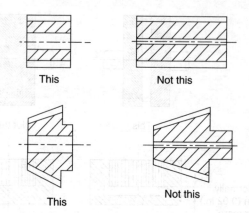

FIGURE 13.62 Avoid wide-face narrow-bore gears.

13.3.7.2 Dimensional Factors

The dimensional variations in gears result in noise, vibration, operational problems, reduced load-carrying capacity, and reduced life. The AGMA provided dimensional accuracy levels in terms of quality numbers that apply to gears of varying size, diametral pitch, and type (see AGMA Standard 390.03). Control of accuracy to high AGMA quality levels requires control of the environment and all machining conditions and necessitates secondary machining operations such as grinding, shaving, and lapping.

13.4 DESIGN FOR GRINDING

13.4.1 SURFACE GRINDING

Parts are produced by surface grinding to the required dimensions and surface finish. Horizontal-spindle reciprocating table machines are used for low-quantity production. Small hand-operated tool room surface grinders are ideally suited for single parts such as dies, molds, gauges, and cutting tools. Large-volume production is most effectively carried out on the vertical-spindle rotary-table machines where automatic part loading and unloading is possible. Carbon and alloy steel grades and other high tensile strength metals are ground with aluminum oxide wheels. CI, soft brass, bronze, aluminum, copper, plastics, and other nonmetallic materials are ground with silicon carbide wheels. Hardened tool-and-die steels use CBN as an abrasive material.

13.4.1.1 Design Recommendations

Surface grinding is a necessary step in finishing a part and therefore consideration must be given to the design so that the grinding operation can be performed easily:

1. For large-volume production, the part should be designed such that it can be ground on the vertical-spindle surface grinders.
2. The part should not have any surface higher than the surface to be finished (Figure 13.63).
3. The part should be magnetic or easily clamped in an automatic fixture.
4. For large-volume production that requires form grinding or has projections above the surface to be finished, a horizontal-spindle-powered table surface grinder is used.
5. For wider tolerances and rough surface finish of milling or other machining operations, avoid surface grinding of the part.

Design for Machining

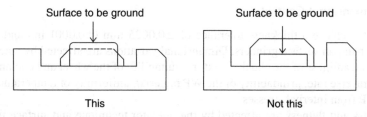

FIGURE 13.63 The ground surface should be higher than other surfaces to avoid wheel obstruction.

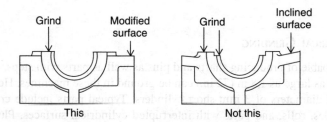

FIGURE 13.64 Design parts so that they can be machined in a single setup without wheel obstruction.

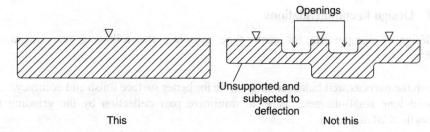

FIGURE 13.65 Avoid openings and unsupported surfaces.

6. Design the part to be held by magnetic chucks; large and flat locating surfaces are the best.
7. Do not specify a better ground finish than necessary because this requires longer machining times.
8. Design the part so that surfaces to be ground are all in the same plane (Figure 13.64).
9. Avoid openings in the flat ground surfaces, as the GW cuts deeper at the edges of interrupted surfaces.
10. Avoid unsupported surfaces that may deflect under GW pressure (Figure 13.65).
11. Whenever grooves or other forms are ground on surface grinders, corner relief is more preferable than sharp corners.
12. Avoid blind cuts where the wheel must be stopped during the cut or reversed with too little clearance provided.
13. Design the part for minimum stock removal by grinding, especially if horizontal-spindle machines are used.
14. Avoid extremely thin sections that cause burning or wrapping of the part.
15. When possible, avoid dissimilar materials that cause wheel loading problems.
16. Indicate clearly on drawings the permitted straightness and parallelism required.

13.4.1.2 Dimensional Control

It is possible to achieve a thickness tolerance of ±0.0025 mm (±0.0001 in.) and flatness of less than one light band on surface grinders. Dimensional variations are affected by machine condition, accuracy, and cleanliness of the chuck or fixture, suitability of the wheel and coolant, wheel speed, depth of cut, traverse rate, grindability of the WP material, uniformity of temperature, and the freedom of the WP from internal stresses.

Straightness and flatness are affected by the operator technique and surface finish is a function of the process time. Small speeds and slower traverse rates favor improved finishes. Fine grit wheels and the use of grinding fluids, slow traverse, diamond dressing of the wheel, good wheel balancing, and the use of hardened work materials of high grindability enhance the quality of surface finish.

13.4.2 Cylindrical Grinding

The process is capable of producing shafts and pins as well as parts with steps, tapers, and ground forms. Diameters as large as 1.8 m (72 in.) can be ground in these machines. However, the process tackles minimum diameters of 3 mm short cylinders. Typical parts include crankshaft bearings, bearing rings, axles, rolls, and parts with interrupted cylindrical surfaces. Plunge-type grinding is limited to ground surfaces shorter than the GW width. Typical output rates range from 10 to 130 pieces/h, with about 60 pieces/h being a fair average for operations involving a single surface or a single cut.

13.4.2.1 Design Recommendations

When designing components for center-type grinding, the following recommendations are followed:

1. Keep the parts as well balanced as possible for better surface finish and accuracy.
2. Avoid long small-diameter parts to minimize part deflection by the grinding forces (length ≤ 20 diameter).
3. Keep profile parts for plunge grinding as simple as possible by avoiding tangents to radii, grooves, angular shapes and tapers, and component radii.
4. Avoid interrupted cuts, as the surface adjacent to interruptions ground deeper than the rest of the continuous surface (Figure 13.66).
5. Undercuts on facing surfaces should be avoided (Figure 13.67).
6. If fillets are used, the designer should consider machining or casting a relief on the WP at the junction of two ground surfaces (Figure 13.68).
7. Center holes on parts held between centers should be made accurately at a 60° angle for accurate cylindrical grinding. They may be lapped in case of precision grinding.
8. Avoid clamping of thin-walled tubular parts by a three-jaw chuck.
9. Minimize the stock removed by grinding.

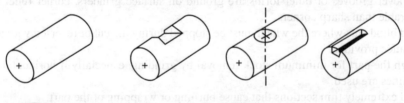

FIGURE 13.66 Avoid interrupted surfaces for better accuracy.

Design for Machining

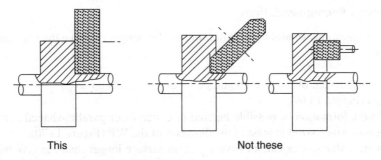

FIGURE 13.67 Avoid undercuts that are costly.

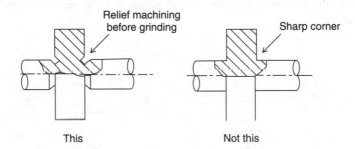

FIGURE 13.68 Machine or cast a relief at the junction of two surfaces before grinding.

13.4.2.2 Dimensional Factors

The accuracy of the final dimensions of machined components reflects the condition of the equipment and the skill of the operator. Worn bearings, centers, machine guide ways, poor coolant action, improper GWs, and the deflection of the WP can adversely affect the finished surface and dimensional accuracy that can be enhanced by

- The use of correct feed and speed
- The use of steady-rest supports for long parts
- Improving the roundness of center holes in the WP

Recommended dimensional tolerances for cylindrically ground parts under normal production conditions are ±0.0125 mm for diameter, parallelism, and roundness. A surface roughness of 0.2 μm is obtained under normal conditions, which can be improved to 0.05 μm under tight machining conditions.

13.4.3 CENTERLESS GRINDING

Centerless grinding is used for solid parts with diameters as small as 0.1 mm and as large as 175 mm. Parts as short as 10 mm and as long as 5 m are also centerless-ground. Pins, shafts, and rings with close tolerances for outside diameters, precise roundness, and smooth finishes are possible. The process is suitable for mass production when short pieces are machined by the through feed method. Using automatic magazines or hoppers parts are machined at a rate of 6000 pieces/h. Infeed grinding can be performed at a rate of 30–240 pieces/h. In other words, production rate ranges from 1 to 9 m/min per pass and one to six passes may be used. Brittle, fragile, and easily distorted parts and materials are more suitable for centerless grinding than center-type.

13.4.3.1 Design Recommendations

The following suggestions should be kept in mind by the designer to take the maximum advantage of the process:

1. The ground surface of the WP should be its largest diameter to permit the through-feed operation (Figure 13.69).
2. To avoid the formation of possible tapered or concave- or parallel-shaped surfaces, keep the ground surface sized at least to the diameter of the WP (Figure 13.70).
3. Parts of irregular shapes cannot have a ground surface longer than the GW width unless the shape permits a combination of through-feed and infeed grinding (Figure 13.71).
4. Avoid grinding the ends of infeed centerless-ground parts (Figure 13.72).
5. Avoid fillets and radii and instead use undercut or relief surfaces (Figure 13.68).
6. For the form infeed method, keep the form as simple as possible to reduce wheel dressing and other costs.

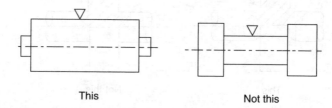

FIGURE 13.69 Avoid through-feed centerless grinding of smaller diameters.

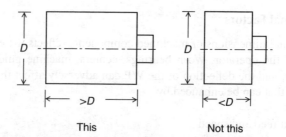

FIGURE 13.70 Avoid shorter WP lengths.

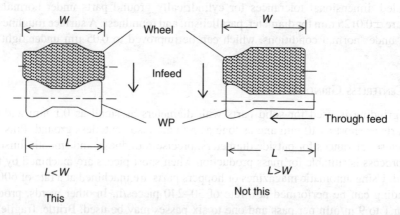

FIGURE 13.71 Avoid long parts of irregular surfaces so that infeed can be applied only to the wheel.

Design for Machining

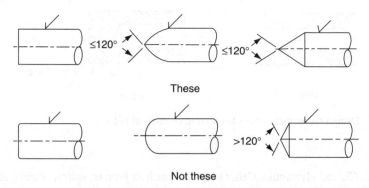

FIGURE 13.72 Avoid grinding the faces of centerless ground parts unless infeed grinding is used and the ends have an included angle of less than 120°.

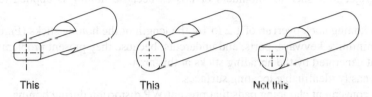

FIGURE 13.73 Avoid unbalanced and end-of-shaft interruptions.

7. For high-accuracy requirements, avoid keyways, flats, holes, and other interruptions to the surface to be ground or make them as small as possible (see Figure 13.66).
8. Avoid unbalanced and end of shaft interruptions (Figure 13.73).

13.4.3.2 Dimensional Control

The condition of the equipment, such as the wheel-spindle bearings, the use of proper wheels and coolant, and the evenness of the temperature of the WP, machine, and coolant, affects the process accuracy. Typical dimensional tolerances are ±0.0125 mm for diameter and parallelism. Surface roughness $R_a = 0.2$ μm is obtained under normal conditions, which can be improved to 0.05 μm under tight machining conditions.

13.5 DESIGN FOR FINISHING PROCESSES

13.5.1 Honing

The main purpose of honing is to generate an accurate surface configuration and an improved surface finish. It is used primarily for inside diameters and flat surfaces. The bore diameter ranges from 2.4 mm to 1.2 m, and the length of bore varies between 1.6 mm and 9.1 m. External cylindrical honing can tackle parts of 6.3 mm diameter by 14.7 m long to 450 mm in diameter by 9.1 m long. Flat surfaces of area less than 645 mm² and spherical diameters ranging from 3 to 300 mm can also be honed. The nature of a cross-hatched honed surface finish is efficient in moving bearing applications, such as automotive-cylinder bore or a valve body bore. In such applications, the peaks carry the load of the mating parts and valleys provide reservoirs for the lubricating oil. The absence of surface damage by heat is another process advantage. Gear teeth, races of ball and roller bearings, crank-pin bores, drill bushings, gun barrels, piston pins, and hydraulic cylinders are often finished by honing. Honing can be used economically; manual honing suits one-of-a-kind or very limited quantity. When honing is applied in mass production, millions of parts are produced per year using the general-purpose horizontal or vertical honing machines. The most common materials to be

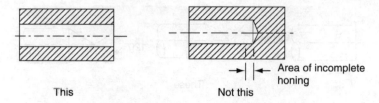

FIGURE 13.74 Design recommendations for honing of internal holes.

honed are steel, CI, and aluminum. Other materials such as bronze, stainless steel, and plastics can be honed at lower speeds.

When designing the WP for honing operation, the following guidelines are recommended:

1. Avoid projections such as shoulders or bosses because honing is applied to the entire surface.
2. Allow a honing stick overrun of 1/2 to 1/4 the length of the honing sticks (Figure 13.74).
3. Avoid/minimize keyways, ports, and undercuts, because they present problems due to the undercut generated by the abrading sticks at this area.
4. Provide easily identifiable locating surfaces.
5. Provide convenient clamping pads that prevent WP distortion during honing.

Bores of automotive engine blocks can be honed to geometric tolerances of roundness, straightness, and size control within 0.008 mm. On hardened steel bores or outside diameter, an accuracy of 0.008 mm can be also produced. Honing produces surface finishes of 0.25–0.4 μm depending on the grit size, coolant used, honing speed, and the generated forces by the honing process.

13.5.2 Lapping

Lapping is normally used to produce one or more of the following:

- An improved surface finish
- Extreme product accuracy
- Correction of minor imperfections in shape
- Precise fit between mating surfaces

Valve spools, fuel injector plungers, seal rings, piston rods, valve stems, cylinder heads, and spherical valve seats are typical applications. Holes or pins as small as 0.8–300 mm in diameter can be lapped. Flat WPs having an area from 6 to 1300 cm² can be lapped successfully. For one-of-a-kind fabrication, hand lapping is used. Automatic lapping machines are available for mass production at high rates of 3000 piece/h. Steel and CI are most frequently used by lapping. Glass, aluminum, bronze, magnesium, plastics, and ceramics can also be finished by lapping. Soft materials are less satisfactory for lapping, as the abrasive grits are embedded in the lapped surface.

When designing a part to be lapped, the following factors should be considered:

1. Avoid shoulders, projections, or interruptions.
2. When two opposite sides of a WP must be machined to a highly refined parallelism, the two surfaces should extend beyond the other surface of the WP.

Normal tolerances of lapped surfaces are as follows:

- Diameter or other dimensions: ±0.0006 mm
- Flatness, roundness, or straightness: ±0.0006 mm
- Surface roughness: R_a = 0.1–0.4 µm

13.5.3 Superfinishing

Superfinishing is utilized for surface refining and improving geometric characteristics of the WP. It can be applied to inside and outside diameters, conical surfaces, spherical surfaces, flats, flutes, keyways, and recesses. Typical parts include pistons, piston rods, shaft-sealing surfaces, crank pins, valve seats, bearing recesses, and steel-mill rolls. When using automatic part handling systems, production rate reaches 240 pieces/h. The process tackles any ferrous and nonferrous metallic parts. It is most frequently used to hardened and ground alloy-steels. The degree of surface finish depends on the grit size, applied force, relative surface speed, condition of the coolant used, and superfinishing time. Generally, a surface finish of 0.025–0.075 µm is common. Due to the small amount of stock removal capability, refinement of geometric tolerances and control in size is limited by superfinishing.

13.6 DESIGN FOR CHEMICAL AND ELECTROCHEMICAL MACHINING

13.6.1 Chemical Machining

In CHM, parts are machined over their entire surface area. CHM finds applications in the aerospace industry, where large surface areas require a small depth of material removal and other parts require weight reduction. When selective machining in required, maskants are used to protect areas that do not require machining. Scribe-and-peel maskants are used for cuts as deep as 13 mm. With silk screen masks; the depth is limited to 1.5 mm, but more accurate details are possible. Photo-resist provides more accurate details where the depth of etch is limited to 1.3 mm. Photochemical blanking produces very intricate blank shapes from sheet metals by covering the sheet with a precisely shaped mask made using photographic techniques and then removing the unmasked metal by chemical dissolution action. Photochemical milling produces very intricately shaped blanks used for electric motor laminations, shadow masks for color television, fine screens, printed circuit cards, and so on. Sheet thickness ranges from 0.0013 to 3 mm with a common range of 0.0025–0.8 mm. Any material that can be chemically etched or dissolved can be chemically machined. Many ferrous and nonferrous metals are chemically machined. Material grain size, rolling direction, hardness, freedom from inclusions, and surface quality must be considered.

13.6.1.1 Design Recommendations

When designing a part to be chemically machined, the designer should consider the following points:

1. The smallest hole size or slot width that can be produced by chemical blanking is 1.5 times the metal thickness (Figure 13.75).
2. The minimum land width (Figure 13.76) should be twice the depth of cut but less than 3.18 mm.
3. The radius and the undercut produced by CHM (Figure 13.77) equals the depth of cut.
4. With CHM, sharp corners can be rounded to about 0.7–1.0 mm radius.
5. The normal taper for blanked parts from one side is one-tenth of the blank thickness.
6. The normal taper for parts blanked from two sides is 0.05 of the depth of etch.

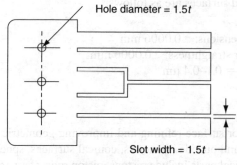

FIGURE 13.75 Maximum hole diameter and slot width by CHM.

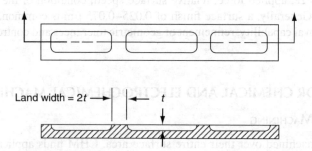

FIGURE 13.76 Minimum land widths.

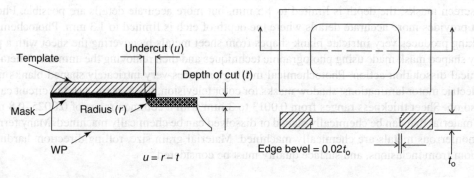

FIGURE 13.77 The radius of undercut and edge bevel.

7. Avoid part design that has deep, narrow cavities, or folded metal seems to prevent any entrapped chemical solutions.
8. Parts machined by masking techniques should be flat.
9. When machining aluminum, the designer should specify the grain direction and minimize machining across the grains.
10. Provide excess material at the periphery of the part for trimming after chemical machining.
11. Sharp internal corners are not produced by CHM, where a radius of 0.5–1 times the depth of etch should be allowed.
12. External corners may have a radius of one-third of the depth of the etch (Figure 13.78).

Design for Machining 563

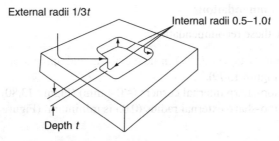

FIGURE 13.78 Minimum radii of CHM corners.

13.6.1.2 Dimensional Factors and Tolerances

Achieving close tolerances by CHM and PCM depend on the following factors (Bralla, 1999)

- The accuracy of the part's dimensions before CHM and PCM
- The accuracy of the mask
- The undercut allowance
- Uniformity of etching

Tolerance of the length and width of chemically machined parts are as follows:

- With scribe-and-peel maskants and cuts to 1.3 mm deep: ±0.38 mm
- With scribe-and-peel maskants and cuts over 1.3 mm deep: ±0.64 mm
- With silk screen maskants: ±0.25 mm
- With photoresist mask and a depth 1.0 mm aluminum alloy: ±0.2 mm for other materials and thicknesses

Surface finish produced by chemical machining depends on factors that include:

- The initial surface finishes of the WP
- The etchant used
- WP material
- Depth of etching
- Heat treatment condition of the WP

Normal surface finish R_a is as follows:

- Aluminum (up to 6.3 mm depth of etch): 2.25 μm
- Magnesium: 1.25 μm
- Titanium: 0.63 μm
- Steel: 1.75 μm

13.6.2 Electrochemical Machining

ECM is suitable for materials and shapes that are difficult to machine by conventional methods. ECM finds applications in machining jet engine parts, nozzles, cams, forging dies, and other contoured shapes. ECM is generally used for difficult-to-machine conductive materials such as hardened steel, including alloy and tool steel, nickel alloys, cobalt alloys, tungsten, molybdenum, zirconium, and other refractory metals. The process is suitable for quantity production, as the tool requires special design and testing. However, the tool life is infinite and requires little maintenance. Due to the high machining speeds attainable (25 times as fast as EDM), the process is used at machining high removal rates.

13.6.2.1 Design Recommendations

Designers should follow these recommendations:

1. Avoid the formation of tapers (0.1%) in hole drilling and apply tool insulation that stops stray machining (Figure 13.79).
2. Avoid specifying too-sharp internal corners (<0.4 mm) (Figure 13.80).
3. Avoid specifying too-sharp external radius (0.05 is minimum) (Figure 13.81).

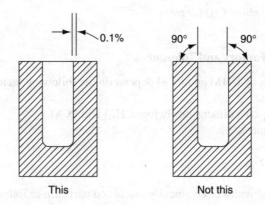

FIGURE 13.79 Allow a taper on the sidewalls of ECM cavities.

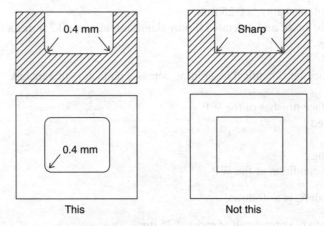

FIGURE 13.80 Minimum corner radii for ECM cavities.

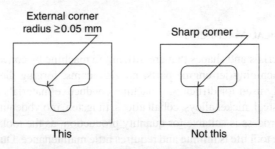

FIGURE 13.81 Allow a radius of 0.05 mm or more for external corners.

Design for Machining

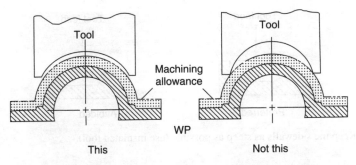

FIGURE 13.82 Allow sufficient machining allowance on castings and forgings. (From Yankee, H. W., *Manufacturing Processes*, Prentice Hall, New York, 1979. With permission.)

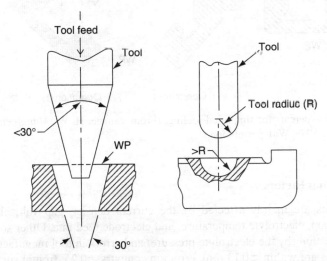

FIGURE 13.83 The electrode tool is less than the true shape of the profile. (From Yankee, H. W., *Manufacturing Processes*, Prentice Hall, New York, 1979. With permission.)

4. When specifying irregular shapes, the designer should allow for ample deviations from the nominal shape to minimize the trial and error development of the tool.
5. Consider an over cut or sidewall tolerance (±0.03 to ±0.76 mm), which can be controlled by adjusting the feed tool rate to within ±0.127 mm.
6. Avoid specifying roundness errors less than 0.013 mm and runout of less than 0.001 mm/mm of hole depth.
7. Hole size of 0.76–100 mm can be achieved.
8. Consider the maximum aspect ratio of 180:1.
9. The rough contour to be machined should conform as closely as possible to the profile of the final designed shape.
10. The designer should provide a uniform machining (stock) allowance so that a constant surface finish will be achieved across the entire machined area (Figure 13.82).
11. The tool is not the exact mirror image of the required shape (Figure 13.83).
12. Electrode tools are designed such that when the final depth is reached, the final size and finish of the shape is achieved.
13. Sidewalls should be as steep as possible (Figure 13.84).
14. To avoid complex tooling requirements, avoid undercuts into shapes machined by ECM (Figure 13.85).

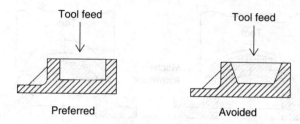

FIGURE 13.84 Keep the sidewalls as steep as possible (use insulated tool).

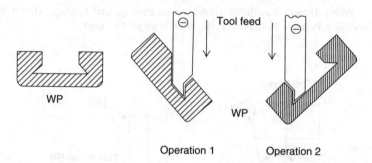

FIGURE 13.85 Avoid undercuts for simplified tooling. (From Yankee, H. W., *Manufacturing Processes*, Prentice Hall, New York, 1979. With permission.)

13.6.2.2 Dimensional Factors

Dimensional variations are mainly affected by the current density, gap volt, electrolyte flow, electrolyte concentration, electrolyte temperature, and electrode feed rate. Other sources of errors include machine deflection (by the electrolyte pressure) and errors in tool manufacturing. Normal dimensional tolerances are within ±0.13 mm, errors in contours ±0.25, frontal surface roughness $R_a = 1.6$ μm, and surface roughness R_a at the sidewalls in about 3 μm.

13.6.3 ELECTROCHEMICAL GRINDING

ECG is used for materials that are difficult to machine conventionally. Fragile materials as well as stress- and burr-free components are good candidates for ECG. Steels (HRC > 60), stainless steels, high-nickel alloys, and WC are typical materials. Materials that are fragile or susceptible to heat damage or distortion from normal grinding stresses are tackled by ECG. Examples are aircraft, honeycomb materials, surgical needles, thin walled tubes, and laminated materials. ECG is used for moderate and high-level production, where the rate of material removal is one to five times the rate attainable by conventional grinding.

13.6.3.1 Design Recommendations

The product designer should consider the following if the part is to be machined by ECG:

1. Allow 0.75–1.0 mm for inside radii.
2. Specify more liberal tolerances if the groove is deep.
3. Electrochemical action concentrated near the WP corners will round them by 0.13 mm radius.
4. For better accuracy, allow for a final nonelectrolytic pass (with the current switched off) over the surface.

Design for Machining

13.6.3.2 Dimensional Factors

As the case of ECM, normal tolerances for ECG are as follows:

- Specific dimensions: ±0.025 mm
- Contours: ±0.13 mm
- Surface finish (plunge-grinding carbide): 0.25 μm
- Surface finish (traverse-grinding carbide): 0.40 μm
- Surface finish (steel): 0.75 μm

13.7 DESIGN FOR THERMAL MACHINING

13.7.1 Electrodischarge Machining

EDM is a thermal machining process that removes the WP material by the erosive action of electrical discharges that melt and evaporate the WP material. The process produces intricate shapes in hard materials that are difficult to machine conventionally. EDM is burr-free process that eliminates the need for secondary operations. Stamping, extruding, wire drawing, die casting, and forging dies and plastic molds are common applications. It is used for one-of-a kind or job-lot quantities.

The low cutting rate of EDM (8 cm³/h) is compensated for by the ease of electrode tool machining from a highly machinable material. Wire EDM cuts at a rate of 130–140 cm²/h in 5 cm thickness using good flushing conditions. EDM machines any conductive material regardless of its hardness. Hardened steel and carbides are the most commonly machined materials. PCD used in form tools and other cutting tools is currently machined by EDM.

13.7.1.1 Design Recommendations

Designers should adhere to the follow guidelines:

1. It is not economical to use EDM for obtaining an ultrafine finish. This requires many passes, low current, and slower MRRs.
2. Design the part so that the machining allowance is as small as possible.
3. Design the part so that as much of the machining is made by a conventional method or another manufacturing method rather than by the slow EDM process (Figure 13.86).
4. Design the part so that several parts can be machined simultaneously or a single part can have several EDM operations simultaneously.
5. Design the part so that the tool electrodes are produced at low cost.
6. Avoid thin and fragile electrode sections, especially in graphite.

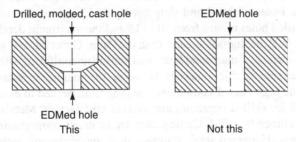

FIGURE 13.86 Perform maximum conventional machining, molding, or casting before the slow EDM process.

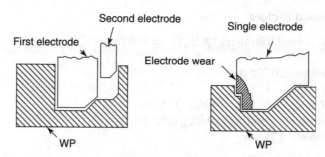

FIGURE 13.87 Complex shapes require special or multiple electrodes. (From Yankee, H. W., *Manufacturing Processes*, Prentice Hall, New York, 1979. With permission.)

7. Complex cavities can be produced by several simple electrode shapes (Figure 13.87).
8. The minimum radius obtainable for internal corners will be equal to the overcut (0.1 mm).
9. It is advisable to specify a cavity tolerance that allows a taper angle of 2–20 min/side.
10. Enlarging or reshaping through holes rather than blind holes permits easier dielectric flow supply.
11. Consider a finishing pass at low removal rate to minimize the adverse effects of the white layer formed on the machined surface.

13.7.1.2 Dimensional Factors

A surface finish of 0.8 μm is obtainable in die sinking by EDM. Better finishes ($R_a = 0.4$ μm) are possible when using orbiting or rotating tool electrodes. Generally, the surface finish depends on the spark energy. A finish cut may be required at low pulse current, short pulse duration, and a high discharge frequency. Under such conditions, a surface roughness of $R_a = 0.4$ μm is possible. Dimensional accuracy in EDM die sinking depends on the accuracy of tool electrode manufacture, electrode wear, quality of the power supply, previous stresses existing in the WP material, and the operator skill. Tolerance of ±0.05 to ±0.13 mm is possible in a single cut, and ±0.005 mm with multiple cuts. A dimensional tolerance of ±0.0025 mm is possible with wire EDM.

13.7.2 Electron Beam Machining

EBM utilizes an electron beam that impinges on an area of 0.32–0.46 mm² and has a power density of 15×10^6 W/mm² to melt and evaporate the WP material. It is used for fine cutting and drilling in any material. Holes and slots of few thousands of an inch and very precise contour are possible. EBM is used for drilling accurate holes for diesel fuel injection, gas orifice, wire drawing dies, and sleeve valve holes. Holes of 0.013 and slots having 0.025 mm width can be cut. The depth-to-diameter ratio of drilled holes ranges from 10 to 15 and the maximum depth is 6.4 mm. A taper of 1–2° is expected for WP thicknesses greater than 0.13 mm. Cratering and spattering occur near the hole entrance. The surface of machined holes and slots is nonuniform, with a HAZ of 0.25 mm depth. Cutting rates are rapid for thin materials. However, the volumetric removal rate ranges from 0.8 and 2 mm³/min. The high cost of equipment, the long time needed to evacuate the machining chamber, and the need for skilled operators are adverse cost factors. Metals, ceramics, plastics, and composites are machined by EBM. Cutting rates are inversely proportional to the melting and evaporating temperatures. Hardened steel, stainless steel, molybdenum, nickel, cobalt, titanium, tungsten, and their alloys are all machined by EBM.

Design for Machining

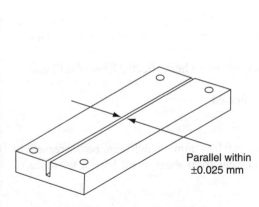

FIGURE 13.88 Parallelism of EBMed parts.

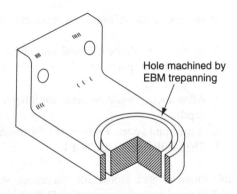

FIGURE 13.89 Trepanning of large-diameter holes. (From Yankee, H. W., *Manufacturing Processes*, Prentice Hall, New York, 1979. With permission.)

In EBM designers should follow these recommendations:

1. Keep the part size as small as possible so that many parts can be loaded to the vacuum chamber of the electron beam machine.
2. Avoid specifying internal corners less than 0.25 mm.
3. Do not exceed the maximum cut by 6.3 mm; thinner WP cuts faster with less sidewall taper.
4. Allow for the surface effects, which may be undesirable and may require a secondary operation. A tolerance of ±10% should be allowed on hole diameters and slots. The normal surface roughness R_a ranges between 0.5 and 2.5 µm.
5. The edge of slot walls can be held parallel to a tolerance of 0.05 mm (Figure 13.88).
6. Hole entrance angles can be kept between 20° and 90° to the WP surface.
7. Large diameter holes (>beam diameter) are produced by the trepanning method (Figure 13.89).
8. Blind cuts can be made by switching off the power when reaching the required depth.

13.7.3 Laser Beam Machining

LBM is generally used for micromachining of thin parts that are difficult to cut using conventional methods. Holes smaller than 1.3 mm diameter in 5 mm thickness are produced; larger holes can be machined by trepanning. Other laser applications include cutting out blanks from sheet metals or other materials up to 13 mm thickness, slitting, trimming, and perforation. The minimum hole diameter is about 0.005 mm but 0.13 mm is more common. Slits and profile cuts as narrow as 0.4 mm are normally achieved. Laser beam holes deeper than 0.5 mm suffer from a taper, nonuniform diameter, and roundness errors. Cratering at the entrance surface and a narrow damaged layer of 0.13 mm is also common. Due to the high cost of equipment used, higher production quantities and the use of difficult-to-machine material favors the use of the process. The most practical materials for LBM are ceramics, glass, carbides, and some aerospace alloys. Other materials such as copper, aluminum, gold, and silver are not suitable candidates because of their high reflectivity and thermal conductivity. Plastics, rubber, beryllium, zirconium, stainless steel, tungsten, CI, brass, molybdenum, cloth, cardboard, wood, and composites have been machined successfully by LBM.

Designers should follow these recommendations:

1. Surfaces should be dull and unpolished.
2. Through-cut parts should be thin to reduce the time required, taper, and surface irregularities.
3. Allowances should be made for a taper of 3° per side and a heat-affected layer of 0.13 mm depth.
4. A minimum corner radius of 0.01 mm/min is expected.
5. The ideal aspect ratio is 4:1.

Hole-diameter and slot-width tolerances should be ±0.025 mm, which can be increased to ±0.1 mm using assisting gasses. Diameters of blanks cut from sheet metal have a tolerance of ±0.13 mm.

13.8 DESIGN FOR ULTRASONIC MACHINING

USM is used to machine irregular holes in thin sections or shallow, irregular cavities in hard, brittle, and fragile materials. Holes as small as 0.08 mm can be drilled; the maximum possible size is 90 mm. Hole depth ranges between 25 and 50 mm. Larger holes can be machined by RUM using a coring or a trepanning method. Machined surfaces do not suffer from thermal damage and they are burr-free. Cavities machined by USM suffer from an overcut and a sidewall taper. USM is a slow machining process that can be used when other conventional processes are not suitable. Cutting rates range from 0.03 to 4 cm^3/min and tool wear ranges from 1:1 to 1:200 of WP material. USM is most advantageous for hard, brittle, nonconductive materials. Materials having a hardness HRC > 64 are best suited for the process and a hardness of HRC < 45 is not recommended for USM.

When designing parts to be machined by USM, the designer should follow the below mentioned points:

1. Shallow holes and cavities (depth <2.5 diameter) are more suitable than deep ones.
2. Through holes or holes through passages for abrasive slurry are preferred to blind holes (Figure 13.90).
3. A backup plate should be used when machining through holes/cavities in brittle materials to prevent chipping of edges (Figure 13.91).
4. Allow for a taper of 0.05 mm/mm (depending on the WP material) when machining deep holes (Figure 13.92). This taper can be reduced by the proper tool design or using a finishing pass.
5. Do not specify sharp corners at the bottom of blind holes because tool wear is concentrated at such corners of the tool (Figure 13.93).
6. Allow for an overcut, which equals the tool diameter plus twice the abrasive grain diameter.

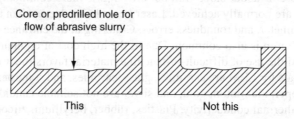

FIGURE 13.90 Provide through passage for the abrasive slurry.

Design for Machining

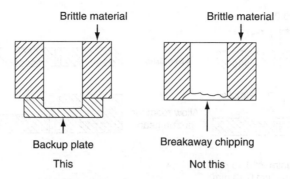

FIGURE 13.91 Avoid breakaway chipping at the exit surface of USM cavities.

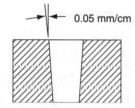

FIGURE 13.92 Allow taper for sidewalls of USM cavities.

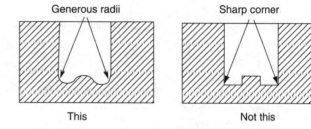

FIGURE 13.93 Allow generous radii at machined corners.

The following factors affect the dimensional accuracy of USM machined parts:

- The amount of overcut
- Tool wear
- Machine tool rigidity
- Abrasive grain size
- Abrasive wear
- Tool design
- Slurry conditions

Recommended dimensional tolerances range between ±0.013 and ±0.025 mm. A surface roughness of 0.25–1.0 μm is achievable.

13.9 DESIGN FOR ABRASIVE JET MACHINING

AJM is widely used for cutting, drilling, slotting, trimming, etching, cleaning, deburring, and stripping. The process is also applicable to machining heat-sensitive components. The minimum slot width machinable with AJM is 0.13 mm with sidewall taper that increases with the stand-off distance. The choice of AJM does not depend on the production quantity. Tooling costs are low, which makes the process suitable for small quantities. Removal rates of 0.016 cm^3/min are possible. It is the most suitable process for machining hard, fragile, heat-sensitive materials. Typical materials include ceramic, glass, porcelain, sapphire, quartz, tungsten, chromium/nickel alloys, hardened steels, and semiconductors such as germanium, silicon, and gallium.

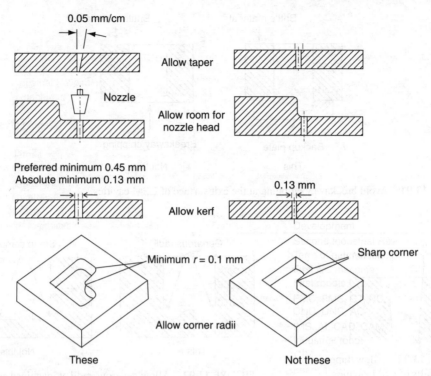

FIGURE 13.94 Design recommendations for AJM.

Designers intending to use AJM should consider the following allowances in their design (Figure 13.94):

1. The taper of the sidewalls of cuts should be at least 0.05 mm/cm of depth.
2. Allow access room for the jet nozzle.
3. Allow for a kerf, which should be at least 0.13 mm (0.45 mm is preferred).
4. Provide corners of at least 0.1 mm.

Normal tolerances for dimensions are from ±0.05 to ±0.03 mm; the surface finish can be held between 0.25 and 1.3 μm depending on the machining conditions.

13.10 REVIEW QUESTIONS

1. State the general principles of manufacturing components at the minimum cost.
2. What are the general rules adopted when designing for manufacturing?
3. List the general guidelines used when designing for economic machining.
4. State the design recommendations adopted when designing for:
   - Turning parts
   - Drilling holes
   - Reaming and boring
5. Explain what is meant by dimensional factors and tolerances of milled parts.
6. Specify the design rules adopted for shaping, planing, and slotting of parts.
7. Describe the design recommendations that should be followed when broaching splines.
8. Explain the principles of economic design for thread cutting and gear machining.

9. Mention the rules followed by the designer when machining WPs on surface and cylindrical grinding machines.
10. Summarize the design guidelines that should be considered when performing lapping, honing, and superfinishing.
11. Show using line sketches the major design factors for ECM and CHM.
12. Explain what is meant by:
    - Design for EDM
    - Design for LBM
    - Design for USM

REFERENCES

American Gear Manufacturing Association (AGMA) Standard 390.03.
Bralla, J. (1999) *Design for Manufacturability Handbook*, 2nd Edition, McGraw Hill, New York.
Yankee, H. W. (1979) *Manufacturing Processes*, Prentice Hall, New York.

14 Accuracy and Surface Integrity Realized by Machining Processes

14.1 INTRODUCTION

The quality of a machined surface is becoming important to satisfy the increasing demands of component performance and reliability. Machined parts used in military, aerospace, and automotive industries are subjected to high stresses, temperatures, and hostile environments. The dynamic loading and design capabilities of machined components are limited by the fatigue strength of the material, which is commonly linked to the fatigue fractures that always nucleate on or near the surface of the machined components. Stress corrosion resistance is another important material property that can be directly linked to the machined surface characteristics. When machining any component, it is necessary to satisfy the surface technological requirements in terms of high product accuracy, good surface finish, and a minimum of drawbacks that may arise as a result of possible surface alterations by the machining process. The nature of the surface layer has a strong influence on the mechanical properties of the part.

Any machined surface has two main aspects—the first aspect is concerned with the surface texture or the geometric irregularities of the surface, and the second one is concerned with the surface integrity, which includes the metallurgical alterations of the surface and surface layer, as shown in Figure 14.1. Surface texture and surface integrity must be defined, measured, and controlled within specific limits during any machining operation.

14.2 SURFACE TEXTURE

Surface texture is concerned with the geometric irregularities of the surface of a solid material, which is defined in terms of surface roughness, waviness, lay, and flaws, as described in Figure 14.2:

1. *Surface roughness* consists of the fine irregularities of the surface texture, including feed marks generated by the machining process.
2. *Waviness* consists of the more widely spaced components of surface texture that may occur due to the machine or part deflection, vibration, or chatter.
3. *Lay* is the direction of the predominant surface pattern.
4. *Flaws* are surface interruptions such as cracks, scratches, and ridges.

Stylus contact type instruments are widely used to provide numerical values of surface roughness in terms of the arithmetic average (R_a) or centerline average (CLA), the root mean square (R_q), and the maximum peak-to-valley roughness (R_{max}), as shown in Figure 14.3. Other methods of surface characterization include microphotography and scanning electron microscopy.

The arithmetic average or CLA is determined as follows:

$$R_a = \frac{1}{L} \int_{x=0}^{x=L} |y| \, dx$$

where L is the sampling length and y is ordinate of the profile from the centerline shown in Figure 14.3a.

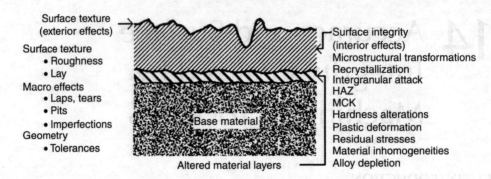

FIGURE 14.1 Surface technology by machining.

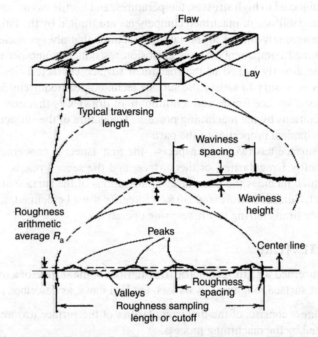

FIGURE 14.2 Surface texture. (From Surface Texture [Surface Roughness, Waviness, and Lay], ANSI/ASME B 46.1, American Society of Mechanical Engineers, 1985. With permission.)

The root mean square roughness is calculated as follows:

$$R_q = \left(\frac{1}{L} \int_{x=0}^{x=L} y^2 \, dx \right)^{1/2}$$

This can be approximated by the following equation (Figure 14.3b):

$$R_q = \sqrt{\frac{(y_1 - Y_M)^2 + (y_2 - Y_M)^2 + \cdots + (y_N - Y_M)^2}{N}}$$

The maximum peak-to-valley roughness (R_t or R_{max}) is the distance between two lines parallel to the mean line that contacts the extreme upper and lower points on the profile within the roughness sampling length (Figure 14.3c).

Accuracy and Surface Integrity Realized by Machining Processes

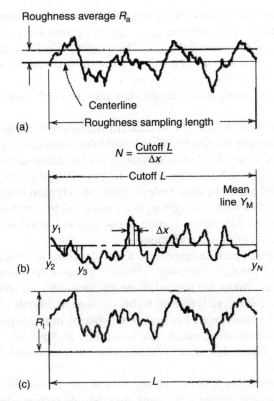

FIGURE 14.3 Commonly used surface roughness symbols. (a) Average roughness R_a, (b) root mean square roughness (R_q), (c) maximum peak-to-valley roughness height (R_t or R_{max}). (From Surface Texture [Surface roughness, waviness, and lay], ANSI/ASME B 46.1, American Society of Mechanical Engineers, 1985. With permission.)

Enhanced surface texture specifications are essential to improve fatigue strength, corrosion resistance, appearance, and sealing. Typical applications of such surfaces include antifriction and journal bearings, food preparation devices, parts operating in corrosive environments, and sealing surfaces.

14.3 SURFACE QUALITY AND FUNCTIONAL PROPERTIES

The quality of surface finish affects the functional properties of the machined parts as follows:

1. *Wear resistance.* Larger macro irregularities result in nonuniform wear of different sections of the surface where the projected areas of the surface are worn first. With surface waviness, surface crests are worn out first. Similarly, surface ridges and micro irregularities are subjected to elastic deformation and may be crushed or sheared by the forces between the sliding parts.
2. *Fatigue strength.* Metal fatigue takes place in the areas of the deepest scratches and undercuts caused by the machining operation. The valleys between the ridges of the machined surface may become the focus of concentration of internal stresses. Cracks and microcracks (MCK) may also enhance the failure of the machined parts.
3. *Corrosion resistance.* The resistance of the machined surface to the corrosive action of liquid, gas, water, and acid depends on the machined surface finish. The higher the quality

of surface finish, the smaller the area of contact with the corrosive medium, and the better the corrosion resistance. The corrosive action acts more intensively on the surface valleys between the ridges of micro irregularities. The deeper the valleys, the more destructive will be the corrosive action that will be directed toward the depth of the metal.

4. *Strength of interference.* The strength of an interference fit between two mating parts depends on the height of micro irregularities left after the machining process.

Figure 14.4 shows the ANSI Y14.36 (1978) standard symbols used for describing part drawings or specifications. These include the maximum and minimum roughness, maximum waviness, and the lay for machined parts. Table 14.1 shows the symbols used to define surface lay and its direction. Accordingly, a variety of lays can be machined, including parallel, perpendicular, angular, circular, multidirectional, and radial ones. The same table also suggests a typical machining process for each produced lay. Figure 14.5 shows the surface roughness produced by common production methods. Several machining processes that employ cutting, abrasion, and erosion actions are also included and compared to some metal forming applications.

The accuracy of machined parts indicates how a part size is made close to the required dimensions. Produced accuracy is normally expressed in terms of dimensional tolerances. Each machining process has its own accuracy limits that depend on the machine tool used and the machining conditions. Tolerances required for materials that are highly engineered, heavily stressed, or subjected to unusual environments are closely related to surface roughness. In this regard, closer dimensional tolerances require very fine finishes, which may necessitate multiple machining operations that raise the production cost. Figure 14.6 shows typical surface roughness and dimensional tolerances for machining operations.

Theoretically, surface roughness in milling and turning can be calculated as a function of the feed rate, tool nose radius, end cutting edge angle, and the side cutting edge angle. However, the actual surface roughness may be higher, due to the formation of the built-up edge (BUE) and the possible tool wear. Table 14.2 summarizes different factors that affect the surface roughness for different machining operations.

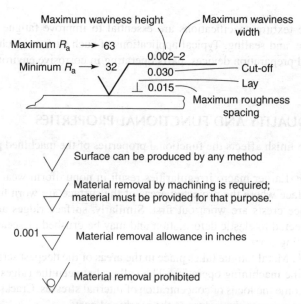

FIGURE 14.4 Surface texture symbols for drawings or specifications. (From Surface Texture [Surface Roughness, Waviness, and Lay], ANSI/ASME B46.1, American Society of Mechanical Engineers, 1985. With permission.)

TABLE 14.1
Symbols Used to Define Lay and Its Direction

Symbol	Meaning	Operation
=	Lay approximately parallel to the line representing the surface to which the symbol is applied	Shaping vertical milling
⊥	Lay perpendicular to the line representing the surface to which the symbol is applied	Horizontal milling
X	Lay angular in both directions to the line representing the surface to which the symbol is applied	Honing
M	Lay multidirectional	Grinding
C	Lay approximately circular relative to the center to which the symbol is applied	Face turning
R	Lay approximately radial relative to the center to which the symbol is applied	Lapping
P	Lay particulate, nondirectional, or protuberant	ECM, EDM, LBM

14.4 SURFACE INTEGRITY

Surface integrity is defined as the inherent condition of a surface produced in a machining or other surface generating operation. Surface integrity is concerned primarily with the host of effects that a manufacturing process produces below the visible surface. During machining by conventional methods, the pressure exerted on the metal by cutting and frictional forces, heat generated, and plastic flow change the physical properties of the surface layer from the rest of metal in the part. Similarly, thermal machining by EDM and LBM is accompanied by material melting, evaporation, resolidification, and consequently, the formation of a heat-affected layer. Machining by chemical and EC processes does not impose thermal changes to the WP. However, the surface suffers from several other effects such as pits and intergranular attack (IGA).

As a result of some machining processes, the thickness of the altered layer may reach a considerable value during rough machining operations. The mechanical, thermal, and chemical properties of the WP material determine the extension of surface effects and thickness of the altered layer. Surface alterations have a major influence on the material performance, especially when high stresses or severe environments are used.

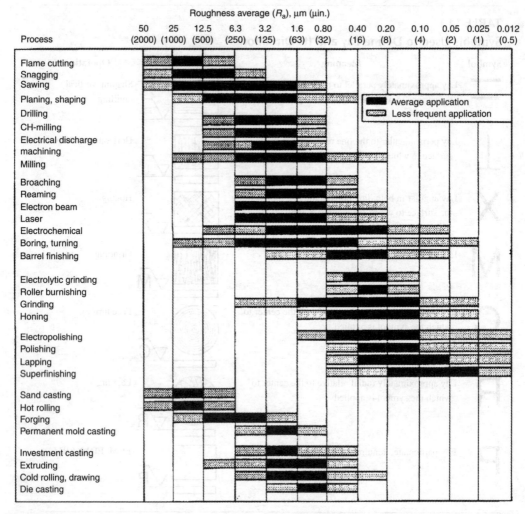

FIGURE 14.5 Surface roughness produced by common production methods. (From Surface Texture [Surface Roughness, Waviness, and Lay], ANSI/ASME B 46.1, American Society of Mechanical Engineers, 1985. With permission.)

The nature of the surface layer has in many cases a strong influence on the mechanical properties of the machined part. This association is more pronounced in some materials and under certain machining operations. Typical surface integrity problems are as follows:

1. Grinding burns on high-strength steel of landing-gear components
2. Untempered martensite (UTM) in drilled holes
3. Stress corrosion properties of titanium by the cutting fluid
4. Grinding cracks in the root section of cast nickel–based gas turbine buckets
5. Lowering of fatigue strength of parts processed by EDM or ECM
6. Distortion of thin components
7. Residual stress induced in machining and its effect on distortion, fatigue, and stress corrosion

The subsurface characteristics occur in various layers or zones, as shown in Figure 14.1. The subsurface altered material zone (AMZ) can be as simple as a stress condition different from that in the body of the material or as complex as a microstructure change or IGA.

Accuracy and Surface Integrity Realized by Machining Processes

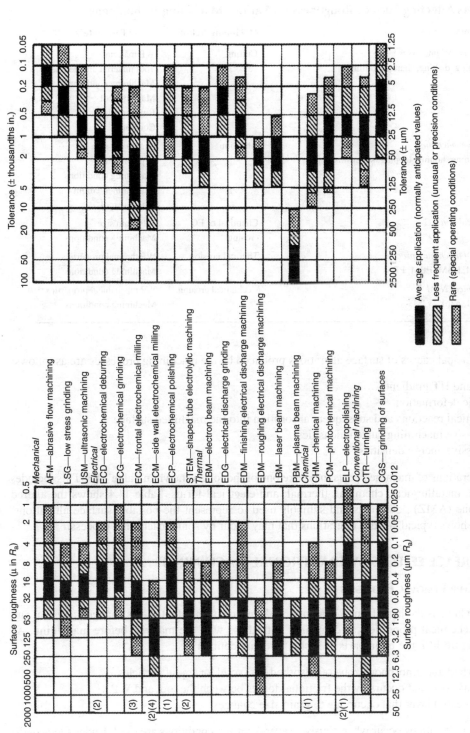

FIGURE 14.6 Different tolerances achieved by machining methods: (1) depends on state of starting surface, (2) titanium alloys are generally rougher than nickel alloys, (3) high-current density areas and (4) low-current density areas. (From *Machining Data Handbook*. Vol. 2, Metcut Research Association, Machinability Data Center, Cincinnati, OH, 1990.)

TABLE 14.2
Factors Affecting Surface Roughness for Various Machining Technologies

Machining Process	Machining Action	Parameters
Chip removal processes Turning, drilling, shaping, milling	Cutting	WP material Tool material and geometry Machining conditions Machine tool Built-up edge Coolant
Abrasive machining Grinding, honing, lapping, superfinishing	Abrasion	Grain-type size Type of bond Machining conditions Machining medium Machine tool
Chemical and EC	Chemical or EC erosion	WP grain size Machining conditions
Thermal machining process EDM, LBM, EBM, PBM	Thermal erosion	WP thermal properties Machining conditions
Mechanical NTM USM, AJM, WJM	Mechanical erosion	WP mechanical properties Machining conditions

The principal causes of surface alterations produced by the machining processes are as follows:

1. HTs and HT gradients
2. Plastic deformation
3. Chemical reactions and subsequent absorption into the machined surface
4. Excessive machining current densities
5. Excessive energy densities

Surface alterations of machined parts may take one of the forms shown in Table 14.3. This includes mechanical, metallurgical, chemical, thermal, and electrical forms. Table 14.4 shows the altered material zone (AMZ) definitions and symbols used to represent the possible surface alterations. Table 14.5 shows typical surface alterations that may occur by some machining process.

14.5 SURFACE EFFECTS BY TRADITIONAL MACHINING

14.5.1 CHIP REMOVAL PROCESSES

The surface layer is the layer from the geometrical surface inward that shows changed physical and sometimes chemical properties, as compared with those with the material before machining. As shown in Figure 14.7, the main parts of such a layer are defined as:

1. Adsorbed and amorphous zone of adsorbed gas, solid, or liquid particles
2. Fibrous zone that occurs by the frictional forces between the tool and WP
3. Compressed layer that occur due to grain size changes

Surface layer alterations occur when abusive (severe) cutting conditions are used. Under such conditions, high temperature (HT) and excessive plastic deformation is promoted. Figure 14.8 shows the surface alterations produced from drilling with dull tools where cracks, UTM at the surface, and a softer overtempered zone below the untempered surface layer are clear. Figure 14.9 shows the surface

TABLE 14.3
Forms of Surface Alterations by Machining

Form	Description
Mechanical	Plastic deformations (as a result of hot or cold working)
	Tears and laps and crevice-like defects (associated with "built-up edge" produced in machining)
	Hardness alterations
	Cracks (macroscopic and microscopic)
	Residual stress distribution in surface layer
	Process inclusions introduced
	Plastically deformed debris as a result of grinding
	Voids, pits, burrs, or foreign material inclusions in surface
Metallurgical	Transformation of phases
	Grain size and distribution
	Precipitate size and distribution
	Foreign inclusions in material
	Twinning
	Recrystallization
	UTM or OTM
	Resolutioning or austenite reversion
Chemical	IGA
	Intergranular corrosion (IGC)
	Intergranular oxidation (IGO)
	Preferential dissolution of microconstituents
	Contamination
	Embrittlement due to chemical absorption of elements, such as hydrogen, chlorine, and so on
	Pits or selective etch
	Corrosion
	Stress corrosion
Thermal	HAZ
	Recast or redeposited material
	Resolidified material
	Splattered particles or remelted metal deposited on surface
Electrical	Conductivity change
	Magnetic change
	Resistive heating or overheating

produced by face milling of Ti-6Al-4V (aged, 35 HRC). For gentle machining conditions, a slight white layer and no changes in microhardness occur. In abusive machining conditions, a white layer of 0.01 mm and a plastically deformed zone of 0.04 mm are visible. Figure 14.10 shows typical residual stresses in milling operations, which tend to be compressive. Residual stresses and part distortion are closely related, such that the greater the area under the residual stress curve, the greater will be the distortion of the machined WP.

14.5.2 Grinding

Using severe grinding conditions, the process becomes more likely to produce surface damage. In this regard, low-stress grinding (LSG) of AISI 4340 steel produced no visible surface alterations, as

TABLE 14.4
Altered Material Zone (AMZ) Definitions, Symbols, and Examples

AMZ Definition	Symbol	AMZ Definition	Symbol
Cracks. Narrow ruptures or separations with depth-to-width >4:1 that alter the continuity of the surface		**Craters.** Surface depressions with rough edges approximately round or oval and shallow	
Hardness alterations. Changes in hardness as a result of heat, mechanical deformation, or chemical changes		**HAZ.** A layer subjected to sufficient thermal energy that causes microstructure alterations or microhardness alterations	
Inclusions. Small particles in the surface layer in an object, may be either foreign or a part of the normal composition of the material		**IGA.** A form of corrosion or attack in which preferential reactions are concentrated at the surface grain boundaries	
Laps, folds, or seams. Defects in the surface from continued plastic working overlapping surface		**Low-stress surface.** A surface containing a residual stress less than 138 MPa or 10% of tensile stress, whichever is greater, at depths below the surface greater than 0.025 mm	
Metallurgical transformation. Micro structural changes resulting from external influences		**Pits.** Shallow depressions such as craters with depth-to-width ratio less than 4:1 or localized selective etching or corrosion that result in holes or pockets left by the machining process	
Plastic deformation. Microstructural changes as a result of exceeding the yield point of the material		**Recrystallization.** The formation of a new strain-free grain or crystal structure from that existing in the material before machining, usually as a result of plastic deformation and subsequent heating	
Recast material. Occurs when some materials become molten and are then resolidified		**Redeposited material.** When some material is removed from the WP in the molten state, and then prior to solidification is attached to the surface (splattered metal)	
Remelted or resolidified material. The portion of the surface that becomes molten, but is not removed from the surface prior to solidification		**Residual stresses.** Those stresses that are present in the material after all external influences are removed	
Selective etch. A process or attack in which preferential reactions are concentrated on certain constituents of the base material corrosion			

shown in Figure 14.11a, compared to abusive grinding, shown in Figure 14.11b. It is therefore clear that abusive grinding produces UTM of 0.03–0.13 mm deep with a hardness of 65 HRC. Below this layer, an overtempered martensite (OTM) zone having a hardness of 46 HRC is clear. The hardness returns back to its normal value at a depth of 0.3 mm below the surface as shown in Figure 14.12a. Abusive grinding produces residual stresses within the altered layer. As shown in Figure 14.12b,

TABLE 14.5
Examples of Surface Alterations for Some Machining Operations

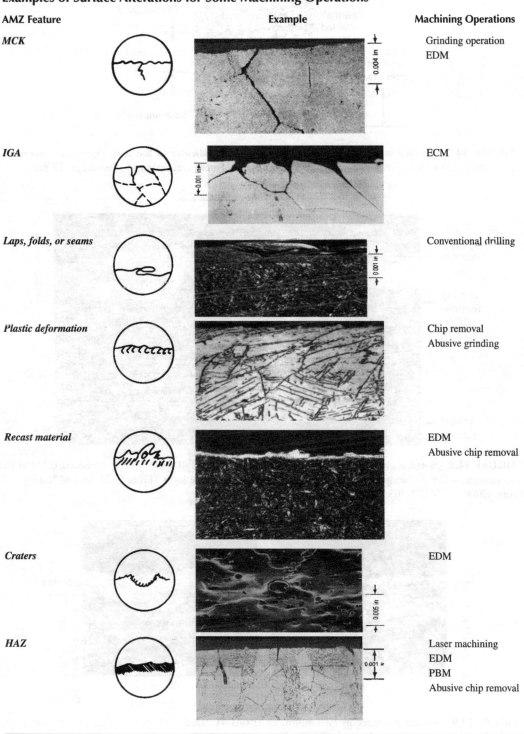

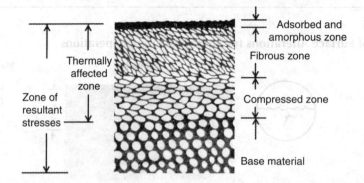

FIGURE 14.7 Surface layer after machining, section perpendicular to the tool. (From Kaczmarek, J., *Principles of Machining by Cutting, Abrasion, and Erosion*, Peter Peregrines Ltd., Stevenage, 1976.)

FIGURE 14.8 Surface alterations produced from drilling with dull tools, where a cracked UTM at the surface and softer overtempered zone below the untempered surface layer. (From Field, M. and Kahles, J.F., *Ann. CIRP*, 20(2), 153–163, 1971. With permission.)

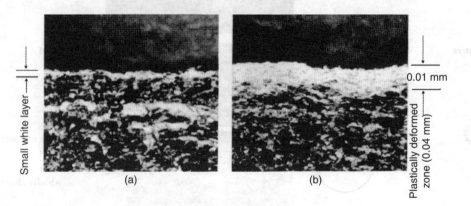

FIGURE 14.9 Surface produced by face milling of Ti-6Al-4V (aged, 35 HRC): (a) gentle conditions, slight white layer, no changes in hardness and (b) abusive machining conditions, white layer of 0.01 mm and plastically deformed zone of 0.04 mm are visible. (From Field, M. and Kahles, J.F., *Ann. CIRP*, 20(2), 153–163, 1971. With permission.)

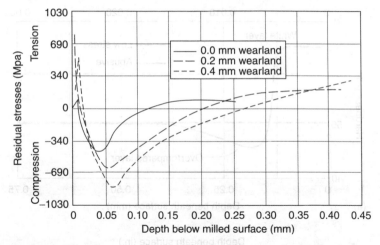

FIGURE 14.10 Residual stresses from surface milling of 4340 steel (quenched and tempered to 52 HRC). (From *Machining Data Handbook*, Vol. 2, Metcut Research Association, Machinability Data Center, Cincinnati, OH, 1990.)

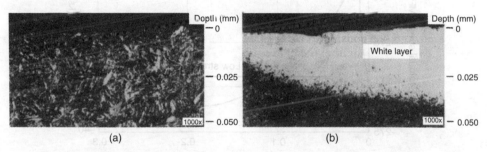

FIGURE 14.11 Surface characteristics produced by surface grinding of AISI 4340 steel: (a) LSG—no visible alterations and (b) abusive grinding. (From Field, M., Kahles, J.F., and Cammett, J.T., *Ann. CIRP*, 21(2), 219–238, 1972. With permission.)

LSG produced a surface of low compressive stress compared to the abusive grinding condition. Additionally, abusive grinding seriously reduces the fatigue strength, as shown in Figure 14.12c. Table 14.6 shows the effect of some machining methods on the fatigue strength together with the percentage change with respect to gentle grinding. Accordingly, the endurance limit of 4340 steel has been decreased by 12% after polishing, compared to gentle grinding. In case of abusive grinding, there is a tendency to form batches of UTM or OTM on the surface, which is associated with a significant drop in the fatigue strength. Shot peening is used to improve the fatigue strength and stress corrosion properties of most structural alloys that are subjected to high stresses and severe environments. It improves the properties of metals that have been machined and that tend to have degraded fatigue strength and other mechanical properties, as shown in Table 14.7.

14.6 SURFACE EFFECTS BY NONTRADITIONAL MACHINING

NTM methods have been introduced due to components' complex shape and fine finish requirements. It is used whenever the conventional machining processes are not able to cope up satisfactorily with the enhanced properties of difficult-to-machine materials. The principal mechanism of material removal may be chemical during the ECM and CHM processes. The melting and boiling

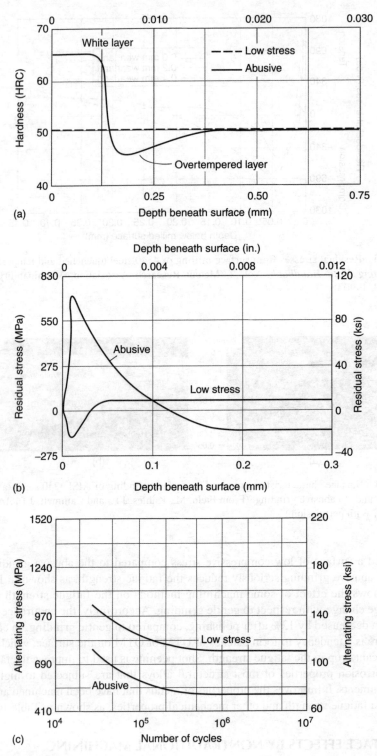

FIGURE 14.12 Surface effects by LSG and abusive grinding of AISI 4340 steel: (a) hardness (b) residual stresses and (c) fatigue strength. (From Field, M., Kahles, J.F., and Cammett, J.T., *Ann. CIRP*, 21(2), 219–238, 1972. With permission.)

TABLE 14.6
Effect of Machining Method on Fatigue Strength

Alloy	Machining Operation	Endurance Limit in Bending, 10^7 Cycles (MPa)	Change Compared to Gentle Grinding (%)
4340 steel, 50 HRC	Gentle grinding	703	—
	ELP	620	−12
	Abusive grinding	430	−39
Ti-6Al-4V, 32 HRC	Gentle grinding	430	—
	Gentle milling	480	+13
	CH-milling	350	−18
	Abusive milling	220	−48
	Abusive grinding	90	−79
Inconel 718, aged 44 HRC	Gentle grinding	410	—
	ECM	270	−35
	Conventional grinding	165	−60
	EDM	150	−63

Source: Field, M. and Kahles, J. F., *Ann. CIRP*, 20(2), 153–163, 1971. With permission.

TABLE 14.7
Effect of Shot Peening on Fatigue Strength of Machined Parts

		Endurance Limit in Bending, 10^7 Cycles (MPa)		
Alloy	Machining Operation	Before Shot Peening	After Shot Peening	Percentage Increase
4340 steel, 50 HRC	Gentle surface grinding	703	772	110
	Abusive surface grinding	430	630	146
	Electropolished	620	660	106
Inconel 718, solution treated, and aged 44 HRC	EDM roughing	170	540	317
	EDM finishing	170	480	282
	ECM	285	560	196
	ELP	290	540	186

Source: Koster, W. P., Gatto, L. E., and Cammett, J. T., Influence of Shot Peening on Surface Integrity of Some Machined Aerospace Materials, Proceedings of 1st International Conference on Shot Peening, France, Pergamon Press, 287–293, 1981.

temperatures, heat of fusion, and the specific energy of the machined components play major roles in thermal machining processes. In other cases, the mechanical characteristics are the controlling factors in the removal mechanism associated with mechanical NTM processes.

Some processes combine more than one removal action, such as the double one in ECG and electroerosion dissolution machining (EEDM) and the triplex effect in electrochemical discharge grinding (ECDG). Such processes are generally classified on the basis of the main removal mechanism contributing to the bulk of material removal as a chemical or a thermal machining process (El-Hofy, 2005). Wear resistance, the contact pressure, stress concentration, type of fit, and the corrosion resistance properties determine the performance of any machined component by NTM methods.

14.6.1 Electrochemical and Chemical Machining

In ECM, considerable variations are possible in surface finishes produced due to WP characteristics and machining conditions. Crystallographic irregularities in crystal lattices, such as voids, dissociation and grain boundaries, differing crystal structures and orientations, and locally different alloy composition produce an irregular distribution of current density, thus leaving microscopic peaks and valleys that produce surface roughness (König and Lindenlauf, 1978). Figure 14.13 shows the mechanism of surface roughness generation in an alloy machined by ECM. For nonpassivating electrolytes (NaCl), the reduction in electrolyte concentration and the increase of its temperature improves the quality of surfaces. For passivating electrolytes ($NaNO_3$), a low electrolyte concentration and a rise in its temperature increase the formation of a protective layer that causes deterioration to surface quality. Further increase in current density breaks up this layer, so that a smoother surface is produced. Surface roughness depends on the structure of the WP material. In this regard, the more fine-grained and homogeneous the structure, the better is the surface quality. The roughness obtained at greater grain sizes was possibly due to the reduced number of grain boundaries present on such surfaces. Surface finish also deteriorates with increasing grain size at lower flow velocities. The different levels of surface roughness R_a and dimensional tolerance that can be obtained by different ECM applications are shown in Figure 14.6.

In CHM, etching rates of 0.025 mm/min with tolerances of ±10% of the cut width (±0.01 to ±0.04 mm) can be achieved, depending on the WP material and depth of etch. The final surface roughness is influenced by the initial WP roughness. It increases as the metal ion concentration of the enchant rises. Table 14.8 shows the dependence of surface roughness on the WP material and production method. In PCM, cutting rates of 0.01–0.2 mm/min are possible. Tolerance of ±15% of the material thickness can be obtained, leaving a surface roughness similar to that of CHM.

Electropolishing (ELP) is a finishing process with metal removal rates between 0.013 and 0.038 mm/min. Because the erosion of the surface asperities takes place at a faster rate than the erosion in the valleys, the surface can be smoothed without considerable material removal. Surface roughness R_a of 0.1–0.8 μm, depending on the initial surface roughness, can be obtained. However, under special machining conditions, a roughness of 0.025–0.05 μm is possible.

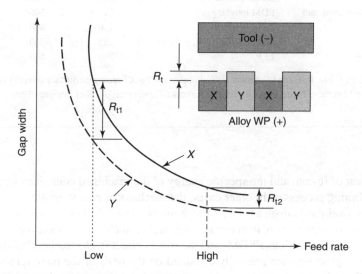

FIGURE 14.13 Surface roughness generation in ECM.

TABLE 14.8
Surface Roughness Achieved in CH-Milling after Removing 0.25–0.40 mm from the Surface

Material	Form	Surface Roughness (R_a) (µm)
Aluminum alloys	Sheet	2.0–3.8
	Casting	3.8–7.6
	Forging	2.5–6.4
Magnesium alloys	Casting	0.75–1.4
Steel alloys	Sheet	0.75–1.5
Nickel alloys	Sheet	0.75–1.0
Titanium alloys	Sheet	0.20–0.8
	Casting	0.75–1.5
	Forging	0.38–1.0
Tungsten	Bar	0.50–1.0
Beryllium	Bar	3.8–6.4
Tantalum	Sheet	0.25–0.5
Columbium	Bar	1.0–1.5
Niobium	Sheet	1.0–1.5

Source: Wilkinson, B. H. and Warburton, P., *Iron and Steel Report* 94, 215–220, 1967. With permission.

ECM, CHM, and ELP produce surfaces that are free from metallurgical alterations. However, under certain conditions, selective etching (SE) or IGA may occur, as shown in Figures 14.14 and 14.15. Surface softening occurs in most materials machined by ECM as well as CHM, as shown in Figures 14.16 and 14.17. Accordingly, the surface is about five points HRC lower in hardness than the interior to approximately 0.05 mm in depth. This softening may be severe enough to affect the fatigue strength and other mechanical properties of metals and may necessitate postprocessing.

As shown in Table 14.6, CHM of Ti-6Al-4V resulted in an 18% drop in fatigue strength, while ECM of Inconel 718 produced a 35% drop in endurance limit compared to gentle grinding. Generally, ECM, CHM, and ELP produce stress-free surfaces. Additionally, the decrease in their fatigue strength resulted from the absence of the compressive stress associated with surface grinding. For CHM of Ti-6Al-4V, the endurance limit is 350 MPa compared to 460 MPa in case of LSG. Table 14.7 shows how shot peening raises the endurance limit of parts machined by ECM, CHM, and ELP. In ECG, the surface is similar to that obtained using metallographic polishing, free from process-induced residual stresses, and there is no heat-affected layer.

14.6.2 Thermal Nontraditional Processes

14.6.2.1 Electrodischarge Machining

In this process, the machined surface consists of a multitude of overlapping craters that are formed by the action of microsecond duration spark discharges. These craters depend on the thermal properties of the material, the composition of the machining medium, the discharge energy, and the pulse on-time, as shown in Figure 14.18. The integral effect of many thousands of discharges per second leads to the formation of the corresponding profile with specified accuracy and surface finish, as shown in Figure 14.6. The peak-to-valley roughness is usually represented by the depth of the resulting crater. The maximum depth of the damaged layer is usually 2.3 times the surface roughness R_a.

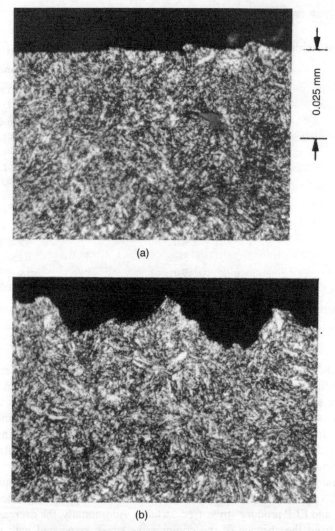

FIGURE 14.14 Surface generation of 4340 steel machined by ECM: (a) gentle conditions (high current density) and (b) abusive conditions (low current density). (From Koster, W.P., et al., *Surface Integrity of Machined Structural Components*, AFML-TR-70-11, Metcut Research Associates, Cincinnati, OH, P2, 1970.)

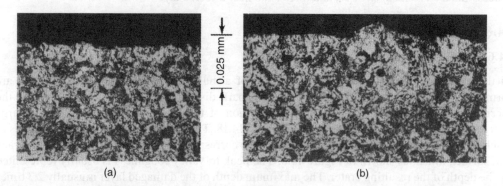

FIGURE 14.15 Surface generation of 4340 steel machined by CHM: (a) gentle conditions and (b) abusive conditions. (From Field, M. and Kahles, J.F., *Ann. CIRP*, 20(2), 153–163, 1971. With permission.)

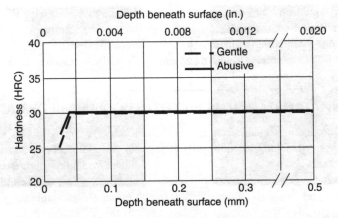

FIGURE 14.16 Hardness of machined surfaces of 4340 steel machined by ECM for gentle and abusive conditions. (From Koster, W.P., et al., *Surface Integrity of Machined Structural Components*, AFML-TR-70-11, Metcut Research Associates, Cincinnati, OH, P2, 1970.)

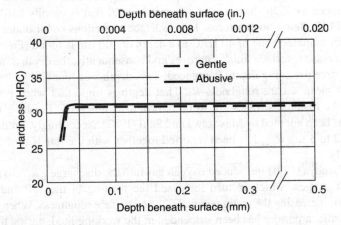

FIGURE 14.17 Hardness of machined surfaces of 4340 steel machined by CHM: (a) gentle conditions and (b) abusive conditions. (From Field, M. and Kahles, J.F., *Ann. CIRP*, 20(2), 153–163, 1971. With permission.)

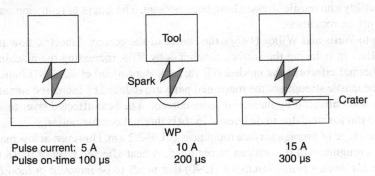

FIGURE 14.18 Effect of pulse characteristics on EDM crater size.

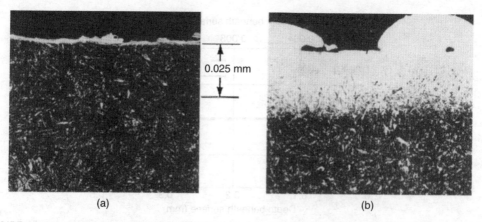

FIGURE 14.19 Surface characteristics of AISI (quenched and tempered 50 HRC) after (a) finish EDM and (b) rough EDM. (From Field, M. and Kahles, J. F., *Ann. CIRP*, 20(2), 153–163, 1971. With permission.)

According to Rajurkar and Pandit (1984) and Delpretti (1977), the maximum peak-to-valley height was considered to be 10 times the average roughness R_a. Surface roughness increases linearly with metal removal rate, and graphite electrode, produce rougher surfaces than the metal ones.

Surfaces produced by EDM have recast splattered metal that is usually hard and cracked, as shown in Figure 14.19. Below such a recast layer, surface alterations occur under roughing conditions where an overtempered zone of 40 HRC to a depth of 100 µm is found (Figure 14.20). As the pulse energy is decreased, surface finish improves and consequently, the depth at which the surface craters disappear (free polishing depth) is reduced. This depth was found to lie between three and six times the root mean square roughness R_q. That depth is important when polishing dies and molds and when the residual stresses are to be removed. A quick EC finishing technique, using mate electrode, has been adopted by Masuzawa and Saki (1987). Accordingly, a reduction of surface roughness from 22 to 8 µm R_{max} has been reported together with the removal of the heat-affected layer (Figure 14.21).

Koneda and Furuoya (1991) introduced oxygen gas into the discharge gap provided extra power by the reaction of oxygen, which in turn increased the melting of the WP and created greater expulsive force, thus increasing the metal removal rate and surface roughness. When EDM is used for cusp removal, the silicon powder has been suspended in the working fluid, during the stage of finish EDM. Consequently, surface roughness is changed from 45 to 10 µm R_{max} as shown in Figure 14.22. In contrast, the matte appearance of the ED-machined surfaces has been found satisfactory in some applications of texturing described by Aspinwal and Wise (1992) and Amalnik, et al. (1997).

Abusive machining conditions have a minor effect on the static mechanical properties of machined materials. However in case of EDM followed by stress relief heat treatment, a marked decrease in ductility and tensile stresses have been noticed. The extent of reduction was found to be a function of surface roughness.

According to Wells and Willey (1977), the choice of the correct dielectric flow in the gap has a significant effect in reducing the surface roughness by 50%, increasing the machining rate, and lowering the thermal effects in the eroded WP surface. Benhaddad et al. (1991) indicated that, for Al/Li alloys, the tensile strength of the machined parts are reduced by increased surface roughness. This reduction was enhanced by increased pulse current. The heat-affected layer reached 200 µm compared to 30 µm for steel due to difference in their thermal conductivities.

In EDG, the range of average surface roughness is 1.6–3.2 µm. However, at low machining rates of 0.03 cm³/h, a roughness of 0.4 µm can be obtained. A heat-affected layer of 0.013 mm has been reported in the *Machining Data Handbook* (1990) that needs to be removed or modified to ensure the best surface integrity. In electrodischarge sawing (EDS), surface roughness for high feeds, is

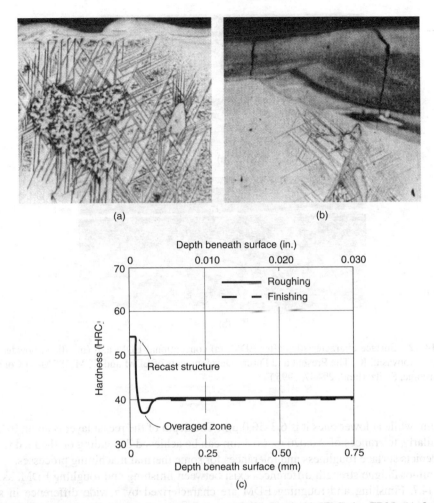

FIGURE 14.20 Surface characteristics of cast Inconel 718 (aged, 40 HRC) by EDM: (a) finish EDM, (b) rough EDM, and (c) microhardness alterations. (From Field, M. and Kahles, J. F., *Ann. CIRP*, 20(2), 153–163, 1971. With permission.)

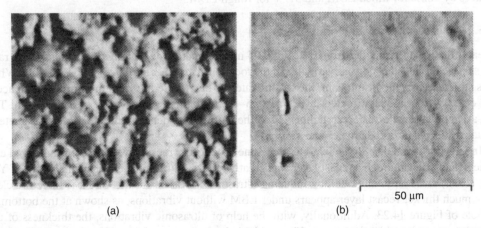

FIGURE 14.21 Use of ECM mate electrode to finish SKS 3 tool steel machined by EDM: (a) after EDM and (b) after ECM finishing. (From Masuzawa, T. and Saki, S., *Ann. CIRP*, 36(1), 123–326, 1987. With permission.)

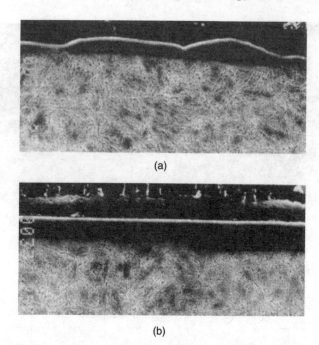

FIGURE 14.22 Surface characteristics after EDM: (a) conventional EDM and (b) silicon powder mixing EDM. (From Kobyashi, K., The Present and Future Developments of EDM and ECM, ISEM-XI Conference, EFPL, Lausanne, Switzerland, 29–47, 1995.)

10–12.5 µm, while at lower ones it is 6.3–10.0 µm. The depth of the recast layer is about 0.025 mm deep. Similarly, tolerances of ±0.076 to ±0.4 µm can be achieved depending on the feed rate. Figure 14.6 depicts surface roughness and tolerances for some thermal machining processes.

Very minor fatigue strength differences exist between finishing and roughing EDM, as shown in Table 14.7. Finishing and roughing EDM are characterized by a wide difference in surface roughness levels (1.25–5 µm R_a) and wide differences in recast layer thicknesses (5–125 µm). High cycle fatigue strength, however, is nearly the same for finishing as for roughing by EDM. The fatigue strength of Inconel 718 alloy machined by EDM reported a reduction by 63% compared with gentle grinding. As shown in Table 14.7, using shot peening has raised the fatigue strength of EDM surfaces by 282% for finish EDM and 317% for rough EDM.

14.6.2.2 Laser Beam Machining

A heat-affected layer ranging from 0.025 to 0.05 mm is created, depending on the machining rate. The surface produced is normally rough and the location accuracy of ±0.025 mm is possible. When a gas stream assists the laser beam, the cutting rate has been raised to 7500 mm/min with an accuracy of ±0.1 mm, a heat-affected layer of 0.025–0.25 mm, and a roughness of 3.2–6.3 µm. The heat-affected layer, the surface roughness, and the cutting rate depend on the process parameters, including the gas nozzle diameter and gap.

In LBM of Inconel 718, the heat-affected zone produced a recast layer at the entrance and exit of the hole produced. Such a recast layer was greatly reduced during the experimental work of Yue et al. (1996) and Lau et al. (1990) by inducing ultrasonic vibrations to the WP. It is therefore clear that a much thicker recast layer appears under LBM without vibrations, as shown at the bottom of the hole of Figure 14.23. Additionally, with the help of ultrasonic vibrations, the thickness of the recast layer is reduced by four times. Figure 14.24 shows the boundary of HAZ with and without WP vibration.

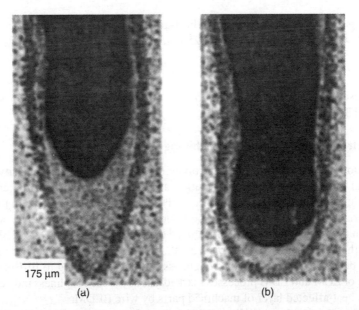

FIGURE 14.23 LBM drilling: (a) without vibrations and (b) with ultrasonic vibrations. (From Yue, T.M., Chen, T.W., Man, H.C., and Lau, W.S., *Ann. CIRP*, 45(1), 169–172, 1996. With permission.)

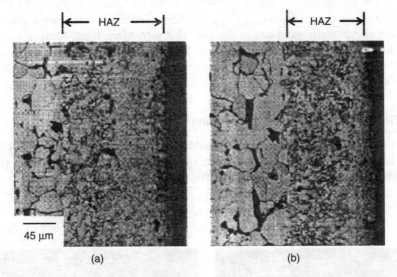

FIGURE 14.24 Laser beam machining: (a) without vibration and (b) with ultrasonic vibration. (From Yue, T.M., Chen, T.W., Man, H.C., and Lau, W.S., *Ann. CIRP*, 45(1), 169–172, 1996. With permission.)

14.6.2.3 Electron Beam Machining

In EBM, the quality of the surface produced is influenced by the thermal properties of WP and pulse energy or charge. There is a thin layer of recast or heat-affected material on the cut surface that should be removed or modified. The heat-affected layer increases with pulse energy and therefore the use of short pulses limits the extent of the heat-affected zone. Typical tolerance is about ±10% of the slot width with heat-affected layer of 0.25 mm is produced.

14.6.2.4 Plasma Beam Machining (PBM)

The depth of fused layer ranges from 0.81 to 4.67 mm and the average roughness R_a is usually between 0.8 and 3.2 μm.

14.6.2.5 Electroerosion Dissolution Machining

Surface generation takes place through electrolytic dissolution of varying intensity depending on the gap size and the consequent crater formation at random locations over the entire machined surface. Saushkin et al. (1982) reported that electrolysis is apparently localized in the proximity of the pits of the craters that are soon made smooth, probably as a result of the HT of the metal and the electrolyte in this zone. The general appearance of the machined surface by EEDM constitutes less turbulence than that of EDM. The crater depth, volume, diameter-to-depth ratio, and surface roughness have been calculated using the roughness profiles. According to El-Hofy (1992, 1996), the combination of ECM and EDM processes markedly reduced the roughness indices together with the absence of a heat-affected layer of machined parts by wire EEDM.

The combination of ECM and EDM in EEDM die sinking produces intrinsic effects on surface roughness, shape geometry, and subsurface layers. The molten crates on the WP surface formed by severe thermal erosive pulses are quenched very rapidly in the surrounding cold electrolyte. The surface and subsurface quickly undergo metallurgical phase transformations. The white damaged layer was clear at the machined die surface. The thickness of such a layer is consistently decreasing in the direction of the electrolyte upstream, owing to the continuous attack by ECD. The thickness of the white layer ranged from 0 to 130 μm. Micro- and macrocracks were also observed in the white layer, occasionally extended to the annealed subsurface layer. Most of the MCK were removed by the electrochemical dissolution action. The pitted surface of the die by selective etching is shown in Figure 14.25a; Figure 14.25b shows the white layer by EEDM process at the die bottom section, with clear cracks of micro and macro sizes.

14.6.2.6 Electrochemical Discharge Grinding

The WP, mainly eroded by discharges, is immediately smoothed by mechanical grinding and electrolytic dissolution, thus producing an average roughness of 0.4 μm R_a. The thickness or the altered layer also decreases at lower machining voltages, as well as smaller wheel speeds (Kaczmarek, 1976).

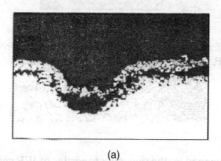
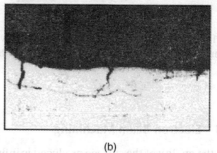

(a) (b)

FIGURE 14.25 EEDM surface layer: (a) pitted surface by selective etching and (b) white layer by EEDM. (From Khayry, A. B. M., *Ann. CIRP*, 39(1), 191–195, 1990. With permission.)

14.6.3 Mechanical Nontraditional Processes

In USM, the material removal rate and surface finish depend on the feed force, oscillation amplitude, WP material, and grain size and type of abrasives used. Surface roughness increases with oscillation amplitude and grain size. Rougher surfaces are produced when machining glass than when machining hard alloy steels. Generally, the more brittle the material, the greater the resulting surface roughness, as the cavity left by each particular grain is deeper. As the amplitude of vibration is raised, the surface becomes rougher, because the individual grains are pressed further into the surface of the WP. Smoother finish can also be achieved when the viscosity of the liquid carrier is lowered. The roughness in the sides of cut is much higher than that in the bottom. This results from the sidewalls being scratched by grains entering and leaving the machining zone. As a rule, USM produces an average surface roughness R_a of 0.51–0.76 μm, out-of-roundness 10 μm, a taper of 0.005 mm/mm, and production tolerances of ±0.005 to ±0.025 mm. The process leaves no heat-affected zone, and no chemical or electrical alterations. Shallow compressive residual stresses occur at the surface, which promote an increase in the high-cycle fatigue strength of the work material (Rooney, 1957).

In AJM, surface roughness depends on WP material, grit size, and type of abrasives. A material with a high removal rate results in large surface roughness. For this reason, fine grains are used for machining soft metals to obtain the same roughness in hard ones. The decrease of surface roughness with a smaller grain size is related to the reduced depth of cut and the undeformed chip cross section. In addition, the larger the number of grains per unit volume, the larger the number of grains that fall on a unit surface area. Generally, surface roughness of 0.15–1.60 μm R_a with tolerance of ±0.05 mm is possible. In AWJM, a carrier liquid consisting of water with anticorrosive additives has much greater density than air, which contributes to higher acceleration of the grains, with consequent larger grain speed and increased metal removal rate. Moreover, the carrier liquid, when spreading over the surface, fills its cavities and forms a film that impedes the striking action of the grains, and hence bulges and tops of the surface irregularities are the first to be affected and the surface quality improves. Experiments showed that water air jet permits one to obtain, as an average, a roughness number higher by one, as compared with the effect of an air jet. In high speed WJM of Inconel, Hashish (1992) concluded that the roughness increases at higher feed rates as well as lower slurry flow rates. The surface produced during WJM is generally a polished, scoured clean one of light peening texture.

In AFM, stock removal rates and surface roughness depend on grit type and extrusion pressure. The process leaves residual stress of depth 0.025 mm. It can be used for quick polishing where a surface of roughness 0.8–7.6 μm R_a finishes to one-tenth of its original value. When the process is used for deburring of stainless steel boles, surface roughness changes from 1.6 to 0.4 μm.

14.7 REDUCING DISTORTION AND SURFACE EFFECTS IN MACHINING

Table 14.9 summarizes the possible surface effects produced by different machining processes of some engineering metals and alloys. Reduction of these effects can be achieved by considering the following recommendations:

1. Chip removal processes
   - Select machining conditions that lead to long tool life and good surface finish.
   - Use sharp tools.
   - Use rigid and high-quality machine tools.
   - Avoid hand feeding during drilling and reaming.
   - Use deburring to remove sharp corners.
   - Use honing to improve surface quality.
2. Abrasive machining
   - Use LSG to remove the last 0.25 mm of material.
   - Dress the grinding wheels frequently.

TABLE 14.9
Summary of Possible Surface Alterations Resulting from Various Material Removal Processes

Material	Conventional		Nontraditional		
	Milling, Drilling, and Turning	Grinding	EDM	ECM	CHM
Nonhardenable 1018 steel	R	R	R	R	R
	PD	PD	MCK	SE	SE
	L and T		RC	IGA	IGA
Hardenable 4340 and D6ac steel	R	R	R	R	R
	PD	PD	MCK	SE	SE
	L and T	MCK	RC	IGA	IGA
	MCK	UTM	UTM		
	UTM	OTM	OTM		
	OTM				
D2 tool steel	R	R	R	R	R
	PD	PD	MCK	SE	SE
	L and T	MCK	RC	IGA	IGA
	MCK	UTM	UTM		
	UTM	OTM	OTM		
	OTM				
Type 410 stainless steel (martensitic)	R	R	R	R	R
	PD	PD	MCK	SE	SE
	L and T	MCK	RC	IGA	IGA
	MCK	UTM	UTM		
	UTM	OTM	OTM		
	OTM				
Type 302 stainless st (austenitic)	R	R	R	R	R
	PD	PD	MCK	SE	SE
	L and T		RC	IGA	IGA
17-4 PH steel	R	R	R	R	R
	PD	PD	MCK	SE	SE
	L and T	OA	RC	IGA	IGA
	OA		OA		
350-grade maraging (18% Ni) steel	R	R	R	R	R
	PD	PD	RC	SE	SE
	L and T	RS	RS	IGA	IGA
	RS	OA	OA		
	OA				
Nickel and cobalt base alloys	HAZ	HAZ			
Inconel alloy 718	R	R	R	R	R
Rene 41	PD	PD	MCK	SE	SE
HS 31	L and T	MCK	RC	IGA	IGA
IN 100	MCK				
Ti-6Al-4V	HAZ	HAZ			
	R	R	R	R	R
	PD	PD	MCK	SE	SE
	L and T	MCK	RC	IGA	IGA
Refractory alloy molybdenum TZM	R	R	R	R	R
	L and T	MCK	MCK	SE	SE
	MCK			IGA	IGA
Tungsten (pressed and sintered)	R	R	R	R	R
	L and T	MCK	MCK	SE	SE
	MCK			MCK	MCK
				IGA	IGA

Note: R = roughness of surface, PD = plastic deformation, L and T = laps and tears, MCK = microcracks, HAZ = heat-affected zone, SE = selective etch, IGA = intergranular attack, UTM = untempered martensite, OTM = overtempered martensite, OA = overaging, RS = resolution or austenite reversion, RC = recast, respattered, vapor-deposited metal.

Source: Field, M., Kahles, J. F., and Cammett, J. T., *Ann. CIRP*, 21(2), 219–238, 1972. With permission.

- Apply cutting fluids.
- Avoid hand wheel grinding.
- Allow proper allowance for subsequent cleanup of cutoff parts.
3. Chemical and electrochemical machining
   - Consider the prior metallurgical condition of the machined surface.
   - Monitor and control the preselected operating parameters.
   - Control the machining current density.
   - Clean the surface to remove electrolytes or etchants.
   - Adopt post treatment (shot peening) to restore fatigue strength after ECM or CHM.
   - Consider the prior metallurgical condition of the surface.
4. Thermal machining
   - Consider the prior metallurgical condition of the surface.
   - Monitor and control the preselected operating parameters.
   - Control the magnitude of the energy impinging on that surface.
   - Clean the surface to remove dielectric fluids, beads, and vapor residue.
   - Apply post treatment to restore fatigue strength after EDM of critical or highly stressed surfaces by:
     a. Removal of layers by LSG
     b. Removal of layers by CHM
     c. Addition of a metallurgical-type coating
     d. Reheat treatment
     e. Application of shot peening

14.8 REVIEW QUESTIONS

1. Differentiate between surface roughness and surface texture.
2. Describe the theoretical roughness that may be generated during turning.
3. Sketch the possible lay that may be formed after milling, shaping, grinding, and honing.
4. State the main reasons behind surface alterations in machining.
5. Explain how the surface properties in EEDM are better than EDM.
6. Explain why ECM and CHM produce low fatigue strength compared to LSG.
7. Differentiate between surface texture and surface integrity.
8. Explain the effect of shot peening on raising the fatigue strength of machined surfaces.
9. State the possible forms of surface alteration after ECM, drilling, and EDM.
10. State the general recommendations that may be considered to reduce the surface alterations in case of EDM, turning, LBM, and ECM.

REFERENCES

Amalnik, M. S., El-Hofy, H., and McGeough, J. (1997) An intelligent knowledge based system for manufacturability evaluation of design for electro discharge texturing, MATADOR Conference, Manchester, pp. 418–424.

Aspinwal, D. K. and Wise, M. L. (1992) Electrical discharge texturing, *International Journal of Machine Tools and Manufacture*, 32(12), 183–193.

Benhaddad, M. A., McGeough, J. A., and Barker, M. B. (1991) Electrodischarge machining of Al-Li alloys and its effect on surface roughness, hardness and tensile strength, *Processing of Advanced Materials*, 6(314), 123–128.

Delpretti, M. R. (1977) Physical and chemical characteristics of superficial layers, *Proceedings of ISEM-5*, 209–212.

El-Hofy, H. (1992) Electroerosion dissolution machining of graphite, Inco 901, 2017 Al and Steels, 5th PEDAC Conference, Alexandria, pp. 489–501.

El-Hofy, H. (1996) Surface generation in nonconventional machining, 6th MDP Conference, Cairo University, Egypt, pp. 203–213.

El-Hofy, H. (2005) *Advanced Machining Processes, Nontraditional and Hybrid Machining Processes*, McGraw-Hill, New York.

Field, M. and Kahles, J. F. (1971) Review of surface integrity of machined components, *Annals of CIRP*, 20(2), 153–163.

Field, M., Kahles, J. F., and Cammett, J. T. (1972) Review of surface integrity of machined components, *Annals of CIRP*, 21(2), 219–238.

Hashish, M. (1992) Machining with High Velocity Water Jets, 5th PEDAC Conference, Alexandria, pp. 461–471.

Kaczmarek, J. (1976) *Principles of Machining by Cutting, Abrasion and Erosion*, Peter Peregrines Ltd., Stevenage.

Khayry, A. B. M. (1990) Die-sinking by electroerosion-dissolution machining, *Annals of CIRP*, 39(1), 191–195.

Kobyashi, K. (1995) The present and future developments of EDM and ECM, ISEM-XI conference, EPFL, Lausanne, pp. 29–47.

Koneda, M. and Furuoya, S. (1991) Improvements of EDM efficiency by supplying oxygen gas into gap, *Annals of CIRP*, 40(1), 215–218.

König, W. and Lindenlauf, P. (1978) Surface generation in electrochemical machining, *Annals of CIRP*, 29(1), 97–100.

Koster, W. P., Gatto, L. E., and Cammett, J. T. (1981) Influence of shot peening on surface integrity of some machined aerospace materials, Proceedings of 1st International Conference on Shot Peening, France, Pergamon Press, pp. 287–293.

Koster, W. P., et al. (1970) *Surface Integrity of Machined Structural Components, AFML-TR-70-11*, Metcut Research Associates, Cincinnati, OH, P2.

Lau, W.S., Lee, W.P., and Pans, S.Q. (1990) Pulsed Nd: YAG laser cutting of fiber composite materials, CIRP, 39(1), 179–182.

Machining Data Handbook (1990) Vol. 2, 3rd Edition, Metcut Research Association Machinability Data Center, Cincinnati, OH.

Masuzawa, T. and Saki, S. (1987) Quick finishing of WEDM products using mate-electrode, *Annals of CIRP*, 36(1), 123–326.

Rajurkar, K. P. and Pandit, S. M. (1984) Quantitative expressions for some of surface integrity in electro discharge machined components, *ASME Journal of Engineering for Industry*, 106, 171–178.

Rooney, R. J. (1957) The Effects of Various Machining Processes on the Reversed-Bending Fatigue Strength of A-110AT Titanium Alloy Sheet, U. S. Air Force Technical Report WADC-TR-57-310, Wright Air Development Center, Wright-Patterson Air Force Base, OH.

Saushkin, B. P. et al. (1982) Special features of combined electrochemical and electroerosion machining of elongate machine parts, *Electrochemistry in Industrial Processing and Biology*, 105(3), 8–14.

Surface Texture (Surface Roughness, Waviness, and Lay) ANSI/ASME B 46.1 (1985) American Society of Mechanical Engineers, NY.

Surface Texture Symbols, ANSI Y14.36 (1978) American Society of Mechanical Engineers, NY.

Wells, P. W. and Willey, P. C. T. (1977) The effect of variation of dielectric flow rate in the gap on wear ratio and surface finish during electrodischarge machining, *Proceedings of ISEM*-5, 110–117.

Wilkinson, B. H. and Warburton, P. (1967) Electrochemical machining-machinability, *Iron and Steel Report*, 94, 215–220.

Yue, T. M., Chen, T. W., Man, H. C., and Lau, W. S. (1996) Analysis of ultrasonic aided laser drilling using finite element method, *CIRP*, 45(1), 169–172.

15 Automated Manufacturing System

15.1 INTRODUCTION

In the early days of the eighteenth century, the first industrial revolution started when attempts were made to substitute the muscle power by mechanical energy. Machine tools such as boring machines, lathes, drill presses, copying lathes, turret lathes, and milling machines were introduced into the production of goods. Geared and automatic lathes were introduced in the 1900s. Mass production techniques and mechanized transfer machines were developed between 1920 and 1940. These systems had fixed mechanisms and were designed to produce specific products. These developments were best represented in the automobile industry, and characterized by high production rates at low cost (Figure 15.1). Since that time, productivity has become a major concern; productivity is often defined as the use of all resources such as materials, energy, capital, labor, and technology, or may be defined as the output/labor hour. It is basically a measure of operating efficiency.

Mechanization related to the first industrial revolution, and reached its peak by 1940s, when most manufacturing operations were carried out on traditional machinery, such as lathes, milling machines, and automatic lathes, which required skilled operators and lacked flexibility. Each time a different product was manufactured, the machine had to be retooled. Furthermore, new products with complex shapes required tedious work from the operator to set the proper processing parameters. Mechanization refers to the use of various mechanical, hydraulic, pneumatic, or electrical devices to run the manufacturing process. In a mechanized system, the operator still directly controls the particular process and checks each step of machine operation.

The next step after mechanization was automation, derived from the Greek word *automatos* (self-acting); this word was first used in 1945 by the U.S. automobile industry to indicate automatic handling and processing of parts in production machines.

The world is now passing through the second industrial revolution, with the fantastic advances occurring continuously in the fields of electronics and computer technology. The computer is substituting for the human brain in controlling machines and industrial processes. A major breakthrough in automation was the invention of the first digital electronic computer (1943), followed by the first prototype of a numerically controlled machine tool (1952). Since this historic development, rapid progress has been made in automating most aspects of manufacturing, including the introduction of computers to enhance automation using CNC, adaptive control (AC), industrial robots, and computer integrated manufacturing systems (CIMSs), including CAD, computer aided engineering (CAE), and CAM (Figure 15.1).

The manufacturing situation today made the mass production of any component economically possible; however, industry in many cases demands variety in products in small lots. The economical production methods suitable for smaller lots should be followed. Further, higher accuracies are required at lower cost. To meet these requirements, there is a rapidly growing need for improved communication and feedback between manufacturing and design processes, integrating them into a single system capable of being optimized as a whole. The use of computerized integrated manufacturing (CIM) is the answer to meet these requirements and objectives. Computers therefore have an important role to play, especially in job shop and batch production manufacturing plants, which

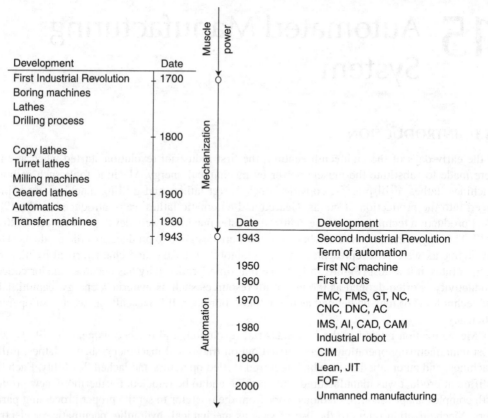

FIGURE 15.1 Mechanization and automation of MS.

constitute an important domain of the overall manufacturing activity. The traditional job shop and batch manufacturing suffers from drawbacks such as:

- Low equipment utilization
- Long lead times
- Inflexibility to market needs
- High inventory
- Dependence upon highly skilled operators
- Poor quality control
- Increased indirect cost

It is estimated that in traditional batch production, only 5–10% of the time is used on machines and the rest is spent in moving and waiting. Out of the total time on the machine, only 30% is machining time, the rest being for positioning, gauging, and idling (Jain and Gupta, 1993). These shortcomings can be overcome through the use of the following:

- Material handling equipment
- Feedback systems and continuous flow process
- Computers for process control and data collection, planning, and decision making to support manufacturing activities

It must be understood that a computer cannot change the basic metal working processes. It can only influence their control and their sequences so that down-time is kept to a minimum. It provides quick reflexes, flexibility, and speed to meet the desired results. The computer enables detailed analysis and accessibility to accurate data necessary for the integration of the manufacturing system (MS). For a plant to produce diversified products of best quality at enhanced productivity and lower prices, it would be essential that all elements of manufacturing (design, machining, assembly, quality assurance, management, and material handling) be computer-integrated, both individually and collectively (Jain and Gupta, 1993).

This chapter emphasizes the importance of flexibility in machine, equipment, tooling, and production operations to be able to respond to market and to ensure on-time delivery of high-quality products to attain customer satisfaction. Important developments during the past three decades have had a major impact on modern manufacturing, among which are group technology (GT), cellular manufacturing, FMSs, and just-in-time (JIT) production.

15.2 MANUFACTURING SYSTEMS

The MS takes inputs and produces products for the customer (Figure 15.2). It is a complete set of elements that include machines, people, materials, information, handling, equipment, and tooling. The system output may be consumer goods or services to user or inputs to some other processes. The materials are processed within the system and gain value as they are passed from machine to machine. MSs are very interactive and dynamic. An MS should be designed and integrated for low cost, superior quality, and on-time delivery.

It should be an integrated whole that is composed of integrated subsystems, each of which interacts with the entire system. System operation requires information gathering and communication within the decision-making processes that are integrated into the MS. Each company will have many differences resulting from discrepancies in subsystem combinations, people, product design, and materials (Degarmo et al., 1997).

MSs differ in structure or design. They may be classified into the following types:

1. *Job shop.* In this type, a variety of products are manufactured, which result in small lot sizes; often these products are one of a kind. It is commonly done by specific customer order. Because the plant must perform a wide variety of manufacturing processes, general-purpose production equipment is required. In a job shop, workers must have relatively high skill levels to perform a range of work assignments. Job shop products include space vehicles, aircrafts, machine tools, special tools, and equipment. Figure 15.3 shows the functional or process layout of a job shop. Forklifts and handcarts are used for material handling.
2. *Flow shop.* This type has a product-oriented layout composed mainly of a flow line. When the volume gets very large, as in assembly lines, it is called mass production. Specialized equipment, dedicated to the manufacture of a particular product, is used. This system

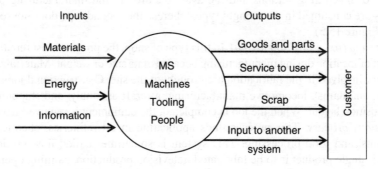

FIGURE 15.2 Manufacturing system.

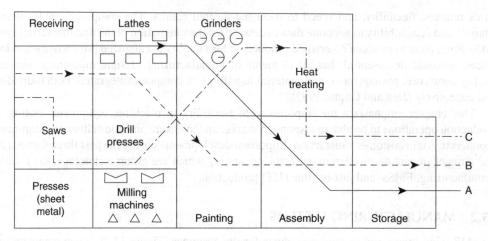

FIGURE 15.3 A Job shop MS—functional or process layout. (From Degarmo, E. P., Black, J. T., and Kohser, R. A., *Materials and Processes in Manufacturing*, Prentice Hall, New York, 1997.)

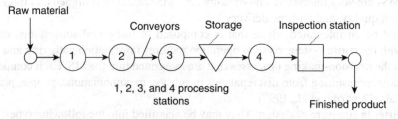

FIGURE 15.4 Schematic of an automated production line (flood shop MS).

may be designed to produce a particular product of a family of products, using special-purpose machines of high investment, rather than general-purpose ones. Figure 15.4 shows an automated production flow line consisting of a number of machines or stations arranged according to a certain configuration and linked to each other by conveyors, to direct the part from one machine to other. The manual skill level tends to be lower than in a job shop. The time item spent at each station is fixed and equal (balanced). Automated production flow lines are adapted for production at reduced labor cost, increased production rate, and reduced inventory cost. In this respect, two types of automated production lines (transfer units) are differentiated: straight and circular. The choice of either type depends upon the number of machining stations and the available area in the manufacturing plant. More stations are consumed in the straight type, whereas the second type does not require large areas (Figure 15.5).

3. *Project shop (fixed-position layout).* In this type of shop, the product must remain in a fixed position or location during manufacturing because of its size or weight. Materials, machines, and labor involved in the fabrication are brought to the site. Constructional jobs (buildings, bridges, and dams), locomotive manufacturing, aircraft assembly, and shipbuilding use a fixed-position layout. When the job is completed, the equipment is removed from the site.

4. *Continuous process.* This process finds application in oil refineries, chemical processing plants, and food processing. This system is sometimes called flow production, if a complex single product is to be fabricated (television production, canning operations, and

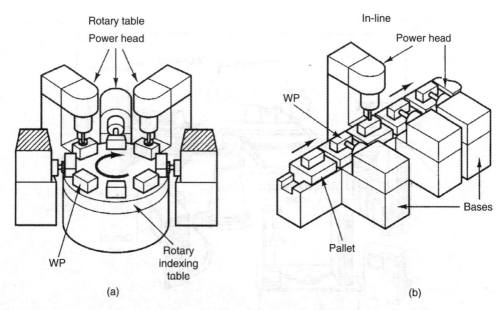

FIGURE 15.5 Rotary (a) and line (b) transfer machines. (Courtesy of Heald Machine Company.)

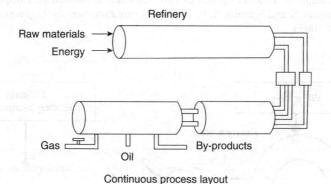

FIGURE 15.6 Continuous process layout of a refinery. (From Degarmo, E.P., Black J.T., and Kohser, R.A., *Materials and Processes in Manufacturing*, Prentice Hall, New York, 1997.)

so on). Continuous processes are usually easy to control and efficient, and are the simplest systems; however, they are the least-flexible systems (Figure 15.6).

5. *Cellular manufacturing*. This process is intended for producing parts one at a time in a flexible design. The cell capacity (cycle time) can be altered quickly to respond to rapid changes in market demand, thus allowing more product variety in smaller quantities, which is highly desirable. Figure 15.7 illustrates an example of a cellular unit, comprising two machine tools, automated part inspection, and a serving robot.

Flexible cells are typically manned, but unmanned cells are beginning to emerge with a robot replacing the worker (Figure 15.8). WPs are placed on the rotary feeder, and are loaded and unloaded one by one to the CNC machine by the robot. Machining is performed according to NC command data (cutting information) stored in advance in the CNC machine tool memory.

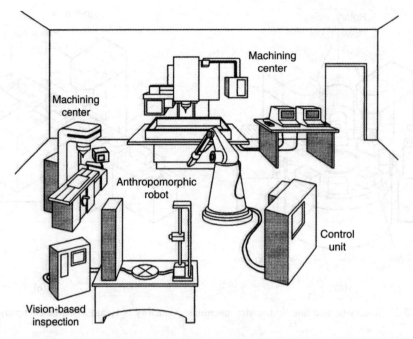

FIGURE 15.7 Example of FMC composed of two machines, automated part inspection and a serving robot. (From Kalpakjian, S. and Schmidt, S. R., *Manufacturing Processes for Engineering Material*, Pearson Education, Inc., NJ, 2003.)

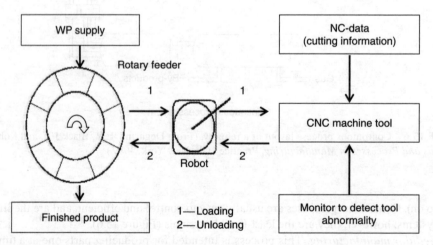

FIGURE 15.8 Unmanned manufacturing cell.

For best results, it is not only recommended to automate the local environment of machine tool, but also to automate the global environment, including the following activities:

- Management and provision of resources
- Preparation and transportation of WPs
- Supply and evolution of production data
- Inspection of WPs and machine tools

Automated Manufacturing System

The flexibility is acquired when a computer is applied to control the mentioned global environment of production.

The flexible manufacturing cell (FMC) process offers the following advantages:

- Flexibility for varied and small quantity production
- Manned and unmanned operation
- Automatic operation of numerous processes
- Simple setup
- Easy operation
- Integration to FMS
- High operation economy
- Immediate stop if necessary
- Operation reliability depending on an adequate supply of NC command data

15.3 FLEXIBLE AUTOMATION-FLEXIBLE MANUFACTURING SYSTEMS

In earlier times, the automation of manufacturing processes was limited to mass production (fixed automation), which was feasible only for a large number of parts.

Another field of automation is the FMS, which is a highly automated MS comprising a collection of production devices, logically organized under a host computer and physically connected by a central transporting system. It has been developed to provide some of the economics of mass production to small-batch manufacturing.

The main advantage of an FMS is its high flexibility in terms of small effort and short time required to manufacture a new product. It is an alternative that fits in between the manual job shop and hard automated transfer lines (Figure 15.9). It is best suited for applications that involve an intermediate level of flexibility and productivity (El-Midany, 1994).

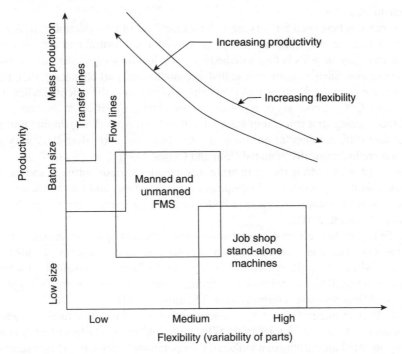

FIGURE 15.9 Flexibility against production rate of different MSs (intermediate flexibility/productivity of FMSs).

FMS can be regarded as a system that combines the benefits of two other systems: (a) the highly productive but inflexible transfer lines, and (b) job shop production, which fabricates a large variety of products on stand alone machines, but inefficient. In FMS, the time required for changeover to a different part is very short (1 min), thus making the quick response to product and market-demand variations a major benefit of FMS.

15.3.1 Elements of Flexible Manufacturing System

The basic elements of an FMS are as follows:

- Workstations
- Automated handling and transport of materials and parts
- Control systems

Workstations are arranged to achieve the greatest efficiency in production, with an orderly flow of materials, and products through the system. The type of workstation depends on the type of production. For machining operations, these usually consist of a variety of three- to five-axes machining centers and CNC machines (milling, drilling, and grinding). They also include other equipment, such as that for automated inspection (including coordinate measuring machines (CMMs)), assembly, and cleaning. For sheet-metal forming, punching, and shearing and forging, the workstations of an FMS incorporate furnaces, trimming presses, heat-treatment facilities, and cleaning equipment.

Machine tools may be equipped either with turret or tool changers for supplying desired tools for machining. A turret has a shorter cycle time and is preferred when turning small components; indexing tools with a turret is the fastest method. For larger components with longer cycle times, an automatic tool changer is required where a bigger magazine can be used (Jain and Gupta, 1993).

Handling and transport are very important for the system flexibility. Material handling is controlled by a central computer and is performed by automated vehicles, conveyors, and various transfer mechanisms.

Industrial robot is best used for serving several machines in a production cell. A robot should not be located in front of the machine tool, as it prevents manual control and supervision. However, the automatic changing of WPs is best satisfied by using handling equipment that is built integral with the machine tool. Such equipment is called a computerized part changer (CPC). It is installed physically separate from the machine tool to eliminate vibration to the machine when the parts are being handled by CPC. It consists of a portal, shuttle carriage, vertical slide, and gripping unit. The jaws of the gripping unit can clamp external and internal parts. The gripping unit axis can be positioned in four different angular positions: 0°, 90°, 180°, and 270°. Both the rotary motion and gripping action are hydraulically actuated (Jain and Gupta, 1993).

Computer control of FMS is the main brain that includes various software and hardware. This subsystem controls the machinery and equipment in workstations and transportation of materials in various stages. It also stores data and provides communication terminals that display the data (Kalpakjian and Schmidt, 2003).

Because FMS involves a major capital investment, efficient machine utilization is essential; machines must not stand idle. Consequently, proper scheduling and process planning are crucial. Scheduling for FMS is dynamic, unlike that in job shops, where a relatively rigid schedule is followed. Dynamic scheduling is capable of responding to quick changes in product type and is thus responsive to real-time decisions (Kalpakjian and Schmidt, 2003).

In brief, FMS is an integrated system of computer-controlled machine tools and other workstations with an automated flow of information, WPs, tools, and so on. Control of FMS is achieved by computer-implemented algorithms that make all the operational decisions. This system is arranged so that the automated production of a group of complex WPs in any lot size, particularly small and medium batches, is possible (Figure 15.9). The FMS is usually planned by simulated techniques in which a model is drawn. This model is an idealized representation of the components, internal relations, and characteristics of a real-life system. Analysis of model behavior shows ways for improving the system by carrying out necessary changes (Jain and Gupta, 1993).

FMS offers the advantages of part cost reduction, throughput time reduction, increasing flexibility toward changes of the product mix, reduced inventory and lead times, and increased productivity. This system usually incorporates features like adaptive controls, tool breakage detectors, and a tool life monitoring system. A major drawback of FMS is the unknown availability and reliability of the planned system (Jain and Gupta, 1993).

15.3.2 Limitations of Flexible Manufacturing System

The use of FMS is hindered due to the following reasons
- High programming cost
- Smaller degree of sophistication of fabrication and assembly processes
- Unavailability of reliable feedback devices for tool wear and breakage

15.3.3 Features and Characteristics

FMS features and characteristics are summarized as follows:

- FMS offers immerging cost and quality benefits for most engineering sectors requiring batch production.
- Batch production using conventional machine tools necessitates a minimum number of similar components to be produced economically. There is no batch size limitation in case of FMS; consequently, there is no need to lock up the money in extensive stocks of finished parts. The work in progress is reduced considerably and the inventory cost is therefore eliminated.
- It is possible to produce at random all varieties of products planned by a firm. FMS is capable of quickly responding to any design changes or market demands in the product.
- FMS are usually equipped with robots or handling equipment. Software is developed to integrate CNC control and the handling systems. All necessary tools can be stored in a magazine.
- All part programs of different models are stored in the system memory. The system identifies the model program to be produced.
- FMS can be conceived in multiples of 15–20 minute operations. If a certain operation takes longer time, multiples of similar machines can be installed in the line.
- Extensive use of touch triggers is made to minimize the operator's, intervention in the line.
- Industrial robots are used for material handling (loading and unloading), inspection activities, and assembly operations.

15.3.4 New Developments in Flexible Manufacturing System Technology

These developments greatly and dramatically boost the capabilities of FMS:

- Computerized tool setting station
- Establishing tool informations like tool lengths, tool offsets, and so on by linear variable displacement transducer (LVDT)
- Automatic tool changer
- Monitoring of tool life
- Providing tool breakage detectors
- Providing tool compensation system and AC
- Increased use of robots and handling system
- Application of laser and fiber optic technology to check bore diameters and part surface location

- Spindle probes to check WP features like bore diameter and hole pattern location
- Improved software
- Fault analysis (vision system for online quality control)
- Swarf and coolant control
- Computerized simulation to establish efficiencies and programming facilities

15.4 COMPUTER INTEGRATED MANUFACTURING

CIM is a recent technology that has been in development and trial since the 1990s. It comprises a combination of software and hardware for product design, production planning and control, production management, and so on, in an integrated manner. It is a methodology and a goal rather than merely an assemblage of equipment and computers. Its effectiveness greatly depends on the use of large-scale integrated communications systems involving computers, machines, and their controls.

As with traditional manufacturing approaches, the purpose of CIM is to transform product designs and materials into sellable goods at a minimum cost in the shortest possible time. The CIM begins with the design of a product (CAD) and ends with the manufacture of that product (CAM). With CIM, the usual split between CAD and CAM is supposed to be eliminated (Degarmo et al., 1997).

CIM differs from the traditional job shop in the role the computer plays in the manufacturing process. CIMSs are basically a network of computer systems tied together by a single integrated database (DB). Using the information in the DB, a CIMS can direct manufacturing activities, record results, and maintain accurate data. CIM is the computerization of design, manufacturing, distribution, and financial functions into one coherent system (Degarmo et al., 1997).

CIM is an attempt to integrate the many diverse elements of discrete parts manufacturing into one continuous process-like stream. It can result in an increased manufacturing productivity and quality and reduced production cost. It employs FMS, which saves the manufacturer from replacing equipment each time a new part has to be fabricated. The current equipment can be adopted to produce a new part, as long as it is in the same product family, with programmable software and some retooling. Thus, this system has the ability to switch from component to component with no down-time for change over. This system requires NC lathes, machining centers, punch presses, and so on, which have the ability to be readily incorporated into multimachine cell or fully integrated manufacturing system (IMS).

Multispindle CNC machines with greater horsepower, stiffness, and wider speed ranges are important for CIM. Automatic tool changers (to change the tool in the spindle and to renew dulled tools) are a must. A robot for handling WPs is another important machine tool peripheral essential for the CIMS.

CIM is a very powerful concept and has the potential for great benefits; however, it is not easy to implement. Like any powerful and complex tool, it can be dangerous and costly if not implemented properly.

The main tasks involved in CIM can be separated into four areas (Figure 15.10):

1. *Product design*, for which interactive CAD system allows drawing, analysis, and design to be performed. The computer graphics are useful to get the data out of the designers mind to be ready for interaction (Figure 15.11).
2. *Manufacturing planning*, where the computer-aided process planning (CAPP) helps to establish optimum manufacturing routines and processing steps, sequences, and schedules.
3. *Manufacturing execution*, in which CAM identifies manufacturing problems and opportunities. Intelligence in the form of microprocessors is used to control machines and material handling and to collect data controlling the current shop floor (Figure 15.11).
4. *Computer-aided inspection (CAI) and computer-aided reporting (CAR)*, so as to provide a feedback control loop (Figure 15.10).

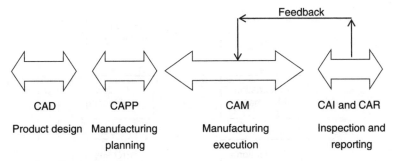

FIGURE 15.10 Main tasks of CIM.

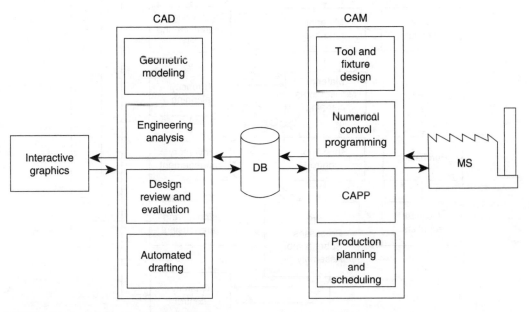

FIGURE 15.11 Database to CAD/CAM.

Computer integration of these four tasks provides the most current and accurate information about manufacturing, thereby permitting better and tighter control, and enhances the overall quality and efficiency of the entire system. Improved communication among these activities results in enhanced productivity and accuracy if the designer considers limitations and manufacturing problems and vice versa. The availability of current data permits instantaneous updating of production-control data, which in turn permits better planning and more effective scheduling. All machines are fully utilized, handling time is reduced, and parts move more efficiently through production. Workers become more productive and do not have to waste time in coordination and searching of previous data.

Figure 15.12 illustrates a block diagram of activities for a typical CIMS. With the introduction of computers, changes have occurred in the organization and execution of production planning and control through the implementation of material requirement planning (MRP), capacity planning, inventory management, shop floor control, and cost planning and control. Engineering and manufacturing DBs contain all the information needed to fabricate the components and assemble the products (Figure 15.12). The design engineering and process planning functions provide the inputs for the engineering and manufacturing DB, which includes all the data on the product generated during design, such as geometrical configurations, part lists, and material specifications (Figure 15.12).

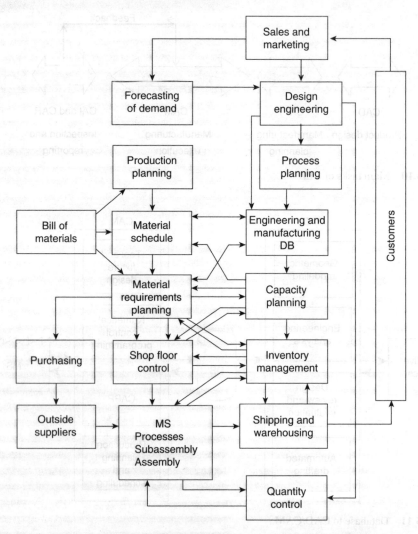

FIGURE 15.12 Cycles of activities in CIMS. (From Degarmo, E. P., Black J. T., and Kohser, R. A., *Materials and Processes in Manufacturing*, Prentice Hall, New York, 1997.)

The bill of material is a key part of the DB. Figure 15.11 shows how a CAD/CAM DB is related to the design and manufacturing. Included in the CAM is a CAPP module that acts as an interface between CAD and CAM (Degarmo et al., 1997).

Capacity planning is performed in terms of labor or machine hours available. The master schedule is transformed into material and component requirements using MRP. These requirements are then compared with available plant capacity. If the schedule is incompatible with capacity, adjustments must be made either in the master schedule or in plant capacity. The possibility of adjustments in the master schedule is indicated by the arrow in Figure 15.12, leading from capacity planning to the master schedule. The term "shop floor control" (Figure 15.12) refers to a system for monitoring the status of manufacturing activities and the plant floor and reporting the status to management so that effective control can be exercised. Cost planning and control system consists of a DB to determine the expected costs to manufacture a product. It also consists of cost collection and analysis software to determine what the actual costs of manufacturing are and how such costs compare with expected ones (Degarmo et al., 1997).

In order that a computer successfully controls a given production line, a DB has to be built up that provides the computer with the necessary information based on a deep understanding of the manufacturing process and the entire system. Normally, the FMS is managed by a computer through a distributed logic architecture operating at several levels.

A machine may have a dedicated program for its own functioning. It must simultaneously obtain commands from many computers that provide necessary information desired for functioning of the complete system. All such activities have to be properly coordinated. An intermediate level of control could be provided to take care of material handling. Parts in such a system will be loaded according to instructions from a supervisory computer. The computer also dispatches tooling according to the expected tool life for each tool and takes desired action at appropriate time. It also generates a variety of related reports like tool life data and production data.

CIM has the advantage of possessing intelligence to maximize the process performance, provided that all the parameters can be measured in real time. It has been established that in metal cutting, every parameter related to machining can be determined if the shear angle is known. It is possible to feed into the computer all the data to compute shear strain, strain rate, flow stress, coefficient of friction at the chip–tool interface, unit horsepower and the total power being consumed, cutting temperature, and so on (Jain and Gupta, 1993).

The intelligent machine tools confined in the CIMS result in the following merits.

- Increased accuracy
- Reduced scrap
- Reduction in manned operation
- Increased predictability of machine tool operation
- Less skill for setting up and operation on the machine
- Reduced machine down-time
- Reduced production cost
- Increased machine throughput
- Reduced setup time
- Reduced tooling due to better operation planning
- Increased range of materials and part geometries
- Increased quantity and quality of information exchange between the machine control and part designer and between human and machine tool control

CIM technology offers the following advantages:

- High rates of production with high precision
- Remarkable flexibility for producing diverse components in the same setup
- Easy and quick manipulation of software
- Uninterrupted production with less supervision or work handling
- Economical production, even in case of moderate batch sizes
- Drastic reduction of lead times
- Drastic changes in product design
- Integrating and fine-tuning of all factory functions, such as raw and semifinished materials flow, tooling, metal cutting, and inspection

15.4.1 Computer-Aided Design

A major element of a CIMS is CAD system that involves any type of design activity that makes use of a computer to create, develop, analyze, optimize, and modify an engineering design.

The design-related tasks performed by a CAD system (Figure 15.11) are:

- Geometric modeling
- Engineering analysis
- Design review and evaluation
- Automated drafting

Geometric modeling is the most important phase of the design process, through which the designer constructs the graphical image of the object on the cathode ray tube (CRT).

Engineering analysis may involve stress calculations, finite element analysis for heat transfer computations, or the use of differential equations to describe the dynamic behavior of the system. Generally, general-purpose programs are used to perform these analyses.

Design review and evolution techniques check the accuracy of the CAD design. Other features of these techniques are checking and animation, which enhance the designer's visualization of the mechanism and help to avoid interference.

Automated drafting involves creation of hard-copy engineering drawings from the CAD DB. It is also capable to perform dimensioning automatically, generate cross-hatching, scale drawings, develop sectional views, enlarge a view for details, perform rotation of parts, and perform transformations such as oblique, isometric, and perspective views (Degarmo et al., 1997).

15.4.2 COMPUTER-AIDED PROCESS PLANNING

CAPP uses computer to determine how a part is to be made. If GT is used, parts are grouped into part families. For each part family, a standard process plan is established and stored in computer files and then retrieved for new parts that belong to that family.

For a manufacturing operation to be efficient, all its diverse activities must be planned and coordinated; this task has traditionally been done by process planners. Process planning involves selecting methods of production, tooling, fixtures, machinery, sequence of operations, standard processing time for each operation, and methods of assembly. These choices are all demonstrated on a routing sheet (Table 15.1). When performed manually, this task is labor-intensive, time-consuming, and also relies heavily on the process planner's experience.

TABLE 15.1
Sample Routing Sheet in CAPP

Routing sheet — Part name: Valve body
Customer's name: Midwest Valve Co.
Quantity: 15 — Part no.: 302

Operation No.	Operation Description	Machine
10	Inspect forging, check hardness	RC tester
20	Rough machine flanges	Lathe no. 5
30	Finish machine flanges	Lathe no. 5
40	Bore and counter bore	Boring mill no.1
50	Turn internal grooves	Boring mill no. 1
60	Drill and tap holes	Drill press no. 2
70	Grind flange endfaces	Grinder no. 2
80	Grind bore	Internal grinder no. 1
90	Clean	Vapor degreaser
100	Inspect	Ultrasonic tester

Source: From Kalpakjian, S. and Schmidt, S. R., *Manufacturing Processes for Engineering Material*, Pearson Education, Inc., NJ, 2003.

Automated Manufacturing System

These route sheets may include additional information regarding materials, estimated tooling time for each operation, processing parameters, and other details. It travels with the part from one operation to another.

CAPP is an essential adjunct to CAD/CAM, although it requires extensive software and good coordination with CAD/CAM as well as other aspects of IMSs. CAPP is a powerful tool for effective planning and scheduling operations. It is particularly effective in small-volume, high-variety parts production requiring machining, forming, and assembly operations.

Process planning activities are a subsystem of CAM (Figure 15.11). Several functions can be performed using these activities, such as capacity planning for plants to meet production schedules, inventory control, and purchasing (Kalpakjian and Schmidt, 2003).

CAPP offers many advantages that can be summarized as follows:

- The standardization of process plans improves productivity, reduces lead times and costs of planning, and improves the consistency of product quality and reliability.
- Process plans make use of GT to retrieve plans to produce new parts.
- Process plans can be modified to suit specific needs.
- Neat and legible routing sheets can be prepared more quickly.
- Many other functions, such as cost estimation and work standards, can be incorporated into CAPP.

15.4.3 Computer-Aided Manufacturing

CAM involves the use of computers to assist in all phases of manufacturing a product, including process and production planning, scheduling, manufacturing, quality control, and management. CAM is another major element of CIM. Because of the increased benefits, CAD and CAM are integrated into CAD/CAM systems. This integration allows the transfer of information from the design stage to the planning stage for manufacturing of the product without the need to manually reenter data on part geometry. The DB developed during CAD is stored, and then processed further by CAM into necessary data and instructions for operating and controlling production machinery and material handling equipment and for performing automated testing and inspection (Kalpakjian and Schmidt, 2003).

An important feature of CAD/CAM integration in machining is the capability to describe the cutting tool path for various operations such as NC turning, milling, and drilling. The programs are computer-generated and can be modified by the programmer to optimize the tool path, and to visually check for possible tool collisions with clamps or fixtures or for other interferences. The tool path can be modified at any time to accommodate other shapes to be machined.

The tasks performed by a CAM system (Figure 15.11) are as follows:

- NC or CNC scheduling
- Production planning and scheduling
- Tool and fixture design
- CAPP

NC can use special computer languages. Today APT and COMPACT II are the two most common language-based computer-assisted programming systems used in the industry. These systems take the CAD data and adapt them to the particular machine control unit to make the part (Degarmo et al., 1997).

15.5 LEAN PRODUCTION – JUST-IN-TIME MANUFACTURING SYSTEMS

Regardless of all that has been written about CIM, this technology is not widespread. What is called "lean production" appears to be more important to the future rather than CIM. It is evident that unless a company first adopts the approach of lean production, the conversion to CIM is likely to fail.

In this approach, the functions of the production system such as production control, inventory control, quality control, and machine tool maintenance are first to be integrated. Lean production has been developed and practiced by Toyota (the Toyota system) instead of CIM (Degarmo et al., 1997).

Integration of the production system functions into the MS requires commitment from top-level management and communication with everyone, particularly manufacturing. Total employee and union participation is absolutely necessary but it is not usually the union leadership or the production workers who raise barriers to integrated manufacturing production system (IMPS). It is those in middle management who have the most to lose in this systems-level change. In this respect, here are the preliminary steps:

1. All levels in the plant, from the production workers to the president, must be educated in the IMPS philosophy and concepts.
2. Top management must be totally committed to this venture and everyone involved must be motivated.
3. Everyone in the plant must understand that cost, not price, determines the profit. Customer wants low cost, superior quality, and on-time delivery.
4. Everyone must be committed to the elimination of waste to reduce cost; this is fundamental for getting lean production.

15.5.1 Steps for Implementing the IMPS Lean Production

Many companies have implemented IMPS (lean production) by converting a factory from job shop-flow MS to a true IMS. The following steps are to be followed:

1. *Build foundation by forming U-shaped cells*—This is done to replace the production job shop. Restructure and reorganize the FMS, composed of cells that fabricate families of parts. Creating cells is the first step in designing an MS in which inventory control and quality control are integrated.
2. *Rapid exchange of tooling and dies (RETAD)*—Everyone on the plant floor must be informed to reduce setup time by using single-minute exchange of die (SMED). A setup reduction team assists workers and foremen and demonstrates a project on the plant's worst setup problem. Reducing setup time is critical to reducing lot size.
3. *Integrating quality control*—A multifunctional worker can do more than operating machines. The worker is also an inspector who understands process capability, quality control, and process improvement. In lean production, every worker has the responsibility and the authority to make the product right the first time and every time and the authority to stop the process when something is wrong. The integration of quality control into the MS considerably reduces defects while eliminating the inspector. Cells provide for integration of quality control.
4. *Integrating preventive maintenance and ensuring machine reliability*—Installing an integrated preventive maintenance program makes machines operate reliably; moreover, it gives the chance to the workers to maintain equipment properly.
5. *Leveling and balancing final assembly*—This is done by producing a mix of final assembly products in small lots. Each process, cell, and subassembly has essentially the same cycle time as the final assembly.
6. *Linking cells*—Integration of production control is realized by linking the cells, subassemblies, and final assembly elements, utilizing Kanban. All linked cells, processes, subassemblies, and final assemblies start and stop together in a synchronized manner. Thus, the integration of production control into the MS is realized.
7. *Integrating the inventory control*—The inventory levels are directly controlled by the people on the floor through the control of Kanban. This is the integration of the inventory control to reduce work in progress (WIP). The minimum level of WIP is determined by the percentage of defectives, the reliability of the equipment, and the setup time.

8. *Integrating the suppliers*—This involves educating and encouraging suppliers (vendors) to develop their own lean production system for superior quality, low cost, and rapid on-time delivery. They should deliver material to the customer when needed, where needed, without inspection.
9. *Automating and robotizing*—Solve problems by converting manned to unmanned cells, initiated by the need to solve problems in quality and reliability, and to eliminate bottlenecks.
10. *Computerizing the whole production system*—Once the MS has been restructured into JIT MS and the critical functions well integrated, the company (MS) will find it expedient to restructure the rest of the company. It is basically restructuring the production system (PS) to be as waste-free and efficient as the MS.

15.5.2 Just-in-Time and Just-in-Case Production

As previously mentioned in lean production, JIT production concept was first implemented by Toyota in Japan under the name Kanban (visible record), to eliminate wastage of materials, machines, capital, manpower, and inventory throughout the MS. The JIT philosophy is summarized as follows (Degarmo et al., 1997):

> Produce and deliver finished goods just-in-time to be sold, subassemblies just-in-time to be assembled into finished goods, fabricated parts just-in-time to go into finished good, fabricated parts just-in-time to go into subassemblies, and purchased materials just-in-time to be transformed into fabricated parts.

To be more specific, JIT seeks to achieve the following goals (McMahon and Browne, 1998):

- Zero defects
- Zero setup time
- Zero inventories
- Zero handling
- Zero breakdowns
- Zero lead time
- Lot size of one

To achieve this goal, all elements of excess should be eliminated. Large safety stocks, long lead times, long setting times, large queues at machines, high scrap and rework levels, machine breakdowns, and so on should also be eliminated. JIT is not, therefore, a simple off-the-shelf solution to all manufacturing problems. If JIT is realized in the firm, the unnecessary inventories will be completely eliminated, making stores or warehouses unnecessary, inventory cost will be diminished, and the ratio of capital turnover will be increased. Consequently, JIT is sometimes called zero inventories, material as need, stockless production, or demand scheduling (El-Midany, 1994).

In traditional manufacturing, the parts are made in batches placed in inventory, and used whenever necessary. This approach is known as the just-in-case (JIC) or push system, meaning that the parts are made according to a schedule and are kept in inventory to be used if and when they are needed. In contrast, JIT manufacturing is a pull system, meaning (as previously mentioned) that parts are produced to order and the production is matched with the demand for the final assembly of products.

In JIT, parts are inspected by the worker and used within a short period of time. Accordingly, the worker maintains a continuous production control, identifies defective parts immediately, and produces quality products. Implementation of the JIT requires that all manufacturing aspects be continuously reviewed and monitored so that all operations and resources that do not add value are eliminated.

Because the basic promise of JIT is to produce the kind of units needed in the quantities needed at the time needed, the system should depend on smoothing (leveling) of the MS, so it is necessary to eliminate fluctuation in the final assembly. This is called leveling or balancing the final assembly. The object of JIT is to make the same amount of product part everyday. Balancing is making the output from the cells equal to necessary demand for the parts downstream. In summary, small lot sizes, made possible by setup reduction within the FMCs, single-unit conveyance within the cells, and standard cycle times are the keys to accomplish a smoothed MSs (Degarmo et al., 1997).

Advantages of Just-in-Time

- Low inventory cost
- Fast detection of defects in production and hence low scrap loss
- Reduced need for inspection and reworking of parts
- Production of high-quality parts at low cost

Implementation of JIT, as compared to FMS (Kalpakjian and Schmidt, 2003) realizes:

- Reduction of 20–40% in production cost
- Reduction of 60–80% in inventory
- Reduction up to 90% in rejections
- Reduction up to 90% in lead times
- Reduction up to 50% in scrap, rework, and warranty cost
- Increase of 30–50% in direct labor productivity
- Increase of 60% in indirect labor productivity

15.6 ADAPTIVE CONTROL

AC systems for machine tools are a logical extension of CNC systems. The part programmer sets the processing parameters, based on the existing knowledge of the WP material and various data on the particular manufacturing process. In CNC machines, these parameters are held constant during a particular process cycle. In AC, on the other hand, the system is capable of automatic adjustments during processing, through closed-loop feedback control. It is therefore readily appreciated that this approach is basically a feedback system. A schematic of a typical AC configuration for a machine tool is shown in Figure 15.13. Accordingly, AC represents a process control that operates in addition to the CNC position or servo control system.

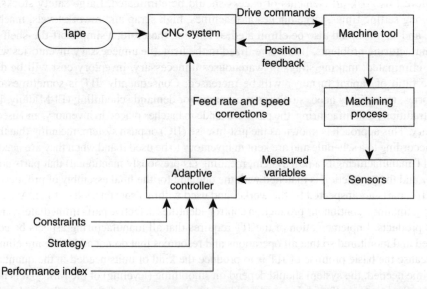

FIGURE 15.13 Typical adaptive control configuration for a machine tool. (From Koren, J., *Computer Control of Manufacturing Systems*, McGraw-Hill, Tokyo, 1983. With permission.)

Automated Manufacturing System

In manufacturing, several AC systems or strategies are distinguished (*Metals Handbook*, 1989):

1. *Adaptive control with optimization (ACO)*, in which an economic index of performance is used to optimize the process using online measurements. This strategy may involve maximizing material removal rate or improving surface quality.
2. *Adaptive control with constraints (ACC)*, in which the process is controlled using online measurements to maintain a particular process constraint (force, power, temperature, and so on). Referring to Figure 15.14, if the cutting force and hence the torque increases excessively (Figure 15.14a, b), the AC system changes the speed or the feed (cutter travel), to lower the cutting force to an acceptable level (Figure 15.14c). Without AC or without direct intervention of the operator (in case of conventional machining), high cutting forces may cause the tools to chip or break, or the WP to deflect or distort excessively. As a result the accuracy and surface finish would deteriorate.
3. *Geometric adaptive control (GAC)*, in which the process is controlled using online measurements to maintain desired dimensional accuracy or surface finish (Figure 15.15).

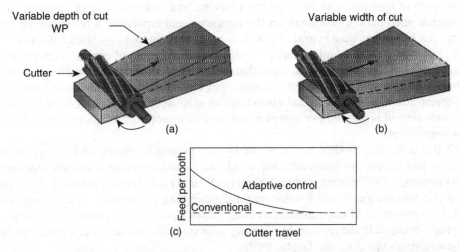

FIGURE 15.14 Adaptive control in milling: (a), (b) increase of cutting force with cutter travel, and (c) AC is used to decrease the feed with cutter travel. (From Koren, J., *Computer Control of Manufacturing Systems*, McGraw-Hill, Tokyo, 1983. With permission.)

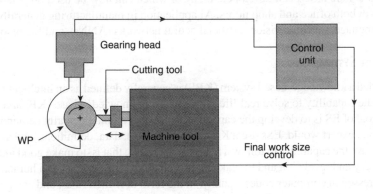

FIGURE 15.15 In-process inspection of WP diameter in turning. (From Kalpakjian, S. and Schmidt, S. R., *Manufacturing Processes for Engineering Material*, Pearson Education, Inc., NJ, 2003.)

Response time must be short for AC to be effective. Currently, all AC systems are based on either ACC or GAC, because the development and proper implementation of ACO is complex. ACC is well-suited for rough cutting, whereas GAC is used for finishing operations.

Integration of AC into CAD/CAM/CIM systems

This integration is an important issue in the future development of AC systems, as well as their role in the CIM hierarchy. Such issues are extremely important for unmanned manufacturing and will require additional research to extend the current understanding of AC system. Important issues include the interface between CAD and CAM and the application of expert systems (ESs) and other methods from artificial intelligence (AI) to AC systems, as well as process monitoring and diagnostics (*Metals Handbook*, 1989).

15.7 SMART MANUFACTURING AND ARTIFICIAL INTELLIGENCE

AI is the basic tool for smart manufacturing (SM). AI is an area of computer science concerning systems that exhibit some characteristics that are usually associated with intelligence in human behavior, such as learning, reasoning, problem solving, and understanding of language. Its goal is to simulate such human endeavors on the computer, and represents a technique for solving problems in a better way than is available with conventional computer programs (CCPs). A CCP typically relies on an algorithmic solution in which a finite number of explicit steps produce the solution of a specific problem. These algorithms work fine for scientific or engineering calculations that are numeric in nature to produce satisfactory answers. In contrast, AI uses a heuristic—a rule-of-thumb search. It should be understood that an exhaustive search can only be for relatively simple, well-defined problems. For complex and uncertain problems, exhaustive search routines become impractical.

A CCP is difficult to modify, but AI is usually easy to modify, update, and enlarge. In a CCP, information and control are integrated, but in AI, the control structure is usually separate from domain knowledge. CCP is often primarily numerical, but AI is concerned primarily with processing symbolic information in which some meaning other than a numerical value is attached to a symbol. The symbols in AI may represent a concept about a process or a condition related to it; the AI programs manipulate the relationships among such symbols, and arrive at logical conclusions from these relationships (Jain and Gupta, 1993).

AI is having a major impact on all steps of the manufacturing cycles, including design, automation, production planning, scheduling, and the overall economics of manufacturing operations. AI programs find also applications in diagnosis, monitoring, analysis, interpretation learning, consultation, instruction, conceptualization, prediction, debugging, and repair. AI packages costing approximately a few thousand dollars have been developed, many of which can now be used on personal computers for application in both office and shop floors. AI application in manufacturing generally encompasses ESs, natural language, machine vision, artificial neural networks (ANNs), and fuzzy logic.

15.7.1 Expert Systems

An ES also called a knowledge-based system (KBS) is generally defined as an intelligent computer program that has the capability to solve real-life problems using knowledge base (KB) and inference procedures. The goal of ES is to develop the capability to conduct an intellectually demanding task in the way that a human expert would. ESs use a KB containing facts, data, definitions, and assumptions.

They also have the capability to follow a heuristic approach; that is, to make good judgments on the bases of discovery and revelation and to make high-probability guesses, just as a human expert would. The KB is expressed in computer codes, usually, in the form of if-then rules, and can generate a series of questions; the mechanism for using these rules to solve problems is called an inference engine. ESs can also communicate with other computer software packages (Kalpakjian and Schmidt, 2003).

To construct ESs for solving complex design and manufacturing problems encountered, needed elements include:

- A great deal of knowledge
- A mechanism for manipulating the knowledge to create solutions

Because of the difficulty involved in modeling the many years of experience of a team of experts and the complex inductive reasoning and decision-making capabilities of humans, including the capability to learn from mistakes, the development of KBSs require much time and effort. ESs operate on a real-time basis, and their short reaction times provide rapid responses to problem. The programming languages most commonly used are C++, list processing (LISP), and programming logic (PROLOG). A significant development is ES-software shells, or environments (framework). These shells are essentially ES outlines that allow a person to write specific applications to suit special needs. Writing these programs requires considerable experience and time (Kalpakjian and Schmidt, 2003).

Several ESs have been developed to be used in:

- Problem diagnosis in machines and equipment
- Modeling and simulation of production facilities
- CAD, process planning, and production scheduling
- Management of a company's manufacturing strategy

15.7.2 Machine Vision

In systems that incorporate machine vision, computers, and software-implementing AI are combined with cameras and other optical sensors. These machines then perform operations such as inspecting, identifying, and sorting parts and guiding intelligent robots (Figure 15.16); in other words, operations that would otherwise require human involvement and intervention.

15.7.3 Artificial Neural Networks

ANNs are used in applications such as noise reduction in telephones, speech recognition, and process control in manufacturing. For example, they can be used to predict the surface finish of

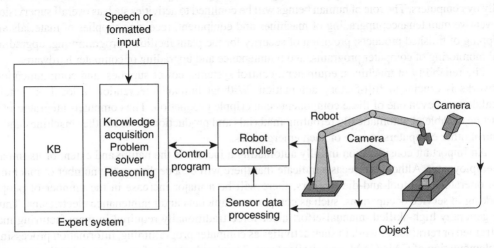

FIGURE 15.16 ES as applied to an industrial robot guided by machine vision. (From Kalpakjian, S. and Schmidt, S. R., *Manufacturing Processes for Engineering Material*, 2003, Pearson Education, Inc., NJ, 2003.)

machined WPs on the basis of input parameters such as cutting force, torque, acoustic emission, and spindle acceleration. However, this field is still under development.

15.7.4 Natural-Language Systems

These systems allow a user to obtain information by entering English language commands in the form of simple typed questions. Natural-language software shells are used in the scheduling of material flow in manufacturing and analyzing information in DBs. Major progress is being made to develop computer software with speech-synthesis and voice-recognition capabilities, thus eliminating the need to type commands on the keyboard.

15.7.5 Fuzzy Logic (Fuzzy Models)

Fuzzy logic is an element that has important application in control systems and pattern recognition. It is based on the observation that people can make good decisions on the basis of nonnumerical information. Fuzzy models are mathematical means of representing vagueness and imprecise information (hence the term "fuzzy"). These models have the capability to recognize, manipulate, interpret, and use data and information that are vague or that lack in precision with fuzzy-logic methods including reasoning and decision making at a level higher than ANNs. Typical concepts used in fuzzy logic are: *few, more or less, small, medium, extreme,* and *almost all*.

Fuzzy-logic technologies and devices have been developed and successfully applied in areas such as intelligent robotics, motion control, image processing and machine vision, machine learning, and design of intelligent systems. Some applications of fuzzy logic include automatic transmissions of lexus cars, automatic washing machines, and helicopters that obey vocal commands (Kalpakjian and Schmidt, 2003).

15.8 FACTORY OF THE FUTURE

The trend toward the automated factory seems unavoidable in modern industries. The integration of many new techniques adopted in MSs, such as CNC, machining centers, FMS, robots, material-handling equipment, and automatic warehouses, together with CAD, CAM, CAPP, and GT software have made the factory of the future (FOF) closer to a reality. All manufacturing, material handling, assembly, and inspection will be done by computer-controlled machinery and equipment. Similarly, activities such as the processing of incoming orders, production planning and scheduling, cost accounting, and various decision-making processes performed by management will also be done automatically by computers. The role of human beings will be confined to activities such as overall supervision; preventive maintenance; upgrading of machines and equipment; receiving supplies of materials and shipping of finished products; provision of security for the plant facilities; programming, upgrading, and monitoring of computer programs; and maintenance and upgrading of computer hardware.

The reliability of machines, equipment, control systems, power supplies, and communications networks is crucial to full-factory automation. Without human intervention, a local or general breakdown in even one of these components can cripple production. The computer-integrated FOF must be capable of automatically rerouting materials and production flows to other machines and to control other computers in case of such emergencies.

An important consideration in fully automating a factory is the nature and extent of its impact on employment. Although forecasts indicate that there will be a decline in the number of machine-tool operators and tool-and-die workers, there will be a major increase in the number of people working in service occupations, such as computer technicians and maintenance electricians. Thus, the generally high-skilled, manual-effort labor force traditionally required in manufacturing may be trained or retrained to work in such activities as computer programming, information processing, implementation of CAD/CAM, and similar tasks. The development of more user-friendly computer software is making these tasks much easier.

In this respect, it should be recognized that the designation of "world-class," like "quality," is not a fixed target for the manufacturing country or the company to reach, but rather a moving target, rising to higher and higher levels as time passes. Manufacturing organizations must be aware of this moving target and plan and execute their programs accordingly (Kalpakjian and Schmidt, 2003).

15.9 CONCLUDING REMARKS RELATED TO AUTOMATED MANUFACTURING

1. Installations of FMSs are very capital-intensive; consequently, a thorough cost/benefit analysis must be conducted before a final decision is taken. This analysis should include:
   - Capital cost, energy, materials, and labor.
   - Market analysis for which products are to be produced.
   - Anticipated fluctuations in market demand and product type.
   - Time and effort required for installing and debugging the system. An FMS can take 2–5 years to install and at least 6 months to debug.
2. Although FMS requires few, if any, machine operators, the personnel in charge of the total operation must be trained and highly skilled. These include manufacturing engineers, computer programmers, and maintenance engineers.
3. The most effective FMS applications have been in medium-volume, high variety batch production (50,000 units/year). In contrast, high-volume, low-variety parts production is best attained from transfer machines.
4. CIMSs have become the most important means of improving productivity, responding to changing market demands and better controlling manufacturing operations and management functions. Regardless of all written about CIM, this technology is not so widespread as that of lean production.
5. CAM is often integrated with CAD to transfer information from the design stage to planning stage and to production; that is, CAD/CAM bridges the gap from design to production.
6. Advances in manufacturing operations, such as CAPP, computer simulation of manufacturing processes, GT, cellular manufacturing, FMSs, and JIT manufacturing contribute significantly to the improvement in productivity.
7. Significant advances have been made in material handling, particularly with the implementation of industrial robots and automated conveyors.
8. The FOF appears to be theoretically possible. However, there are important issues to be considered regarding its impact on employment.
9. In the FOF, many of the functions of production system are integrated into the MS. This requires that the job shop MS is replaced with a linked cell or MS. The functions of production control, inventory control, quality control, and machine tool maintenance are the first to integrate.

15.10 REVIEW QUESTIONS

1. Describe the difference between mechanization and automation. Give some typical examples of each.
2. Explain the difference between hard and soft automation. Why they are named as such?
3. Describe the principles and purpose of adaptive control. Give some applications in manufacturing in which you think it can be implemented.
4. Differentiate between ACO, ACC, and GAC.
5. What are the benefits and limitations of FMS?
6. Draw a sketch to show the idea of:
   - An MS
   - A flow-line manufacturing cell

- Unmanned FMC
- GAC for turning operation
7. What are the components of an MS?
8. List the benefits of CIMS.
9. Describe the principles of FMS. Why does it require a major capital investment?
10. What are the benefits of JIT production? Why it is called a pull system? What is a push system?
11. Differentiate between JIT and JIC production.
12. What is meant by the term "FOF"? Explain why humans will still be needed in the FOF.
13. What is Kanban? Why was it developed?
14. Describe the elements of AI. Why is machine vision a part of it?
15. Explain the principles of CAM, CAPP, and CIM to an older worker in a manufacturing facility who is not familiar with computers.
16. What is lean production? Numerate and explain the main steps toward lean production.

REFERENCES

Degarmo, E. P., Black, J. T., and Kohser, R. A. (1997) *Materials and Processes in Manufacturing*, 8th Edition, Prentice Hall, New York.

El-Midany, T. T. (1994) *Computer Automated Manufacturing and Flexible Technologies*, 1st Edition, El-Mansoura University, Egypt.

Jain, R. K. and Gupta, S. C. (1993) *Production Technology*, 13th Edition, Khanna Publishers, New Delhi.

Kalpakjian, S. and Schmidt, S. R. (2003) *Manufacturing Processes for Engineering Material*, 4th Edition, Prentice Hall Publishing Co., New York.

Koren, J. (1983) *Computer Control of Manufacturing Systems*, 1st Edition, McGraw-Hill, Tokyo.

McMahon, C. and Browne, J. (1998) *CADCAM—Principles, Practice and Manufacturing Management*, 2nd Edition, Addison-Wesley, Reading, MA.

Metals Handbook (1989) Machining, Vol. 16, 9th Edition, ASM International, Materials Park, OH.

Midwest Valve Company, Detroit, MI.

Index

A

Abrasive
    ac motors, 45, 300
    electrodischarge grinding, 516
    flow machining, 581
    grain, 123, 125, 396, 570, 571
    grit, 130, 432
    jet machining, 3, 10, 392, 395, 521, 571
    material, 554
    slurry, 410–413, 415, 429, 496, 497, 500, 519, 521, 570
    water jet machining, 392, 402, 405
Accuracy
    levels, 554
    of machined parts, 55, 298, 579
Adaptive control, 603, 620, 621, 625
Agility, 363, 369
Alloy steel, 166, 342, 391, 435, 554
Aluminum oxide, 122, 125, 554
Anodic dissolution, 392, 433, 445, 449
APT, 317, 334–338, 362, 617
Architecture, 345, 346, 367, 615
Arithmetic
    average, 575, 576
    progression, 31, 32
Art to part, 362, 367, 369
ASA, 185, 545
ASCII, 308, 309, 316
Aspect ratio, 433, 565, 570
Automatic
    bar, 169, 236, 237, 244
    full, 60
    lathes, 233, 241, 243, 246, 254, 261, 284, 603
    multispindle, 2, 6, 235–237, 241, 256–260, 266, 268, 270, 281, 282, 284, 529, 530
    screw machine, 6, 241–245, 248, 251, 253, 265, 271, 273, 274, 276, 283, 284
    semi, 243, 246, 254, 261, 266, 283, 284
    single-spindle, 236, 241, 244, 261, 266, 283, 284
    Swiss-type, 252–254, 270, 276, 278, 283, 284
    tool changer, 303, 304, 343, 363, 610
Automatically controlled, 7, 22
Automation, 7, 11, 68, 112, 233–235, 341, 370, 603, 604, 609, 622, 624, 625
Auxiliary shaft, 242–250, 253, 283

B

Ball screw strut, 352, 354, 368
Bar work, 217, 223, 226, 228
Base line, 320
BCD, 308, 343
Bearing mounting, 24, 25
Bevel gear
    forming, 190
    generation, 190, 206

Bifurcated, 299, 352–354, 356–358, 369
Binary, 308, 310, 311, 341
Boring
    machines, 14, 109–111, 365, 458, 536
    mills, 60, 69, 70
    operation, 109
    tools, 108, 109
Brake, 47, 48, 469
Broach
    pull, 115
    push, 115
    tool, 113
Broaching
    external, 112
    internal, 112, 115, 117, 455, 549
    machine, 113, 116–119
Built up edge, 578, 582, 583
Burnishing, 115, 144, 183, 207, 211, 212, 215, 229, 526, 552, 580

C

CAD/CAM, 4, 8, 317, 339, 340, 361, 409, 613, 614, 617
Calibration, 352, 353, 356, 366, 369, 374
Camshaft, 236, 241, 243–247, 250, 254, 255, 260, 263, 265, 283, 284
Canned cycle, 316, 322, 326, 332, 343
Carbide tools, 103, 224, 302, 454, 550
Carbon steel, 23, 201, 517, 552
Carborundum, 122
Carriage, 18, 20, 62, 63, 65, 66, 69, 219, 221, 248, 262, 263, 610
Cast iron, 16, 51, 57, 361
Cathode, 434, 445, 448, 471, 472, 486, 491, 512, 616
Cemented carbides, 122, 302
Centerline average, 575
Ceramics, 9, 86, 99, 109, 119, 142, 144, 367, 391, 394, 403, 411, 412, 421, 431, 443, 476, 481, 490, 506, 514, 520, 527, 560, 568, 569
Character, 142, 311, 341
Chasers
    circular, 160, 169–171
    radial, 160, 170, 171
    tangential, 160, 161, 171
Chasing, 69, 159, 163, 164, 168, 179
Chatter, 14, 103, 142, 145, 299, 575
Chemical
    dissolution, 434, 561
    etching, 434
    machining, 3, 434, 510, 561–563, 581, 590
    milling, 434, 435, 439, 440
    processes, 392
Chip breakers, 114
Chromium, 144, 452, 504, 510, 513, 514, 516, 571
Chucking work, 219, 226–228, 232

627

Circular interpolation, 297, 305, 307, 313, 320, 322, 325, 327, 329, 330, 340, 341, 343
CLDATA, 335, 337, 340, 362
Closed frames, 13–15
Closed loop, 297, 298, 301, 345, 620
Clutches, 46, 47, 182, 220, 247, 250
Coated tools, 321
Cobalt, 116, 302, 408, 452, 504, 563, 568, 600
Collect chuck, 220, 221, 244, 246, 248, 253, 259
Compact II, 617
Compensation
    cutter diameter, 316, 319
    tool length, 316, 343
    tool nose radius, 329–331
Composites, 398, 403, 410, 433, 481, 568, 569
Computer
    -aided design, 4, 339, 615
    -aided engineering, 603
    -aided inspection, 612
    -aided manufacturing, 4, 339, 617
    -aided process planning, 612, 616
    -integrated manufacturing, 5, 603, 612
    numerical control, 4, 285, 287, 281, 293, 295, 297
    program, 7, 316, 317, 337, 622
Computerized part changer, 610
Constant surface speed, 329, 341
Contour machining, 367, 433
Contouring, 295, 296, 334, 340, 341, 361, 363, 367, 433
Controller, 286, 298, 355, 361, 362, 620, 623
Conventional machines, 4, 286, 287, 289, 363–365
Conversational MDI, 308
Coolant
    control, 612, 335
    pressure, 351
    supply, 136
Copper, 28, 71, 124, 148, 164, 398, 431, 443, 448, 452, 460, 465, 470
Coring, 570
Corundum, 122
Counterboring, 59, 70–73, 76, 109, 110, 154, 294, 535
Countersinking, 59, 70–73, 76, 154, 394, 306, 387, 535, 551
Coupling, 30, 39, 47, 161, 400, 417
Cross slide, 62, 63, 65, 110, 163, 174, 217–219, 221, 223, 227, 238, 243, 248, 257, 264–268, 270, 281, 388
Cubic boron nitride, 86
Current
    ac, 414
    dc, 392, 489
    density, 434, 449, 450, 566, 581, 590, 592, 601
    efficiency, 450
Cutting
    condition, 155, 509, 582
    energy, 120, 485
    fluid, 133, 168, 499, 506
    forces, 11, 13, 14, 66, 83, 321, 371, 496, 505, 524, 527, 538, 543, 550, 551, 553, 621
    of rocks, 398
    operations, 207, 234, 264, 382, 496, 497, 510
    orthogonal, 382, 383
    rate, 409, 451, 469, 486, 567, 596

    temperature, 505, 615
    time, 39, 270
    tool material, 503

D

Database, 362, 612, 613
dc
    motors, 43, 69, 106, 354
    power supply, 434, 445, 453
Deburring, 9, 124, 395, 399, 403, 446, 448, 489, 525, 549, 571, 581, 599
Deep hole drilling, 2, 74, 77, 447, 508
Degrees of freedom, 345, 349
Depth of cut, 60, 83, 99, 101, 106, 121, 133, 135, 154, 159, 160, 163, 189, 194, 229, 332, 371, 412, 488, 556, 561, 562, 599, 621
Dexterity, 366, 369
Diamond tool, 130, 131
Die
    head, 160, 163, 171
    self opening, 160, 170, 171, 179
    sinking, 466, 514, 516, 586, 568, 602
Dielectric
    fluid, 467
    flushing, 469
    liquids, 514, 516
    supply, 469
Dies and molds, 594
Dimensional accuracy, 108, 143, 144, 168, 172, 412, 554, 557, 568, 571, 621
Direct numerical control, 4, 286
Dividing head, 94, 95, 98, 155
DNC, 4, 285–287, 309, 340, 341, 604
Dome, 348, 352, 354, 359, 361
Drill press, 74–76, 165, 169, 170, 536, 616
Drilling
    jig, 80–82, 155, 536
    operations, 315, 359, 538
    tool, 7, 70, 77, 305, 586
Drilling machine
    gang, 155
    multispindle, 47, 76, 77
    single spindle, 76
Dry cutting, 501, 502, 506–509
DXF, 339, 409

E

ECG process, 454, 489
ECM
    advantages and disadvantages, 489
    applications, 447, 590
    equipment, 449
    process, 433, 445, 447
EDM
    electrodes, 455
    machine, 4, 463, 489, 490, 516, 567
    milling, 466
Electrical stepless speed drive, 42, 43

Index

Electrochemical
    deburring, 581
    discharge grinding, 589, 598
    grinding, 392, 453, 566, 581
    honing, 392
    machining, 3, 446, 447, 512, 561, 563, 601, 602
Electrode
    material, 462–464
    wear, 462, 466–468, 568
Electrodischarge
    grinding, 516
    machining, 3, 6, 514, 567, 591, 601, 602
Electroerosion dissolution machining, 589, 601, 602
Electrolysis, 445, 489, 513, 598
Electrolyte
    feeding, 450
    passivating, 590
    temperature, 556
    velocity, 449
Electrolyzing current, 433, 434, 445, 451
Electromagnetic field, 520
Electron beam machining, 4, 470, 471, 568, 581, 597
Electropolishing, 144, 580, 581, 590
Environmental impacts, 8, 10, 495, 499–501, 510, 515
Etch
    factor, 437, 439
    rate, 439–441, 443, 510
Etchant, 392, 434–442, 445, 446, 497, 510–512, 563, 601
Evaporation, 392, 579

F

Face milling, 82, 84, 86, 90, 91, 105, 296, 305, 324, 325, 540, 583, 586, 587
Face plate, 23, 184
Factory of the future, 624
Faraday's laws, 445, 489
Feed gearbox, 29, 30, 37, 38, 57, 63, 75, 88, 107
Finish turning, 61, 328, 331–333
Fixed block format, 311
Fixed zero, 293, 294
Fixtures, 83, 118
Flexibility, 194, 363, 609
Flexible manufacturing systems, 609–611
Floating zero, 293
Flushing, 459, 467
Follower rest, 62, 66, 67, 115, 154
Fuzzy logic, 622, 624

G

G codes, 311, 319, 322, 329, 362
Gap
    voltage, 463, 490, 566
    width, 451, 590
Gear
    broaching, 188–189
    burnishing, 211, 215
    finishing, 207
    forming, 185, 189, 190, 208
    grinding, 212–214
    hobbing, 6, 7, 190, 193–196, 204, 205, 215
    lapping, 214, 215
    milling, 86, 184, 508
    production, 183, 215
    shaping, 183, 189, 198, 199, 201, 202, 208, 215
    shaving, 207, 209–211
    spur, 6, 45, 46, 96, 181, 183–185, 188–190, 192, 193, 202, 209, 211, 214, 215, 552
    train, 69, 75, 98, 161, 162, 179, 191, 195–198, 243, 244
Geometric
    adaptive control, 621
    progression ratio, 32, 33, 63
Grain size, 122, 123, 125, 131, 137, 392, 396, 403, 432, 434, 499, 561, 571, 582, 583, 590, 599
Graphite, 16, 402, 412, 463–465, 467, 468, 470, 490, 509, 527, 567, 594, 601
Grinding
    action, 135, 453
    centerless, 131, 137–141, 155, 178, 557, 558
    conventional, 177, 485, 566, 589
    creep feed, 132
    cylindrical, 2, 7, 124, 126, 133, 135–138, 155, 290, 291, 556, 573
    machines, 6, 46, 119, 126, 131–133, 136, 137, 177, 214, 290, 453, 499, 573
    operations, 121, 135–137, 506
    plunge, 132, 133, 159, 176, 178, 556, 567
    process, 119–121, 124, 130, 207, 214
    surface, 124, 127
    wheel, 117, 122–125, 127, 130, 131, 212, 213, 403, 453
Grooving, 148, 244, 264, 328, 329, 332, 529
Group technology, 605
Guideways
    ball bearing, 22
    dovetail, 18, 20
    externally pressurized, 19, 22
    rolling friction, 21, 22, 57
    sliding friction, 18
    vee, 18

H

Hardening
    case, 52, 53, 214, 552
    induction, 20
    surface, 470, 477, 478, 481, 482
Hardness
    Brinell, 51
    Knoop number, 122, 13
    Rockwell, 23
Headstock, 62, 69, 110, 143, 163, 218–220, 226, 242, 244, 252–255, 257, 260, 264, 277
Heat
    affected zone, 141, 596, 598, 599
    generation, 485
Helix angle, 70, 71, 94, 98, 173, 174, 185, 188, 191, 193, 202, 208, 209, 546, 553
High speed
    machining, 504, 505, 508
    steel, 167, 509
High spindle speed, 27, 262, 366

Horizontal milling, 52, 82, 84, 86, 90, 184, 579
Hot hardness, 507
Hybrid machining, 396, 602
Hydraulic
  motor, 43, 44, 300, 301
  pump, 44, 56, 399
  tracer, 9, 65, 237, 238
Hydrogen, 440, 446, 510, 512, 583

I

Ignition lag, 463
Imperial units, 314
Incremental positioning, 293, 294, 322, 329, 340
Indexing
  head, 181
  trips, 185, 186
Input signal, 458
Integrated circuit, 4, 285, 341
Intensifier, 402
Interlocking cutter, 202, 203, 207
Interpolation
  circular, 297, 305, 307, 313, 320, 322, 327, 329, 330, 340, 341, 343
  cubic, 296, 297
  linear, 296, 313, 320, 322, 327
  parabolic, 297
Iron and steel, 51, 164, 512, 591, 602

J

Jet
  cutting, 9, 400, 405, 492
  velocity, 393, 395, 398, 409, 488
Jog, 315, 316
Just-in-time, 605, 617, 619, 620

K

Kanban, 618, 619, 626
Knurling, 59, 63, 221, 226, 233, 244, 264
Konvoid generators, 202, 204
Kopp variator, 41

L

Lapping
  equalizing, 148
  machines, 7, 148–153, 560, 661
  operation, 207, 215
Laser beam
  cutting, 481
  drilling, 479
  machining, 3, 10, 475, 478, 516, 518, 569, 581, 596, 597
Lathe
  automatic turret, 166, 169
  capstan, 170, 217, 219, 221–223, 225, 227, 229, 231, 234
  dog, 66
  facing, 60, 69, 70, 154
  machine, 3, 15, 59, 160, 179, 298
  manual turret, 166, 169
  plain turning, 60, 69
  vertical turret, 70

M

M codes, 322, 323, 329
Machinability
  of materials, 371, 477
  rating, 552
Machine tool
  drives, 28, 42, 45
  frame, 16, 18, 57
  gearbox, 35, 210
  guideways, 18, 19, 23, 57
  manual, 7, 286
  motors, 45, 46
  spindles, 23, 24, 26, 27, 33, 298, 299
  structures, 13, 15–17, 57, 299
  testing of, 53
  traditional, 5
Machinery, 1, 55, 212, 450, 603, 610, 616, 617, 624
Machining
  allowance, 1, 159, 207, 209, 226, 371, 498, 527, 545, 565, 567
  by cutting, 159, 417, 492, 498, 501, 510, 524, 528, 586, 602
  centers, 4, 9, 111, 285, 306, 610, 612, 624
  nontraditional, 1, 4, 5, 9, 391, 392, 490, 510, 587
  productivity, 9, 171, 172, 190
  traditional, 5, 7, 9, 10, 498, 582
Magic three, 314, 341
Magnetic
  chuck, 121, 132, 153, 555
  tape, 286, 298, 308, 309
Magnetostrictor, 417
Maintenance
  corrective, 56
  preventive, 53, 56, 57
Management, 362, 495, 525, 605, 608, 612–614, 617, 618
Mandrel, 55, 66, 67, 142, 191
Manufacturing
  cell, 608, 609, 625
  cellular, 605, 607, 625
  smart, 622
Maskant, 436, 438, 439, 489, 561, 563
Material handling, 605, 610–612, 615, 617, 624, 625, 604
MDI, 308, 316
Mechanical abrasion, 392, 397, 434, 453
Mechanism
  apron, 63
  of material removal, 491, 587
  quick return, 48–50, 57
  reversing, 45, 46, 57, 165
Metallic bond, 125, 144
Micromachining, 456, 467, 470, 475, 478, 479, 481, 569
Milling
  climb, 83, 155
  conventional, 83, 285, 345
  down, 83, 84
  end, 82, 184, 541, 542

Index

face, 82, 84, 86, 90, 91, 105, 296, 305, 324, 325, 540, 583, 586, 587
gear, 86, 184, 508
horizontal, 82, 84, 86, 90, 184, 579
operations, 82, 84, 86, 173, 295, 305, 312, 390, 539
vertical, 84, 85, 87, 110, 579, 583
Modal, 322, 323, 329
Moving target, 626

N

NC control, 289, 294–296, 314, 361, 482
Nickel, 414, 439, 443, 452, 476, 481, 504, 510, 514, 516, 563, 566, 568, 571, 580, 581, 591, 600
Nitric acid, 510
Norton gearbox, 29, 32, 38, 57
Neural network, 622, 623

O

Octahedral hexapod, 346, 349, 351, 363–365, 370
Octahedron, 345, 350, 351, 354
Open
 frames, 13, 14
 loop system, 301
Overcut, 432, 448, 451, 465, 468, 570, 571
Override, 314, 315
Oxygen, 506, 513, 594, 602

P

Parallel action, 257, 258
Parallel kinematics, 345
Parity check, 316, 341
Part programming, 316, 318, 334, 337, 339, 341
Peck drilling, 322, 324, 332
Photochemical machining, 434, 441, 443, 581
Photoresist, 438, 439, 442, 561, 563
Pick-off gears, 29, 30, 38, 57, 174, 243, 244, 270, 282
Plain milling cutters, 84
Planing, 3, 6, 7, 9, 48, 99–101, 154, 155, 542, 572
Plasma
 arc, 485, 486, 488
 beam machining, 3, 581, 598
 channel, 485, 514
 jet, 487
Plastic forming, 1
Plate jig, 80
Pocket milling, 324, 326
Point-to-point, 294
Pole-changing motor, 36, 37
Polishing, 3, 141, 144, 394, 396, 399, 400, 526, 580, 581
Precision
 adjustment, 525
 cutting, 408,
 drilling, 470
 grinding, 124, 367, 509, 556
 machine tool, 23, 26
Printed circuit boards, 435
Process
 accuracy, 394, 559
 capabilities, 142, 391, 396
 characteristics, 392, 397, 402, 445, 454, 470, 475, 485

control, 445, 604, 620, 623
parameters, 4, 391, 460–462, 501, 596
Product design, 6, 8, 10, 497, 539–542, 544, 566, 605, 612, 613, 615
Production
 batch, 30, 38, 76, 603, 604, 611, 625
 interchangeable, 217
 lean, 617–619, 625, 626
 lot, 52, 60, 76, 80, 86, 103, 153, 160, 184, 233
 mass, 59, 68, 77, 80, 89, 112, 113, 119, 150, 154, 200, 212, 233, 236, 237, 256, 271, 276, 283, 287, 398, 446, 525, 557, 559, 560, 603, 605, 609
Program datum, 321
Programmable logic controller, 362
Progressive action, 236, 239, 240, 256–260
Pulleys, 41, 42, 111, 529
Pulse
 current, 568, 593, 594
 duration, 470, 471, 474, 475, 480, 482, 568
 energy, 474, 475, 478, 480, 594, 597
 frequency, 464, 474, 475, 478–480
 generator, 455, 458, 460, 462–464, 468, 490
 on-time, 591, 593
Punched tapes, 286, 308, 309

Q

Quality control, 604, 612, 617, 618, 625

R

Rack and pinion, 75, 101, 102, 106, 181
Rake angle, 70, 83, 100, 103, 116, 120, 169, 170
Rapid prototyping, 496
RC circuit, 460–462, 490
Reaming
 allowance, 74
 operation, 111, 262
Recast
 layer, 435, 455, 464, 471, 479, 594, 596
 structure, 464, 595
Reference plane, 323, 324
Relief angle, 71, 163
Removal rate
 material, 82, 176, 302, 320, 599, 621
 volumetric, 11, 465, 474, 568
Robots, 299, 453, 362, 603, 604, 611, 623–625
Roller box, 299
Rotary ultrasonic machining, 433
Rough turning cycle, 331, 333

S

Saddle, 66, 88, 110, 111, 161, 217–219, 221, 223, 232, 345, 402, 492
Sapphire, 392, 393, 398, 401, 405, 412, 433, 571
Sawing, 250, 453, 500, 508, 580, 594
Scheduling, 610, 613, 617, 619, 622–624
Scribing templates, 438, 440
Setting angle, 60, 101, 130
Shaping machine, 6, 49, 201, 215, 373
Shielded plasma, 486, 487

Short circuit, 446, 449, 458
Slotting, 7, 13, 14, 85, 86, 99–102, 154, 190, 233, 446, 533, 542–544, 571, 572
Slurry supply system, 431
Sodium chloride, 512
Sodium nitrate, 512
Spark
 erosion, 6, 455
 gap, 458, 463, 464, 516
 machining, 492
Speed
 chart, 35, 37
 gearbox, 29, 35–37, 39, 62, 88, 104, 105, 110
Sphere drive, 352–357, 361
Spindle
 bearing, 12, 23–28, 559
 mounting, 57, 136
Spot facing, 59, 71, 73, 110, 154, 233, 306, 540
Staggered, 84, 114
Stand off distance, 392
Steady rest, 62, 66, 67, 74, 77, 154, 557
Stepless speed, 28, 40–44, 57
Stick slip effect, 19, 21, 300
Straight cut, 295, 296, 305, 340, 341, 343
Structural diagram, 35, 36
Structural steels, 52, 57
Structure
 cast and fabricated, 17
 welded, 16, 57
Strut assembly, 354
Superfinishing, 3, 9, 145, 146, 561, 580
Surface
 effects, 569, 579, 582, 587–589
 finish, 168, 592
 grinding, 6, 7, 31, 120, 121, 124, 127, 132, 135, 155, 384, 390, 554, 587, 589, 591
 integrity, 4, 6, 8, 10, 11, 455, 504, 505, 575
 layer, 211, 575, 579, 580, 582–584, 586, 596, 598
 texture, 6, 8, 460–462, 545, 575–578, 580, 601, 602
System
 expert, 622, 623
 knowledge based, 601, 622
 pull, 619, 626
 push, 619, 626
 vision, 612

T

Tape
 magnetic, 286, 298, 308, 309
 paper, 298
 reader, 7, 285, 287, 288, 297, 298, 316, 341
Tapping cycle, 322, 324
Telescopic struts, 349–351, 369
Texturing, 9, 457, 594, 601
Thermal
 effect, 411, 485, 594
 machining, 4, 9, 579, 582, 589, 596, 601
 properties, 474, 582, 591, 597
Thinning of parts, 436

Thread
 broaching, 175
 English, 163
 external, 61, 160, 163, 168, 174, 178, 224, 551
 form, 157, 159, 160, 175, 176, 178, 508, 550
 grinding, 6, 168, 175–179, 550, 551
 internal, 59, 61, 74, 109, 110, 163, 164, 174, 175, 177, 550, 551
 metric, 157, 159, 162
 milling, 160, 172–175, 178, 179, 551
 pitch, 65, 160, 165, 166, 176
 tapered, 174, 175, 179, 529
 tapping, 164, 167
Threading
 conditions, 168, 172
 cycle, 170, 329, 333
Three jaw chuck, 226, 556
Time
 idle, 185, 225, 232, 253, 270, 283, 399
 machine handling, 255
 machining, 9, 10, 18, 39, 40, 189, 200, 205, 207, 209, 223, 225, 226, 260, 268, 281–283, 289, 340, 432, 433, 442, 467, 530, 555
 production, 11, 39, 40, 112, 217, 224, 225, 242
Titanium, 116, 302, 367, 410, 443, 452, 476, 516, 563, 568, 580, 581, 591, 602
Tool
 diameter, 317, 318, 361, 411, 432, 491, 570
 electrode, 451, 455, 461, 463–466, 514, 516, 567, 568
 feed rate, 59, 282, 490, 535
 geometry, 101, 113, 223, 267, 270, 340, 532
 holder, 104, 106, 131, 217, 222, 226–228, 230, 303, 307, 308, 358, 458
 insulation, 447, 564
 life, 37, 40, 116, 168, 170–172, 204, 211, 212, 302, 371, 501, 503–506, 509, 551, 563, 599, 611, 615
 magazine, 303
 material, 1, 11, 30, 31, 101, 154, 189, 224, 268, 270, 282, 302, 371, 391, 411, 432, 448, 463, 503, 506, 507, 516, 535, 582
 offset, 303, 305, 316, 611
 oscillation, 410, 430, 491
 shape, 71, 445, 519
 steel, 86, 123, 150, 409, 470, 563, 595, 600
 wear, 109, 168, 169, 178, 314, 316, 329, 359, 371, 432, 434, 446, 451, 453, 455, 459, 460, 463, 476, 491, 504, 535, 538, 542, 550, 571, 578, 611
Tooling layout, 217, 226, 228–230, 267, 268, 277, 280, 281
Tracer device, 65
Traditional
 grinding, 435
 machining, 5, 7, 9, 10, 582
 manufacturing, 612, 619
Trimming, 4, 394–396, 484, 562, 510
Trip-dogs, 135, 241, 247, 249, 250
Tripod, 345, 346
Truing and dressing, 122, 130, 131
Tungsten, 147, 392, 394, 395, 408, 439, 452, 463, 471, 475, 486, 487, 563, 568, 569, 571, 591, 600
Turbine blade, 363, 366, 447

Turning center, 285, 302, 307, 328, 329
Turret
   head, 70, 220, 223, 226, 245–247, 250–252, 268, 273, 275
   hexagonal, 217–219, 221–223, 225, 226
   screw automatics, 235
   slide, 217–219, 222, 242, 250, 251
   square, 62, 65, 217, 218, 221, 225–227
Twist drill, 62, 70–72, 77, 79, 86, 94, 231, 256

U

Ultrasonic
   assisted ECM, 433
   machining, 3, 92, 393, 410, 433, 519, 750, 581
   sinking, 433
   vibration, 433, 434, 596, 597
Undeformed chip, 60, 599

V

Vaporization, 392, 455, 470, 471, 477, 502, 503, 516

W

Water jet, 3, 8, 392, 397, 402, 405
Wear compensation, 18, 20, 21, 108, 109, 466, 467
Welding, 1, 2, 20, 52, 364, 367, 368, 411, 470, 475, 477, 478, 481, 482
Wheel
   balancing, 129, 336
   bond, 120, 124
   dressing, 213, 558, 589
   loading, 130, 555
   marking, 125
   mounting, 128
   single rib, 175–179
   skip rib, 176, 177, 179
   truing, 130, 131, 135
Wire
   cutting, 468, 469, 497, 514
   EDM, 489, 514, 515, 567, 568
Word address format, 311, 312
Work holding, 12, 79, 226, 289, 299, 321, 322, 542
Workpiece material, 9
Workstation, 395, 396, 476, 610
Worm wheel, 181, 190–194, 208, 315, 244